Electron Microscopy and Analysis 2003

Other titles in the series

The Institute of Physics Conference Series regularly features papers presented at important conferences and symposia highlighting new developments in physics and related fields. Previous publications include:

Electron Microscopy and Analysis 2003

Proceedings of the Institute of Physics Electron Microscopy
and Analysis Group Conference,
University of Oxford, 3–5 September 2003

Edited by
Stephen McVitie and David McComb

Institute of Physics Conference Series Number 179

CRC Press
Taylor & Francis Group
Boca Raton London New York

CRC Press is an imprint of the
Taylor & Francis Group, an **informa** business

CRC Press
Taylor & Francis Group
6000 Broken Sound Parkway NW, Suite 300
Boca Raton, FL 33487-2742

First issued in paperback 2019

ISBN-13: 978-0-7503-0967-7 (hbk)
ISBN-13: 978-0-367-39453-0 (pbk)

**Visit the Taylor & Francis Web site at
http://www.taylorandfrancis.com**

**and the CRC Press Web site at
http://www.crcpress.com**

British Library Cataloguing in Publication Data

A catalogue record for this book is available from the British Library.

Library of Congress Cataloging-in-Publication Data are available

Preface

The 2003 biennial meeting of the Electron Microscopy and Analysis Group of the Institute of Physics (EMAG 2003) was held in the Examination Schools of the University of Oxford from 3 to 5 September. Prior to the conference a one day Advanced School was held in the Department of Materials, University of Oxford, in which a variety of imaging techniques were presented and discussed. A trade exhibition was integrated into the conference site giving delegates the opportunity to discuss recent developments in equipment. Furthermore, a series of lunchtime workshops were organised by some of the exhibitors to promote and educate delegates in a variety of techniques and instrumentation. Delegates were accommodated in Keble and Magdalen colleges. Social events organised during the conference included an exhibitors' buffet in the Examination Schools, and a superb conference dinner set in Balliol college.

The conference attracted 170 delegates from 15 countries and the conference theme was toward objective oriented microscopy. A full program covering many areas of microscopy and microanalysis took place during the three day meeting. As usual the highlights of the conference were the plenary lectures in which recent developments in key areas are reviewed. In the first, David Muller described using imaging and spectroscopy to characterise individual atoms and clusters in electronic materials and devices. John Spence covered tomographic diffractive imaging for nanostructures. Finally, John Pethica detailed atom resolved atomic form microscopy and energy dissipation.

In addition to the plenary lectures, there were 10 invited oral presentations and 67 contributed oral presentations, which ran in two parallel sessions. Furthermore, 66 posters were presented at the conference throughout the three days. These proceedings comprise 113 papers, beginning with the plenary papers and followed by the remaining papers, ordered in the session that they appeared at the conference. The papers were submitted at the conference in camera-ready form and refereed independently during the conference where possible. We are indebted to the efforts of many of the conference delegates who kindly gave up their valuable time to help in this process. It is safe to say that without their efforts it would not have been possible to produce these proceedings so promptly. We hope that readers of these proceedings see this volume as a valuable snapshot of microscopy and microanalysis in the UK and beyond at the time of writing. Finally, we are extremely grateful to local helpers, Rebecca Nicholls and Zhongfu Zhou, who maintained good spirits among the editors and provided enthusiastic and invaluable assistance in the proceedings office during the conference.

Stephen McVitie, David McComb

Organising Committee

The conference was organised by the Electron Microscopy and Analysis Group of the Institute of Physics in association with the Royal Microscopical Society and the Institute of Materials.

Chairman of the Scientific Committee
Dr R M D Brydson (Department of Materials, University of Leeds)

Scientific Program Organiser
Dr A Bleloch (Cavendish Laboratory, University of Cambridge)

Honorary Editors
Dr S McVitie (Department of Physics and Astronomy, University of Glasgow)
Dr D McComb (Department of Materials, Imperial College, London)

Exhibition Organiser
Dr R Doole (Department of Materials, University of Oxford)

Local Arrangements
Prof A Petford-Long (Department of Materials, University of Oxford)

Advanced School Director
Prof D J Cockayne (Department of Materials, University of Oxford)

Conference Co-ordinator
Ms J Bolfek-Radovani (Conference Office, The Institute of Physics)

Electron Microscopy and Analysis Group Committee 2002–2003
Dr R M D Brydson (Department of Materials, University of Leeds)
Dr P D Brown (School of Mechanical, Materials, Manufacturing Engineering and Management, University of Nottingham)
Dr A Bleloch (Cavendish Laboratory, University of Cambridge)
Dr S Galloway (Gatan UK, Oxford)
Dr D McComb (Department of Materials, Imperial College, London)
Prof A Petford-Long (Department of Materials, University of Oxford)
Dr T Rong (School of Engineering, Metallurgy and Materials, University of Birmingham)
Dr R Shipley (FEI UK Ltd., Cambridge)

Contents

Section 7: Advances in Nanoanalysis

Section 8: Advances in Imaging

Section 9: Sample Preparation and Nanofabrication

Section 10: Surfaces and Interfaces

Section 11: Nanomaterials

Inst. Phys. Conf. Ser. No 179: Section 1
Paper presented at Electron Microscopy and Analysis Group Conf. EMAG2003, Oxford, 2003
©2003 IOP Publishing Ltd

1

Imaging and spectroscopy of individual atoms, clusters and interfaces in electronic materials and devices

D. A. Muller, J Grazul

Applied and Engineering Physics, Cornell University, Ithaca, NY 14853

Abstract. The smallest features on transistors used in integrated circuits today have approached atomic dimensions: the SiO_2 gate oxides are between 5 and 7 silicon atoms thick and the concentration of dopant atoms has increased to the point that electrically inactive dopant clusters as small as two atoms are common enough to affect device performance. We have used atomic-resolution STEM with single atom sensitivity to identify the size, structure and distribution of clusters responsible for the saturation of charge carriers and address the question of how many atoms are needed before the gate oxide loses its bulk properties.

1. Introduction

The narrowest feature on an integrated circuit is currently the gate oxide. At the start of this century, gate oxides thinner than 2 nm were being used in some commercial integrated circuits. Between 2004 and 2008, if silicon dioxide is still to be used, then the projected gate oxide thickness will be less than 1 nm, or 5 silicon atoms across as shown by the industry roadmap - ITRS 2000. At least two of those five atoms will be at the silicon/oxide interfaces. The interfacial atoms have very different electrical and optical properties from the desired bulk silicon dioxide yet comprise a significant fraction of the dielectric layer. This fundamental problem has also become a very practical one. Fig. 1 illustrates how dramatically the oxide thickness has shrunk. It is now not only technologically possible to produce metal oxide semiconductor field effect transistors (MOSFET) with gates shorter than 50nm and SiO_2 gate oxides less than 1.3nm thick (Timp et al 1998), but what was a research demonstration 6 years ago, is now a commercial process. Such a thin gate oxide is required to improve the drain-current response of the transistor to the applied gate voltage (allowing lower voltages to be used). Since power dissipation currently limits the scale of integration, lowering the power supply voltage becomes the key to increasing integration and improving IC performance. Therefore, the performance of the gate oxide, becomes the key issue that would disable the manufacture of very large-scale integrated circuits. Since almost all practical alternatives to SiO_2 that provide a higher dielectric constant or a reduced leakage current involve a few layers of SiO_2 (Wallace et al 2002), it is crucial to the future of large-scale integration to discover the practical limits on the thickness of the SiO_2 gate oxide.

A second set of challenges facing device designers is the positioning and activation of dopant atoms. As devices shrink (and all the dimensions must scale accordingly), the dopant concentrations in the source and drain must increase to maintain good contact to

2

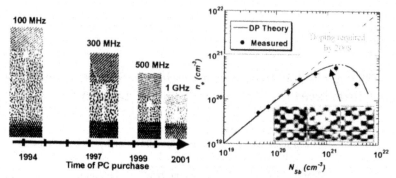

Figure 1. Left:The relentless march to zero thickness. SiO_2 Gate oxide thickness used in personal computers. For more details about modern gate oxides see Baumann et al 2000. Right: Density of free carriers (n_e) as a function of Sb dopant concentration (n_{Sb}). The peak carrier concentration (…) still falls below the 2008 semiconductor roadmap requirements. Inset: ADF STEM images of the Sb dopant clusters responsible for the decrease in carrier activation.

the transistor. Such concentrations can already exceed the solid solubility limits and are rapidly approaching the 1% level. Fig. 1 shows that the dopant concentration cannot be increased without limit, and in fact reaches a maximum at a point below that need for the 2008 generation of devices (Citrin et al 1999, Gossmann et al 2000). Whether this limitation is intrinsic or can be overcome by clever processing is a matter of great concern. In fact, some of the motivation for decreasing the gate oxide thickness has been the difficulty in producing, shallow doped junctions with high conductivity – the transistor drive current (which determines the switching speed) depends inversely on a product of the junction depth and the gate oxide thickness. Establishing the limits for both dimensions sets a fundamental limit on the smallest (and hence indirectly the fastest) a conventional MOSFET can operate.

In this paper we discuss the use of scanning transmission electron microscopy (STEM), for measuring the physical and electronic properties of ultra-thin gate oxides and dopant clusters. After an introduction to annular dark field (ADF) imaging, the imaging of individual dopant atoms and clusters is dicussed. Electron energy loss spectroscopy (EELS) of Si/SiO_2 interfaces follows. Finally, some challenges for replacement gate dielectrics are discussed.

2. Annular Dark Field STEM Imaging

The bulk of the work reported in this paper was performed on a JEOL 2010F electron microscope was fitted with an analytical (C_s=1 mm) polepiece, JEOL and Fischione ADF detectors, Gatan imaging filter and BF/ADF detectors. Room environment proved to be a critical factor in obtaining clear stable images and reducing the drift to a point that high-quality EELS signals could be obtained with a 0.2 nm probe size (Muller et al 2001a). Early EELS measurements were also made on the Cornell VG-HB501 dedicated STEM, which has higher gun brightness and a factor of two improvement in energy resolution.

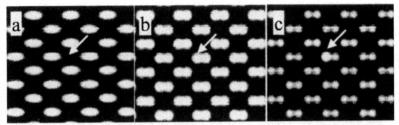

Figure 2. Multislice simulations (with frozen phonons) for a JEOL 2010F (Cs=0.5 mm, 200 kV beam, 1.2Å information limit, 45Å thick slab, a single Sb atom placed at a depth of 22Å). (a) In HRTEM, atoms are dark. Arrow marks the position of the Sb atom. (b) The noise-free electron phase shows the atoms as bright. (c) In ADF-STEM the atoms are bright, and the Sb atom is the brightest. The Sb contrast is 0.1% for HRTEM, 5% for electron phase and 65% for ADF-STEM.

In a STEM, the atomic-resolution image is formed by scanning an atomically small electron beam and collecting the signal on an annular dark field detector. The image can also be used to align and locate the small probe needed for electron-energy-loss spectroscopy. This makes it possible to measure the chemical composition and bonding of an interface with sub-nanometer spatial resolution (Batson 1993, Kaiser et al 2002, Muller et al 1999).

Apart from the manner in which the image is formed (scanning versus parallel illumination), the main difference between ADF and HRTEM (high-resolution transmission electron microscopy) imaging is the contrast mechanism. For very thin specimens, HRTEM imaging is primarily a coherent, phase-contrast imaging technique while ADF imaging is an incoherent, amplitude-contrast technique (Black et al 1957, Pennycook 1989, Silcox et al 1992). The incoherent nature of the ADF image (which measures the square of the electron wave function) largely removes the contrast reversals present in phase-contrast HRTEM. ADF imaging also obtains higher resolution at a cost of reduced contrast. Consequently, raw ADF images always look more 'blurry' than HRTEM images, which in turn are artificially sharpened by the microscope (The HTREM CTF removes the lower frequencies, while the ADF CTF enhances them). These models are no longer qualitatively correct for thicker specimens where multiple scattering is significant, but they still serve as a useful guide to the differences in image formation.

The simulated images of a single antimony dopant atom in silicon by ADF-STEM but not HRTEM (figure 2) emphasizes the higher atomic number contrast and point-point resolution of incoherent imaging (roughly a factor of 40% better). The antimony dopant atoms show up roughly nine times brighter than a silicon atom would – the scattering cross section scales roughly as $Z^{1.7}$, so ADF imaging is sometimes known as "Z-Contrast" imaging, although not all contrast effects are related to atomic number alone.

The effects of crystal orientation and strain fields can be pronounced (Perovic et al 1993), so it is always important to check the ADF result with an analytical technique such as EELS (e.g. Citrin et al 1999). More often than not, a bright line at an interface in an ADF image is likely to be a strain field, rather than evidence of dopant segregation, and as the effects are strongly thickness dependent, great care should be taken to check results for different sample thicknesses and camera angles, in addition to EELS.

3. Imaging Dopant Atoms and Clusters

Many of the properties of materials are controlled by the distribution and motion of low concentrations of impurity atoms. Techniques such as x-ray absorption spectroscopy or nuclear magnetic resonance can determine the average local environment of impurities in some cases. This is insufficient information, however, if there are several different possible local environments. Imaging can characterize the environment around each impurity individually, yielding the maximum possible information. Single-atom imaging has also taken on direct technological relevance. As silicon transistors continue to shrink in size, whether or not a device works will depend on the position of just a few dopant impurity atoms (Packan 1999). Moreover, as the concentration of such impurities grows larger, they form nanoclusters, and the free carrier concentration as a function of impurity concentration saturates (Williams et al 1982). This may impose a fundamental limit on future generations of Si technology (Gossmann et al 2000).

Two generic models for electrically deactivating defect structures have been proposed for Group V dopants in Si (i.e., electron donors). The first consists of an *extrinsic* defect formed between 1 and 4 dopant donor atoms surrounding a Si vacancy - Sb_nV(Fair et al 1973), while the second is *intrinsic*, consisting of only 2 donor atoms bound to reconfigured Si with no vacancies (Chadi et al 1997). Determining which type of deactivating defect is formed in Si will establish how the fundamental limits to the free-electron concentration are reached.

This requires imaging and resolving individual dopant atoms that are still buried in their bulk-like environment. As a starting point, single atoms on surfaces have been imaged previously. The earliest such report, by Muller 1957, reports imaging of atoms on a metal surface using a field-ion microscope. Imaging heavy atoms on an amorphous supporting film was one of the first applications of annular dark-field scanning transmission electron microscopy (ADF-STEM) by Crewe et al 1970. Scanning electron microscopy has been very successful in imaging dopant atoms on III-V semiconductor surfaces which rearrange to remove dangling bonds, but has been less useful for silicon

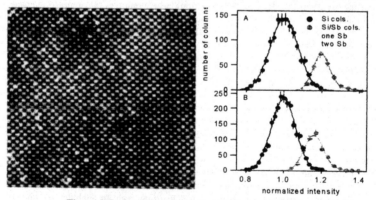

Figure 3. Left: ADF-STEM image of Sb-doped Si. The undoped region shows atomic columns of uniform intensity. The brightest columns in the doped region contain at least one Sb atom. The image is smoothed and background subtracted. Right: Intensity histograms from two images. Intensities are normalized to the average unoccupied column intensity. Images A and B are 15 and 23 Å thick, respectively.

where the defect states pin the Fermi level(Ebert 2001, Castell et al 2003). Atoms on a surface, however, behave very differently from atoms in the bulk. With different local environments, coordination numbers, strains, relaxations and screening responses, states that may have been active in the bulk, may lie midgap on a surface – or visa-versa.

Because the electrons pass completely through the sample, STEM sees a 2D projection of the 3D sample structure. So STEM can, in principle, see dopants buried in a device, but it is only recently that bulk samples have been made thin enough, and with smooth enough surfaces, that individual dopant atom became visible (Voyles et al 2002). Previously, the signal from one dopant atom was easily swamped by surface roughness in mechanically thinned bulk samples. STEM works best with heavier atoms, but it is predicted that individual dopants, as light as boron, should be detectable at low temperatures(Hillyard et al 1995). Recent advances in instrumentation may push things further. In principle, channeling of the electron beam means the signal is a function of depth, so adding some three-dimensional information to the image (Vanfleet et al 1998) should be possible. Signal-to-noise ratios aren't yet good enough for this, but aberration corrected STEM optics with huge increases in resolution, contrast and signal may make this practical (Batson et al 2002). Electron spectroscopy from a single column of dopant atoms is already possible (Kaiser et al 2002). In a microscope with a monochromator, core and optical spectroscopy from a single dopant atom, which would indicate its identity and its electrical activity, becomes a distinct possibility.

We have recently demonstrated ADF-STEM images of Si that show quantifiable contrast individual Sb atoms still bonded to their crystalline Si host using atomic-resolution ADF-STEM (Voyles et al 2002). Samples with an Sb concentration of 9.35×10^{20} cm^{-3}, measured by RBS, were prepared by low-temperature molecular beam epitaxy (Gossmann et al 1993). Figure 3 is an ADF-STEM image of a $\langle 110 \rangle$ cross section of the sample. In the substrate region, where no Sb was deposited, all of the atomic columns are similar in intensity. In the region doped with Sb, some atomic columns are much brighter than the surrounding columns. These columns contain at least one Sb atom.

The intensity histograms of the images, shown in Figure 3, demonstrate that we have single Sb sensitivity. For random substitution of Sb atoms in the Si lattice the number of Sb atoms in a column is given by the binomial distribution. The intensity distribution of unoccupied columns is well-fit by a Gaussian, the width of which measures the residual effects of sample surface roughness. The occupied column distribution is fit by two Gaussians, the second at twice the excess intensity of the first. We identify the peaks as singly and doubly Sb-occupied columns, respectively. The areas under the peaks are within experimental error of the expected binomial distribution (Voyles et al 2002), consistent with a random distribution of Sb on Si sites, imaged with single Sb sensitivity and 97±3 % Sb detection efficiency. While single atoms have been previously imaged on surfaces, we believe this is the first time single atom sensitivity in the bulk has been quantitatively demonstrated.

The key to single-atom detection is good sample preparation. We prepared cross-sections by double-wedge mechanical polishing using an Allied High Tech Inc. TechPrep/MultiPrep polishing wheel and diamond coated plastic lapping paper, finishing with a 0.02 μm colloidal silica polish. The polishing results in samples with a highly damaged surface layer, which is quickly oxidized in air. The oxide is stripped off using a weak HF dip (30 s in 100:1 HF), leaving a sample with regions that are <40 Å thick, have <1 Å rms roughness, and no amorphized surface layer. Ion milling must be avoided entirely, as it produces a damage layer which is not removed by the etch.

STEM images such as those of Fig. 3 or on the inset of Fig. 1 contain information for identifying different deactivating dopant clusters. Our ability to identify particular defect structures is limited by viewing the structure in projection. Thus, for example, the ⟨110⟩ projections used here allow us to see only three of the four Sb atoms in a Sb_4V defect, but such an image is also consistent with one of the projections from a Sb_3V defect. The electrical activity of a particular Sb atom cannot be firmly established for the same reason: what appears to be an inactive cluster-pair in the image could actually be two active Sb atoms at different depths. To overcome the projection problem, we have generated full three-dimensional structures using the image data and Monte Carlo simulation. Starting with the Sb-occupied column positions from the images, we populated each column with Sb atoms at random depths. Nearest-neighbor Sb atoms were not allowed, consistent with x-ray absorption measurements (Citrin et al 1999). The percentage of Sb atoms potentially involved in any one of the defect structures, and therefore the electrically inactive fraction, was then computed, subject to ± 3% error from counting statistics.

We find that defects with two Sb next-nearest neighbor atoms comprise 30 and 33% of the Sb in images A and B. Both values are in excellent agreement with the Hall measurements of the electrically inactive fraction of dopants (30%). The three and four atom Sb_3V and Sb_4V clusters are found to comprise only 1% of the Sb atoms. From these results, we conclude that the primary deactivating defect in these samples contains only two Sb atoms. The paired defects outnumber the 3- and 4-atom clusters by roughly 50:1. That the smaller defects dominate suggests it will be much harder to control this clustering by optimizing the processing conditions.

4. Mapping Si/SiO₂ Interface States with EELS

Even if the interface structure was known of Si/SiO_2, the connection between the physical arrangement of atoms at the interface and their electrical properties is neither direct nor obvious. Instead, it is far simpler to measure the averaged electronic states that directly determine the electrical properties of the interface. This has been done with atomic-scale electron energy loss spectroscopy (Batson 1993, Muller et al 1999), which maps the unoccupied electronic density of states by site, atom-column and atomic species. These measurements give localized information about both chemical composition and electronic properties. The scattering geometry of the microscope is optimized so that the EELS measurement is dominated by dipole transitions from core states to unoccupied final states, and can selectively resolve spectra from a particular column of atoms, and so therefore comparable to a local densities of unoccupied states. Interactions between the excited electron and the hole left behind in the core level can distort the shape of the spectrum. The resulting exciton on the Si L edge is sufficiently pronounced that the spectrum no longer reflects the ground state density of states (DOS) (Neaton et al 2000). In contrast, the core hole on the O-K edge only weakly perturbs the spectrum (Neaton et al 2000), so it can be used to map out the O p-DOS. Furthermore, the O-K edge produces a more localized signal than the Si-L edge, making it easier to isolate the interface states (Muller et al 1999), although these can be detected using the Si-L edge (Batson 1993).

Atomic-scale EELS from atomically abrupt interfaces [001] reveals 1-2 monolayers of suboxide with a distinct fingerprint on the O-K edge (Muller et al 1999). The different shapes for the interfacial and bulk O EELS spectra are used map out the number of bulk and interfacial states in rougher oxides. The width of the interfacial layer (0.7 nm total – 0.35 nm per interface) is seen to be constant with oxide thickness (Fig. 4). Consequently, a 0.7 nm thick gate oxide would not retain any bulk-like electronic structure. It is important to note that EELS (like the electronic structure changes it directly reflects) is also affected by second neighbor changes. *That is, the electronic width of the interface is always larger than the physical width.*

Our finding of an interfacial layer can be understood using the moments theorem (Cyrot-Lackmann 1968, Neaton et al 2000). Such an analysis shows the number of states in the local energy gap is proportional to the number of O 2nd neighbors around a given O atom, and the bulk SiO_2 gap is only obtained for a fully-coordinated 2nd neighbor shell (Muller et al 2001b, Neaton et al 2000). An oxygen atom at a perfectly abrupt Si/SiO_2 interface must be missing at least half its O neighbors – if it wasn't then it would not be the last atom in oxide! Since the O-O distance is about 0.27 nm, this is the minimum electronic width the interface can be before bulk-like properties are attained. However, the 50% bond-density mismatch between Si and SiO_2 adds a monolayer of suboxide (another 0.16 nm). Taking in to account the bond angle distortions, and the presence of 2 interfaces (top and bottom), this puts a fundamental limit of 0.7 - 0.8 nm on the oxide thickness in order for bulk SiO_2 properties to be achieved, suggesting there is little room to improve on our experimental results. Interface roughness increases the minimum usable thickness for present commercial growth techniques and severely impacts the performance of transistors using gate dielectrics with effective oxide thicknesses less than 1.2 nm. As most replacement gate dielectrics contain subnanometer-thick silicon oxide layers, similar physical constraints apply.

In the search for replacements for the SiO_2 gate oxide the materials considered must at least satisfy four important constraints: (1) the material must have a higher dielectric constant, so a larger capacitance can be obtained with a thicker layer; (2) low leakage current means a large band offset (>1eV) so materials like TiO_2 or $SrTiO_3$ are unsuitable; (3) the material should not react with silicon to form SiO2 or silicides – even a 1nm thick interlayer is thicker than what could be done with SiO_2 alone; and (4) a low density of defects states (< $10^{10}e/cm^2$ of fixed and trapped charges) is needed to prevent reductions in mobility and turn-on voltage. The ultimate test is in the construction of a transistor that has a higher drive current than that of an SiO_2-based device, as well as a leakage current that is less. While reducing the leakage current has proved to be relatively easy, no device produced to date has yet shown an improvement in drive current over that of the best SiO_2-based devices (Wallace et al 2002).

The problem is essentially criterion 4 as a flatband or threshhold voltage shift of even a few tens of millivolts impacts device performance. With proper control of vacancies and other point defects, the fixed charge might be reduced to a level competitive with the best SiO_2 gate oxides (10^{10} e/cm^2 or better).

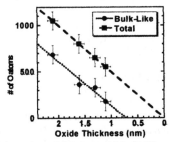

Figure 4. The number of O atoms with bulk-like and interfacial bonding arrangements (from EELS) plotted as a function of gate oxide thickness. There is no more bulk-like silicon dioxide when the gate oxide is less than 0.7 nm thick. Only the interface states remain.

5. Summary

The electronic structures of gate oxides have been mapped at the atomic scale using EELS. The finite width of the interface states, observed both in experiment and ab-initio simulations places a fundamental limit of 0.7 nm on the minimum thickness for an bulk-like SiO_2 gate oxide. Most candidate replacement dielectric incorporate SiO_2 layers – some accidentally and some by design. There is considerable room for improvement in the design of replacement gate dielectrics, and spatially resolved EELS can play a useful role in guiding and elucidating this process. ADF imaging has been demonstrated to provide quantitative information on the distribution of dopant atoms and nature of small defect clusters. The dominant deactivating cluster in Sb-doped silicon at high concentrations is from paired dopants, which outnumber 3 and 4 atom clusters by roughly 50:1.

References

Batson P E. 1993 Nature **366**, 728

Batson P E et al. 2002 Nature **418**, 618-620

Baumann F H et al. 2000 Mat. Res. Soc. Symp **611**, C4.1.1 - C4.1.12

Black G & Linfoot E H. 1957 Proc. Roy. Soc. (London) **A239**, 522

Castell M R et al. 2003 Nature Materials **2**, 129-131

Chadi D J et al. 1997 Phys. Rev. Lett. **79**, 4834-4837

Citrin P H et al. 1999 Phys. Rev. Lett. **83**, 3234-3237

Crewe A V et al. 1970 Science **168**, 1338-1340

Cyrot-Lackmann F. 1968 J. Phys. Chem. Solids **29**, 1235 -- 43

Ebert P. 2001 Current Opinions in Solid State and Materials Science **5**, 211 - 250

Fair R B & Weber G R. 1973 J. Appl. Phys. **44**, 273-279

Gossmann H-J et al. 2000 Mat. Res. Soc. Symp. **610**, B1.2.1 - B1.2.10

Gossmann H-J et al. 1993 J. Appl. Phys. **73**, 8237-8241

Hillyard S E & Silcox J. 1995 Ultramicroscopy **58**, 6-17

Kaiser U et al. 2002 Nature Materials **1**, 102-105

Muller D A & Grazul J. 2001a J. Elec. Microsc. **50**, 219-226

Muller D A & Neaton J D. in *Fundamental Aspect of Silicon Oxidation* (ed. Chabal, Y.) 219 - 246 (Springer, New York, 2001b).

Muller D A et al. 1999 Nature **399**, 758-761

Muller E W. 1957 J. Appl. Phys. **28**, 1-6

Neaton J B et al. 2000 Phys. Rev. Lett. **85**, 1298 -1301

Nellist P D & Pennycook S J. 1996 Science **274**, 413-415

Packan P A. 1999 Science **285**, 2079-2081

Pasquarello A et al. 1996 Phys. Rev. **B53**, 10942

Pennycook S J. 1989 Ultramicroscopy **30**, 58-69

Perovic D D et al. 1993 Ultramicroscopy **52**, 353-359

Silcox J et al. 1992 Ultramicroscopy **47**, 173

Timp G et al. 1998 IEDM Technical Digest **San Francisco, 6-9 Dec**, 615-618

Vanfleet R R et al. in *Characterization and Metrology for ULSI Technology: 1998 International Conference* (eds. Seiler, D. G. et al.) 901-905 (1998).

Voyles P M et al. 2002 Nature **416**, 826-829

Wallace R M & Wilk G D. 2002 Mat. Res. Soc. Bulletin, 186-191

Williams J S & Short K T. 1982 J. Appl. Phys. **53**, 8663-8667

Inst. Phys. Conf. Ser. No 179: Section 1
Paper presented at Electron Microscopy and Analysis Group Conf. EMAG2003, Oxford, 2003
©*2003 IOP Publishing Ltd*

Tomographic Diffractive Imaging for Nanoscience and Biology

J C H Spence

Department of Physics and Astronomy, Arizona State University, Tempe, Az. USA 85287-1504.

Abstract. Progress in the lensless reconstruction of aberration-free three-dimensional images from microdiffraction electron and X-ray diffraction patterns is reviewed. The hybrid input-output algorithm of Fienup has been successful recently in phasing both experimental X-ray and electron diffraction patterns from isolated nanostructures, including a single nanotube. The use of iterative phasing methods is proposed to reduce the number of HREM images needed at high tilts for cryomicroscopy of proteins. A new iterative algorithm has been described which does not require an independent estimate of the boundary of the unknown object.

1. Introduction

Nanoscience (described by Nobel Laureate Raul Hoffman as "the next industrial revolution") requires three-dimensional internal views of nanostructures at atomic resolution. A similar capability is desired in structural biology, however few proteins diffract to atomic resolution, and radiation damage is dominant. (The atomic structure of the twenty constituent amino acids is, however, known, and these can be distinguished at about 0.25nm resolution). Cryomicroscopy of monolayer crystals addresses the important problem of solving, for example, membrane proteins important for drug delivery, which are very difficult to crystallize in three dimensions. Materials scientists (including polymer scientists) have been slow to adopt the tomographic methods of cryomicroscopy in biology, and only recently have we seen the first sub-nanometer tomographic TEM images of catalyst particles [1]. In biology one must take advantage of crystallographic redundancy (or image many randomly oriented identical particles) to deal with the radiation damage problem. Then the chief difficulty is that of recording high resolution images of, say, a crystalline monolayer at large angles of sample tilt, in order to minimize the "missing cone" of data. In materials science, imaging the same nanoparticle at atomic resolution through the wide range of orientations needed for tomography is almost as demanding. In this paper we describe approaches to atomic resolution tomography which are based on the collection of diffraction data alone for this purpose. Such images can be expected to be diffraction limited, un-aberrated, and, for the imaging of two-dimensional proteins, can be formed at very large numerical aperture, thereby improving resolution for a given damaging radiation dose. We give recent examples of this "coherent diffractive imaging" (CDI) method using both electrons in a TEM and soft X-rays.

2. Theory.

Over the past fifty years, three new ideas have emerged among different disciplines, which, taken together, now provide us with a working solution to the phase problem for non-periodic objects. This has made possible the recent atomic-resolution reconstruction of an image of a single carbon nanotube from an electron diffraction pattern [2], and three-dimensional X-ray imaging [3]. In the X-ray community, the early suggestion of Sayre that Bragg diffraction undersamples the transform of the autocorrelation function of a molecule in a crystal (which therefore differs from a Patterson function) was the first of these ideas [4]. In the electron microscopy community, where real-space images are readily available, the question arose in the early seventies as to whether an algorithm which iterates between measured diffracted intensities and the corresponding experimental image intensity might be used to phase both, and so provide super-resolution [5]. This work came to the attention of experts in signal processing, who were able to derive from this algorithm an improved version (including feedback) which required only diffracted intensities and a rough estimate of the boundary (support) of the object to solve the phase problem [6]. The development of this iterative algorithm was the second breakthrough. (A "real" (weak phase) object are both assumed in early work). The resulting "hybrid input-output" (HIO) algorithm was used to analyze the aberrations of the Hubble space telescope, and to provide the first inversion of an experimental optical speckle pattern to an image [7]. At the same time, a considerable literature was generated in the image processing community both analyzing the uniqueness and convergence properties of the HIO algorithm [8] [9], and exploring other related approaches based on analyticity [10]. Finally, a powerful insight was provided by applied mathematicians who pointed out that the HIO iterations can be interpreted as Bregman projections between constrained sets (see [11] for a review), leading to the development of improved algorithms [12, 13]. An important review, which attempted to integrate the approaches of the optical and crystallographic communities, appeared in 1990 [14]. Here the connection was made between the "solvent-flattening" or "density-modification" techniques of crystallography and the compact support requirements of the HIO algorithm. The importance of fine sampling of the diffraction pattern was recognised at an early stage in Bates' work [15], and has lead to the method being referred to as "oversampling", since the Shannon sampling is half the Bragg sampling in one dimension. We point out, however, that the sampling required in the HIO algorithm is simply that required by the Shannon sampling theorem (with the autocorrelation of the object treated as a "band-limit"), so that the HIO sampling is optimal, rather than oversampled. The relationship between the number of unknown phases, the number of equations available and the sampling interval was analysed in detail in 1998 [16].

Until recently, these theoretical advances were not accompanied by corresponding experimental advances, and, like economics, image processing had acquired the reputation of something of a dismal science. For the X-ray community, a major breakthrough therefore occurred in 1998, with the first inversion of an experimental diffraction pattern to an image of a non-periodic object [17] at about 100nm resolution. Since then we have seen experimental three-dimensional reconstructions at higher resolution [3]. Since the early Gerchberg-Saxton papers, little experimental research was undertaken using electron beams based on the HIO algorithm until the recent reconstructions of small holes [18] and the atomic-resolution image of a nanotube mentioned earlier [2]. Looking back, we see how crucial results in one field were ignored by others for decades, how important are interdisciplinary review articles, and how powerful is the recent synergy in this field between astronomers, X-ray crystallographers, the signal processing community, applied mathematics and electron microscopy.

The HIO algorithm works by iterating between real and reciprocal space, applying known information in each domain. In real space, this is the support of the object and the sign of the scattering potential. In reciprocal space it is the known Fourier Modulus. For real objects a triangular-shaped support, larger than the object, will usually produce convergence. Details of the algorithm are given in [19]. Figure 1 shows how the autocorrelation function of the object (the transform of the diffracted intensity) acts as a "bandlimit", and so defines the sampling interval of the scattered intensity. Phase information is encoded in this intensity and may be extracted by the HIO algorithm if this sampling is used, since it ensures that the number of Fourier equations is at least equal to the number of unknown phases [16].

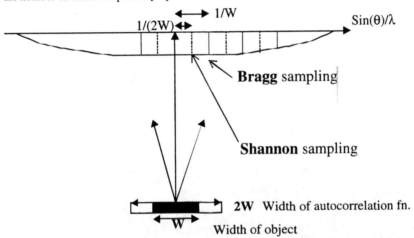

Figure 1. The autocorrelation function (of width 2W) acts as a bandlimit for a molecule of width W to define the Shannon sampling interval required of the Fraunhoffer diffraction intensity shown.

3. Experimental CDI with electrons.
Since the earliest days of electron microscopy (TEM) there has been a continuous effort to take advantage of both real-space and diffraction pattern information, since both are easily obtained by the instrument from a common area in the "selected area diffraction" mode. However almost all this work has been devoted to crystals (to which HIO is not readily applied, but see section 4), and these efforts have had limited success. Since the current resolution of the instrument in image mode is about one Angstrom [20], one might inquire why CDI is pursued. The reasons have to do with the need for tomographic atomic-resolution images of nanostructures, and with the practical difficulties of obtaining the needed images (which show a projection of the structure) at very high

angles of sample tilt. If a coherent probe larger than an isolated nanoparticle is used, the vibrational and electronic stabilities are much less for CDI than for HREM imaging.

The first attempt at CDI based on the HIO algorithm was reported   recently [18]. A special opaque mask was created by lithography, containing irregularly shaped holes of about 50nm diameter. The coherence width of the field-emission electron beam spanned both holes, and the intention was first to reconstruct images of the holes from the Young's fringe-like diffraction pattern. In a second step, it was planned to fill the holes with nanoparticles of magnesium oxide (or molecules), which would be reconstructed at atomic resolution. To reduce penetration of the opaque areas of the mask, the field-emission TEM was operated at low voltage (30kV). The resulting reconstructed CDI images showed clear images of the holes at about 2nm resolution, rather less than that of the TEM in conventional imaging mode. The reasons for this were traced to inelastic scattering in the walls of the holes, poor diffraction-mode alignment and the use of a CCD detector rather than image plates. Simulations were included demonstrating the atomic-resolution capability of electron CDI, and the possibilities of phasing coherent nanodiffraction patterns from complex objects (to allow for multiple scattering) were explored by using prior knowledge of the probe wavefield as a support [21].

Since that work there has been a dramatic advance. Using an image-plate detector (to avoid the blooming of the CCD), Zuo et al [2] have obtained the striking image of a single nanotube at atomic resolution shown in figure 2(b) from the nanodiffraction pattern shown in figure 2(a). From this they obtain the diameter, number of walls (2) and chirality of the tube. The tube lay across a hole in a carbon film larger than the fully-coherent electron beam, and it would appear that compact support was obtained only in the dimension across the tube. The use of any physical supporting medium (such as amorphous carbon) is found to cause the HIO algorithm to fail due to the strong electron interaction with it, so that the challenge for this field is to find ways of supporting isolated objects, perhaps though the use of laser tweezers [22].

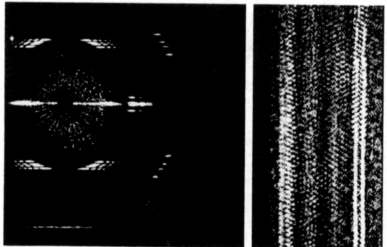

Figure 2. Electron diffraction pattern from a double-walled nanotube at left (a) and reconstructed image (b) at right. Phases of the diffraction pattern were found by HIO [2].

4. Iterative methods for tomographic imaging of proteins.

Recently, a way has been found to apply the HIO algorithm to two-dimensional protein crystals in cryomicroscopy (TEM), in order to greatly reduce the number of high-resolution tilted images needed for tomographic reconstruction [23]. In conventional cyromicroscopy, one HREM image is recorded for each diffraction pattern, and the diffraction patterns are much easier to record, particularly at high sample tilts, than the images. We have applied a one-dimensional support condition normal to the slab to assist with phasing along the reciprocal space rods. Some phases are then needed to link the rods, but these can be obtained from a few images at small tilts. Figure 3 shows the principle of the method, figure 4 results from a simulation for a small molecule, and figure 5 simulated results for the protein lysozyme. Figure 6 shows how noise affects the results. The correlation coefficient between the true lysozyme density and the HIO estimate is given as a function of iteration for different noise levels and ranges of image tilts. The jumps in the HIO error occur when switching between HIO and error-reduction algorithms. The algorithm rapidly converges in the presence of noise with few images. We note that the HIO error in the error-reduction algorithm is monotonic with CC, so that it provides a good predictor of convergence for unknown structures, for which CC is not known.

5. Imaging nanostructures with soft-X-rays.

Similar methods may be applied to the phasing of transmission X-ray "speckle" patterns. In recent work, we have attacked the problem of estimating the support of the object from the known support of its autocorrelation function. First results were encouraging, allowing images of 50nm diameter gold balls to be reconstructed from transmission X-ray patterns recorded at 588 eV [24]. More recently we have incorporated the "support-finder" into the HIO algorithm itself. The boundary of the autocorrelation function is taken as the first estimate of the support of the object, and improved estimates are obtained with each iteration of the HIO algorithm. The resulting "shrink-wrap" algorithm is thus capable of solving the phase problem for an isolated object with no a-priori knowledge of the object, except the sign of the scattering potential and a knowledge of the scattered intensity. This new algorithm has been applied to experimental soft-X-ray patterns from similar clusters of gold balls lying on a transparent silicon-nitride membrane [25]. The extension of the method to complex objects is possible for objects separated into two parts [26], however this requires a rather accurate knowledge of the object support. Tomographic soft-Xray reconstructions of lithographed objects have also been achieved recently [3].

A new analysis of the dependence of radiation damage on resolution and dose, based on both electron spot-fading, X-ray crystallography and X-ray zone-plate imaging is in progress – preliminary results have been published [27]. In this work we attempt to span the resolution ranges of these three techniques, and to consider any radiation-damage advantages of tomographic diffractive imaging for biology.

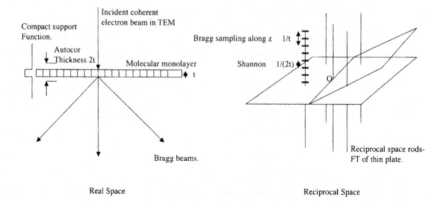

Real Space   Reciprocal Space

Figure 3. Geometry of "oversampling" for a two-dimensional crystalline slab. The compact support constraint is applied in real space normal to the molecular monolayer. In reciprocal space the intensities along the relrods are assumed known from a diffraction pattern tilt series. Known phases are supplied on a few planes approximately normal to the relrods. The HIO-algorithm then reconstructs the phases along the relrods.

Figure 4. Isopotential view of the electrostatic potential of TCNE as reconstructed by the HiO algorithm from simulated diffracted intensities (to 0.35 Å) combined with 3 images (to 0.35 Angstroms). The electron beam runs across the page. The computational supercell is shown, the size in the z (horizontal) direction is three times the thickness of the monolayer, providing compact support. The size in the x and y direction is one unit cell (0.97nm).

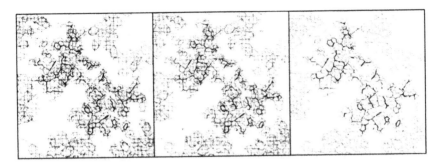

Figure 5. A section of the unit cell of lysozyme normal to z, comparing model and retreived structures. Left: The model. Center: Reconstructed using HIO, with noise-free amplitudes and image phases up to 15 degree tilt angles. 105 iterations. Right: Similar reconstruction, with noise added to structure factors corresponding to R=25% and using image phases up to 15 degree tilts. These are part of a three-dimensional reconstruction.

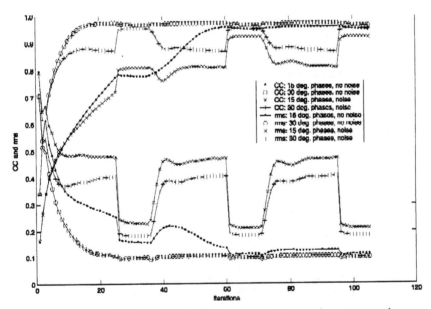

Figure 6. Cross-correlation coefficient CC (upper curves) measuring agreement between model lysozyme potential and the HIO estimate as a function of iteration number. The HIO error is also shown (lower curves). Calculations include starting phases obtained from images at tilt angles up to 15 or 30 degrees, with or without noise as shown in the key. Jumps in lower curves are due to switches between algorithms.

Acknowledgement: This work was supported by ARO award DAAD190010500 and DOE award DE-AC03-76SF00098.

16

References

1. Midgley, P.A., M. Weyland, J.M. Thomas, and F.G. Johnson, Chem. Commun., 2001. **2001**: p. 907.
2. Zuo, J.M., I.A. Vartanyants, M. Gao, M. Zhang, and L.A. Nagahara, Science, 2003. **300**: p. 1419.
3. Miao, J., T. Ishikawa, E.H. Johnson, B. Lai, and K. Hodgson, Phys Rev Letts, 2002. **89**: p. 088303.
4. Sayre, D., Acta Cryst, 1952. **5**: p. 843.
5. Gerchberg, R. and W. Saxton, Optik, 1972. **35**: p. 237.
6. Fienup, J.R., T.R. Crimmins, and W. Holsztynski, J. Opt Soc Am, 1982. **72**: p. 610.
7. Cederquist, J.N., J.R. Fienup, J.C. Marron, and R.G. Paxman, Optics Letters, 1988. **13**: p. 619.
8. Seldin, J. and J.R. Fineup, J. Opt Soc Am, 1990. **7**: p. 412.
9. Barakat, R. and G. Newsam, J Math Phys, 1984. **25**: p. 3190.
10. Liao, C., M. Fiddy, and C. Byrne, J. Opt Soc Am, 1997. **A14**: p. 3155.
11. Stark, H., *Image Recovery: Theory and applications.* 1987, New York: Academic Press.
12. Bauschke, H., P.I. Combettes, and R. Luke, J. Opt Soc Am, 2002. **19**: p. 1334.
13. Elser, V., J. Opt. Soc. Am., 2003. **20**: p. 40.
14. Millane, R., J. Opt Soc Am, 1990. **7**: p. 394.
15. Fright, W. and R. Bates, Optik, 1982. **62**: p. 219.
16. Sayre, D., H. Chapman, and J. Miao, Acta Cryst., 1998. **A54**: p. 232.
17. Miao, J., C. Charalambous, J. Kirz, and D. Sayre, Nature, 1999. **400**: p. 342.
18. Weierstall, U., Q. Chen, J. Spence, M. Howells, M. Isaacson, and R. Panepucci, Ultramic., 2001. **90**: p. 171.
19. Fienup, J.R., Applied Optics, 1982. **21**: p. 2758.
20. Spence, J.C.H., *High Resolution Electron Microscopy.* 3rd edition ed. 2003, New York: Oxford University Press.
21. Spence, J., U. Weierstall, and M.Howells., Phil Trans., 2002. **360**: p. 875.
22. Weierstall, U., J. Spence, and G. Hembree, Optics Express, 2003. **In press.**
23. Spence, J., U. Weierstall, T. Fricke, R. Glaeser, and K. Downing, J. Struct Biol, 2003. **in press.**
24. He, H. Marchesini, S. Howells, M., Weierstall, U. Hembree, G., Spence, J.C.H. Acta Cryst. (2003) A59, p. 143
25. Marchesini, S. He, H., Chapman, H. Hau-Riege, S., Noy, A., Howells, M., Weierstall, U., Spence, J. Phys. Rev. B. (2003). In press.
26. Fienup, R. J. Opt Soc Am. (1987) A4, p. 118.
27. Howells, M. Chapman, H. Hau-Riege, S. London, R. , Szoke, A., Howells, M. He, H. Padmore, H. Rosen, R., Spence, J, Weierstall, U. Optics Express, 2003. In press.

Inst. Phys. Conf. Ser. No 179: Section 2
Paper presented at Electron Microscopy and Analysis Group Conf. EMAG2003, Oxford, 2003
©2003 IOP Publishing Ltd

Electron Energy Loss Spectroscopy of Extended Defects

U Bangert[1], A Gutierrez-Sosa[1], A Harvey[1], R Jones[2], C J Fall[2], A Blumenau[2,3], R Briddon[4]

[1]Department of Physics, UMIST, Manchester M60 1QD, UK
[2]School of Physics, University of Exeter, Exeter EX4 4QL, UK ·
[3]Department of Physics, Faculty of Science, University of Paderborn, Warburgerstr 100, D-33098 Paderborn, Germany
[4]Department of Physics, University of Newcastle upon Tyne, Newcastle NE1 7RU, UK

Abstract. The advent of cold field emission scanning transmission electron microscopes with high mechanical and energy stability and a monochromacity of 0.3 eV has made spectroscopy of sub-nm size features possible. High spectrometer dispersions (0.01 eV/channel) furthermore allow electron energy loss spectroscopy (EELS) to be carried out at low energy losses, i.e., in the bandgap regime of semiconductors with bandgaps down to 2 eV. In this paper the technical procedures and the evaluation of low loss spectra, as well as limits imposed by energy resolution, point spread function and delocalisation are highlighted. The 'spectrum imaging' technique employing the UHV ENFINA EELS detector and software package has proven invaluable for the detection of the subtle changes involved with dislocations: these may go unnoticed in the balance between local stability and spectrum statistics, which this kind of highly spatially resolved spectroscopy entails. However, in contrast to results from singular spectra, a spectrum map captures systematic changes over an extended area in direct correlation with the microstructural feature. EELS results of dislocation obtained of GaN are presented. Supported by *ab initio* calculations [1,2], changes to the density of states can be observed in low loss spectra maps at individual dislocations at bandedge. Preliminary results on pre-bandedge absorption of basal stacking faults in SiC are also reported.

1. Introduction

The electronic structure of extended defects in electronic and opto-electronic materials is of great interest and has been the focus of a large body of work over the past decade, and the issue is still controversial. Concerning ourselves first with GaN, predictions based on bandstructure calculations of various dislocation types and core structures, stating that all dislocations give rise to gap states (e.g. Fall et al, Phys Rev B 65, 245304 (2002)), are to date accepted. Despite the high threading dislocation densities optical devices of GaN are believed to work because of carrier localisation due to inhomogeneities of the GaInN alloy, which constitutes the active device region [e.g., Duxbury et al, 2000]. Thus carriers would not reach the dislocations. How dislocations affect the long term performance, however, is still in question. A substantial body of information on separate aspects of electronic activity of dislocations has been accumulated so far: there is agreement that dislocation cores are negatively charged in n-material (undoped) (e.g. Saarinen, Cherns, both 2003 ONR workshop *unpublished*, Cai and Ponce 2002) and that negative, neutral or positive charge arises in semi-insulating (Zn-doped) and p-type (Mg) material (Srinivasen et al 2002). Contact AFM has shown that charged dislocations are electrically non-conducting, whereas non-charged screw dislocations can be conducting (Simpkins and Yu 2003 ONR workshop *unpublished*). More significant than that of pure dislocation cores is believed to be the effect of point defects interacting with extended defects on the electronic properties in GaN materials: It has been variously reported that V_{Ga} may be trapped in the strain field of dislocations (e.g. Elsner et al 1998), and recent experiments have shown that V_{Ga} are indeed trapped at grain boundaries and hence at edge dislocations (Saarinen 2003 ONR workshop *unpublished*). Other point defect interactions include formation of traps resulting from impurity segregation to dislocations (Look and Fang 2003 ONR workshop *unpublished*), substitution of nitrogen by oxygen in open core screw dislocations (Browning and Arslan 2003 ONR workshop *unpublished*) and dopant segregation (e.g. Mg) to inversion domain boundaries (Northrup 2003 ONR workshop *unpublished*). Threading dislocations with *a*-Burgers' vectors in MOCVD GaN have proven optically active in cathodo luminescence, *c* and *c+a* dislocations, on the contrary, have not (Strunk 2003 ONR workshop *unpublished*, Albrecht et al 2002). Furthermore, stacking faults induce a luminescence band (Skromme et al 2003 ONR workshop *unpublished*)

In SiC a distinct relationship between dislocation and device performance has been identified: thermal and electrical stresses induce slip of partial dislocations, which in turn leads to device degradation. Both, stacking faults and partials are optically and electrically active (Skowronski, ONR workshop 2003 *unpublished*, Lendenmann et al 2001, Liu et al 2002, Galeckas 2002, Bergman et al 2003 ONR workshop *unpublished*). A model for partial dislocation slip has been suggested, whereby due theoretically predicted gap states of the Si partial recombination enhanced glide might occur (Blumenau 2003 ONR workshop *unpublished*). The electrical and optical activity of basal stacking faults is explained by the fact that atomic planes of the 3C variant embedded in 4H (or 6H) SiC introduce quantum wells, to which carriers are lost (Glembocki et al, Skromme et al, both 2003 ONR workshop *unpublished*).

Spatially resolved electron energy loss spectroscopy (EELS) is one of the few experimental techniques to give direct information about electronic levels associated with dislocations. The technique has become refined enough to give information of electronic states down to 2 eV (above valence band). The energy loss in the region following the zero loss peak (ZLP) tail up to 10 eV arises from inter- and intraband scattering. Given sufficient energy resolution and a fast enough decaying point spread function (PSF), the low loss intensity in wide-gap semiconductors can be observed to drop to zero before the inter/intra band scattering events set in, and the first subsequent rise corresponds to the bandgap.

This paper sets out to examine the possibility of measuring bandgap states in wide gap semiconductors and to reconcile the measurements with calculations of the joint density of states (JDOS) and of low EEL spectra from first principle methods. It furthermore investigates, whether the JDOS of different dislocations types can be discerned unambiguously, and how much can be revealed about the nature of electronic states at dislocations.

2. Experimental

The experimental details including measurement procedures and sample geometry considerations, have been published elsewhere (Sosa et al 2002, Bangert et al 2002). It should be re-emphasised that high spectrometer dispersions (0.02 or 0.01 eV/ch) are required. An energy resolution of 0.3 eV, as obtainable with a cold field emission source, enables pre-bandedge features of down to approximately 0.2 eV below conduction band to be detected. With a spatial resolution of not much lower then 10 nm this is not an atomic resolution technique, however, it provides a highly localised absorption spectroscopy. The method, as we currently employ it, does not yield quantitative results, the prospects of extracting those depend strongly on the model used for background extraction from the spectra and, on the theory side, on how accurately the transition matrix elements can be established. In principle, however, quantitation ought to be possible. For background extraction, subtracting a measured ZLP from the low loss spectrum seems to be the obvious way, however, a pure ZLP, i.e., without specimen, does not produce the same shape as the ZLP of a low loss spectrum. Ideally a low loss spectrum of a very large bandgap material in a region of identical thickness to that of the material of interest ought to be subtracted from the spectrum of interest; this is, however, hardly ever possible! Deconvolution of an experimental PSF or ZLP from the low loss spectra has been carried out (Rafferty and Brown xxxx). This method, however, invariably introduces artefacts. Constructing a theoretical ZLP tail has been attempted, using e.g., the Fowler-Nordheim distribution and theoretical or measured spectrometer point spread functions (PSF) (Bangert et al 1997). This has proven cumbersome. Furthermore, fitted curves to the ZLP tail, using, e.g., Bessel functions, Lorentzians, exponentials or polynomials have been used by various researchers (Reed et al) with limited success, as the shape of the fitted curve depends strongly on the fitting regime, and might not produce appropriate backgrounds in case of smaller band gaps. We found that a double exponential and the GATAN 'background removal' routine work well for medium and large bandgaps, they are however, not quantitative. Else, for large bandgaps (e.g. diamond) no background subtraction at all is required.

3. Spectrum Imaging

The spectrometer employed for the low EELS is a UHV Enfina (Gatan), which allows spectrum mapping or 'spectrum imaging' (SI). Spectrum mapping greatly facilitates the visualisation of defects and enabling the detection of marginal spectral differences in on- and off-defect locations, which are unrecognisable using point spectra measurements. This is due to the fact that SI maps may contain many spectra e.g., along a dislocation line. This is equivalent to accumulating a statistical data set. To 'enhance the evidence', spatially correlated trends can be used. To combine statistics, spatial resolution requirements, microscope and spectrometer stability, spectrum maps of between 100 and 400 pixels were usually taken with pixel frame lengths of 5-10 nm, dwell times of 0.1-0.3 s and up to 10 read-outs per pixel.

Figure 1 shows a STEM BF image of an edge dislocation and the area (white frames), in which spectrum maps were taken. The intensity maps represent the normalised

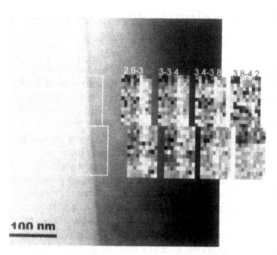

Fig. 1   BF STEM image of an edge dislocation. White frames indicate areas, where spectrum maps were taken. To the right are intensity maps corresponding to different energy windows (see text)

integrated counts in the respective energy window (shown by the numbers above the maps) for the spectrum in each pixel as grey value. The band of relative brighter intensity running from top to bottom through the maps, corresponding to the counts in energy slices 2.6-3 eV and 3-3.4 eV, show that the edge dislocation exhibits enhanced absorption at these energies. The band of bright contrast subsides at energies above 3.4 eV. This means that the dislocation has band gap states below the conduction band edge. The intensity values are normalised (explanation below), so that effects arising from sample thickness variations are cancelled. Measurements on screw dislocations yield similar pronounced intensity at the dislocation line, although the intensity profile as a function of the energy varies slightly. In order to investigate the fingerprint of a particular core structure, individual spectra must be scrutinized and compared to theoretical low loss spectra.

4. Calculations of low EEL spectra

Rather then extracting the JDOS from the measured spectra, we took the approach to calculate low loss spectra from first principle methods, and compare them to the measurements. This way spectrum processing was kept at an absolute minimum and artefacts introduced to the experimental spectra as a result of deconvolution routines are avoided. The theoretical method has previously been explained in great detail (Fall et al 2002a, Fall et al 2002b). In essence *ab initio* calculations using the local density approximation (AIMPRO code) are carried out: a cluster or supercell containing atoms in the required positions is defined and for these the bandstructure is calculated. The transition matrix elements are summed over all the bands and integrated over the Brillouin zone. This yields the imaginary part of the dielectric function, ε_2. The real part, ε_1, and thus the complete dielectric function, ε, is found via Kramers-Kronig transformation. The loss signal then is proportional to $Im\{-1/\varepsilon\}$.

Figure 2 shows calculated low loss spectra for a neutral and a negatively charged edge dislocation as well as for a full-core and a Ga-core screw dislocation. It should be noted that the detailed loss intensity depends on the e-beam direction with respect to the dislocation line. Since we look at cross-sections with the e-beam orthogonal to the dislocation line, the relevant curves are the dotted and the dash-dotted lines. All dislocation core structures show gap states, although they clearly differ from each other, the charged edge exhibits an intensity tail protruding deep into the bandgap with a peak at a deep level. The full core screw similarly tails into the gap and exhibits a peak deep in the gap. The Ga-core screw appears to have states all through the gap.

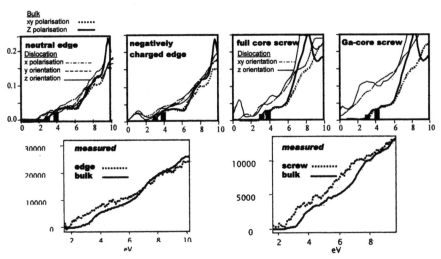

Figure 2. Theoretical (top row) and measured low loss spectra for edge and screw dislocations (bottom row)

5. Ratio method and difference spectra applied to dislocations in GaN

The shaded boxes in the upper panels indicate the appropriate energy windows for intensity integration and normalisation of the measured spectra. One window is set before and another just after the bandedge. From the calculations it would be predicted that the intensity ratio resulting from lower and higher energy window is higher at the dislocation then in the bulk. As the lower window is stepped through the bandedge towards the higher energy window, the ratio should change, e.g., in case of the charged edge dislocation it should gradually assume lower values, which are eventually indistinguishable from the bulk. This trend is confirmed by the intensity maps in fig.1 of edge dislocation and surrounding bulk.

The nature of the dislocation cannot be determined from the intensity maps alone. In order to compare the spectrum shape with the calculations spectra along the dislocation core were summed and a proportion of a bulk spectrum was then subtracted. The fraction of the bulk spectrum to be subtracted was established by estimating the ratio of dislocation core volume and surrounding un-dislocated material in the supercell used for the calculations. The bottom panel in fig.2 shows a difference spectrum for the edge dislocation in fig.1 and for a screw dislocation evaluated in the same way. The current background extraction routine using a fit to the tail of the ZLP does not allow us to extract scattering intensity below 2 eV with sufficient accuracy, and hence the deep gap states cannot be detected. Therefore it is difficult to distinguish between neutral and charged state of the edge dislocation. However, in the energy region between 3 and 4 eV the difference spectrum agrees better with that of the calculated negatively charged edge. For the screw dislocation the situation is clearer and the Ga-core can be ruled out. These results are in support of current statements concerning pure dislocations in n-type (un-doped) GaN.

6. Intensity maps of stacking faults in SiC
It has been observed that basal plane stacking faults in 4H-SiC produce a luminescence

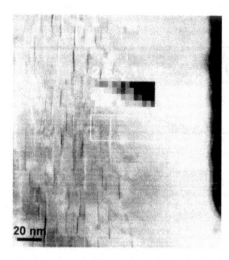

Figure 3. Basal plane stacking faults in SiC with spectrum map region (white frame) and intensity map.

band at 2.6-2.9 eV, whereas the bounding partials luminesce with energies below 2 eV (Galeckas et al 2002). Due to our restrictions we will not easily observe the partial gap states, however, the gap states giving rise to the stacking fault luminescence ought to be detected, in particular in a geometry, where the stacking fault is viewed edge-on, thus providing active volume throughout the entire thickness of the thin TEM foil. Figure 3 shows a STEM BF image of a stacking fault band, which was introduced by Ar-implantation. The white frame is the region, of which a spectrum map was taken. Two stacking faults intersect this region vertically in the left half. The two bright lines in the intensity map representing integrated counts in the energy slice 2.6-3.1 eV clearly show correlation with the stacking faults, in agreement with current findings.

7. Conclusions

Low EELS spectrum mapping, applied to wide bandgap materials, reveals density of states in the bandgap at extended defects in agreement with current beliefs. The nature of dislocation states, for example, of a negatively charged edge and a full core screw dislocation, could be identified, and the existence of states responsible for stacking fault luminescence in SiC could be proven.

References
Albrecht M, Strunk H P, Weyher, Grzegory I, Porowski S, Wosinski T 2002 J. Appl. Phys. 92(4) 2000
Bangert U , Gutierrez-Sosa A, Harvey A J, Jones R, Fall C J 2002 J. Appl. Phys 93(5) 2728
Bangert U, Harvey A J, Freundt D, Keyse R 1997 Ultramicroscopy 68 173
Cai J, Ponce F A 2002 Phys. Stat. Sol. A 192 407
Duxbury N, Bangert U, Dawson P, Thrush E J, Jacobs K, Van der Stricht W, Moerman I 2000 Appl. Phys. Lett. 76(12)1600
Elsner J, Jones R, Sitch P K, Porezag V D, Elstner M, Frauenheim T, Heggie M I, Oeberg S, Briddon P R 1998 Phys. Rev. B 58 12571
Fall C J, Jones R, Briddon P R, Blumenau A T, Frauenheim T, Gutierrez-Sosa A, Bangert U, Mora A E, Steeds J W, Butler J E 2002a Phys. Rev. B 65 205206
Fall C J, Jones, Briddon P R, Blumenau A T, Frauenheim T, Heggie M I 2002b Phys. Rev. B 65 245304
Galeckas A, Linnros J, Pirouz P 2002 Appl. Phys. Lett. 81(5)
Gutierrez-Sosa A, Bangert U, Harvey A J, Fall C J, Jones R, Briddon P R, Heggie M I 2002 Phys. Rev. B 66 035302
Lendenmann H, Dahlquist F, Johannson N, Soderholm R, Nilsson P A, Bergmann J P, Skytt P 2001 Mater. Sci. Forum 353-356, 727
Liu J Q, Chung H, Kuhr T, Li Q, Skowronski M 2002 J. Appl. Phys. 80 2111
Raffery B, Brown L M 1998 Phys. Rev. B 15 10
Reed B W, Sarikaya M 2002 Ultramicroscopy 93(1) 25
Srinivasen S, Cai J, Contreras O, Ponce F A, Look D C, Molnar R J 2002 Phys. Stat. Sol. C0 508

Inst. Phys. Conf. Ser. No 179: Section 2
Paper presented at Electron Microscopy and Analysis Group Conf. EMAG2003, Oxford, 2003
©2003 IOP Publishing Ltd

TEM assessment of As-doped GaN epitaxial layers grown on sapphire

M W Fay, I Harrison,[1] E C Larkins,[1] S V Novikov,[2] C T Foxon[2] and P D Brown

School of Mechanical, Materials, Manufacturing Engineering and Management

[1]School of Electrical and Electronic Engineering

[2]School of Physics

University of Nottingham, University Park, Nottingham NG7 2RD, UK

Abstract. TEM investigations of As-doped GaN layers grown by plasma-assisted molecular beam epitaxy on sapphire substrates reveal the presence of extensive regions of cubic stacking disorder within the hexagonal GaN matrix. Electron energy loss spectroscopy suggests the localization of As within grains immediately below domains containing stacking disorder, and additionally at the layer surface. This suggests that localised strain plays a role in the formation mechanism of the stacking faults.

1. Introduction

There is currently considerable theoretical and experimental interest in As-doped GaN. Three main reasons motivate such investigations. Firstly, the difference in the native crystal structures of GaAs and GaN and the large difference in their lattice parameters leads to a strong negative bowing of the Ga(As,N) band gap as a function of composition [1,2]. Secondly, As-doped GaN shows very strong room temperature blue emission at ~2.6eV, which raises the potential of using this material in blue light emitting diode (LED) applications [3-5]. Thirdly is the possibility of As-stimulated growth of zincblende GaN in a controlled fashion [6]. In this paper, we report on a TEM investigation of As-doped GaN films grown by PA-MBE that exhibit strong blue emission.

2. Experimental

As-doped GaN layers of varying thicknesses used in this study were grown on sapphire at a temperature of ~ 800°C, as described in detail elsewhere [7]. Active nitrogen for the growth was provided by an Oxford Applied Research (OAR) CARS25 RF activated plasma source. The nitrogen flux was about 3×10^{-5} mbar beam equivalent pressure (BEP). Arsenic, in the form of dimers (As_2) produced using a two-zone purpose made cell, provided a flux of about 7×10^{-6} mbar BEP. Prior to layer growth, the sapphire substrates were exposed to nitrogen at a temperature of ~800°C for 30 minutes in the same MBE reactor. Cross-sectional specimens for TEM were prepared by sequential mechanical polishing and dimpling, followed by argon ion milling to electron transparency. These specimens were examined using a JEOL 4000fx operating at 400kV equipped with a Gatan Imaging Filter (GIF) for chemical analysis. X-ray diffraction techniques were also used to study the structural properties of this sample set.

3. Results and discussion

X-ray diffraction studies previously confirmed that the As-doped GaN samples contained a mixture of phases, i.e. {0001} oriented hexagonal GaN, {111} oriented cubic GaN and {111} oriented cubic GaAs with respect to the growth plane [8]. The proportion of cubic GaN decreases rapidly with increasing film thickness suggesting that the majority of these domains are close to the epilayer/substrate interface. From the X-ray intensity of the GaAs peaks relative to GaN, it was concluded that GaAs constituted ~0.03% of the total volume of these films, with this fraction being roughly constant for films of different thickness. The volume fraction of cubic GaN was similarly determined to be 4-9% for samples with epilayer thicknesses ranging from 1 μm to 2.5 μm.

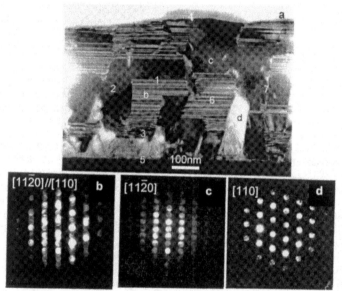

Figure 1. Focused probe diffraction patterns obtained from the areas labelled in (a) the bright field image for (b) a region containing stacking faults; (c) the wurtzite GaN<11$\bar{2}$0> matrix and (d) a zincblende GaN<110> grain.

Cross-sectional TEM investigations revealed regions of stacking faults, up to 200nm wide on the scale of the layer subgrains (Fig. 1a). These stacking fault domains were established above the highly faulted regions near the substrate and were laterally defined by threading dislocations. Focused probe rather than selected area diffraction techniques were required to analyze these layer regions individually. Thus, converged probe diffraction patterns from the banded regions containing stacking faults confirmed the presence of both hexagonal and cubic GaN, with the orientational relationship

GaN<11$\bar{2}$0>$_{hex}$ // GaN<110>$_{cubic}$ (Fig 1). However, no evidence for the presence of GaAs was obtained from these diffraction patterns. A high density of grains of single crystal cubic GaN were identified near the epilayer/substrate interface, whilst regions free of stacking faults away from the substrate corresponded to hexagonal GaN. The tilted bright field image presented within Fig 1 emphasises the layer grain structure and the associated distribution of stacking faults. High resolution TEM imaging of mixed phase material regions confirmed the presence of basal plane cubic stacking disorder. The cubic stacking sequence was predominantly observed to be only one or two monolayers thick.

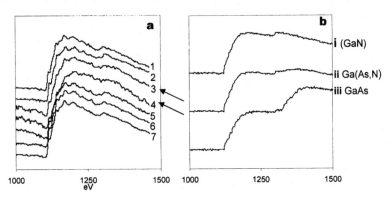

Figure 2. (a) Background subtracted EEL spectra, taken from various regions of the sample as shown in Fig. 2. (1) above stacking faults; (2) at the centre of a stacking fault free grain; (3) directly below a domain containing stacking faults; (4) a surface step feature; (5) at the GaN/sapphire interface; (6) at the centre of a grain containing stacking faults and (7) at a dislocation in the middle of the sample. The Ga $L_{2,3}$ edge at ~1100eV dominates these spectra. The Ga L_1 edge at 1298eV is present in all cases. Spectra 3 and 4 show an additional contribution due to the As $L_{2,3}$ edges. This feature takes the form of a gentle hump beyond 1298eV (arrowed). (b) Reference simulated spectra corresponding to (i) GaN, (ii) GaAs$_{0.25}$N$_{0.75}$, (iii) GaAs

Electron energy loss (EEL) spectroscopy was used to profile the distribution of As within these samples. Individual features were investigated for As content using a focused probe of ~20nm and even though the overall As content was extremely low it was identified and found to be concentrated at specific regions within the layer. Fig. 2 compares portions of the EEL spectra acquired from specific features within the sample shown in Fig 1. The spectra shown are dominated by the Ga $L_{2,3}$ edges and contain the position of the characteristic As $L_{2,3}$ edges which in two cases (Fig 2a$_3$, 2a$_4$) reveals the presence of low As content. The background prior to the Ga $L_{2,3}$ edges at 1115eV has been subtracted. Particular care was taken due to the overlap between the Ga L_1 edge at 1298eV, which is a hydrogenic edge, and the As $L_{2,3}$ edges at 1323 eV, which have a less clearly defined onset. In the absence of As, the Ga L_1 edge appears as a step-like feature, with a steady reduction with increasing energy (Fig. 2b$_i$). The presence of localised higher concentrations (>5%) of As would be expected to result in a visible change in the shape of this Ga L_1 tail beyond 1298eV, with a hump-like feature becoming more pronounced for increasing As content (Fig 2b$_{ii}$). Such As-containing features were only reproducibly observed in spectra obtained from two types of feature within the sample, either directly below the grains containing stacking faults (Fig 1a$_3$), or at step edge features on the surface of the GaN layer (Fig 1a4). This distribution suggests that As is mediating the formation of stacking faults within the hexagonal GaN matrix whilst acting as a surfactant. However, in view of the projected volume of material analyzed it is not possible to determine from EEL spectroscopy whether these regions of As content are due to nanoscale GaAs grains, or a Ga(N,As) alloy.

Layer growth in the presence of As may result in the development of cubic GaN material in two ways. With a high concentration of As at the surface of the sample during growth, the arrival of Ga would initially tend to form cubic GaAs. It is suggested that, as the Ga-N bond is stronger than the Ga-As bond, the displacement of the As atom by N could then promote the localised formation of cubic GaN that subsequently propagate sideways on the basal plane, being confined by the subgrain boundary structures. Alternatively, the incorporation of As onto N sites would result in localised lattice strain which may introduce basal plane GaN stacking faults. It is noticed that As was detected at concentrations >5% in regions immediately below such domains containing stacking disorder, and hence it is considered that the high density of basal plane GaN stacking faults observed within these samples is more likely to have formed as a consequence of the increased lattice distortion in these grains. The lateral propagation of these monolayer thick stacking disorders is still crystallographically driven and this inference is supported by the observation that regions of stacking faults appear near the surface even though the surface layer itself is not flat. It is noted that stacking faults have recently been observed at the location of a GaAs interrupt within GaN [9], however, the study presented here is, to our knowledge, the first such TEM study on GaN layers grown with constant As flux.

4. Summary

The presence of As during the growth of hexagonal GaN by PA-MBE induces the formation of a high density of GaN stacking faults within the hexagonal GaN matrix. No planar GaAs stacking faults were identified. The limited precision of the analytical techniques used prohibit the unambiguous identification of a Ga(As,N) alloy within the faulted regions.

References

[1]   Weyers M, Sato M and Ando H, Jpn J Appl Phys., Part 2 **31**, L853 (1992)
[2]   Sakai S, Ueta Y and Terauchi Y, Jpn J Appl Phys, **32** 4413 (1993)
[3]   Pankove JI and Hutchby JA, J. Appl. Phys. **47** 5387 (1976)
[4]   Li X, Kim S, Reuter EE, Bishop SG and Coleman JJ, Appl Phys Lett **72** 1990 (1998)
[5]   Winser AJ, Novikov SV, Davis CS, Cheng TS, Foxon CT and Harrison I, Appl. Phys. Lett., **77**, 2506 (2000).
[6]   Cheng TS, Jenkins LC, Hooper SE, Foxon CT, Orton JW, Lacklison DE, Appl Phys Lett **66** 1509 (1995)
[7]   Foxon CT, Novikov SV, Cheng TS, Davis CS, Campion RP, Winser AJ, Harrison I, J. Crystal Growth **219** 327 (2000).
[8]   Andrianov AV, Novikov SV, Li T, Xia R, Bull S, Harrison I, Larkins EC and Foxon CT, phys. stat. sol. (b) **238**, No. 1, 204 (2003)
[9]   Kim H, Andersson TG, Chauveau J-M and Trampert A, Appl. Phys. Lett. **81** 3407 (2002)

Inst. Phys. Conf. Ser. No 179: Section 2
Paper presented at Electron Microscopy and Analysis Group Conf. EMAG2003, Oxford, 2003
©2003 IOP Publishing Ltd

Investigating the Graphitization of Carbon using Analytical FEGTEM

H R Daniels, A P Brown, B Rand and R Brydson
[1]Institute for Materials Research, University of Leeds, Leeds, LS2 9JT, UK.

Abstract. A combination of high resolution electron microscopy, electron diffraction, energy filtered imaging and electron energy loss spectroscopy on a field emission TEM has been used to study the microstructural development in graphitising carbons as a function of heat treatment.

1. Introduction

The graphitization of carbon is an important, yet sometimes difficult, microstructural process occurring during the production and processing of a large range of materials including carbon fibres and tapes, carbon nanotubes, carbon composites, carbon coatings and dispersed precipitates in cast irons and mild steels. Graphitization of carbon leads to large changes in mechanical, thermal and electronic properties and there is a need to develop accurate spatially resolved imaging and spectroscopic methods for the determination of the degree of graphitic character as a function of heat treatment, both to study the nature of the graphitization process itself as well as to extract information on the resultant physical properties of the material's microstructure.

2. Method

Samples of a petroleum-based pitch (Aerocarb 80) were heat treated in an inert atmosphere at a series of temperatures between 200 °C and 2730 °C. Powders were initially ground and examined by X-ray diffraction (XRD) and He density measurements. Powders were then dispersed onto holey carbon support films and thin areas examined in a FEI CM200 FEGTEM fitted with a GIF 200 using HRTEM, selected area electron diffraction (SAED), EELS and EFTEM.

3. Results and Discussion

XRD, SAED and HRTEM imaging do provide a (well documented) means of tracking the graphitisation process. XRD can provide averaged values for the (002) interlayer spacing and also, from the widths of diffraction peaks, the crystallite dimensions (L_a and L_c). SAED produces spatially resolved data on the (002) spacing, its variation and also the angular deviation of basal planes from a perfectly planar configuration. HRTEM can image (002) basal planes directly and also provide a direct indication of crystallite size [1]. Generally the interlayer spacing decreases monotonically with increasing heat treatment temperature, whilst the crystallite size correspondingly increases. Some important findings from these comparative studies include: the presence of significant structural heterogeneity within samples heat treated at a particular temperature (that appeared to be correlated with the presence of S and Si heteroatoms which are known to promote cross-linking and inhibit graphitization) and, secondly, slight variations in the monotonic trend at intermediate temperatures (ca. 1500 °C) which may be linked to the loss of heteroatoms and subsequent "puffing" of graphene layers.

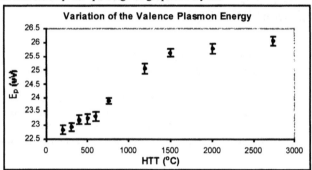

Figure 1. Variation of the volume plasmon energy with heat treatment temperature.

EELS measurements can also provide important additional information on the graphitisation process. Figure 1 shows the variation in the volume plasmon energy of the graphitisation series as a function of heat treatment temperature. Each data point is the average of at least ten measurements. In a free electron model, the volume plasmon energy scales as the square root of the valence electron density and accordingly this data shows an excellent linear correlation with the square root of the He density measurements. The plasmon energy thus provides a sensitive measure of the degree of graphitic character and it is possible to use an EFTEM technique to image variations in graphitic character within complex specimens such as carbon-carbon composites and graphitic nodules in medium carbon cutting steels. Here EFTEM images are acquired using two 3 eV windows centred at 22 eV and 27 eV in the plasmon region, the ratio of these two images providing high resolution (1.6 nm) graphitisation maps free from diffraction (i.e. orientation) contrast and thickness effects [2].

Another feature in the EELS low loss spectrum is the $\pi \rightarrow \pi^*$ transition peak at ca. 6.5 eV which represents a single electron transition that exhibits significant collective (plasmon-like) character. As the degree of graphitic character increases, the proportion of planar sp^2-bonded carbon should increase. The $\pi \rightarrow \pi^*$ peak intensity should therefore provide a measure of $\%sp^2$-bonded carbon when normalized to the data for 2730 °C (i.e. assuming 100% sp^2-bonded carbon at this temperature). Figure 2 shows the variation in relative $\pi \rightarrow \pi^*$ peak intensity of the graphitisation series as a function of heat treatment temperature. Again this data is an average of multiple measurements and care was taken to ensure that, under the experimental conditions employed (near parallel illumination and a collection angle, $\beta = 1.7$ mrads), this peak intensity was independent of the crystallographic orientation of the sample area [1, 3]. The data show a gradual increase in $\%sp^2$ carbon content with temperature and an apparent dip at 1500 °C corresponding the buckling or puffing of the basal planes due to loss of volatiles.

The carbon K-edge in the EELS core loss region can also provide important information on the local carbon environment during the graphitisation process. Figure 3 shows the evolution of the carbon K-ELNES for the graphitisation series as a function of heat treatment temperature.

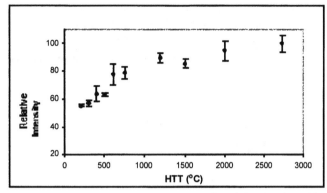

Figure 2. Variation of the relative ($\pi \rightarrow \pi^*$) peak intensity with heat treatment temperature.

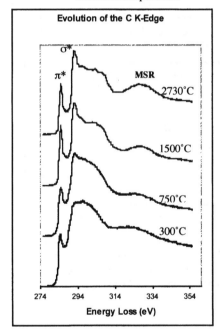

Figure 3. Evolution of C K-ELNES with heat treatment.

These spectra were measured using parallel illumination and $\beta = 1.7$ mrads, where spectra are independent of crystallite orientation [4]. The C K-ELNES exhibits a number of important features: firstly the 1s $\rightarrow \pi^*$ peak at 285 eV, the 1s $\rightarrow \sigma^*$ region commencing at ca. 292 eV and a broad multiple scattering resonance (MSR) centred at ca. 330 eV. The exact energy position of the MSR above the edge onset is known to be proportional to the inverse square of the next nearest neighbour bond length around the carbon atom. Thus it is possible to derive the C-C bond length in the graphene sheets. Figure 4 shows the calculated variation in bond length as a function of heat treatment temperature and this correlates well with XRD and SAED data.

Figure 3 shows an increase in π^* peak intensity as graphitisation proceeds owing to the increasing proportion of sp^2-bonded carbon, as well as a significant sharpening of the σ^* structure due to increasing long range order. The relative intensity of the 1s $\rightarrow \pi^*$ peak, normalized to the intensity within a 20 eV window covering both the π^* and σ^* features, can be used to extract the %sp^2-bonded carbon in a similar fashion to the low loss data in figure 2. Figure 5 shows the results of this analysis of the C K-ELNES. The trend in figure 5 is very different to that in figure 2, notably the higher values of sp^2 content derived from the core loss data. One possible reason for this is due to the significant hydrogen content present in the pitches which will result in significant C-H bonds giving rise to a 1s $\rightarrow \sigma^*$ (C-H) feature between 287-289 eV. Deconvoluting the effect of this additional intensity contributing to the 1s → π* peak we obtain the lower curve in figure 5 which is in considerably better agreement with the sp^2 contents shown in figure 2.

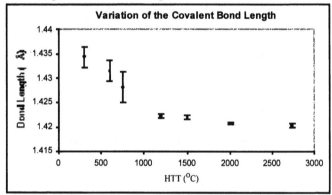

Figure 4. Variation of the C-C covalent bond length (Angstrom) as a function of heat treatment temperature.

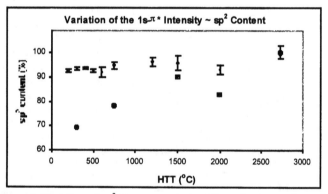

Figure 5. Variation of sp^2 C content derived from the 1s → π* peak intensity as a function of heat treatment (upper curve with error bars); lower curve (dots) includes a correction for C-H bonding.

4.   Conclusion

EELS analysis of a series of graphitising carbons heat treated at different temperatures can provide quantitative information on the graphitisation process complementary to that derived from XRD, He density measurements, SAED and HRTEM.

5.   References

[1]   Daniels H R, PhD Thesis, IMR, University of Leeds, 2003.
[2]   Daniels H R et al., Ultramicroscopy 96, 547-558, 2003.
[3]   Daniels H R et al., submitted to Carbon
[4]   Daniels H R et al., Ultramicroscopy 96, 523-534, 2003.

Inst. Phys. Conf. Ser. No 179: Section 2
Paper presented at Electron Microscopy and Analysis Group Conf. EMAG2003, Oxford, 2003
©2003 IOP Publishing Ltd

3D analysis of semiconductor structures using HAADF STEM tomography

T J V Yates, L Laffont, M Weyland, D Zhi and P A Midgley

Department of Materials Science and Metallurgy, University of Cambridge, Pembroke Street, Cambridge, CB2 3QZ, UK.

Abstract. The high spatial resolution and atomic number contrast available using high angle annular dark field scanning transmission electron microscopy (HAADF-STEM) tomography makes it an ideal technique for the 3D analysis of physical science specimens and of semiconductors in particular. In this work, we have begun to apply this form of tomography to investigate Si/Ge quantum dots to reveal their 3D morphology (in order to identify the major growth facets) and, in the longer term, to reveal the 3D distribution of Si and Ge within the dot itself.

1. Introduction

The drive in the semiconductor industry for smaller and increasingly complex devices is linked with the ability to analyse the morphology, chemistry and electronic structure of each successive generation. Shrinking feature sizes and increasing complexity are challenging traditional methods for characterising nanoscale structures because the 2D projection obtained from a conventional electron microscopy image may not reveal a 3D object's true structure. Electron tomography, using either HAADF-STEM images or energy-filtered TEM (EFTEM) images has been seen as one way to overcome this limitation [1]. By rastering the beam and using a high angle annular detector, STEM HAADF images can be formed using electrons scattered to high angles. To a good approximation, these images are incoherent in nature (thus limiting the effects of Fresnel diffraction and other interference phenomena) and exhibit strong atomic number, Z-contrast, which classically approaches a Z^2 dependence. The spatial resolution of an HAADF image is given approximately by the size of the probe.

Most of our work to date using HAADF tomography has concentrated on qualitative structural characterisation. In this paper we start to consider how best to reveal quantitative compositional information contained within the reconstructed 3D object, in particular the germanium variation in a Si/Ge quantum dot. To do this, we need to determine how the reconstruction intensity is affected by both elastic and inelastic (phonon) scattering, how dynamical effects disturb the intensity distribution of the reconstruction and lastly how the probe shape varies through the volume of the object - initial experiments have indicated how the central maximum of the probe shape remains remarkably constant over a large depth of focus [1].

2. Principles of tomography

Electron tomography consists of several key stages: the acquisition of a tilt series of 2D projections, the alignment of these projections about a common tilt axis and, tomographic reconstruction [1]. A conventional tilt series is acquired by tilting the specimen about a single axis at regular angular increments, correcting specimen position and focus, before acquiring an image. Whilst the resolution of the 3D reconstruction is maximised by acquiring the largest number of images over the greatest angular range, the extent of the tilt series is presently constrained by specimen shadowing and the increased projected thickness of most specimens at high tilt angles. In addition, for some samples beam damage associated with prolonged exposures can be highly detrimental to the fidelity of the 3D reconstruction.

3. STEM HAADF tomography of a Si/Ge quantum dot

A specimen comprising two layers of capped (buried) and one layer of uncapped Si/Ge quantum dots (QDs) was grown using conventional MBE methods on a Si (001) substrate. The foil was prepared using standard cross-sectional techniques, with final thinning carried out using a Gatan PIPS. The specimen was mounted in a high-tilt tomography holder (Fischione model 2020) with the glue line perpendicular to the tilt axis, in order to minimise shadowing at high tilts, and plasma cleaned to reduce possible contamination. Using a FEI Tecnai F-20 FEG-(S)TEM with automated acquisition software (FEI Xplore 3D), a STEM-HAADF tilt series was acquired from a region containing a capped and uncapped quantum dot, with a HAADF inner-radius ~50 mrad, sufficiently large to minimise the effects of diffraction contrast and strain. A total of 73 images were acquired at 2° increments over the range −70° to +74°. The aligned tilt series was reconstructed using an iterative reconstruction algorithm (SIRT) [2].

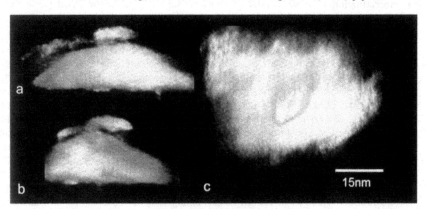

Figure 1. Three mutually perpendicular views (voxel projections) of the reconstructed uncapped Si/Ge quantum dot. The 'dirt' on the top of the dot is an artefact of the sample preparation process.

Voxel projections of the SIRT reconstruction of the uncapped QD are shown in Figure 1; some 'dirt' is clearly evident on the surface of the uncapped dot and has arisen from the ion thinning. It is clear from the reconstruction that a large proportion (approximately 50%) of the uncapped QD was removed during specimen preparation. The limited tilt angle will introduce some elongation into the reconstruction but its appearance has been minimised using the SIRT routines.

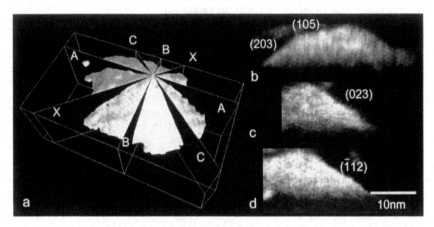

Figure 2. (a) Perspective view of the QD. The sections delineated are shown in (b), (c) and (d) in which the index of individual facets have been tentatively assigned. Section XX defines the boundary between facets seen in AA and BB.

Figure 2(a) shows a perspective view of the same QD with sections delineated that are shown in Figures 2(b), (c) and (d). These sections help to display the morphology of the dot and reveal some of the faceting. For a dot of this size the facets are not well established and through AFM experiments are known to show rather rounded features. Nevertheless, we have been able to tentatively ascribe Miller indices to some of the facets revealed within the sections. The major facets appear to be of type {203} with more minor facets of type {105} and {$\bar{1}12$}. Further work is required however to be certain that the indexing is correct.

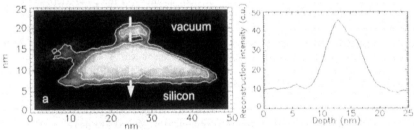

Figure 3. (a) Mean of 13 slices (3.5 nm) through the reconstruction of the uncapped quantum dot showing the variation in reconstruction intensity. The additional structures at the top, and to the right, of the dot are artefacts arising from the 'dirt' on the sample surface. (b) A line trace, an average over 13 voxels (3.5 nm) through the centre of the slice, as marked by the dotted line. The asymmetry of the intensity distribution indicates a germanium maximum near the dot surface. A 3D spatial resolution of about 1-2 nm is seen in the reconstruction.

34

4. Compositional variation within the quantum dot

As seen in the previous figure, it is only by taking sections through the 3D object that the true structure and internal composition of the dot can be revealed. Figure 3(a) shows a section through the dot parallel to the section marked AA in Figure 2. The intensity in a slice from the reconstruction through the centre of the uncapped dot may, in principle, be related directly to composition - the increased intensity towards the upper surface implies that atomic species of higher atomic number, namely germanium in this case, is concentrated in that region. The line trace in Figure 3(b) is an average over 13 voxels (3.5 nm) through this region, the centre of which is marked by the dotted line in Figure 3(a). The sampling volume is approximately 3.5 nm × 3.5 nm × (the sampling depth).

5. Discussion

The observed variation seen in Figure 3 is unlikely to be an artefact of reconstruction. It occurs along the direction of the highest resolution in the reconstruction, the intensities in the reconstruction are constrained to the intensities in the original images and the response of the HAADF detector, for sensible integration times, gives image intensities reproducibly to within 2%. However, as yet we have no proven method for converting these intensities quantitatively to composition, although the compositional simplicity of this system makes it ideal for assessing any quantification methods. Although 'frozen phonon' multislice calculations can be performed at every tilt this will be an extremely time consuming operation and impractical for most data sets. As such a way to analyse the data in a much simpler fashion is sought [3]. One (grossly simplified) method is to ignore the effects of dynamic scattering and of the sample's crystallinity, assuming both to be 'averaged out' over a large tilt range. In essence we assume the sample is composed of an ensemble of independently scattering atoms and only the atomic scattering factors (elastic and thermal diffuse, TDS) need be calculated. Using parameterised atomic scattering factors [e.g. 4,5], initial quantification of the QD intensities did not result in sensible values for the Ge composition. This may be partly because of these gross assumptions made and partly because of the poor accuracy of the parameterisations for $s>2$ Å^{-1}. It seems clear that either more complex theory is needed to better fit the data or that experimental calibration is achieved using Si:Ge samples of known composition.

In conclusion, HAADF STEM tomography can be used in principle to reveal the internal structure and composition of semiconductor specimens. We have reconstructed a Si/Ge quantum dot, revealing faceting and apparent compositional fluctuations but more work is required to quantify the 3D variation of the dot's composition.

References

[1]   Midgley P A and Weyland M 2003 Ultramicrosc. 96(3-4) 413-431.
[2]   Gilbert P F C 1972 Proc. R. Soc. Lond. B. 182 89-102.
[3]   Liu C P, Preston A R, Boothroyd C B and Humphreys C J 1998 J. Microsc. 194 171-182.
[4]   Peng L-M, Ren G, Dudarev S L and Whelan M J 1996 Acta Cryst. A52 456-470.
[5]   Rez D, Rez P and Grant I 1994 Acta Cryst. A50 481-497.

Acknowledgements

We would like to thank EPSRC, FEI Company, the Royal Society, the Royal Commission for the Exhibition of 1851 and the Isaac Newton Trust for financial support.

Inst. Phys. Conf. Ser. No 179: Section 2
Paper presented at Electron Microscopy and Analysis Group Conf. EMAG2003, Oxford, 2003
©*2003 IOP Publishing Ltd*

The Non-Uniform Composition Profile in Ge(Si)/Si(001) Quantum Dots

C Lang, D Nguyen-Manh, and D J H Cockayne

Department of Materials, Oxford University, Parks Road, Oxford, OX1 3PH,UK

Abstract. The composition profile of a pyramid-shaped Ge(Si)/Si(001) quantum dot has been calculated using a combination of molecular static relaxations and a Monte Carlo process. The strain field and the displacement field of the non-uniformly alloyed quantum dot have been calculated using finite element analysis and compared to the case of a uniform alloying profile. TEM image simulations based on the displacements calculated in the finite element analysis have been performed. It is found that the differences in the contrast between uniformly and non-uniformly alloyed pyramid-shaped quantum dots are not significant.

1. Introduction

During the layered growth of lattice-mismatched semiconductors, a strain-induced transition from planar growth to 3D island growth (Stranski-Krastanow growth [1]) can occur leading to the formation of quantum dots (QDs). Their small size (nm in dimension) leads to quantum confinement of the electrons, resulting in electronic and optical properties which show promise for use in a wide range of devices ranging from semiconductor lasers to quantum computers . These properties of the QDs are controlled by their size, shape and composition. The growth of Ge(Si) on Si(001) has been extensively studied because of its importance as a model system for the growth of other, more complicated, lattice mismatched semiconductors.

Whereas the shape and size of Ge(Si)/Si(001) QDs can be determined from AFM [2] and TEM [3] to a high precision, an accurate measurement of the composition and composition distribution remains unsolved. Several different TEM methods (EFTEM [4], EDX [5], plan-view TEM [6]) have been applied. All methods consistently show a significant non-uniformity in the composition profile but although several details of the composition profile have been resolved, a coherent model of the alloying profile, taking into account all different driving forces, remains elusive. During growth Ge(Si)/Si(001) QDs undergo a shape transition from pyramids to domes at a certain critical size [3]. Measuring the composition profile is especially difficult for the pyramid-shaped dots because of their smaller size. Until an experimental method for studying the profile is identified, modelling the composition distribution is a good alternative. This paper summarises the key results of modelling work undertaken to determine the composition profile of pyramid-shaped Ge(Si)/Si(001) quantum dots and shows how these results can be linked to plan-view TEM experiments via TEM image simulations. Image simulations have been successfully used to study dome-shaped islands [6]. Their application to pyramid-shaped islands poses several problems, which will be analysed.

2. Monte Carlo simulations of the composition profile

The formation of a non-uniform composition profile in a QD is essentially driven by the lowering of the free energy of the QD, $E_{formation}$, given by four terms [7]:

$$E_{formation} = E_{surface} + E_{strain} + E_{mix} + E_{chem}$$

which are, in order, the surface energy, the strain energy, the alloy mixing energy and the chemical potential of the reservoir (i.e. wetting layer). By redistributing the atoms within the island, the mixing energy, the surface energy and the strain energy can all be lowered, whereas the chemical potential does not change for a fixed average composition. The geometry assumed in this paper is a QD in the form of a square based pyramid with the sides having normals parallel to the <105> directions and placed on a Si substrate with an (001) surface. This geometry is in agreement with the experimental work of Ross et al [3] for small QDs. The substrate is a rectangular Si block with a square base of 32.6nm side length, and a height of four layers, each consisting of 7200 atoms. The QD in our modelling work has a base length of 11.1nm and containing 3268 atoms and an average composition of two thirds Ge and one third Si. A uniform starting composition profile is assumed and the energy of the system is minimized computationally. The method used is a combination of a Monte-Carlo process and molecular static relaxations.

CUTTING PLANES

Figure 1. A plot of the composition profile as calculated in the Monte Carlo simulations with Ge rich areas bright and Si rich areas dark (a) from the bottom and (b) from the top of the QD

The composition distribution obtained after the redistribution is shown in figure 1 with the amount of segregation indicated by greyscales where black represents pure Si and white represents pure Ge. It is found that there is a Ge rich layer at the QD surface. This is energetically favourable because Ge has a lower surface energy than Si. Only in the corner of the pyramid is a high Si content detected, even at the surface. This is due to a high strain concentration at this location, which is relieved by diffusion of Si there. At the interface between the QD and the Si substrate the composition distribution shows a ring of high Si content. This is in agreement with findings by Sonneth et al [8] who explained this effect as being due to a stress concentration in this region. Careful analysis of the layers parallel to the substrate surface reveals a redistribution of Ge to the top of the QD, which is consistent with experiment results from Liao et al. [6] for dome-shaped islands.

3. Finite Element Analysis

Finite element analysis (FEA) has been used successfully previously to calculate the strain field of dome-shaped quantum dots and to perform TEM image simulations that correspond well to experimental results [6]. In this work we apply a similar method to analyse the strain field of a non-uniformly alloyed pyramid-shaped quantum dot using the finite element package STRAND7 [9] and the composition profile shown in figure 1. STRAND7 allows the user to define a thermal expansion coefficient α for each material.

Furthermore the temperature can be defined at each node, which determines the expansion in combination with α. This can be used to introduce a strain term into the FEA, which is equivalent to the strain due to the lattice mismatch between Si and Ge in the atomistic simulations. To clarify this, two examples are given:

(i)   A pure Ge QD on a Si substrate: The lattice mismatch between Ge and Si is 4% and therefore α is set to be 0.04 for Ge and 0 for Si. Because the material in the QD is pure Ge QD, all node 'temperatures' are set to 1.

(ii)   A non-uniform composition profile: To transfer the composition profile from the atomistic simulations the local Ge content near each node is mapped to node 'temperature' in an interval between 0 and 1 with 0 being pure Si and 1 being pure Ge. This number is equivalent to the node temperatures in the FEA.

The finite element software calculates the strain energy density at each position. This value can be plotted, and represents the strain energy distribution. The strain energy density of a uniformly alloyed pyramid-shaped quantum dot is shown in figure 2. It is found that the strain field changes considerably when the non-uniform composition distribution from figure 1 is used.

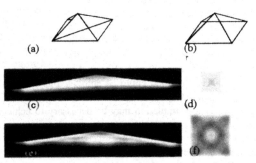

Figure 2. sketch of the (a) diagonal cutting plane and (b) base cutting plane. The strain maps for the uniform composition are shown in (c) and (d) and for the non-uniform composition in (e) and (f).

The Si rich ring observed in the composition profile in figure 1 is found to result in a low strain area in the strain map, whereas a large strain concentration can be seen at the pyramid edge. The displacements calculated in the FEA are next used to simulate TEM images.

4. TEM Image Simulations

The aim of the image simulations is to show whether pyramid- and dome-shaped QDs can be distinguished in on-zone axis multiple-beam bright field TEM images and to test if this technique can be used to distinguish between uniform and non-uniform alloying in pyramid-shaped QDs, as was successfully shown for dome-shaped QDs [6]. In dome-shaped QDs a Maltese cross type contrast was observed inside the quantum dot and the splitting of the bars of the cross was found to indicate non-uniform alloying [6]. It has also been found that the contrast of the QDs varies strongly with the effective extinction distance.

38

We have simulated a pyramid-shaped QD with a base length of 80nm over a range of substrate thicknesses for the case of a uniform composition profile and the case of the non-uniform composition profile obtained from the Monte Carlo simulations. Representative results are presented in figure 3 and show that there is no Maltese cross like contrast in the case of pyramid-shaped QDs. There are differences between the images of uniformly and non-uniformly alloyed QDs but the differences found are smaller than in the case of the dome-shaped QDs and we expect they will be more difficult to detect in a TEM experiment. However, the contrast of a pyramid-shaped QD is significantly different from dome-shaped QDs so that

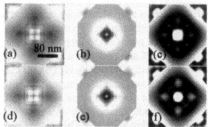

Figure 3. Image simulations are shown for different substrate thicknesses for the uniform (first row) and non-uniform (second row) composition. (a) and (d) 56nm substrate thickness, (b) and (e) 66nm substrate thickness, (c) and (f) 80nm substrate thickness.

this plan-view method could be employed as a tool to distinguish between dome- and pyramid-shaped QDs and to determine their relative distributions on a plan-view TEM sample. Image simulation may be necessary to achieve this, and also to determine the orientation of the pyramid, since the actual shape cannot be seen in the TEM image.

5. Conclusions

We have shown that the three dimensional composition distribution of a pyramid-shaped Ge(Si)/Si(001) QD can be calculated by a combination of a MC process and atomistic relaxations. The strain map of the QD, calculated using finite element analysis shows significant differences between the case of a uniformly and a non-uniformly alloyed QD. Subsequently, we have performed image simulations and have shown that the differences in the TEM contrast are small in the case of pyramid-shaped QDs. Therefore this method is not expected to be suitable to distinguish between different composition profiles in the case of pyramid-shaped quantum dots. However, with supporting image simulations, it may be possible to distinguish between pyramid- and dome-shaped QDs on a plan-view TEM sample.

References:

[1] Stranski I N and Krastanow L 1939 Akad. Wiss. Wien Kl IIB 146 797
[2] Goldfarb I Hayden P T Owen J H G and Briggs G A D 1997 Phys Rev Lett 78 3959
[3] Ross F M Tromp R M and Reuter M C 1999 Science 286 1931
[4] Liao X Z Zou J Cockayne D J H Wan J Jiang Z M and Wang K L 2002 Phys Rev B 65 153306
[5] Havey A Davock H Dunbar A Bangert U and Goodhew P J 2001 J Phys D:Appl Phys 34 636
[6] Liao X Z Zou J Cockayne D J H Jiang Z M and Wang K L J 2001 Appl Phys 90 2725
[7] Tersoff J 1998 Phys Rev Lett 81 3183
[8] Sonnet P and Kelires P C 2002 Phys. Rev. B 66 205307
[9] For details see: http://www.strand7.com

Inst. Phys. Conf. Ser. No 179: Section 2
Paper presented at Electron Microscopy and Analysis Group Conf. EMAG2003, Oxford, 2003
©2003 IOP Publishing Ltd

The structure of uncapped, buried and multiple stacked Ge/Si quantum dots

D Zhi, M Wei[1], D W Pashley[1], B A Joyce[2], M Weyland, L Laffont, T J V Yates, A C Twitchett, R E Dunin-Borkowski, P A Midgley

Department of Materials Science and Metallurgy, University of Cambridge, Cambridge, CB2 3QZ, UK

[1]Department of Materials, Imperial College, London, SW7 2BP, UK

[2]Department of Physics, Imperial College, London, SW7 2AZ, UK

Abstract. In the present study, several different multiple-layered Ge/Si (001) QD structures with the final QD layer uncapped have been fabricated using gas source molecular beam epitaxy (GSMBE). The different characterisation techniques, including atomic force microscopy, diffraction-contrast transmission electron microscopy, high-resolution lattice imageing, analytical electron microscopy, and electron tomography, have been applied to characterise the morphology, structure and chemistry of both surface and buried QDs. The size, the shape and the number density of Ge islands show drastic changes when altering parameters such as growth temperature, Ge coverage and Si spacer layer thickness within Ge/Si multilayers. The atomic scale details obtained by HRTEM are combined with the results of field-emission gun scanning transmission electron microscopy (FEG-STEM), electron tomography to provide information on the structure of the QDs. All these structure information about QDs are discussed in relating to GSMBE growth conditions.

1. Introduction

The formation of self-organized islands has become a fascinating subject in the field of semiconductors in recent years (Mo et al 1990). These islands have the potential to become ideal quantum dots and recently room temperature lasing has been reported in InAs dots on GaAs substrates (Heinrichsdorff et al 1997). Ge self-assembled islands have also been of considerable interest due to the possibility of introducing quantum effects within a well-developed silicon technology (Schittenhelm et al 1998). In contrast to InAs dots, however, Ge/Si islands normally have a larger size and a lower number density, which prevent them from acting as quantum dots in optoelectronic devices. Low temperature growth and vertically stacked dot layers, which results in size ordering, have been used to solve these problems. Ge islands grown on Si(001) substrates in the Stranski–Krastanov growth mode are known to have various sizes and shapes corresponding to their growth temperatures. Low temperature growth was found necessary for fabricating a high density of small Ge islands which have pyramidal and hut-cluster shapes with facets, whereas at higher growth temperatures, large dome-shaped islands with steeper facets tend to be formed. It is also necessary to introduce size ordering in some way, since the size distribution of the small Ge islands is quite large as a result of their instability (Kamins *et al* 1997).

The formation process and size ordering of Ge dots, produced by gas source molecular beam epitaxy (GSMBE), have been studied thoroughly by AFM and photoluminescence (Mirua *et al* 2000). Here, we have studied further the structure properties and formation mechanism of Ge islands as a function of growth temperature, Ge coverage, and Si spacer layer thickness within Ge/Si multilayers by the means of different microscopy techniques.

2. Experimental

All samples were grown by GSMBE using Si_2H_6 and GeH_4 as source gases. The pressures in the Si_2H_6 and GeH_4 source lines were fixed at 5 and 10 Torr, respectively. Following the deposition of a Si buffer layer on the Si(001) substrate, a Ge layer was grown to determine the critical thickness of island formation by reflection high energy electron diffraction (RHEED). The Ge layer was then covered by a thick Si layer, in order to smoothen the growth front for subsequent growth. Having obtained this RHEED calibration, two to ten periods of Ge/Si multilayers with Ge nominal coverage slightly greater than the critical layer thickness were grown at a temperature between 525 and 750°C. The Si spacer layer thickness was between 80 and 130 nm for each growth temperature, which was confirmed to be thick enough to eliminate the strain effects of one Ge layer on the next. The last Ge layers of all these structures were uncapped and were first studied by tapping mode AFM.

Both plan-view and cross-sectional TEM specimens were made using conventional sample preparation method, including mechanical polishing, dimpling, and precise ion-milling. Diffraction contrast TEM was performed with a JEOL 2000FX microscope, and HRTEM lattice images were recorded using both JEOL 2010 and JEOL 4000EX microscopes. High angle annular dark field (HAADF) imaging was performed using a Tecnai F20 microscope.

3. Results and discussions

Figure 1 shows AFM images of uncapped Ge islands grown at different growth temperatures. In these images, the morphological features and the density of Ge islands show drastic changes with varying temperature. At the lowest temperature (525°C), there is quite a high number density of hut-cluster shaped islands, together with a much smaller number of pyramidal islands. The width and height of the hut-clusters each has a narrow distribution, whereas the length of the clusters varies between 30 and 60 nm. When the growth temperature goes up to 600°C, the growth mode changes to a bimodal one consisting of pyramid-shaped islands with {105} facets and dome-shaped islands with steeper facets (Figure 1b). In this sample, the pyramid-shaped islands have a considerably higher number density than that of the dome-shaped ones and also have a much larger size distribution due to the instability. On the other hand, dome-shaped islands are in fact almost identically sized. The highest temperature (750°C) gives rise to the same bimodal growth mode as in the case of 600°C; however, the size of both types of islands becomes too large for the lateral quantum effect to be expected (Figure 1c).

Both AFM and plan-view TEM images were used to study two uncapped Ge dot samples grown at the same temperature (600°C) and with different nominal Ge coverage (3.8 and 7.7 ML). For the sample with low Ge coverage, only pyramidal islands were observed with an average base length of about 48nm and a low dot number density, as shown in Figure 2(a). Plan-view TEM image in Figure 2(b) shows one of these dots grown with low Ge coverage. However, with the increasing Ge coverage, a bimodal size distribution was observed and the smaller pyramidal islands were gradually replaced by larger dome-shaped ones with an average base size of 67nm, as shown by AFM in Figures 2(c) and by plan-view TEM in Figure 2(d).

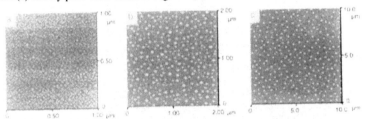

Figure 1. AFM images of Ge islands grown at different growth temperatures; (a) 7.2 ML of Ge at 525°C, (b) 5.8 ML at 600°C, (c) 5.5 ML at 750°C.

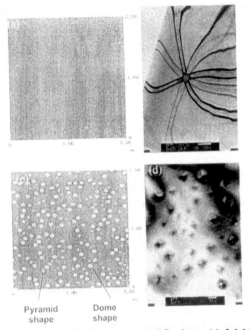

Pyramid   Dome
shape   shape

Figure 2. (a) AFM image of uncapped Ge dots with 3.8 ML Ge coverage, (b) bright field TEM image of a faceted Ge dot in a sample with 3.8 ML Ge coverage, (c) AFM image of uncapped Ge dots with 7.7 ML Ge coverage, and (d) bright field TEM image of uncapped Ge dots with 7.7 ML Ge coverage.

Figure 3 shows the HRTEM and HAADF images taken from a two-layer stacked Ge QD specimen with a spacer thickness of 80nm. Apparently, the uncapped surface dot in Figure 3(a) has a larger height than the buried one in Figure 3(b), which might be attributed to the interdiffusion during the Si encapsulation overgrowth process. It was also observed that, with this large spacer thickness, the surface dot and the buried dot were not vertically aligned, as illustrated by the HAADF image in Figure 3(c). Further study by HAADF STEM tomography suggested the facet nature of uncapped dot and an increased Ge content towards the upper surface.

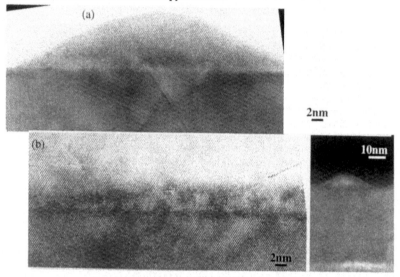

Figure 3. (a) HRTEM image of an uncapped Ge dome-shaped dot, (b) HRTEM image of a buried Ge dot, and (c) HAADF image showing both surface dot and buried dot.

4. Conclusions

The morphology and structure of multiple-layered Ge quantum dots were characterised in terms of different growth conditions, using AFM, TEM, HRTEM and HAADF. The drastic change on the morphology, and number density of Ge dots were observed with both increasing growth temperature and Ge coverage. Bimodal dot distribution (pyramid and dome) and larger size were determined with the dots of higher growth temperature or more Ge coverage. Decreased height was determined by both HRTEM and HAADF with the buried Ge dots, due to possible interdiffusion during encapsulation. Chemical composition variation was observed with both uncapped and buried dots.

References

Mo Y W, Savage D E, Schwartzentruber B S and Lagally M G 1990 Phys. Rev. Lett. **65**, 1020

Heinrichsdorff F, Mao M H, Kirstaedter N et al 1997 Phys. Lett. **71**, 22

Schittenhelm P, Engel C and Findeis F 1998 J. Vac. Sci. Technol. B **16**, 1575

Kamins T I, Carr E C, Williams R S and Rosner S J 1997 J. Appl. Phys. **81**, 211

Mirua M, Hartmann J M, Zhang J, Joyce B A and Shiraki Y 2001 Thin Solid Films **369**, 104

Inst. Phys. Conf. Ser. No 179: Section 2
Paper presented at Electron Microscopy and Analysis Group Conf. EMAG2003, Oxford, 2003
©2003 IOP Publishing Ltd

Image interpretation in Lorentz transmission electron microscopy

S McVitie

Department of Physics and Astronomy, University of Glasgow, Glasgow
G12 8QQ, United Kingdom.

Abstract. Lorentz microscopy can be used to measure the induction distributions in magnetic thin film systems. By using a micromagnetic simulation package it is possible to calculate the form of Lorentz microscopy images. In doing so it is apparent that great care must be taken in interpreting experimental images if quantitative information on the magnetisation and induction distributions is to be extracted.

1. Introduction

Lorentz microscopy is a branch of electron microscopy in which the Lorentz force between the electron beam and the sample is utilised to generate images which allow observation of the magnetic structure of materials [1]. Transmission electron microscopy (TEM) is very useful in investigating thin film structures of technologically relevant materials such as spin valves and spin tunnel junctions which display giant magnetoresistance [2, 3]. Generally it is the magnetisation structure of a sample that is of interest and although it is the induction that is the relevant quantity in Lorentz microscopy, for many samples interpretation of the magnetisation from the induction distribution is quite straightforward. However if quantitative measurements are to be made then it is important that any differences between the magnetisation and induction distributions are well understood to avoid misinterpretation. In this paper such examples are considered. This is achieved through the use of micromagnetic simulation software and calculations of induction distributions from various simulations.

2. Lorentz microscopy

There are a number of different methods that can be used in the TEM to reveal magnetic structure which fall into the category of Lorentz microscopy. The intention in this paper is to concentrate on the quantitative aspects of Lorentz imaging and so the differential phase contrast (DPC) mode of scanning TEM is discussed [1]. In this imaging mode a focussed beam of electrons pass through the sample and are locally deflected through the Lorentz interaction to give a net deflection angle. Maps of components of the deflection angle which is proportional to integrated induction components are made by using a segmented quadrant detector situated in the far field. An important factor to note is that scan coils ensure that any displacement of the beam on the detector is due only to magnetic deflections and not scanning of the beam. Only the bright field cone is incident on the detector and by taking difference signals from opposite segments of the detector, induction maps may be displayed. An example of the induction maps obtained from a lithographically fabricated thin film element of permalloy is shown in Figures 1(a) and (b). Additionally the inferred magnetisation distribution of the element is shown in Figure 1(c). The element shown here has a flux closure structure and so the magnetisation and induction maps should be expected to be very similar to each other.

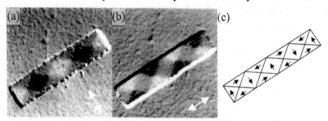

Figure 1. (a) and (b) are examples of DPC images obtained from a 39 nm thick NiFe element with in-plane dimensions 1000 nm × 200 nm. The arrows indicate the direction of induction mapped. (c) shows the domain structure in the element inferred from (a) and (b).

In cases where the magnetisation distribution gives rise to significant stray fields it is necessary to take great care in analysing the images. This is especially true if absolute values are to be inferred from the induction maps, or even relative values within an image. In order to put this in context the origin of the contrast in the DPC mode must be critically assessed. The quantum mechanical effect on the electron beam of a magnetic sample can be written as a phase difference (the Aharanov-Bohm effect):

$$\Delta\phi = -\frac{e}{\hbar}\int_{-\infty}^{\infty}(\underline{A}\cdot\hat{n})dl \tag{1}$$

where e is the electronic charge, $\underline{A}$ the magnetic vector potential, $\hat{n}$ is the unit vector along the electron path and $\hbar$ is Planck's constant divided by 2π. The integral is performed along the electron path with phase differences being relative to a beam along the same path with no magnetic interaction. It should be noted that only components of the magnetic vector potential along the electron path make any contribution to the phase. By taking the gradient of the phase perpendicular to the electron path the following equation is obtained:

$$\nabla\phi = \frac{e}{\hbar}\int_{-\infty}^{\infty}(\underline{B}\times\hat{n})dl \tag{2}$$

where $\underline{B}$ is the magnetic induction. The identity $\underline{B} = \nabla\times\underline{A}$ has been used to relate the phase gradient to the induction. The right hand side of Equation (2) is identical to that which can be calculated from the classical deflection of the electron beam due to the Lorentz force except for a constant factor [1].

Both the classical and quantum mechanical interpretation are valid here although fundamentally the phase information is the best that can be recovered from Lorentz imaging. From equation (1) this is seen to relate to the component of the magnetic vector potential along the electron path. Generally the magnetisation distribution is of greatest interest so equation (2) which relates directly to the DPC images is more useful even though this gives information on the induction rather the magnetisation. Part of the problem with the induction interpretation is that induction is not necessarily local to the sample, unlike the magnetisation. A question therefore arises as to what quantity, if any, that is local to the sample can be deduced from the images i.e. the phase information. It turns out that this quantity is the curl of the magnetisation [4]. For thin film samples at normal incidence to the electron beam it is only the out of plane component of the curl that contributes to phase differences. Indeed the phase can be written as a convolution of this curl component with a simple geometric function [4].

The increasing power of desktop computers has allowed sophisticated micromagnetic simulation programs to be available to anyone with access to such a machine. One of the most commonly used packages is called OOMMF (object oriented micromagnetic framework) which is provided by NIST [5]. In the next section details of how this simulation package has been used to assist quantification of Lorentz DPC images is discussed.

3. Micromagnetic simulations

The example of the Lorentz images given in Figure 1 appears relatively straightforward as the element possesses a flux closure structure in which the magnetisation and the induction distributions are very similar. Using the micromagnetic simulation package it is possible to simulate the induction distribution from the magnetisation components output from the simulation [4]. However care must be taken in how this calculation is performed as discrepancies can arise between the predicted and expected structures. To illustrate this a 20 nm thick element of permalloy with in-plane dimensions of 1000 nm × 200 nm was used for the initial study. For the simulation the element needs to be broken down into discrete cells which are assumed to be uniformly magnetised and the program performs an energy minimisation for the given situation that is input. The cell size chosen here was 5 nm × 5 nm × 20 nm with last parameter being through the thickness. It is assumed that the magnetisation will not vary through the thickness of the film. The in-plane components of magnetisation simulated for this element are shown in Figure 2. Here the magnetisation distribution does not possess flux closure and indeed has a net moment along its long in-plane axis. The structure is shown schematically in Figure 2(c). As discussed in the previous section the out of plane component of the curl of the magnetisation can be considered to form the basis for the calculation of the Lorentz image. This is shown in Figure 2(d).

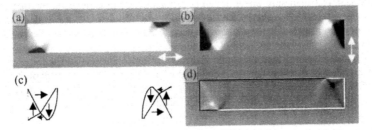

Figure 2. (a) and (b) Shows the in plane-magnetisation components of a 20 nm thick NiFe element with dimensions 1000 nm × 200 nm. The arrows indicate the direction of magnetisation mapped. (c) A schematic representation of the magnetisation distribution within the element. (d) The out of plane component of the curl of the magnetisation for the element.

46

4. Lorentz microscopy calculations

Calculation of Lorentz images are made possible by first determining the electron phase from the Aharonov-Bohm effect in equation (1). As noted in the previous section the phase can be considered as a convolution of the out of plane magnetisation curl and a simple geometric function [4]. This suggests that it may be simple to use Fourier transforms to calculate the phase effects, and then to simply differentiate the phase image to get the components of integrated induction. Whilst we have used this in the past it is important that certain considerations be taken into account in order that quantitative information can be extracted from the simulations and also experimental images. These factors are discussed in more detail now.

A phase calculation has been proposed based on a Fourier series method [6] and Lorentz images have been determined using such calculations [7]. Our image calculations are based on this method. Using this Fourier series representation of the magnetisation is a quick and reasonably efficient method. However one must be careful in ensuring that the magnetisation and the resulting phase are consistent with each other. For most simulations, magnetisation patterns are considered which are either periodic or can be considered to be an isolated element as in the example shown in Figure 2. In the case of Figure 2 an area of "free space" has been created around the element which means we have effectively isolated elements, assuming periodic boundary conditions are considered. A problem can arise with phase calculations in that while periodic magnetisation distributions can be easily produced, the resulting phase from such distributions may not necessarily be periodic. In fact the Fourier series representation of the phase omits the constant component of the magnetisation arguing that it is not physically important. However this term is important if it is non-zero. If there is a non-zero constant magnetisation term the calculated phase is incorrect and this means that the phase gradient calculation will be wrong by a constant term. The problem can be overcome by making a composite and larger magnetisation distribution created by suitably tiled images that has no constant magnetisation term. A similar effect has been proposed in the phase reconstruction from Lorentz TEM images [8]. This however increases the image size fourfold which can be impractical. A solution is to include an additional term in the phase calculation to account for constant magnetisation component [4]. This is added to the Fourier series calculation which does not include this component. Calculation of the term can be written as due to a magnetic flux term by re-writing equation (1) [7]. The reason for including it in this way is that a constant in-plane magnetisation gives no out of plane magnetisation curl even though it has an affect on the phase of the beam.

It should be noted that one could also use the curl of the magnetisation convoluted with a simple geometric function as discussed in section 2. As this involves Fourier transforms it is necessary to suitably tile images to make a composite to ensure that the phase is periodic as in reference 8. The implementation of the Fourier series method has been adopted as it has been deemed easiest to use.

5. Results

To illustrate the importance of the phase calculations considered in the previous section results are now given for the simulation shown in Figure 2 together with some simple comparison magnetisation distributions. The results are shown in Figure 3. The systems for comparison are; (i) an infinitely long strip with the same width and thickness as the element in Figure 2 and magnetised along its length, (ii) an element with the same dimensions as that in Figure 2 but uniformly magnetised along its long in-plane axis and (iii) the magnetisation configuration of the element in Figure 2.

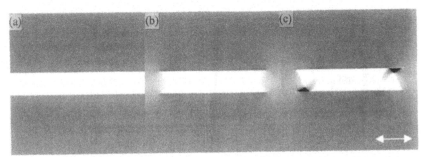

Figure 3. (a) In-plane induction (and magnetisation) component of a long strip uniformly magnetised along its length. The element width and thickness correspond to the element shown in Figure 2. (b) Calculated integrated induction component (DPC image) along the long in-plane axis for uniformly magnetised element with same dimensions as that in Figure 2. (c) Calculated integrated induction component (DPC image) for the element shown in Figure 2 for the component along the long in-plane axis using the OOMMF simulation.

The examples shown are a useful comparison for Lorentz microscopy in that for the three cases, a profile of the magnetisation through the centre of the element across its width will be identical. In each case the magnetisation is uniform and has a value of the saturation induction of the material. It is therefore interesting to compare the calculated integrated induction components along the length of the elements, i.e. a DPC image component. For example the infinitely long strip in Figure 3(a) has the same form for the magnetisation and induction profile as it produces no stray field. In the case of the uniformly magnetised finite sized element in Figure 3(b) there are clearly stray fields visible. This is also true in Figure 3(c) although the integrated stray field effect is rather small compared to the magnetisation, as comparison of Figures 2 (a) and 3 (c) illustrates.

A more useful and quantitative insight can be made by comparing the integrated induction across the centre of the length of each of the elements shown in the DPC images of Figure 3. The results are shown in Figure 4. The units of the integrated induction are in Tesla nanometres (Tnm). For the permalloy films considered here the saturation induction is 1.0 T and so for a uniformly magnetised 20 nm thick film, with no stray fields present, an integrated induction of 20 Tnm is expected. This can be seen to be true for the uniformly magnetised infinitely long strip in Figure 4 in which the integrated induction is a constant 20 Tnm within the element and zero outside. For the finite uniformly magnetised element the integrated induction at the centre is 18.9 Tnm. This reflects the same saturation magnetisation contribution and a stray field contribution in the opposite direction. The integrated stray field outside the element is opposite to the magnetisation as expected and consistent with that of a bar magnet. The induction profile of the OOMMF simulation is similar in form to that from the uniformly magnetised finite element. However even though the magnetisation is saturated at the centre of the element, it is expected that the domain structure seen in Figure 2(c) would lead to a different integrated stray field. This is indeed the case and the integrated induction at the centre of the element here is measured to be 18.2 Tnm.

48

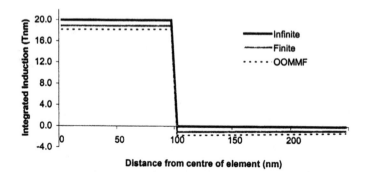

Distance from centre of element (nm)

Figure 4. Integrated induction profiles from the Lorentz DPC images shown in Figure 3. The integrated induction component is given in units of Tesla nanometers (Tnm). The line profiles are taken across the centre of each image, from the centre to 150 nm outside, with the corresponding labels shown.

These results demonstrate that although the domain structure may be obvious from the magnetisation patterns the actual induction distribution from which quantitative measurements are to be made should be treated with care. The examples shown here indicate that for situations where a film is expected to be uniformly magnetised, the induction image does vary significantly due to stray field effects. In the example of the induction linetraces shown in Figure 4 the differences in induction values are of the order of 10% inside the element whereas the equivalent magnetisation traces are identical for all three cases. The integrated stray field effects depend on a number of factors, such as element shape and domain structure, however it is clear that accurate calculation is necessary for quantitative analysis of images.

Acknowledgments
Thanks are due to the UK EPSRC for financial assistance. In addition Mr Gordon White and Dr Patrick Warin are to be thanked for implementing the phase calculations and providing the micromagnetic simulation shown in Figure 2 respectively.

References
[1] Chapman J N and Scheinfein M R 1999 J. Magn. Magn. Mater. 200 729-740
[2] Lim C K, Chapman J N, Rahman M, Johnston A B and O'Donnell D O 2002 J. Phys. D: Appl. Phys. 35 2344-2352
[3] Warot B, Petford-Long A K and Anthony T C 2003 J. Appl. Phys. 93 7287-7289
[4] McVitie S and White G S 2003 submitted for publication
[5] Donahue M J and Porter D G 1999 National Institute of Standards and Technology, Gaithersburg, MD 20899, USA; The current URL: http://math.nist.gov/oommf/.
[6] Mansuripur M 1991 J. Appl. Phys. 69 2455-2464
[7] Haug T, Otto S, Schneider M and Zweck J 2003 Ultramicroscopy 96 201-206
[8] Volkov V V, Zhu Y and De Graef M 2002 Micron 33 411-416

Inst. Phys. Conf. Ser. No 179: Section 2
Paper presented at Electron Microscopy and Analysis Group Conf. EMAG2003, Oxford, 2003
©*2003 IOP Publishing Ltd*

Amorphous ferromagnetic layers for magnetic tunnel junctions

B Warot[a], J Imrie[a], A K Petford-Long[a], M Sharma[b], T C Anthony[b]

[a] Department of materials - University of Oxford, UK

[b] HP Labs, Palo Alto, USA

Abstract : Magnetic tunnel junctions are being actively studied for use as single cells in magnetic random access memory (MRAM). For MRAM application purposes, a narrow distribution of the switching fields of the cells is required. Amorphous ferromagnetic layers used as sense layers in the junctions are predicted to improve the uniformity properties. In this article, we report the results of transmission electron microscopy studies of the crystalline structure of CoFeB and CoFeHfNb ferromagnetic layers, known to be amorphous. The results show the influence of the underlayer and emphasise the importance of deposition order on the crystallographic structure of the layers.

1. Introduction

Magnetic tunnel junctions are being studied for their spin-dependent transport properties – the tunnel magnetoresistance (TMR) effect [1]. In a TMR device, an insulating barrier separates two ferromagnetic layers. One of the ferromagnetic layers, the sense (or free) layer, is free to rotate under the field whereas the other is pinned to an antiferromagnetic layer.

Amorphous magnetic thin films have been paid increasing attention for applications in magnetoresistive structures due to their low coercivity which induce low switching fields and their high electrical resistivity which enables high speed access memories. Indeed, amorphous magnetic materials can achieve lower coercivities than permalloy ($Ni_{80}Fe_{20}$), which is commonly used as the soft ferromagnetic sensing layer in tunnel junctions. Amorphous layers may also help solve the problem of switching field uniformity in Magnetic Random Access Memory (MRAM) devices. The tunnel junctions used as bit cells in MRAM must have sizes close to 0.25 µm or less. The grain structure of the materials has an influence at these dimensions and it is suggested that the use of amorphous materials will improve the uniformity of switching in MRAM devices [2,3].

Boron, niobium and hafnium are known to induce amorphous structure in magnetic layers [4]. For example, CoFeB amorphous layers have been widely studied for spin-valve devices [5,6]. In this study, electron microscopy has been used to study the crystallinity and the texture of magnetic tunnel junctions containing amorphous ferromagnetic sense layers.

2. Experimental details

A magnetic tunnel junction (MTJ) consists of a seed / antiferromagnet (AF) / pinned ferromagnet (F) / insulating barrier layer / free F stack. Four different samples with IrMn as the AF layer and two different F layers ($Co_{77}Fe_1Hf_6Nb_{16}$ and $Co_{72}Fe_8B_{20}$) have been studied. All the layers were deposited by sputtering on a Ta/Ru seed layer deposited on a silicon substrate. The thin alumina layer was formed by aluminium deposition and exposure to oxygen plasma. The samples were annealed for an hour at 250°C. Cross-section TEM samples were prepared by mechanical polishing and the slice was then thinned further using a precision ion polishing system (PIPS). The high-resolution TEM (HRTEM) analysis was carried out using a JEOL 4000EX operated at 400 kV (point-to-point resolution 0.16 nm). Electron energy loss spectroscopy (EELS) compositional mapping of the layers was carried out.

3. Results and discussion

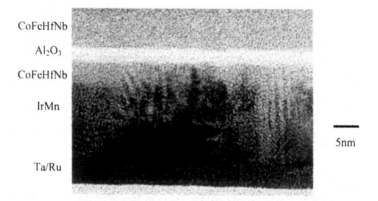

Figure 1 : HRTEM image of a MTJ with CoFeHfNb F layers

Figure 1 shows an HRTEM image of a Ta/Ru/IrMn/CoFeHfNb/Al₂O₃/CoFeHfNb/Ta sample. The layers are labelled on the image. The layers are continuous throughout the sample and the roughness is small. Figure 2a is a HRTEM image of the same sample at higher magnification. The AF and pinned F layers are crystalline and textured. The free F layer is amorphous and this has been observed throughout the sample.

In order to study the influence of the pinned layer on the crystallinity of the free F layer, a 3.5nm thick $Co_{50}Fe_{50}$ layer was used as the pinned F layer (figure 2b). For this sample the free CoFeHfNb layer is nanocrystalline with an average grain size of 2nm. An ion etching step has been performed on the surface of the pinned CoFe layer prior to the barrier deposition. The CoFe surface is rougher than the CoFeHfNb pinned layer surface observed in figure 2a.

CoFeHfNb

Al₂O₃

CoFeHfNb

IrMn

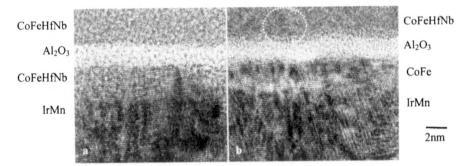

CoFeHfNb

Al₂O₃

CoFe

IrMn

———

2nm

Figure 2 : HRTEM images of a MTJs with (a) CoFeHfNb pinned layer
and (b) CoFe ion-etched pinned layer

The difference in the crystallinity of the CoFeHfNb layer cannot be explained by
"transmission" of the crystalline structure from the pinned to the free layer as the
alumina barrier is too thick. The transition between amorphous and crystalline growth
can be achieved by changing the local temperature. The sputtering method used to
deposit the materials locally increases the temperature because of the high energy ions
arriving on the surface. The dissipation of this energy proceeds on the surface and
through the pinned layers. As the pinned layers for the two samples are different, we
might expect different heat transfer behaviour. However, heat transfer coefficients have
been calculated for the two different pinned layers, CoFe and CoFeHfNb, and are
similar so other parameters need to be taken into account to explain the difference. One
possibility is that the higher roughness for the sample with the CoFe pinned layer could
induce local nucleation sites for nanocrystallites.

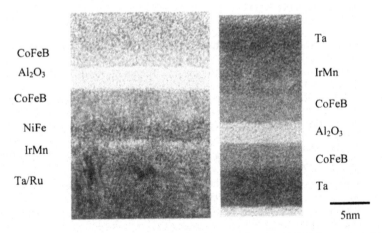

CoFeB

Al₂O₃

CoFeB

NiFe

IrMn

Ta/Ru

Ta

IrMn

CoFeB

Al₂O₃

CoFeB

Ta

———

5nm

Figure 3 : HRTEM images of (a) a MTJ with CoFeB layers and (b) an
inverted MTJ with the same layers

Poor magnetic properties have been measured for the MTJs with CoFeHfNb free layers
which are attributed to the poor intrinsic magnetic properties of the CoFeHfNb material.

Further samples have been deposited using CoFeB as a material for both the free and pinned layers. A NiFe layer has been inserted between the IrMn AF layer and the CoFeB pinned layer to induce exchange coupling between the two layers [5]. Figure 3a presents an HRTEM image of the Ta/Ru/IrMn/NiFe/CoFeB/Al_2O_3/CoFeB sample. The free layer is amorphous and the pinned layer is crystalline and textured. We then deposited another MTJ stack with the same layers but deposited in reverse order so that the free layer is close to the substrate. This is shown on figure 3b and all the layers are amorphous. Chemical mapping on this sample revealed that boron has diffused from the free layer to the Ta capping layer.

The CoFeB material is amorphous when deposited on an amorphous Al_2O_3 layer and crystalline on a textured layer. Crystallinity in the CoFeB cannot be the result of a crystallising process after deposition as a fcc structure would require annealing at high temperature, 400°C [7]. When deposited directly on the Ta layer, the CoFeB layer is amorphous and induces an amorphous texture throughout the sample, which could be due to the diffusion of boron. Moreover, the IrMn layer is amorphous and magnetic measurements have shown that there is no exchange coupling between the IrMn and the CoFeB layers. Exchange has been measured on the crystalline MTJ (figure 3a) confirming the need for a crystalline AF layer to observe exchange [8].

4. Conclusions

Magnetic tunnel junctions with different amorphous sense layers ($Co_{77}Fe_1Hf_6Nb_{16}$ and $Co_{72}Fe_8B_{20}$) have been studied using TEM. A CoFeHfNb layer deposited on a crystalline IrMn layer is crystalline whereas the CoFeHfNb sense layer is amorphous or nanocrystalline depending on the pinned layer. The difference could be explained by different local temperatures due to different heat transfer as well as topological differences of the two pinned layers. A CoFeB layer deposited on a crystalline NiFe layer is crystalline, but is amorphous when deposited on amorphous alumina or textured tantalum.

References

[1]   Moodera J. S. and Kinder L. R. 1996 J.Appl.Phys. 79 4724[

[2]   Kano H, Bessho K, Higo Y, Ohba H, Hahimoto M, Mizuguchi T and Hosomi M 2002 IEEE Trans. Mag.

[3]   Wang D, Qian Z, Daughton J, Nordman C, Tondra M, Reed D and Brownell D. 2001 J.Appl.Phys. 89 6754

[4]   Balogh J, Bujdosó L, Kemény T, Pusztai T, Tóth L and Vincze I 1997 Appl. Phys. A 65 23

[5]   Fujita M, Yamano K, Maeda A, Tanuma T and Kume M 1997 J.Appl. Phys. 81 4909

[6]   Jimbo M, Kurita J, Sakakibara K, Goto T and Shigeoka T 1999 J. Mag, Mag. Mat. 198-199 431

[7]   Jimbo M, Komiyama K, Shirota Y, Fujiwara Y, Tsunashima S and Matsuura M 1997 J. Mag, Mag. Mat. 165 308

[8]   Ritchie L, Liu X, Ingvarsson S, Xiao G, Du Jand Xiao J 2002 J. Mag, Mag. Mat. 247 187

Inst. Phys. Conf. Ser. No 179: Section 2
Paper presented at Electron Microscopy and Analysis Group Conf. EMAG2003, Oxford, 2003
©*2003 IOP Publishing Ltd*

Computer simulation and experimental observation of stripe phases in La$_{1-x}$Ca$_x$MnO$_3$

S Cox, J C Loudon, P A Midgley and N D Mathur

Department of Materials Science and Metallurgy, University of Cambridge, Pembroke Street, Cambridge, CB2 3QZ, UK

Abstract. In this paper we combine low temperature electron microscopy of polycrystalline La$_{1-x}$Ca$_x$MnO$_3$ (0.5≤x≤0.71) with computer simulations of a model system using bi-stripe type sub-units. The results indicate that the observed modulations cannot be described as a random mixture of bi-stripe type sub-units delineated by discommensurations. The modulation is also simulated using the Frenkel-Kontorova model suggesting that electron distribution may vary smoothly (akin to a charge density wave).

1. Introduction

Modulations of wavevector **q** ascribed to charge and orbital order in the mixed-valent pseudo-cubic (orthorhombic) manganites (RE$^{3+}_{1-x}$AE$^{2+}_x$MnO$_3$ with RE=rare earth, AE=alkaline earth) have been observed along **a*** at low temperature using transmission electron microscopy [1-4]. Globally, the wavenumber q of the observed modulations is found to obey the relation $q \approx (1-x)a*$ for a continuous range of values of x≥0.5

Traditionally, the observed modulations have been interpreted [1-4] in terms of a crystalline array of two Mn species, commonly referred to as the idealised cations Mn^{3+} and Mn^{4+}. In this picture, pseudo-cubic {110} planes of coherently (Jahn-Teller) distorted Mn^{3+}O$_6$ octahedra, and similar planes containing undistorted Mn^{4+}O$_6$ octahedra, are present in the ratio (1-x):x and stacked along **a** with their oxygen octahedra sharing corners. However, at a few specific values of x there is now growing evidence to suggest that the repeating units possess different charge distributions (for example [5-10]).

Contrast seen in high resolution electron microscopy (HREM) images has been used to suggest [2-4] that the two types of planes described above can only be stacked in a few specific sequences whose periodicities are integer multiples of a. In this description, the allowed sequences are called "bi-stripes" and the periodicity is an integer multiple of a such that q=**a***/n. In addition, it has been suggested that if the periodicity is not an integer multiple of a, then the modulation is composed, according to the lever rule, of a fine random mixture of the two adjacent integer period sub-units. For example, La$_{1-x}$Ca$_x$MnO$_3$ at x=5/8 has been described [3,4] as a fine random mixture of x=1/2 and x=2/3 type sub-units present in the ratio 1:3. One type of sub-unit is considered to appear in the form of incoherent, atomically sharp, "stacking faults" in a background of the other type of sub-unit [3,4].

Image contrast in HREM varies strongly as a function of sample thickness, defocus and tilt. Quantitative interpretation of such contrast is particularly difficult for the relatively thick (50-100 nm) manganite specimens in which dynamical diffraction will enhance the intensities of the weaker reflections. However, measurements of periodicity by HREM are accurate.

2. Simulation of the random mixture model and comparison with experiment

Throughout this work we consider sub-units obtained from high resolution images in [3]. One period of the modulation at each of the doping levels $x = 1/2$, 2/3 and 3/4 was created from sub-units based on linescans from high resolution electron micrographs [3], and adjusted so that the endpoints of each subunit matched up. Computer simulations were performed to simulate arbitrary compositions of $La_{1-x}Ca_xMnO_3$ ($x>0.5$) as random mixtures of these integer period sub-units, weighted to produce the correct average composition. To simulate diffraction patterns, power spectra were taken

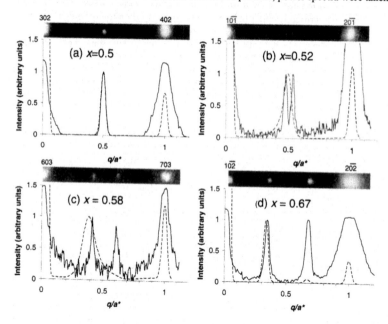

Figure 1 Electron diffraction patterns for **a** x=0.5, **b** 0.52, **c** 0.58 and **d** 0.67 taken at 90 K looking down the [010] zone axis. At intermediate dopings **b**&**c** the corresponding linescans (solid lines) do not match the power spectra (dotted lines) simulated using the random mixture model with x=0.5 and x=0.67 type sub-units. The gaussian convoluted with the simulations in **a-d** was selected in order to match the experimental peak at q/a^*=0.5 in **a**. The least saturated areas of the experimental diffraction patterns were used and a Random Sample Consensus (ransac) method with quadratic equations was used to remove background noise from the experimental linescans. Note that the experimental intensities are affected by dynamical diffraction.

using standard fast Fourier transforms. To achieve self-consistent simulations required ~60000 unit cells (~30 μm in length).

Samples of $La_{1-x}Ca_xMnO_3$ were prepared by repeated grinding, pressing and sintering of La_2O_3, $CaCO_3$ and MnO_2 in stoichiometric proportions. The samples were prepared for transmission electron microscopy by conventional mechanical polishing and argon-ion thinning at liquid nitrogen temperatures. Electron microscopy data at 90 K were taken from polycrystalline samples (grain size ~2μm) using a Philips CM30 microscope. As expected, the wavevectors of the modulations were only ever observed to be parallel, or almost parallel, to a^* as a consequence of the underlying orthorhombic Pnma symmetry.

Four compositions in the range $0.5 \leq x \leq 0.67$ display sharp modulations with wavevector q (Fig. 1). For the two intermediate compositions in this range ($x=0.52$ and $x=0.58$), this is inconsistent with the broad peaks produced by a random mixture of integer period sub-units (Fig. 1b&c). We are able to confirm that $q \approx (1-x)a^*$, but due to small spatial variations in q discussed below, we are unable to rule out the possibility of subtle deviations from this relationship.

3. Simulations using the Frenkel-Kontorova model

The Frenkel-Kontorova model is used to describe systems with competing periodicities [11]. It was used as a possible alternative model to the random mixture model. The experimentally determined value of q sets the number of particles/period to be distributed in each run within a sinusoidal potential of amplitude V and period a. The particles are connected with springs of stiffness K such that only their horizontal separation is taken into account. Each run was initialised in a random configuration and the evolution of the ground state was determined using damped equations of motion. For this an eighth order Runge-Kutta Prince-Dormand solver with ninth order error estimate and adaptive stepsize from the gnu scientific library (http://www.gnu.org/software/gsl/gsl.html) was used. Once the ground state had been determined, a delta function was positioned at the site of each particle. The power spectrum and the value of q were calculated.

It is possible to compare the widths of the peaks in the power spectrum produced by the Frenkel-Kontorova model with the experimental peak widths. However, broadening occurs for a number of different reasons experimentally, so we decided instead to find the standard deviation of q for a number of areas within a grain and compare this to the results from simulating a number of different areas.

Experimentally, the small variations in q within a grain may be mapped with high precision in a dark field image using an objective aperture that includes q and $g-q$ (where g is a reciprocal lattice vector of the unmodulated structure, e.g. a^*). Interference fringes arise with spacing $1/|g-2q|$ from the beating between the two wavevectors. Using this interference method to study 80 distinct 50 nm diameter regions from one grain of $La_{0.48}Ca_{0.52}MnO_3$ we found a mean value of $q=0.446a^*$ with a standard deviation of $0.004a^*$.

In the simulation three thousand runs were performed for each value of $(2\pi/a)^2 V/K$. The standard deviation of the wavevectors was calculated and compared to experiment (Fig. 2). Our results suggest that to achieve a standard deviation as small as $0.004a^*$ with $q=0.446$ places an upper bound of 3% on $(2\pi/a)^2 V/K$. If we take an overly simplistic interpretation and identify the potential V with the high temperature lattice

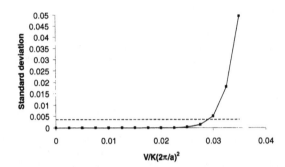

Figure 2 The solid line shows the standard deviation of the peak positions obtained using the Frenkel-Kontorova model simulation, and the dotted line shows the experimental result for the standard deviation. The limiting value of $(2\pi/a)^2 V/K$ is 3%.

configuration, the particles with the valence electrons, and $(2\pi/a)^2 V/K$ as the electron-phonon coupling, then $(2\pi/a)^2 V/K=3\%$ would imply that the electron-phonon coupling is weak. This comes as a surprise since weak electron-phonon coupling is normally associated with the metallic phases of manganites rather than the so-called charge ordered stripe phases.

4. Conclusions

Our results throw doubt on the random mixture model. The results obtained using the Frenkel-Kontorova model indicate that the modulation seen in $La_{1-x}Ca_xMnO_3$ could be a charge density wave.

References

[1] Chen, C H and Cheong, S -W 1996 Phys. Rev. Lett. 76 4042-4045
[2] Chen, C H, Cheong, S -W and Hwang, H Y 1997 J. Appl. Phys. 81 4326-4330
[3] Mori, S, Chen, C H and Cheong, S -W 1998 Nature 392 473-476
[4] Chen, C H, Mori, S and Cheong, S -W 1999 J. Phys. IV France 9 PR10-307-Pr10-310
[5] Wang, R, Gui, J, Zhu, Y and Moodenbaugh, A R 2000 Phys. Rev. B 61 11946-11955
[6] Radaelli, P G, Cox, D E, Capogna, L, Cheong, S -W and Marezio M 1999 Phys. Rev. B 59 14440-14450
[7] Daoud-Aladine, A, Rodríguez-Carvajal, J, Pinsard-Gaudart, L, Fernández-Díaz, M T and Revcolevschi A 2002 Phys. Rev. Lett. 89 097205
[8] Rodríguez-Carvajal, J, Daoud-Aladine, A, Pinsard-Gaudart, L, Fernández-Díaz, M T and Revcolevschi A 2002 Physica B 320 1-6
[9] Williams, A J and Attfield, J P 2002 Phys. Rev. B 66 220405
[10] Ferrari, V, Towler, M D, and Littlewood, P B 2003 arXiv:cond-mat/0304343 v2
[11] Coppersmith S N and Fisher, D S 1983 Phys. Rev. B 28 2566-2581

Inst. Phys. Conf. Ser. No 179: Section 2
Paper presented at Electron Microscopy and Analysis Group Conf. EMAG2003, Oxford, 2003
©2003 IOP Publishing Ltd

Structural and compositional characterization of LiF thin films on Si

A A Suvorova, M Saunders, B J Griffin, and S N Samarin*

Centre for Microscopy and Microanalysis, The University of Western
Australia, 35 Stirling Highway, Crawley WA6009, Australia
*School of Physics, The University of Western Australia, Australia

Abstract. LiF thin films deposited on Si substrates have been found to
exhibit interesting electronic properties that are dependent on the film
thickness. The structural and compositional properties of LiF films have
been studied via a combination of SEM imaging and TEM diffraction and
energy-selected imaging in an attempt to correlate the structural,
compositional and electronic properties of the films.

1. Introduction

Alkali halide films are promising materials for applications in new integrated optical
devices such as waveguides and lasers. Among these materials, LiF is of considerable
interest, having a wide bandgap and extraordinary UV-transparent properties. A number
of experimental and theoretical investigations of the electronic and structural properties
of LiF films have been carried out [1-4]. LiF films have been found to exhibit
exciton/plasmon-assisted secondary electron emission revealed by time-of-flight two-
electron coincidence spectroscopy in reflection mode in combination with EELS [5,6]. It
has been found that secondary emission features are related to the energy losses and their
appearance has been attributed to the exciton/plasmon-assisted processes. However, for a
thin (up to 5nm) LiF film the above mentioned loss-related features do not appear. The
correlation between this emission mechanism and the structure of the LiF films is not
clearly understood.

LiF films of various thicknesses (3-20nm) have been deposited by thermal
evaporation onto Si(100) single crystal substrates. A combination of SEM and TEM has
been used to determine the structural and compositional properties of the films grown
with different film thicknesses.

2. Results and Discussion

2.1. FESEM imaging

FESEM imaging has been applied to study the morphology of the LiF films of various
thicknesses. Charging of the dielectric films makes it difficult to obtain surface
morphology information from conventional, high vacuum SEM. To alleviate this
problem, we have used the in-lens secondary electron detector of a LEO 1555 VPSEM,
operating at low voltage, to observe the surface morphology and thickness dependent
modifications to the surface structure of the LiF films.

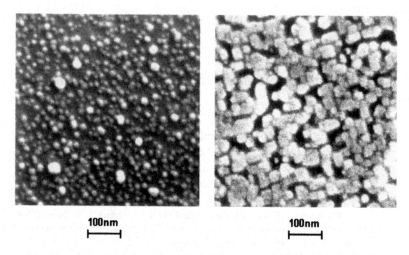

Figure 1. SEM images of "thin" (a) and "thick" (b) LiF films on Si recorded at 3kV, with an in-lens detector, at working distance of 1mm.

As revealed in Fig.1(a), the surface morphology of a 5nm LiF film clearly display an island structure. As the film thickness increases, the islands begin to coalesce, and an ordered morphology is observed at a thicknesses of 10nm, see Fig.1(b). Further increasing in thickness results in almost continuous LiF surfaces, which maintain the surface morphology. The rectangular shape of the islands suggests that they are of single crystal nature.

2.2. High resolution imaging and SAED

High resolution imaging, selected area diffraction and energy-filtered TEM have been carried out on plan view and cross-section samples using a JEOL 3000F FEGTEM equipped with a Gatan Image Filter and digital camera.

The diffraction pattern obtained from a plan view sample at the Si[001] zone axis is shown in Fig.2. Two sets of diffraction spots are observed in addition to those from the Si substrate. First, strong reflections are observed close to the 220 Si spots, which were identified as 220 reflections from the LiF crystals. Second, the diffraction pattern exhibits a complex regular distribution of low intensity reflections resulting from double diffraction.

HREM studies of the LiF/Si system in the [110] projection have shown that the LiF films are polycrystalline with crystallite sizes in the range 3-10 nm depending on film thickness.

The crystallites have rectangular shape and the predominant alighnment of the LiF crystallites is along one of the <110> directions of the Si surface. The LiF crystallites have a preferential orientation relative to the silicon surface that could be determined as LiF(001)[110]//Si(001)[110]. This cube–on–cube orientation relationship is observed throughout the thickness range from 3 to 20nm.

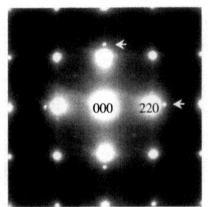

Figure 2. SAED pattern from LiF/Si taken in Si[001] zone axis.

2.3. Energy-filtered imaging

Obtaining Li elemental maps proved difficult due to the low signal from the Li-K edge (at 55eV) and the small particle size, resulting in a poor signal-to-noise ratio. Therefore, low-loss imaging has been used to investigate the compositional variations in the films.

Plasmon-loss imaging is complicated by the overlapping of the Si and LiF signals in the low-loss region. A series of energy-selected images has been acquired across a range of energies from 5eV to 30eV with a 4eV energy width. This range includes the LiF bulk plasmon energy (25.3eV), the Si substrate plasmon energy (16.7eV), and additional energy-loss features such as the LiF surface exciton peak (10.7eV) and LiF bulk exciton peak (13.5eV) [7].

Fig.3 shows a selection of the images from an energy-selected series obtained from an 8nm LiF film. All of the filtered images show contrast associated with the LiF crystallites. This contrast is maximized at an energy loss of ~26eV that is close to the characteristic energy loss of the LiF plasmon (25.3eV). The image at 22eV includes signals from both the LiF particles and the Si substrate, resulting in a greatly reduced contrast. At 18eV, a high signal is obtained from the particle surfaces, possibly as a result of a LiF surface plasmon. As the energy loss is reduced to 10eV, the bulk of the particles again lights up, perhaps from a LiF exciton signal. The resolution obtained in the 10eV image is considerably reduced compared to that seen in the 26eV image, and can be interpreted in terms of the greater delocalisation of the lower energy signal.

Thinner LiF films show considerably less contrast in the 26nm image. This could result from the smaller volume of material giving lower signal. However, it may indicate a reduction in the strength of the plasmon signal as the particle size is reduced (which is supported by the previous electron spectroscopy results [4,5]).

3. Conclusions

SEM analysis has shown that the particle size and morphology of LiF films varies as a function of film thickness. Continuous films can be achieved for thicknesses above 8nm. TEM studies have shown that the LiF films are polycrystalline with crystallite sizes in the range 3-10 nm depending on film thickness. The orientation relationship between the film and substrate was determined to be LiF(001)[110]//Si(001)[110], and is consistent for all film thicknesses.

60

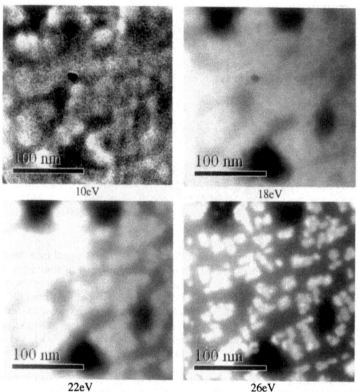

10eV   18eV

22eV   26eV

Figure 3. Series of filtered images taken at various energy loss (10-26eV) with a 4eV slit. 8nm LiF film

Low-loss imaging confirms the presence of LiF, with particle contrast being maximised close to the LiF plasmon energy. The LiF plasmon signal is significantly lower when the film thickness is reduced to ~3nm, an observation that is consistent with previous electron spectroscopy measurements [4,5]. Additional features are observed at other energy-losses that could be associated with surface plasmon and exciton signals.

References

[1] Roy G, Singh G and Gallon T E 1985 Surf. Sci. 152-153 p 1042-1050
[2] Di Nunzio P E, Fornarini L, Martelli S and Montereali R M 1997 Phys.Stat.Sol.(a) 164 p 747-756
[3] Klumpp St, Dabringhaus H 1999 J.Cryst.Growth 204 p 487-498
[4] Golek F and Sobolewski W J 1999 Phys. Stat. Sol. (b) R1 216
[5] Samarin S N, Artamonov O M, Suvorova A A, Kirschner J, Berakdar J and Williams J F submitted Phys.Rev. B
[6] Samarin S N, Artamonov O M, Waterhouse D K, Kirschner J and Morozov A, Williams J F 2003 Rev. Sci. Instrum. vol 74 No 3 p 1274-1277
[7] Golek F and Sobolewski W J 1999 Solid State Communications 110 p 143–146

Inst. Phys. Conf. Ser. No 179: Section 2
Paper presented at Electron Microscopy and Analysis Group Conf. EMAG2003, Oxford, 2003
©2003 IOP Publishing Ltd

Selected techniques for examining the electrical, optical and spatial properties of extended defects in semiconductors

N M Boyall[1], K Durose[1], T Y Liu[2], A Trampert[2], C Liu[3] and I M Watson[3]

[1]Dept of Physics, Univ. Durham, South Rd, Durham, DH1 3LE, UK.
[2]Paul-Drude-Institut für Festkörperelektronik, Hausvogteiplatz 5-7, D-10117 Berlin, Germany.
[3]Institute of Photonics, University of Strathclyde, Glasgow G4 ONW, UK.

ABSTRACT: Three methods in semiconductor defect analysis are described with examples and appraisal of their capabilities: An electron beam induced current method (R-EBIC) is shown to determine the sense of band bending at grain boundaries in the CdTe-CdS system. Cathodoluminescence in the TEM has demonstrated a link between stacking faults and 3.26eV luminescence in GaN. Statistical analysis of spatial pattern applied to 'v-pit' distributions in InGaN structures on sapphire demonstrates the pits to be correlated on the scale of 60-120nm.

1. Introduction

Here three selected methods for assessing the electrical, optical and spatial properties associated with extended defects in semiconductors are described, and examples from appropriate materials systems given. They are:-

a) The 'remote' electron beam induced current mode of the SEM (R-EBIC) in which the fields at grain boundaries are probed. Examples from the thin film solar cell materials system CdTe-CdS are given.

b) Collection of cathodoluminescence *in-situ* in the TEM (TEM-CL) which allows structural defects identified by diffraction contrast to be correlated to optical spectroscopy directly. Results showing luminescence from stacking faults in GaN are described.

c) Statistical analysis of spatial pattern in which the randomness or otherwise of the distribution of defects (e.g. dislocations) can be assessed quantitatively. A study of the dislocation associated 'v-pits' on GaN/InGaN multi-quantum wells is presented.

2. Experimental

Remote-EBIC (R-EBIC) was done in a JSM 848 SEM using a Matelect ISM5 specimen current amplifier. The samples examined were undoped vapour grown CdTe [1], and homogeneous CdS_xTe_{1-x} ($x \approx 0.067$) grown by a self selecting vapour method [2]. Gold contacts or else probes were attached either side of grain boundaries and twins.

Cathodoluminescence was collected from a JEOL 200CX TEM using an Oxford Instruments MonoCL2 for TEM with samples held in a tilt-rotate stage, nominally at

100K. Layers of M-plane oriented ($1\bar{1}00$) GaN/(100) γ-LiAlO$_2$ grown by plasma assisted MBE [3] were examined.

Spatial pattern was assessed using a nearest neighbour distribution parameter, by comparison with a Poisson distribution, and by means of radial correlation and autocorrelation analysis. The methods are described in [4] and briefly in section 5. Here the distribution of 'v-pits' was evaluated for a 14 period $In_{0.09}Ga_{0.91}N$/GaN multi-quantum well grown by MOVPE at 860°C on a GaN buffer on (0001) sapphire [5].

3. R-EBIC of grain boundaries in solar cell materials.

In conventional electron beam induced current (EBIC) carriers excited by the probe beam are collected by contacts close to the generation volume. In the so-called R-EBIC (remote) experiment the contacts are many diffusion lengths removed from the site of generation (fig.1). Contrast arises by two means i) There is a contribution from the beam current. For the convention of a positive current being bright, the contact connected directly to the amplifier will appear dark, and that connected via earth, bright. A uniformly resistive sample acts as a current divider giving a ramp of contrast over the sample. ii) Electron hole pairs generated by the beam may cause a local current in the sample if there is a field, for example as in the case of grain boundary band bending. This gives rise to a displacement current in the amplifier circuit. It can be seen from fig. 1 that the sense of this current (with respect to the sense of the beam current ramp contrast) gives information about the sign of the band bending. For the convention adopted here, and reading left to right, white-black contrast indicates upward band bending, while black-white indicates downward band bending.

First use of the method is reported in [6] and [7], and good explanations of the phenomena is provided by Holt [8] and in [2,9].

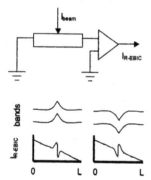

Fig. 1 The experimental configuration used for the R-EBIC experiment. The sample contains a grain boundary between two ohmic contacts. Possible schemes for band bending at grain boundaries and the contrast associated with them are shown. Ramp contrast and the peak and trough contrast may be used to evaluate the sense of the band bending at the grain boundary.

Polycrystalline p-CdTe/n-CdS/ITO/glass solar cells undergo post-growth annealing and also etching during processing and contact formation. The annealing step is known to promote interdiffusion of the CdS window layer with the CdTe absorber; chemical etching of the CdTe surface is known to cause enrichment of the grain boundaries with Te. Without such processing the cell efficiencies are perhaps 2%, but can reach 16.5% afterwards. R-EBIC studies of grain boundaries in bulk CdTe and Cd(STe) give some insight as to how the processing can ameliorate the deleterious effect of grain boundaries.

Fig 2 shows black-white R-EBIC contrast at a randomly oriented grain boundary in CdTe. This corresponds to *downward* band bending such as would cause minority carrier electron capture. Experiments on boundaries decorated with Te inclusions show much reduced contrast at the inclusion sites. Moreover, R-EBIC of boundaries in Cd(STe) crystals shows white-black contrast indicating upward band bending such as would cause minority carrier electron repulsion. Hence both the presence of Te and of the mixed CdS-CdTe phase at grain boundaries would act to increase solar cell performance.

Fig. 2 (left) R-EBIC micrograph of a grain boundary in bulk undoped CdTe.
(below) Linescan of the signal.

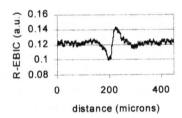

distance (microns)

4. TEM-Cathodoluminescence from stacking faults in M-plane GaN/LiAlO₂

Collection of CL *in-situ* in a TEM offers the possibility to correlate microstructural features (diffraction contrast imaging) directly with spectroscopy and CL imaging. It was first demonstrated by Kingsley [10], and later with scanning beams in a STEM [11] and a TEM with a scanning attachment [12].

M-plane ($1\bar{1}00$) GaN/γ-LiAlO₂ (100) is of interest since layers with this orientation are non-polar and offer the possibility of growing quantum wells free from the spontaneous polarisation that is present in (0001) orientations [3]. LiAlO₂ is tetragonal and hexagonal GaN grows upon it with M-plane orientation (fig. 3) and with low misfit in orthogonal in-plane directions [13] ([$1\bar{1}20$]GaN//[001]LiAlO₂ f=1.4% and [0001]GaN//[010]LiAlO₂ f=0.1%). Trampert [13] demonstrates the presence of stacking mismatch boundaries on the prismatic planes (not fully characterised) and of conventional I₂ stacking fault bundles on the (0001) planes.

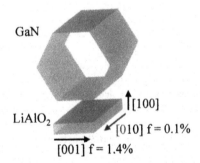

Fig 3. Epitaxial relationship between M-plane - ($1\bar{1}00$) GaN - and (100) γ-LiAlO₂.

The densities of each type of stacking fault are variable and fig. 4a shows an area in which the (0001) stacking faults dominate. A CL spectrum recorded from the stacking fault bundle in area B shows a clear peak at 3.26eV that is not present in the fault-free area A. Recording of a line scan of the intensity at this energy (fig 4b) revealed a strong correlation between the stacking faults and luminescence. Intrinsic stacking faults in the wurtzite lattice comprise a layer of sphalerite phase material. As the band gap of the cubic is marginally lower than the hexagonal phase it is considered that stacking faults may act as type II quantum wells, and this accounts qualitatively for their association with luminescence.

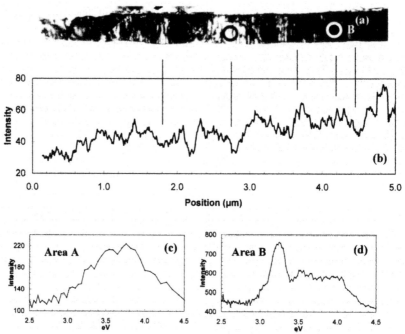

Fig 4. Cross-section TEM and CL spectroscopy and linescan of M-plane GaN on LiAlO$_2$. a) multi-beam TEM image with B=[11-20] - stacking fault bundles are seen end-on, b) CL line-scan with spectrometer centred on 3.26eV(380nm), c) CL spectrum of area A containing a low density of defects, d) CL spectrum of area B containing a high density of stacking faults.

5. Analysis of spatial pattern applied to v-pits in GaN/sapphire

While it cannot be ignored, visual interpretation of spatial pattern is necessarily subjective and must be augmented by quantitative methods. To our knowledge such methods were first applied to dislocations in semiconductors for bulk CdTe - which is prone to polygonisation. Walker [4] used statistical methods to compare the outcomes of different bulk growth processes for CdTe and Palosz [14], the effects of cool-down rate. Durose [2] examined the randomness of twinning in Cd(STe).

Here four methods are applied to evaluate the spatial pattern of hexagonal 'v-pits' on the surface of a GaN/InGaN MQW. For this work the important point is that v-pits are generally considered to be associated with threading dislocations from the underlying GaN/sapphire [15]. The distribution of v-pits in fig. 5a was determined by AFM.

In nearest neighbour analysis a single value distribution parameter is defined as R_n $=2d(n/A)^{1/2}$ where d is the average nearest neighbour distance for a set of n points in an area A [16]. A value of $R_n = 1$ indicates randomness, $R_n < 1$ clustering and $R_n > 1$ regular ordering. The value of 1.09 obtained indicates near-random behaviour.

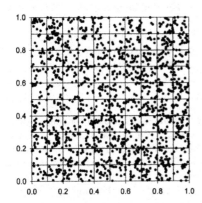

a) Spatial distribution of v-pits on the surface of a GaN/InGaN MQW structure. The unit square is 7.4μm in size, the v-pit density is $2.42 \times 10^9 \text{cm}^{-2}$.

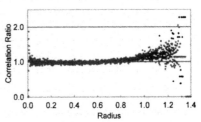

c) Radial correlation function. Deviation from 1.0 indicates excess or deficit clustering.

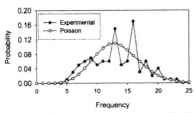

b) Comparison of the spatial distribution over 100 squares with that for a random distribution (Poisson statistics).

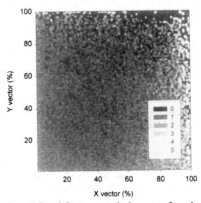

d) Spatial correlation function (autocorrelation). There are no directional trends; the v-pits do not align to crystallographic directions.

Fig. 5 Distribution of v-pits in a 14 period InGaN/GaN MQW structure (MOVPE- grown at 860°C) and its spatial analysis. There is short radius association at distances of 60-120nm, but without crystallographic orientation.

Forming a histogram from the occupancies of the 100 squares in fig. 5a and comparing the result to that expected for a random distribution (Poisson statistics) generates a second test of randomness as shown in fig. 5b. While there are differences they are not striking. Since this test probes the distribution on the scale of one box (0.74μm), it may be concluded that it is random *on that scale*. Indeed χ^2 testing indicates the probability of randomness to be >99%.

Use of a radial correlation function gives an indication of the frequency of associations at all length scales present in the field i.e. from 0-1.41. The frequency for each pit-pit separation radius is calculated and normalised by that expected for a random distribution. Values greater than 1.0 indicate an excess correlation whilst those less than 1.0 a deficit of clustering. In fig. 5c, random behaviour is indicated over most of the range, while the scatter at radii > 1 is due to statistical noise. There is however clustering for radii of 60-120nm and an anti-correlation below 60nm. The former may represent the attraction between dislocations, the latter is the minimum resolvable distance between v-pits which are typically 20-30nm wide.

By reducing all vectors connecting pairs of v-pits to a common origin and normalising by the figure expected for a random field of points, the autocorrelation of fig. 5c was generated. There is no evidence of any preferred crystallographic alignment: the v-pits behave as though they are in an isotropic medium.

6. Conclusions

R-EBIC has been used to determine the qualitative features of the band bending at grain boundaries in the CdTe-Te-CdS system that is successful in that it tallies with the observed efficiency gains from processing. Nevertheless the method remains to be put on a quantitative basis - and its spatial resolution may be limited by the generation volume.

TEM-CL has demonstrated a direct correlation between stacking faults and luminescence at 3.26eV in GaN. Although powerful, signal detection issues make this a difficult method to apply since the luminescence comes from thin sections.

Statistical analysis of spatial pattern has shown that v-pits (threading dislocations) in InGaN MQW's/GaN/sapphire have an excess correlation (are associated) with separations of 60-120nm or less. This is an indication of mosaic structure. The methods described deliver information that is only subjectively determined by eye.

7. Acknowledgements

The authors thank EPSRC for support and Y.J. Sun and O. Brandt for growing the M-plane GaN samples.

8. References

[1] Durose K, Russell G J, Woods J 1985 Journal of Crystal Growth 72 85.
[2] Durose K, Yates A J W, Szczerbakov A, Domogala J, Golacki Z, Swiatek K. 2002 J Phys D: Applied Physics 35 1997
[3] Waltereit P, Brandt O, Trampert A, Grahn H T, Menniger J, Ramsteiner M, Reiche, M and Ploog, K H 2000 Nature 406 865
[4] Walker E M, Buckley D J, Boyall N M, Durose K, Grasza K and Szczerbakow A 2001 Inst Phys Conf Ser No169 215
[5] Deatcher C J, Liu C, Pereira S, Lada M, Cullis A G and Sun Y J 2003 Semicond Sci and Technol 18 212
[6] Russell G J, Robertson M J, Vincent B and Woods J 1980 J Mat. Sci. 15 939
[7] Ziegler E, Siegel W, Blumtritt H, Breitenstein O 1982 Phys. Stat. Sol. A72 593
[8] Holt D B, Raza B, Wojcik A 1996 Materials Science and Engineering B42 14
[9] Russell J D and Leach C 1995 J European Ceramic Society 15 617
[10] Kingsley N 1970 Septieme Congres International de Microscopie Electronique, Grenoble 285
[11] Pennycook S J, Craven A J, Brown, L M 1977 Inst. Phys. Conf. Ser 36 69
[12] Petroff P M and Lang D V 1977 Appl. Phys. Lett. 31(2) 60
[13] Trampert A, Liu T Y and Ploog K H 2002 Proc. 4th Symp. Non-Stoichiometric III-V Compounds, 111
[14] Palosz W, Grasza K, Durose K, Halliday D P, Boyall N M, Dudley M J, Raghotamachar B, Lai C, 2003 Cryst Growth 254(3-4) 316
[15] Liu C, Deatcher C J, Cheong M G and Watson I M 2003 Inst Phys Conf Ser. In press
[16] Waugh D Geography: an integrated approach. 1990 Hong Kong: Thomas Nelson and Sons Ltd

Inst. Phys. Conf. Ser. No 179: Section 2
Paper presented at Electron Microscopy and Analysis Group Conf. EMAG2003, Oxford, 2003
©2003 IOP Publishing Ltd

A room temperature cathodoluminescence study of dislocations in silicon

D J Stowe[a], S A Galloway[b], S Senkader[a], Kanad Mallik[a], R J Falster[c] and P R Wilshaw[a]

[a]*Department of Materials, University of Oxford, Parks Road, Oxford, OX1 3PH, UK*
[b]*Gatan UK, Ferrymills 3, Osney Mead, Oxford, OX2 0ES, UK*
[c]*MEMC Electronic Materials SpA, viale Gherzi 31, 28100 Novara, Novara, Italy*

ABSTRACT: Cathodoluminescence has been used to investigate room-temperature light emission from dislocations generated by ion-implantation and abrasion using silicon carbide paper in standard Czochralski grown silicon wafers. Dislocations in the ion implanted material were generated by implantation of either boron or silicon ions to produce, after suitable thermal annealing, a thin (100 – 200 nm) band of dislocation loops typically 150 nm below the surface. Abrasion created large dislocation tangles up to 15 μm deep. The generation of dislocations by both ion implantation and abrasion led to the observation of room temperature cathodoluminescence peaked at a wavelength of 1154 nm. The luminescence efficiency was found to be lower for the relatively lowly doped samples than in the case of the highly doped samples when similar dislocation densities were present. We attribute the luminescence behaviour to electron-hole recombination at the dislocations themselves and propose a model for this near-band gap luminescence based on one-dimensional energy bands associated with the strain field of dislocations.

1. INTRODUCTION

Bulk silicon (Si) is normally considered a very inefficient light emitter due its indirect band gap. Approaches to produce efficient room temperature light emission from silicon have included the use of quantum confinement, the incorporation of radiative centres such as erbium and the minimisation of the competing non-radiative recombination. However, the development of efficient room-temperature devices has been severely limited, in particular by competing non-radiative recombination processes which normally dominate at room temperature. As a result, there is still opportunity to develop new systems with a view to achieving this goal.

It has recently been reported that samples containing a thin band of dislocations, generated by boron implantation, produced efficient emission when measured by electro- or photoluminescence. In each case a *p-n* junction was present and this was thought to play an important role in the luminescence mechanism [1,2]. However, recent work using cathodoluminescence [3] has shown that the production of dislocations generated via ion-implantation without a *p-n* junction also has the potential to produce relatively efficient room temperature luminescence and thus the *p-n* junction is not critical to the luminescence mechanism. The luminescence observed in all cases case was peaked at 1154 nm with an external quantum efficiency of $2x10^{-4}$ [1] and $5x10^{-6}$ [3]. The present research aims to investigate further this luminescence with a view to understand better the luminescence mechanism and increase its efficiency.

2. EXPERIMENTAL METHOD AND RESULTS

Standard (100) Czochralski-grown (Cz) Si substrates which contain dislocations generated by ion-implantation and also abraded substrates have been investigated. Ion-implantation induced dislocations were created by implantation of either B^+ or Si^+ at a dose of $1x10^{15}$ cm^{-2} into n- and p-type substrates (resistivity ~5 Ωcm), with subsequent annealing at 900°C for 15 mins for B^+ and 10 mins for Si^+ implants. The n-type substrates implanted with B^+ contained a p-n junction. Abrasion using No. 600 SiC paper was performed on both 0.01 Ωcm and ~5 Ωcm p-type Cz wafers at room temperature followed by cleaning using organic solvents. Previous work showed the abrasion to produce disordered dislocation tangles (mainly 1/2<110> burgers vector) up to 15 µm below the wafer surface and associated cracks and fragmentation [4].

Transmission electron microscopy (TEM) was used to characterise the dislocations generated by ion implantation. Representative (110) cross-sectional images are shown in Fig. 1. Figure 1a shows a band of dislocations generated by B^+ implantation and Fig. 1b shows a band of dislocations generated by Si^+ implantation. The implantation and anneal conditions were selected to give very similar dislocation morphologies and densities for each ion species. In this case the B^+ implant and anneal produced a dislocation density of $5x10^{10}$ cm^{-2} and the silicon implant a dislocation density of $6.4x10^{10}$ cm^{-2}, in the form of isolated loops. The dislocations are present in either highly or lowly doped material depending on the implant species. Thus the main difference between the two sets of samples is that the dislocations created by Si^+ implantation are situated in uniformly doped (~5 Ωcm) n- or p-type material, whereas, in the B^+ implanted specimens the doping is much higher and varies across the dislocation band as shown schematically in Fig. 2 for implantation into an n-substrate. Other work [5] on dislocations produced by ion-implantation under similar conditions had shown the burgers vectors of the dislocations to be mainly 1/2<110>, whilst some were 1/3 <111>.

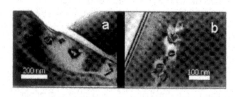
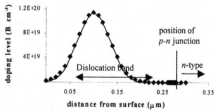

Figure 1. Representative (110) bright field cross-sectional TEM images of a) B^+ implanted sample and b) Si^+ implanted sample.

Figure 2. Doping variation of the B^+ implanted n-type substrate. Position of the dislocation band and the p-n junction are indicated.

Cathodoluminescence was carried out using a Gatan MonoCL3 system and a liquid-nitrogen-cooled Northcoast E0-817L germanium detector attached to a JEOL 6500F scanning electron microscope. The measurements presented here were made with a beam energy of 30 keV and a beam current of 200 nA.

Room temperature CL spectra from a variety of specimens are shown in Figure 3. Unprocessed float zone (FZ) material (p-type, ~3 Ωcm) was found to give no observable emission both before and after the removal of the native oxide layer. Some Cz -wafers (n- and p-type, ~5 Ωcm) also gave no emission whilst others produced very weak emission at sub-band gap wavelengths, the cause of this behaviour in Cz-Si is not understood.

All samples ion-implanted with B$^+$ or Si$^+$ to give a band of dislocations exhibited luminescence with a peak wavelength of 1154 nm (1.07 eV) and a FWHM of 80 nm, Fig. 3a, as reported previously [3]. The peak was the same shape and at the same position as that reported in [1]. No peaks were observed at any other wavelength and no difference was observed between n- and p-type substrates for the excitation conditions used. Defining external quantum efficiency as the fraction of electron-hole pairs generated by the incident electron beam which recombine to give a photon emitted from the sample surface, a value can be estimated using the sensitivity of the detector specified by the manufacturer and the efficiency of the CL system as specified by Gatan. For the B$^+$ implanted sample this is estimated to be about 5×10^{-6} using the excitation conditions stated previously. The efficiency of the silicon implanted specimens (dislocations in relatively low doped material) was a factor of 15 lower than for the more highly doped boron implanted samples.

The CL spectra from specimens containing dislocations generated by abrasion also exhibited a luminescence peak at 1154 nm at room temperature. Luminescence was clearly observed for the 0.01 Ωcm abraded material (intensity ~2.5 times less than the Si$^+$ implanted specimen) whereas for the similarly abraded 5 Ωcm specimen the luminescence (not shown) was very weak and the shape of the emission peak could not be clearly characterised. The shape of the luminescence peak from the highly doped abraded specimen exhibited a significantly different shape to the specimens produced by ion implantation and in this case was broader with FWHM 125 nm, and is extended on the lower energy side of the spectrum, Fig. 3.

3. DISCUSSION

The experimental results show that the presence of dislocations in Si leads to relatively efficient luminescence at room temperature with a peak centred at 1154 nm (1.07 eV) which is 50 meV less than the band gap energy. This is lower in energy than the phonon assisted band to band radiative recombination which has been reported to be peaked at ~1130 nm [6]. We attribute the 1154 nm luminescence to electron-hole recombination at the dislocations themselves.

The radiative recombination is proposed to take place via one-dimensional (1D) energy bands that have been created in potential wells formed by the strain field associated with a dislocation [3]. These 1D-bands have been theoretically predicted to be split off from the conduction and valence band by 37-50 meV [7, 8]. When the thermal energy of carriers in these bands is taken into account, the photons of energy 1.07 eV detected in this work are consistent with radiative recombination between the 1D energy bands.

The complicated dislocation tangles present in the highly doped abraded specimen clearly result in a different shaped spectrum than those observed for the case of isolated dislocation loops created by ion implantation. It is believed that the complicated dislocation tangle results in the significant modification of the local strain field, particularly at the intersection of dislocations. This highly localised change to the strain field may distort the potential well in which the 1D energy bands are produced resulting in a distribution of energy levels from which radiative recombination may occur, potentially leading to a broadening of the luminescence spectrum.

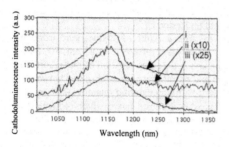

Figure 3. CL spectra at room temperature; accelerating voltage 30 kV, beam current 200 nA. i) B$^+$ implanted specimen, no p-n junction, ii) Si$^+$ implanted specimen p-type wafer and iii) SiC abraded p-type 0.01 Ωcm specimen. Curves are displaced along y-axis for clarity.

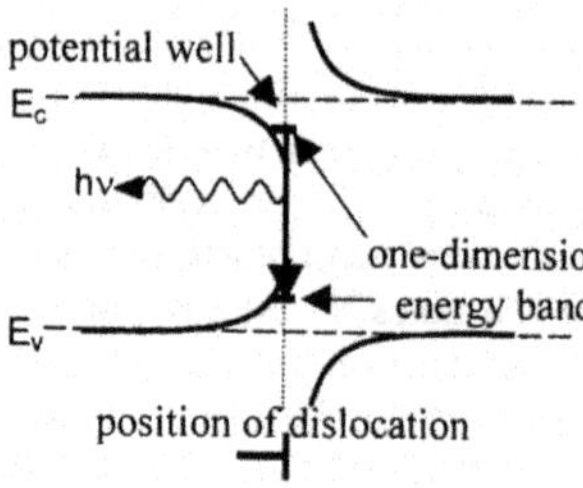

Figure 4. Schematic energy band diagram showing the effect of the strain field on the conduction and the valence bands around a dislocation. The 1D-energy bands between which radiative recombination may occur are indicated.

Thus, we propose that in the case of dislocations where the opportunity for non-radiative recombination via deep states is small, the confinement of electrons and holes in the strain field of a dislocation can lead to relatively efficient, room temperature, radiative electron-hole recombination via the 1D-energy bands.

Acknowledgements: The authors wish to thank Peter Pichler and Oliver Krause at the Fraunhofer Institute, Erlangen, Germany for performing the ion implantations.

References
[1]   W.L. Ng, M.A. Lourenço, R.M. Gwilliam, S. Ledain, G. Shao, K.P. Homewood, Nature 410 (2001) 192.
[2]   M.A. Lourenço, M.S.A. Siddiqui, R.M. Gwilliam, G. Shao, K.P. Homewood, Physica E 16 (2003) 376.
[3]   D.J. Stowe, S.A. Galloway, S. Senkader, Kanad Mallik, R.J. Falster, P.R. Wilshaw *accepted for publication in Proceedings of ICDS-22 in Physica B.*
[4]   R. Stickler and G.R. Booker, Phil. Mag. 91 (1963) 859.
[5]   K.S. Jones and G.A. Rozgonyi. Rapid Thermal Processing Science and Technology (R.B. Fair, ed) (1993) 123-168.
[6]   T.-C. Ong, K.W. Terril, S. Tam, C. Hu. IEEE Electron Dev. Lett. 4 (1983) 460-462.
[7]   J.-L. Farvacque, P. François, Phys. Stat. Sol. B 223 (2001) 635.
[8]   Y.T. Rebane, J.W. Steeds, Phys Rev B 48(20) (1993) 14963.

Inst. Phys. Conf. Ser. No 179: Section 2
Paper presented at Electron Microscopy and Analysis Group Conf. EMAG2003, Oxford, 2003
©2003 IOP Publishing Ltd

Studies of variations in insertion cations in Intergrowth Tungsten Bronzes (ITBs)

J Sloan[*1,2], K L Langley[1], E V Day[1] R R Meyer[1], A I Kirkland[1]

[1]University of Oxford, Inorganic Chemistry , South Parks Road, Oxford, OX1 3QR
[2]University of Oxford, Department of Materials, Parks Road, Oxford OX1 3PH

ABSTRACT: The characterisation of a series of intergrowth tungsten bronzes (ITB) of the form M_xWO_3 (M = K, Rb and Cs) by the technique of focal series restoration is discussed. Preliminary results with regard to the determination of alkali metal concentrations within the hexagonal tunnels formed within the intergrowth phase are presented. The obtained modulus images provide a basis for direct calculation of M_x tunnel occupancy in the ITB structures from both restored modulus images and multislice simulations.

1.   INTRODUCTION

The first report of a tungsten bronze dates back to 1838 when Wöhler reduced a mixture of Na_2WO_3 and WO_3 in hydrogen (Wöhler, 1923). Brightly coloured crystals were formed with electrical conducting properties; the hue and electrical properties resulted in the name of tungsten bronzes. This class of compounds has the general formula M_xWO_3 where the insertion ion, M, is an alkali metal although examples have also been found where M is an alkaline earth, ammonium or rare earth metal ion (Bartha et al., 1995). Bronzes derived from the alkali metals are generally metallic conductors. The physical size of M and magnitude of x also determine which of the three principal structure types is formed (i.e. perovskite, hexagonal or tetragonal tungsten bronzes) or if an intergrowth between distorted ReO_3-type tungsten oxide and hexagonal bronze is obtained (i.e. forming a so-called Intergrowth Tungsten Bronzes (ITBs) (Steadman, 1972).

The present work is concerned with comparing the imaging properties of M_xWO_3 where M = K, Rb, Cs and $x = 0.1$ to 0.33. Here, the effect of varying both the concentration of M and also the scattering power of the insertion ion in the ITB structure is examined using improved resolution afforded by restoring the specimen exit-plane wave. This enhanced resolution also makes it possible to detect the variations in the local structure surrounding the hexagonal tunnel sites occupied by the insertion ions. Additionally, the comparative sensitivity of the restored phase compared to conventional imaging allows the major part of the oxygen sub-lattice to be resolved. These conclusions are supported by multislice simulations of the specimen exit-plane wave.

2.   EXPERIMENTAL

Alkali tungsten bronzes of the form M_xWO_3 (M = K, Rb and Cs) were prepared by heating stoichiometric ground and pelletised mixtures of M_2WO_4 (M = K, Rb and Cs), WO_3 and WO_2 (supplied by Alfa Aesar or STREM, at least 99.99% purity) to 900°C in a Carbolite tube furnace. Suspensions of each preparation were then reground and then pipetted onto 3.05 mm lacey carbon coated copper grids (Agar, 300 mesh). Focal series were obtained from various crystallographic zones using a JEM 3000F FEGTEM (C_s = 0.6 mm at 300 kV) from crystals oriented using selected area diffraction patterns. The images were recorded digitally using a 1024 × 1024 pixel CCD camera mounted axially and primary microscope magnifications of 400kX and 600kX.

The microscope was manually aligned to the coma-free axis and the two-fold astigmatism corrected using on-line diffractograms of the amorphous carbon support film. Focal series of images were recorded with the microscope under automatic control using scripts running under the Gatan Digital Micrograph software. For each orientation a series of 30 images was recorded of a thin crystal edge beginning underfocus with a nominal focal increment of 10 nm between images. A final image at the starting defocus was recorded to assess the focal drift. The image processing techniques subsequently employed to recover the modulus and exit-plane wave function are described elsewhere (Kirkland *et al.*, 1995, 1997, 1999; Meyer *et al.*, 2002).

3. RESULTS AND DISCUSSION

Without the insertion ion (cation dopant), WO_3 forms a covalent array of corner-sharing octahedra based on the ReO_3 structure but with slight distortions from cubic symmetry. With smaller diameter insertion alkali metal cations (i.e. Li^+ and Na^+), M_xWO_3 bronzes form structures approaching the perovskite archetype. With larger cations, such as K^+, Rb^+ and Cs^+, the insertion cations force the host WO_6 network to distort into arrays containing tunnels of triangular and hexagonal cross section resulting in the formation of Hexagonal Tungsten Bronzes (i.e. HTBs; Magneli, 1953). Additionally, at low insertion cation concentration, well ordered intergrowth structures between the parent distorted ReO_3-type WO_3 network and the HTB structures are frequently observed as depicted schematically in Figure 1. A simple scheme for the structural classification of ITBs (Kihlborg, 1979) describes them in terms of the width of the WO_3 slabs and the corresponding width of the of the HTB layer, i.e. a (m,n)-ITB whereby m is the number of chains of octahedra in the HTB block and n indicates the width of the WO_3 layer in octahedra. Thus, the M_xWO_3 ITB depicted in Fig. 1 is a (1,5)-ITB.

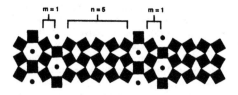

Fig. 1. Generalised structure and naming convention for M_xWO_3 intergrowth tungsten bronzes (ITBs). The generic description is (m,n)-ITB (Kihlborg, 1979). The WO_6 octahedra are represented by shaded squares, the M cations by shaded spheres.

insertion ion concentration (i.e. M_x in M_xWO_3), which can vary considerably within a given HTB or (m,n)-ITB phase and along the 1D tunnels. Given that the tunnel contents can consist of pure elements of variable atomic number (i.e. K (Z = 19); Rb (Z = 37) and Cd (Z = 55)), then the observed contrast ought to be quantifiable. To this end, there have been various attempts to assess the variation of image contrast with tunnel occupancy, for example (Kihlborg and Sundberg, 1988). A key problem encountered here was the difficulty of distinguishing M cations and W cations. In the current investigation, we have attempted to investigate how the restored phase and modulus images computed from focal series of images may be used in conjuction with simulations to distinguish between the different types and relative occupancies of M^+ cations in three different M_xWO_3 preparations.

Upon comparison of restored phase images versus the conventional HRTEM images (i.e. Figs. 2(a), (b) and (c)), the improvement in the available detail is immediately apparent. In particular, the restored phase shows a marked improvement in visibility both in terms of the cation contrast (in this case K^+) and also the oxygen sublattice. In addition, a more complete picture of the cation contrast can be obtained from the restored exit plane modulus (Figs. 3(a) and (b)).

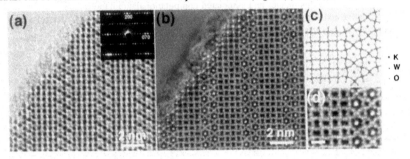

Fig. 2. (a) Conventional HRTEM image and inset electron diffraction pattern obtained from a [001] projection of a (1,5)-ITB corresponding to the composition $\sim K_{0.1}WO_3$. (b) Restored phase image (i.e. with reverse contrast) obtained from the same region as (a) produced from a 20 image focal series. (c) Structure model corresponding to the ReO_3:ITB interface. (d) Detail from (b). Note marked improvement in the visibility both the cation contrast and also the oxygen lattice, both of which are effectively invisible in (a).

In Fig. 3(a) the restored modulus image is shown obtained from a thin crystal of (1,5)-K_xWO_3 aligned parallel to the [001] zone. By comparison with the conventional HRTEM image obtained from the same phase (Fig. 2(a)), the cation contrast is much sharper and, additionally, the cation contrast from in the K_x columns is also clearly visible. It is noteworthy also that there is considerable variation in the latter contrast (see inset, Fig. 3(a)) and differences in the tunnel occupancy are now readily apparent. In Fig. 3(b) a similar result is shown for (1,5)-Cs_xWO_3. In this case the Cs_x contrast is clearly stronger, which should be anticipated from the higher atomic number of Cs, although it goes without saying that this is a thicker wedge crystal with a sharper thickness gradiant. This consideration notwithstanding, it is noteworthy that it is still possible to discern variations in the tunnel occupancy from the restored modulus (see inset, Fig. 3(b)).

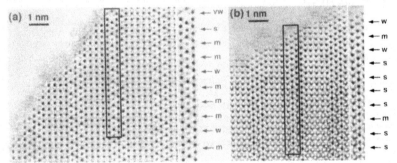

Fig. 3. (a) Restored modulus image produced from an [001] projection of (1,5)-K_xWO_3. The inset detail is magnified from the boxed region with the relative scattering strengths of the K_x columns indicated (vw = very weak, w = weak, m = medium, s = strong). (b) as for (a) but this time the phase is (1,5)Cs_xWO_3.

In Fig . 4, we show simulations of the exit plane modulus produced from a standard multi-slice calculation (Doyle and Turner, 1968) for wedge-shaped crystals of $(1,6)$-K_xWO_3. In the thin (i.e. single unit cell) part of the crystal with $x = 0.35$, the K_x columns are effectively invisible. As the crystal becomes thicker (i.e. four to nine unit cells thickness) the K_x columns can clearly be seen . For the wedge-shaped crystal with $x = 0.70$, the K_x columns are apparent even in the thinnest region of the wedge. These modulus images and associated computational simulations now form the basis of more quantitative comparisons between crystal thickness, ITB- and HTB-M_xWO_3 tunnel occupancy, work currently in progress.

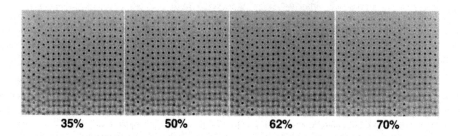

35% **50%** **62%** **70%**

Fig. 4 The simulated restored modulus for wedges corresponding to four different $(1,6)$-K_xWO_3 compositons with the x values indicated as percentage occupancies (unit cell thickness – 1× at top, 9× at bottom) .

4.   CONCLUSIONS

The technique of phase image reconstruction has been used to refine the microstructures of a variety of intergrowth tungsten bronzes of the form M_xWO_3. Restored phase images provide enhancement of both the anion and cation sublattices with respect to conventional HRTEM images whereas the modulus images provide very accurate cation maps making a refiniment of the structures possible on that basis. Additionally the modulus images provide a basis for direct calculation of M_x tunnel occupancy in the ITB structures from both restored modulus images and multislice simulations.

ACKNOWLEDGEMENTS

J.S. is indebted to the Royal Society for a University Research Fellowship.

REFERENCES

Bartha L, Kiss A B, Szalay T 1995 Int. J. of Refractory Metals & Hard Materials, **13**, 77.
Doyle P A and Turner P S 1968 Acta Cryst. **A24**, 370.
Kihlborg L and Sundberg M 1988 Acta Cryst. **A44**, 798.
Kirkland A I, Saxton W O, Chau K-L, Tsuno K and Kawasaki M 1995 Ultramicroscopy **57**, 355.
Kirkland A I, Saxton W O and Chand G 1997 J. Electron Microscopy **1**, 11.
Kirkland A I, Meyer R R, Saxton W O, Hutchison J and Dunin-Borkowski R 1999 Inst. Phys. Conf. Ser. No 161: Section 6.
Magneli A 1953 Acta Chem Scand **7**, 315.
Meyer R R, Kirkland A I and Saxton W O 2002 Ultramicroscopy **92**, 89.
Kihlborg L, 1979, Chem. Sript. **16**, 1.
Steadman R 1972 J. Chem. Soc. Dalton 1271.
Wöhler F 1923 Ann. Chim. Phys., **43**, 23.

Inst. Phys. Conf. Ser. No 179: Section 2
Paper presented at Electron Microscopy and Analysis Group Conf. EMAG2003, Oxford, 2003
©2003 IOP Publishing Ltd

Study of phase transition in filled-skutterudite $PrRu_4P_{12}$ at low temperature by electron diffraction

H Matsuhata†, C H Lee†, C Sekine‡, I Shirotani‡, T Asaka♯, K Kimoto♯, Y Matsui♯ and T Hirayama*

† National Institute of Advanced Industrial Science and Technology(AIST), 1-1-1 Umezono, Tsukuba, 305-8568, Japan
‡ Muroran Institute of Technology, 27-1 Mizumoto, Muroran, 050-8585, Japan
♯ National Institute of Materials Science(NIMS), 1-1 Namiki, Tsukuba, 305-0044, Japan
*Japan Fine Ceramics Center, 2-4-1 Mutsuno, Atsuta-ku, Nagoya, 456-8587, Japan

Abstract. A structural change has been found in $PrRu_4P_{12}$ accompanied by a metal-insulator transition at about 60 K by electron diffraction. At room temperature the structure belongs to space group $Im\bar{3}$, and is a metal. At about transition temperature new and weak reflections appear at h+k+l=2n+1 positions. Space group of the low temperature phase is discussed to be P23 or $Pm\bar{3}$. The origin of the transition was discussed.

1. Introduction

Filled skutterudite compounds RM_4X_{12} (R=La Pr and Ce ; M=Fe, Ru and Os, X=P, As and Sb) [1] have been attracting attention, not only for their capability as candidate materials for thermoelectric devices, but also for various physical properties which originate from the f-electrons of rare-earth elements [2].

$PrRu_4P_{12}$ is a metal at room temperature, and is known to show a metal-insulator transition at about T=60 K [2]. Several studies have been carried out to understand the mechanism of this transition. Powder x-ray measurements made using synchrotron orbital radiation seemed hard to detect a structural change associated with this phase transition [3]. A purely electrical phase transition is deduced to exist. However, PrL_2-edge XANES measurements suggested that Pr atom seemed almost trivalent independent of temperature [4]. Magnetic susceptibility showed no anomaly at the transition temperature. This indicates magnetic ordering is not involved in this transition [4]. Until recently the origin of this metal-insulator transition was uncertain.

Because weak and diffuse intensity is often detected easily by electron diffraction, the possibility of the detection of structure change by electron diffraction remains. In this paper we report on the result of electron diffraction observation at low temperature. We will discuss the structure at low temperature and also the origin of this transition briefly.

2. Experiment

Single crystals of $PrRu_4P_{12}$ were grown by the Sn flux method. The crystals were mechanically polished and ion-milled by standard techniques. The electron microscopy experiments were carried out using JEOL-4000FX in AIST, also Hitach HF-9000 in NIMS. Both microscopes were operated at 100 kV - 300 kV. Oxford double-tilt He-cooling stages

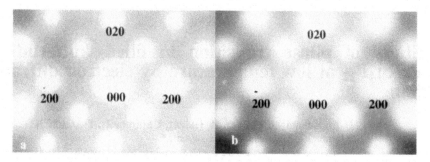

Figure 1. Diffraction patterns at the [001] zone axis. (a) taken at 70 K above the transition temperature , and (b) at about 12 K. The new superlattice reflections appear at the positions; h+k+l=2n+1 at low temperature.

were used. Conventional parallel beam diffraction patterns, convergent beam electron diffraction(CBED) patterns and dark-field images were taken at various temperatures.

PrRu$_4$P$_{12}$ has the body centred cubic structure Im$\bar{3}$(No.204) at room temperature. Pr atoms occupy the corner and the body center points. Ru atoms are at the middle between the two Pr atoms; ie (1/4, 1/4, 1/4) and equivalent positions. Ru atoms are surrounded by distorted octahedron of P atoms [1]. Figure 1a shows the diffraction pattern at the [001] zone-axis taken at 70 K above the transition temperature. A diffraction pattern of typical body center cubic structure is observed. Figure 1b is the diffraction pattern taken at about 12 K. New superlattice reflections are observed at the positions of h+k+l=2n+1 below the metal-insulator transition temperature [5]. These reflections indicate there is a structure change at low temperature. Figure 2a is that observed at the [110] at 12 K. The new reflections are observed also at this zone axis. CBED patterns taken at the [001] zone-axis at room temperature shows 2-fold rotation symmetry in the bright field CBED disc. This symmetry is also observed at about 12K as seen in Fig. 2b.

3. Discussion

The space group of the low temperature phase is deduced to be Pm$\bar{3}$(No.200) or P23(No.195) from extinction rules and the CBED patterns at the [001] zone axis. The superlattice diffraction spots are very weak, and therefore the change in the structure is likely subtle. In order to identify the space group at low temperature, +G/-G experiments [6],[7] were carried out several times along the 200-Kikuchi band, and also at the [110] zone axis orientation. An absence of mirror symmetry should be observed along the 200-Kikuchi bands for P23, whereas mirror symmetry should be observed for Pm$\bar{3}$. At the [110] zone axis an m$_R$ pattern should be observed for P23, whereas a 2$_R$mm$_R$ pattern is for Pm$\bar{3}$. CBED patterns show mirror symmetry along the 200-Kikuchi bands, and 2$_R$mm$_R$ pattern at the [110] zone axis at the low temperature. This indicates that space group at the low temperature is Pm$\bar{3}$(No.200). However, computer simulations of CBED patterns using the dynamical calculations with assumed structures for P23 and Pm$\bar{3}$ show almost identical results for both space groups, if deviations of atoms from the structure Im$\bar{3}$ are small. Therefore, so far the deviations of atoms are very small, the possibility of P23 remains.

The structures at low temperature will now be discussed. In Fig. 3a the positions of the Pr and Ru atoms in the unit cell are illustrated for Im$\bar{3}$ at room temperature. The Pr atoms occupy the corner and the body center points, and the Ru atoms are at the middle between

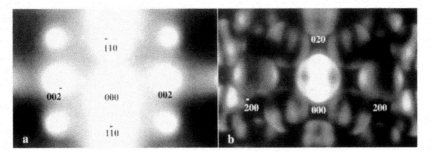

Figure 2. Diffraction patterns taken at about 12K. (a) [110] zone axis orientation. The new superlattice reflections appear at the position h+k+l=2n+1. (b) [001] zone axis CBED pattern. 2-fold rotation symmetry is observed at this orientation, similar to room temperature results.

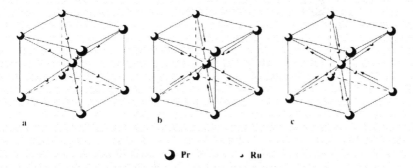

Figure 3. Positions of Pr and Ru atoms in the unit cell. Large spheres denote the Pr atoms located at the corner and the body center positions, while small spheres denote the Ru atoms located between two the Pr atoms. (a) for Im$\bar{3}$, (b) Pm$\bar{3}$, and (c) P23.

two Pr atoms. In Pm$\bar{3}$, the 8 Ru atoms move towards the body center Pr atom, otherwise they move towards the corner Pr atoms. In P23, 4 of the 8 Ru atoms move towards the body center point, and the other 4 Ru atoms move towards the corner points. These two kinds of Ru atoms should move different distances leading to a structure belongs to P23. These deviations of the Ru atoms are shown in Fig. 3b and 3c.

In Im$\bar{3}$ a Pr atom is surrounded by 12 P atoms as seen in Fig. 4a. 4 of the 12 P atoms are located on the plane x=0, other 4 P atoms groups are on the planes y=0, and z=0, respectively. These clusters are located at the corner and the body center. In Pm$\bar{3}$, the interatomic-distance betwen Pr and P, and/or a bond angle P-Pr-P around the corner Pr atom should differ slightly from those of the body center Pr atom. Similarly, Fig. 4b is a cluster in a P23 crystal. The 4 atomic positions of P are twisted along the x, y and z axis, respectively. If the structure is P23, the interatomic-distance between the Pr and P, and/or a bond angle P-Pr-P around the corner Pr atom should be different slightly from those of the body center Pr atom.

In Fig. 5, the domains around the two kinds of Pr atoms in a unit cell are illustrated for Pm$\bar{3}$ and P23. The Pr atoms are located at the centers of the shaded cubics or tetrahedrons, and the Ru atoms are located at the corners of the shaded cubics or tetrahedrons. According to the discussion above, regardless of whether Pm$\bar{3}$ or P23, the volume for the electrons around the Pr atom at the corner point should differ slightly from that of the body center

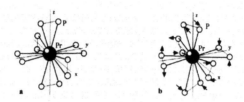

Figure 4. Clusters consist of a Pr and the nearest 12 P atoms. (a) Pm$\bar{3}$, and (b) P23.

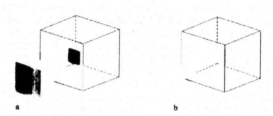

Figure 5. Occupation volumes of Pr atoms at the corner and the body center positions. (a) for Pm$\bar{3}$. Pr atoms are located in the centers of shaded cubics. Nearest Ru atoms are located at the corners of shaded cubics. (b) for P23. Pr atoms are in the shaded tetrahedrons, and those Ru atoms are at the corners.

Pr atom as shown in the Figure. The low temperature superlattice structure consists of the difference in those occupation volumes around the Pr atoms between the corner and the body center positions. Harima and Takegahara calculated the shape of Fermi surface of $PrRu_4P_{12}$ [8]. The shape seems appropriate for the three-dimensional nesting by the appearance of the superlattice reflections. Recently, a precise x-ray diffraction experiment by using synchrotron orbital radiation was carried out. Detailed analysis results will be reported [9].

4. Conclusion

We have detected weak superlattice reflections and subtle structural changes in $PrRu_4P_{12}$ below 60 K associated with the metal-insulator transition. The space group at low temperature was found to be P23(No.195) or Pm$\bar{3}$(No.200). It was shown that there should be a difference in electron-structures around the Pr atoms between the corner and the body-center positions at low temperature.

5. References

[1] Jeitschko W and Braun D 1977 Acta Cryst. **48** 3401-3406
[2] Sekine C Uchiumi T Shirotani I,and Yagi T 1997 Phys. Rev. Letter. **79** 3218-3222
[3] unpublished work, Sekine C and Shirotani I
[4] Lee C H Oyanagi H Sekine C Sirotani I and Ishii M 1999 Phys. Rev. B **60** 13253-13257
[5] Lee C H Matsuhata H Yamamoto A Ohta T Takazawa H Ueno K Sekine C Shirotani I and Hirayama T 2001 J. Phys. C **13** L45-L48
[6] Buxton B F Eades J A Steeds J W and Rackham G M 1976 Phil. Trans. Roy. Soc. London **281** 171-194
[7] Tanaka M Saito R and Sekii H 1983 Acta Cryst **A39** 357-368
[8] Harima H Takegahara K 2002 Physica B **312-313** 843-845
[9] Lee C H et al. will be published

Inst. Phys. Conf. Ser. No 179: Section 2
Paper presented at Electron Microscopy and Analysis Group Conf. EMAG2003, Oxford, 2003
©2003 IOP Publishing Ltd

Understanding gate oxide materials: ELNES of Hf and Zr compounds

D A Hamilton[1], A J Craven[1], M MacKenzie[1] and D W McComb[2]

1. Dept. of Physics & Astronomy, University of Glasgow, G12 8QQ
2. Dept. of Materials, Imperial College London, SW7 2AZ

ABSTRACT: The local coordination environment of the silicates and oxides of Hf and Zr have been investigated using both experimental ELNES data and theoretical modelling of the near-edge structures.

1.   Introduction

One of the biggest challenges currently facing the semiconductor industry is the need to replace SiO_2, or SiON, as the gate oxide in complementary-metal-oxide-silicon (CMOS) devices. As these devices are scaled, all components within the gate stack are reduced in size. In state of the art devices, the SiO_2 thickness is ~ 1.2nm and further reduction will lead to a rapid increase of the tunnelling current and hence unacceptable leakage current in the device.

Using a material with a higher permittivity, k, will allow the same performance but with physically thicker layer thus leading to reduced tunnelling and lower leakage currents. Not only must the layer have a high k but also, ideally, be possible to fabricate devices using standard CMOS processes. The oxides and silicates of Zr and Hf have been examined as candidate materials and the Hf compounds appear to be more suitable. Nevertheless the more studied and very similar Zr compounds can give valuable insights to the properties of the corresponding Hf compounds.

In order to understand fully the behaviour of such materials when incorporated into gate stacks, the bulk materials must first be well understood. To this end, the oxygen and silicon edges of the oxides and silicates of Hf and Zr have been modelled and, where available, compared to experimental results.

2.   Method

The edges were modelled using FEFF8, which is a real space multiple scattering algorithm that is widely used to calculate X-ray absorption near-edge structure (XANES). This algorithm is based on a Green's function formalism of Fermi's golden rule and is the equivalent of KKR band structure calculations provided the multiple scattering expansion converges. The potentials were calculated using a cluster of about 100 atoms and the XANES centred on each unique potential was calculated using a cluster large enough to achieve convergence. In some cases this was up to 400 atoms. Assuming that the dipole approximation is valid in both cases, the XANES data is equivalent to the electron energy near-edge structure (ELNES).

The experimental ELNES data were acquired under several different conditions. Typically spectra were collected on an FEI Tecnai F20, operated at 200 keV, equipped with a Gatan Enfina electron spectrometer. The spectra were recorded with dispersion 0.2eV/ch; probe half angle 9 mrad; collection half angle 10 mrad; integration time 5s. The edges were processed using Digital Micrograph. The background was removed by fitting a power-law model to the pre-edge and subtracting. Several edges were aligned and summed to improve the signal to noise ratio.

3.   Results and discussion

3.1   Oxygen K-edges

Figure 1 shows a comparison of the experimental O K-edges and the calculated XANES for the monoclinic oxides and the silicates of hafnium and zirconium. Here, as in all the figures, the zero of the energy scale is the threshold. The experimental O K-edge from α-quartz is shown for comparison. The O K ELNES of the oxides are both are dominated by two peaks (A and B) within 5eV of the onset. In m-HfO$_2$, the experimental and theoretical data match reasonably well in both peak position and height. While peaks A and B match well in m-ZrO$_2$, peak C appears only in the calculation. The m-ZrO$_2$ K-edge has previously been modelled [1] and provided a better match to experiment. The origin of the discrepancy in the FEFF result is unclear and is currently under investigation. The O K ELNES of m-ZrO$_2$ has been previously discussed by McComb [2]. Peaks A and B were assigned to the interaction of the ligand field split unoccupied Zr d-states with the unoccupied p-states on the oxygen site. Due to the similarity in the electronic structure and ionic radii of Hf and Zr, this argument can be extended to m-HfO$_2$ and the similar structure of its edge can be also be attributed to the mixing of p and d states.

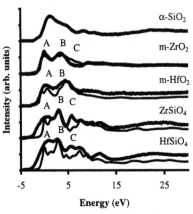

Figure 1. Oxygen K-edges of monoclinic oxides and silicates of Hf and Zr and α-SiO$_2$. Heavy lines indicate experimental data.

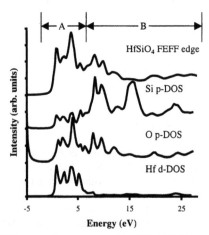

Figure 2. Comparison of modelled HfSiO$_4$ edge and partial DOS.

The O K ELNES of HfSiO$_4$ and ZrSiO$_4$ are also similar to each other. Both show two distinct initial peaks (A & B) followed by a series of smaller peaks (C). These peaks show some similarity to the O K-ELNES of α-SiO$_2$, which is a reflection of the fact that both materials are based on SiO$_4$ tetrahedra. The electronic structure of ZrSiO$_4$ has been reported elsewhere [3] and so HfSiO$_4$ will be considered here, although the similarity of Zr and Hf mean that the origin of the ELNES in each material will be closely related.

Figure 2 shows the modelled O K ELNES of HfSiO$_4$ along with the Hf d-DOS, O p-DOS and Si p-DOS. Comparing the edge with the O p-DOS gives an insight into the origin of the ELNES. The edge can be split into two regions; region A contains four peaks close together and region B contains a number of peaks spread over a larger energy range. The four peaks in region A are clearly associated with corresponding features in O-p DOS. The peaks in the O-p DOS fall at the same energy of four intense peaks in the Hf d-DOS and it can be concluded that, as in the oxides, part of the observed ELNES arises through interaction of the metal d-orbitals with the p-like states on the oxygen. Region B can be attributed to mixing of p-like states on the Si with the p-like states on the O site. The energy shift is likely to be due to the high density of d-states scooping out the p-states and displacing them to either side of the d-band, as described by Muller [4].

3.2   Silicon K-edges

The calculated Si K XANES data for HfSiO$_4$ and ZrSiO$_4$ are shown in Figure 3. As expected there are some similarities between the two plots; both have four peaks (region A) that precede a larger main peak at the start of region B. There are two other intense peaks in region B at 14eV and 22eV. The main differences are the relative intensities of the peaks although in the case of ZrSiO$_4$ there are shoulders on the first two peaks in region B that are less obvious or absent in the HfSiO$_4$ data. The calculated Si K XANES of ZrSiO$_4$ is in reasonable agreement with previously published data [5]. However, the pre-edge features in region A are not apparent in the experimental data. Work is

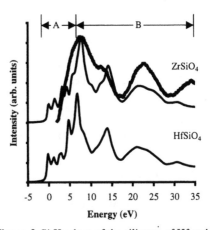

Figure 3. Si K-edges of the silicates of Hf and Zr. Heavy lines indicate experimental data [5].

underway to obtain new experimental data to determine if this discrepancy is due to poor background subtraction on the experimental data or an artefact in the calculation.

Comparing the calculated Si K data for the HfSiO$_4$ edge (Figure 3) with the Si p-DOS and Hf d-DOS (Figure 2), it is clear that region A is associated with the four strong peaks in the metal d-DOS projected onto the silicon p-DOS and region B is associated with the interaction of the Si and O p-like states.

3.3 Silicon $L_{2,3}$-edge

The experimental Si $L_{2,3}$-edges of both HfSiO$_4$ and ZrSiO$_4$ exhibit 5 peaks (A-E) within 12eV of the edge onset followed by a broader peak at about 22eV with some differences in relative intensities in the two spectra. The edge features correspond to peaks in the Si s- and d-DOS as would be expected from the dipole selection rule. The initial peaks (A-E) extend over a much greater energy range than the four-peaked structure observed on the metal d-DOS. The origin of the structure seems to be more complicated than with the K-edges and is probably not simply associated with the interaction with the metal d-states.

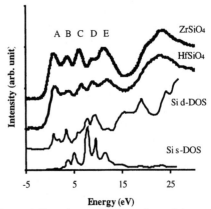

Figure 4. Experimental Si $L_{2,3}$-edges of the silicates of Hf and Zr and calculated DOS.

4. Conclusions

The O K-edges of the monoclinic oxides and the O K-, Si K- and Si $L_{2,3}$-edges of the silicates of Hf and Zr have been presented. In all the cases, the data for Hf and Zr are closely correlated with the major differences being the ratio of peak heights. The experimental and theoretical data, where both are available, show reasonable agreement. Examination of the DOS associated with each edge shows that the K-edges have features within the first 5eV of the edge onset that can be associated with the mixing of the metal d-states with p-states on the excited atom. The origin of the Si $L_{2,3}$ ELNES appears to be more complex.

Acknowledgements

The authors wish to thank Professor J.Hanchar for the provision of the crystalline HfSiO$_4$ sample. This work was supported by the University of Glasgow, EPSRC and SHEFC.

References

[1]   Ostanin S, Craven A J, McComb D W, Vlachos D, Alavi A, Finnis M W and Paxton A T 2000 Phys. Rev. B 62 14728-14735.
[2]   McComb D W 1996 Phys. Rev. B 54 7094-7102.
[3]   Wu Z Y and Seifert F 1996 Solid State Comm. 99 773-778
[4]   Muller D A, Singh D J and Silcox J 1998 Phys. Rev. B 57 8181-8202.
[5]   McComb D W, Brydson R, Hansen P L and Payne R S 1992 J. Phys. Condens. Matter 4 8363-8374.

Inst. Phys. Conf. Ser. No 179: Section 2
Paper presented at Electron Microscopy and Analysis Group Conf. EMAG2003, Oxford, 2003
©*2003 IOP Publishing Ltd*

Analytical Electron Microscope Investigation of Iron within Human Liver Biopsies

A Brown[1], R Brydson[1], C C Calvert[1], A Warley[2], A Bomford[3], A Li[2] and J Powell[2]

[1]Institute for Materials Research, University of Leeds, Leeds LS2 9JT
[2]The Rayne Institute, St Thomas' Hospital, London SE1 7EH
[3]Institute for Liver Studies, King's College Hospital, London SE1 9NN

Abstract. Human liver biopsies from patients with high iron levels (due to haemochromatosis) and normal iron levels have been examined using electron energy loss spectroscopy (EELS) in the transmission electron microscope (TEM). The oxidation state of the iron in individual and clusters of ferritin cores has been estimated using the energy loss near edge structure (ELNES) of the Fe L_3-edge. The preliminary results suggest that some 35 % of the total iron content within a ferritin core is ferrous.

1. Introduction

During the daily human iron cycle any excess iron is temporarily stored in the liver by ferritin molecules before being cycled back into the body. Haemochromatosis is the genetic failure of this iron-cycle that leads to the permanent storage of excess iron in the liver by the continued build up and clustering of ferritin molecules. Ultimately this can lead to cirrhosis of the liver. The ferritin molecule itself is 12 nm in diameter and consists of a crystalline hydrated iron oxide core (6 nm in diameter) and an organic shell that facilitates the movement of iron in and out of the core. The core crystallises predominantly in the ferrihydrite structure [$(FeOOH)_8(FeOPO_3H_2)$, Cowley *et al.* 2000], which should only contain Fe^{3+}. This is significant since Fe^{2+} can catalyse free radical reactions and so may contribute to oxidative tissue damage. Clearly the actual Fe^{2+}/Fe^{3+} content within both a normal and dysfunctional liver is of interest.

The current methods of determining Fe^{2+}/Fe^{3+} ratios can be divided into wet or solid methods. Wet chemical analysis requires separation of the iron bearing fraction and then titration to evaluate the Fe^{2+}/Fe^{3+} ratio, resulting in a loss of co-localisation of the iron within the tissue structure. The solid approaches include Mossbauer Spectroscopy or X-Ray Absorption Spectroscopy. Both of these methods might be described as bulk techniques that certainly do not have the spatial resolution to measure signal from individual ferritin cores (although Kwiatek *et al.* 2001, have used XANES to determine Fe^{2+}/Fe^{3+} ratios in μm-mm sized regions of kidney tissue). Whereas the oxidation state of iron can be studied by electron energy loss spectroscopy (EELS) within a TEM on a single or small group of ferritin cores.

By specifically looking at the shape and intensity of the Fe $L_{2,3}$-edge in the energy loss spectrum an estimate of the Fe^{2+}/Fe^{3+} ratio is possible (Garvie and Buseck 1998 and van Aken and Liebscher 2002). Calvert, (2001) has developed a method that can estimate this ratio by fitting mathematical line profiles under the L_3-edge. In this case, the Fe^{2+} is represented by a peak fitted under the L_3-line at 707.5 eV and the Fe^{3+} by a peak fitted under the L_3-line at 709.5 eV energy loss. Such a technique applied in a microscope allows the examination of ferritin cores within intact tissue sections rather than on isolated material.

Becker et al., (1997) have worked on the co-localisation and concentration of iron in liver from haemochromatosis patients using electron spectroscopic imaging (ESI) but have not done any Fe^{2+}/Fe^{3+} work. Quintana et al., (2000) have looked at this specific question on isolated ferritin from the brain, (showing the presence of some Fe^{2+}). The work undertaken here focuses on the Fe^{2+}/Fe^{3+} ratio from ferrtin cores within relatively intact tissue material (i.e, non-isolated) from both an haemochromatosis and non-haemochromatosis liver.

2. Experimental

Two biopsy specimens were examined, one from a patient with haemochromatosis (with high iron levels) and one control from a patient (with normal iron levels). Electron microscope specimens were conventionally prepared (fixed with glutaraldehyde, post-fixed with osmium tetroxide, dehydrated with ethanol, embedded in resin and sectioned in an ultra-microtome) and examined *unstained (or least altered)*. Or, subsequent to sectioning, were exposed in uranyl acetate and lead citrate and examined *stained*. Synthetic ferrihydrite powder was also examined (as a representative Fe^{3+} standard) and was prepared for TEM by dispersing in acetone and drop casting onto holey carbon film.

The specimens were examined with a Philips CM200 FEG-TEM operating at 197 keV and fitted with a Gatan Imaging filter (GIF 200) and Oxford Instruments UTW ISIS X-ray detector (EDS). Bulk energy loss spectra were taken in diffraction mode (image coupled) with a ~ 0.1 μm diameter selected area diffraction aperture (SAD) inserted, a collection semi-angle of 6 mrad, and convergence semi-angle of ~ 1 mrad. Individual ferritin EEL spectra were taken using a probe approximately 5 nm in diameter (size estimated from the CCD) formed with the microscope in nanoprobe mode but image coupled and the smallest SAD inserted to eliminate stray gun scattering. Here the collection semi-angle is 6 mrad and the convergence semi-angle ~ 6 mrad (defined by using the smallest C2 aperture available, 50 μm).

3. Results

3.1 Haemochromatosis

Clusters of ferritin molecules can be seen in the unstained sections where elemental analysis (EDS and EELS/ESI) confirms the iron and oxygen content of the cores and HREM shows their crystalline nature (Figure 1). Electron diffraction patterns from the clusters indicate that individual cores are in random orientations that can be indexed to the ferrihydrite structure. SAD mode EELS from the ferritin clusters (Figure 2a) reveals an Fe L_3-edge structure with increased intensity at 707.5 eV relative to both the conventionally stained section and the synthetic ferrihydrite standard. Quantification (Calvert 2001) suggests that the unstained (least altered) liver contains some 35 % Fe^{2+} whilst synthetic ferrihydrite and the conventionally stained liver contains only Fe^{3+}.

Figure 1a

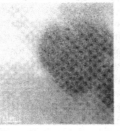

1b

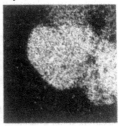

1c

Figure 1. *a*) Bright field image of the unstained haemochromatosis liver section, showing dark clusters within the tissue structure. *b*) Bright field image showing the dark clusters in (*a*) are composed of fine particles (~ 5 nm). *c*) Fe L_3 ESI elemental map showing the co-localisation of iron to the particles suggesting they are ferritin cores.

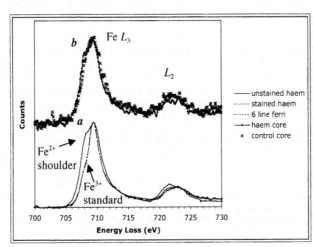

Figure 2. *a*) SAD mode Fe $L_{2,3}$-edges from ferritin clusters in the unstained and stained haemochromatosis liver. There is a clear shoulder at 707 eV on the unstained L_3-edge. *b*) Cold holder nanoprobe Fe $L_{3,2}$-edges from individual ferritin molecules in the haemochromatosis and control livers. Both edges are similar to the unstained edge in (*a*).

However, room temperature nanoprobe EELS from individual ferritin molecules shows significant reduction of the Fe (indicated by a rise in the shoulder of the Fe L_3-edge at 707.5 eV energy loss). Preliminary low temperature (nominally -180 °C) measurements of the Fe L_3-edge from 5 individual cores (figure 2b) prevents this reduction and produces edge shapes similar to that of the selected area method.

3.2 Control

Clusters of ferritin molecules are no longer evident in this section and individual ferritin cores need to be identified for EELS measurement. Preliminary low temperature measurement (under identical collection conditions to those applied to the haemochromatosis specimen) from 2 cores reveals (the crosses in figure 2b) little change in the Fe L_3-edge shape (although possibly less of a shoulder at 707.5 eV) compared to the haemochromatosis specimen.

4. Discussion

Location of iron in the ferritin molecules and their relative position within the biological structure of liver tissue has been demonstrated. It is evident that the cores have the expected ferrihydrite structure. Fe L_3-edge energy loss signals have been obtained from individual and clusters of cores. To a first approximation, the shape of the Fe L_3-edges in both the unstained haemochromatosis and control livers is similar and whilst there is a suggestion of a larger Fe^{2+} shoulder in the haemochromatosis case (figure 2b) further work is required to confirm this. Therefore the Fe L_3-edges in both the unstained haemochromatosis and control are consistent with some 35% Fe^{2+} in the ferritin whilst in the stained (oxidised) liver and the ferrihydrite standard there is only Fe^{3+}.

There are several possibilities for the presence of Fe^{2+}; sample preparation/handling, beam damage or Fe^{2+} is accommodated by the ferritin core. Sample preparation cannot be eliminated, however given that at the very least the osmium tetroxide stain, used in fixation, is an oxidising agent we would not expect reduced Fe. Whilst it is recognised that in room temperature nanoprobe mode the ferritin cores do reduce, (presumably due to ionisation in the beam), it is believed this can be minimised by cooling to liquid nitrogen temperatures. Furthermore when collecting in selected area mode neither the unstained, the stained or the synthetic ferrihydrite show any propensity to beam damage. Therefore it is tentatively suggested that Fe^{2+} is actually present in the ferritin cores. The exact location of the Fe^{2+} is unclear. Pure ferrihydrite cannot accommodate Fe^{2+} directly and is only known to do so by precipitating into a mixture of magnetite and ferrihydrite (Jambor and Dutrizac 1998). However the phosphorus content in ferritin may enable Fe^{2+} accommodation (Chasteen and Harrison 1999) or the Fe^{2+} could be located at the surface of the cores (40 % of Fe atoms in a core of 2100 Fe atoms are predicted to be at the surface [Frankel et al. 1991]).

5. Conclusion

EELS has been used to obtain oxidation state measurements of iron within individual and clusters of ferritin cores in human liver biopsies from both haemochromatosis and non-haemochromatosis livers. It is tentatively suggested that in both livers some 35% of the total iron in the ferritin cores is ferrous.

6. References and Acknowledgements

Becker ALD et al 1997 Micron, Vol 28, No 5, p 349
Calvert et al 2001 Inst. Phys. Conf. Ser. No 168, p 251
Chasteen ND and Harrison PM 1999 Journal of Structural Biology, 126, p 182
Cowley JM et al 2000 Journal of Structural Biology, 131, p 210
Frankel RB et al 1991 Hyperfine Interactions, 66, p 71
Garvie LAJ and Buseck PR 1998 Nature, Vol. 396, p 667
Jambor JL and Dutrizac JE 1998 Chem. Rev. 98, p 2549
Kwiatek WM et al 2001 Journal of Alloys and Compounds, 328, p 276
Quintana C et al 2000 Cellular and Molecular Biology, 46(4), p 807
van Aken PA and Liebscher B 2002 Phys. Chem. Minerals, 29, p 188

The authors are grateful to L. Benning (School of Earth Sciences, University of Leeds) for the preparation of the synthetic ferrihydrite.

Inst. Phys. Conf. Ser. No 179: Section 2
Paper presented at Electron Microscopy and Analysis Group Conf. EMAG2003, Oxford, 2003

Scanning Electron Microscopy of Biomaterials

K J McKinlay,[1] C A Scotchford,[1,2] D M Grant,[1] J M Oliver,[3] J R King[3] and P D Brown[1]

[1]School of Mechanical, Materials, Manufacturing Engineering and Management; [2]School of Biomedical Sciences; and [3]School of Mathematical Sciences, University of Nottingham, University Park, Nottingham, NG7 2RD, UK.

Abstract
A comparison of conventional high vacuum scanning electron microscopy (HVSEM), environmental SEM (ESEM) and confocal laser scanning microscopy (CLSM) in the assessment of cell-material interactions is made. The processing of cells cultured for conventional HVSEM leads to the loss of morphological features that are retained when using ESEM. The use of ESEM in conjunction with CLSM of the labeled cytoskeleton gives an indication of changes to the cell morphology as a consequence of incubation time and substrate surface features.

1. Introduction

The investigation of biological systems on the molecular scale using high vacuum SEM (HVSEM) requires several rigorous processing steps to dehydrate the cells and prevent the sample from charging when exposed to the imaging electron beam [1,2]. Such processing steps can lead to the destruction of finer features and the introduction of artefacts on the cell membrane. In addition, dehydration implements alcohols such as ethanol and acetone and as more polymeric materials are being studied for implant materials, such processing can lead to the degradation of the material. With the advent of environmental SEM (ESEM) [3,4], these dehydration steps can be avoided allowing the morphology of cells to be assessed at high resolution in a state closer to their natural morphology *in vitro*. A comparison of the HVSEM and ESEM techniques in the imaging of cell morphology is discussed here. The advantages of using ESEM to image cell morphology in conjunction with confocal laser scanning microscopy (CLSM) are emphasised. The application of ESEM and CLSM to the assessment of cell response to implant material surface features is also described.

2. Methodology

Primary human osteoblasts (HOBs), derived from femoral head trabecular bone, were seeded at an initial cell density of $2x10^4$ cells/cm^2 onto a variety of substrates sterilised by exposure to ultra-violet radiation. Substrates were incubated in full culture medium

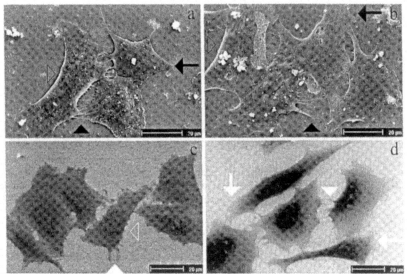

Figure 1: a,b) HVSEM of HOBs cultured on a titania surface and dehydrated; c,d) the equivalent samples imaged using ESEM. Samples were fixed after a,c) 90 minutes and b,d) 6 hours (micron markers = 20μm).

at 37°C and 5% CO_2 for times of 90 minutes and 6 hours before processing for microscopy.

Secondary electron imaging (SEI) using an FEI XL30 field emission gun (FEG)-ESEM was performed in high vacuum and wet (i.e. water vapour imaging gas) modes. A 30 μm incident beam aperture with a spot size of 6 nm and an accelerating voltage of 10 kV were typically used. Secondary electron (SE) and gaseous secondary electron (GSED) detectors were employed to image the cells.

After the selected culture period, samples for ESEM analysis were fixed with 3% glutaraldehyde in 0.1M sodium cacodylate (NaCaco) buffer. Prior to imaging, samples were washed with deionised water to remove residual buffer and prevent interference by salt crystal precipitation. Samples were placed on a copper stub, which was in turn placed in good thermal contact with a Peltier effect cooling stage. The sample chamber was pumped down to a pressure of 6 Torr (above the pressure required for imaging) before flooding with water vapour several times to 10 Torr to ensure saturation of the sample environment with water. The pressure was decreased slowly to the imaging pressure to prevent the sample temperature dropping below 4°C, which would otherwise result in the formation of ice crystals and the subsequent destruction of the cells.

After the selected culture period, samples for HVSEM analysis were fixed with 3% glutaraldehyde in 0.1M NaCaco buffer and post fixed with 1% osmium tetraoxide in Millonig's buffer. Dehydration was implemented using a typical route by submersion in ethanol solutions of increasing concentration before submersion in hexamethyl disilazane (HMDS). The majority of HMDS was removed and the remaining solvent was left to evaporate overnight in air. Samples were mounted on SEM stubs, gold coated and visualised using the FEG-ESEM operated in high vacuum mode.

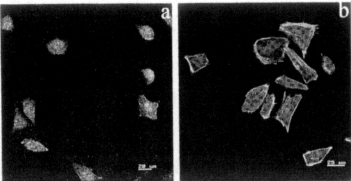

Figure 2: CLSM of HOBs cultured on a titania surface with actin network labeled after a) 90 minutes and b) 6 hours incubation (micron markers = 20 and 25μm, respectively).

After the selected culture period, samples for CLSM were fixed with 4% paraformaldehyde at 4°C and stained with FITC-conjugated phalloidin (10μg/ml,Sigma, Dorset, UK) to label the actin and a 0.5% aqueous solution of propidium iodide (Sigma, Dorset, UK) to label the nucleus of each cell. Samples were then mounted and the cell actin networks were imaged using a Leica TCS 4D confocal laser scanning microscope.

3 Results and Discussion

A comparison of the morphology of HOBs cultured on a smooth titania surface and imaged using HVSEM and ESEM is shown in Figure 1. Both HVSEM and ESEM show that increased incubation time, from 90 minutes to 6 hours, leads to an increase in cell spread as would be expected. However some differences in the detail are evident. In HVSEM, the cell spreading is indicated after 6 hours by broad thin lamellipodia (Figs. 1a,b; black arrows) which show cracking and perforation compared to those cells incubated for just 90 minutes. However, there is very little difference in the morphology of the cells when the two time points are compared with HVSEM. At both time points there is evidence of fine contact processes (Figs. 1a,b; black arrowheads) which have snapped due to the processing are and well defined cell edges (Figs. 1a,b; black-lined arrowheads).

If the images obtained with ESEM are considered, a greater difference between the cells at the two time points is observed. Once again the HOBs are clearly spread to a greater extent after 6 hours and fine contact processes are evident (Figs. 1c,d; white arrowheads). However, the cells are now less precise in their attachment after 90 minutes compared to 6 hours, whereas there was no obvious difference in the shape of the fine processes with incubation time in the dehydrated specimens. The extension of pseudopodia and elongation of the cells after 6 hours incubation is also more obviously seen in the ESEM images (Figs. 1c,d; white arrows). There is also a clearer difference in the edges of the cells (Figs. 1c,d; white-lined arrowheads). After 6 hours the edges of the cells are in general smoother and more clearly defined. The ruffled edges seen in the ESEM image of HOBs incubated for 90 mins are clearly lost during the dehydration processing for HVSEM. This is probably due to membrane shrinkage and the fact that the less well defined processes are not strongly attached. The increased definition of the cell edge seen in the ESEM images after 6 hours incubation could indicate an increased order of the substructure [5].

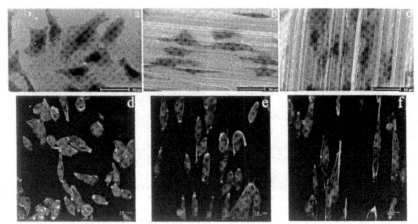

Figure 3: ESEM (a-c) and CLSM (d-f) of HOBs culture for 6 hours on stainless steel surfaces (a,d) polished; (b,e) roughened with P80 SiC and (c,f) roughened with P40 SiC paper.

Surface morphology only provides an indication of changes in the structural organisation of a cell, while labeling and visualisation of the distribution of the cytoskeletal components involved is still required to confirm the state of organisation.

Figure 2 shows equivalent CLSM images of the actin distribution of HOBs cultured for 90 minutes and 6 hours on titania surfaces. The increased definition of the fibrous network and the short fibres extending from the cell body seen in the 6 hour culture but not the 90 minute culture, consistent with the change in cell morphology observed using ESEM, but is not evident in the HVSEM images.

Imaging of cell morphology on roughened surfaces using optical microscopy techniques can prove difficult due to the short focal length, ESEM overcomes this limitation. Figures 3a-c are examples of ESEM images obtained from HOBs cultured on stainless steel surfaces with increasingly roughened surfaces. The overall alignment of the cells with the surface texture is evident, however, the small cell features observed on smooth surfaces are difficult to distinguish from the substrate texture and as such the filopodia and morphology of the cell outline cannot easily be determined. Therefore, CLSM of the cytoskeletal organisation is used to complement the ESEM observations.

The main advantage of using ESEM to image cell cultures grown on biomaterial surfaces lies in the fact that the processing of the samples prior to observation does not destroy features that are indicative of the cell morphology. The processing for HVSEM imaging is found to destroy fine features especially at the early time points.

References

1.   Bozzola J J and Russell L D 1998 *Electron Microscopy: Principles and techniques for biologists*, (Massachusetts: Jones and Bartlett)

2.   Hayat M 2000 *Principles and techniques of electron microscopy: biological applications* (Cambridge: Cambridge University)

3.   Danilatos G D 1993 Micr. Res. Tech., 25 354-361.

4.   Uwins P J R 1994 Mat. For. 18 51-75.

5.   Bray D 2001 *Cell movements: from molecules to motility* (New York: Garland)

Inst. Phys. Conf. Ser. No 179: Section 2
Paper presented at Electron Microscopy and Analysis Group Conf. EMAG2003, Oxford, 2003
©2003 IOP Publishing Ltd

Electron Microscopy of Si-based Photonic Materials

G Shao, M Milosavljević [+], S P Edwards, Y L Chen,
S Ledain, K P Homewood [+] and M J Goringe
School of Engineering (H6),
[+] School of Electronics and Physical Sciences,
University of Surrey, Guildford, Surrey GU2 7XH, UK.

Abstract. Si-based photonic materials have attracted a great deal of research interest due to their compatibility with the well-developed silicon technology. Extensive efforts have been directed towards the synthesis and characterisation of these materials using ion-beam synthesis at the University of Surrey. This paper covers some aspects of the microstructural characteristics of ion beam synthesised silicides such as the semiconducting iron and ruthenium silicides, using analytical transmission electron microscopy.

1. Introduction

Semiconducting metal silicides such as these based on transition metal elements are promising materials for optoelectronic applications due to the main technological advantage for *in situ* synthesis within/over silicon chips [1-6]. These silicides can be fabricated with various methods including ion beam synthesis (IBS) which involves ion implantation and usually subsequent annealing [3-7], and epitaxial growth methods such as molecular beam epitaxy (MBE), solid phase epitaxy (SPE) and reactive deposition epitaxy (RDE) [8-10], as well as physical vapour deposition. IBS has the key advantage of high compatibility with the well-developed silicon technology.

In this work, we discuss microstructural characteristics of IBS semi-conducting $FeSi_2$ and Ru_2Si_3, which are reported to have direct band gaps and hence have the potential for applications in areas such as optical sources and silicon-based optoelectronic components.

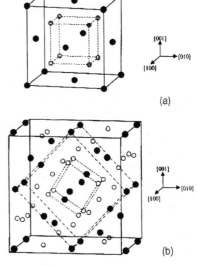

(a)

(b)

Fig. 1 Crystal structure of (a) $\gamma FeSi_2$, and (b) $\beta FeSi_2$. Notice that the atoms joined by dot-dashed and dashed lines highlight a distorted γ-like cell in β.

2. Iron Disilicide

The equilibrium iron disilicide is the $\beta FeSi_2$ phase, and the high temperature equilibrium form is the $\alpha FeSi_2$. A metastable cubic $\gamma FeSi_2$ has also been observed in the system. The

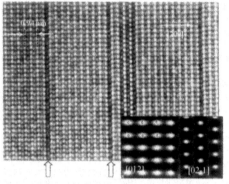

$\beta FeSi_2$ phase (Pearson's symbol oC48; space group Cmca; a=0.9865nm, b=0.7791nm and c=0.7833nm) has a complex crystal structure that can be related to a cubic $FeSi_2$ structure (γ) as is shown in Fig. 1, indicating that β can be considered as a distorted structure of γ (CaF_2 structure, Pearson's symbol cF12; space group Fm3m) [5]. Since the latter is considered as a chemically ordered compound based on the Si unit cell, the relationship in Fig. 1 naturally leads to the well known Type (I) orientation relationship (OR) between Si and $\beta FeSi_2$: $[010>_\beta//<110>_{Si}$ and $(200\}_\beta// \{004\}_{Si}$ (for OR analysis, the b- and c-axes are exchangeable for the β phase) [8,9]. Slight deviation from this OR leads to $[010>_\beta//<110>_{Si}$ and $(202)_\beta// \{111\}_{Si}$, and crystallographically equivalently $[111>_\beta//<110>_{Si}$ and $(110)_\beta //\{111\}_{Si}$, which are referred as *deviated* Type (I) [5]. However, these Type (I) based ORs are dominant only in β crystals on the top of Si wafers [5], and for β crystals embedded inside the Si substrate, various ORs have been observed. These ORs can be related to the Type (I) and Type (II) ORs by combinations of 90°- and 45°-rotations around the axes of the β unit cell [5]. It is worth emphasising that the Type (I) OR is dominant, as long as the β crystals are on the free surface of the Si wafer, with little effect due to the variation of synthesis methods. It has been found recently that embedded β nanocrystals could be thermodynamically less competitive than the metastable γ phase during annealing at high temperatures [11]. This is largely due to the large lattice mismatch in the [100]β direction

Fig. 2 HREM image of a β grain containing 90°-related ODs, taken at [012>β. Insets show the simulated images for the [012] and [021] domains.

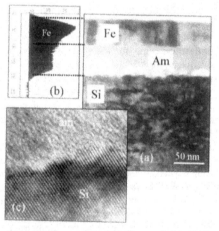

Fig.3 (a) Cross section TEM image (bright field) showing sandwiched structures of Fe, amorphous $FeSi_2$ (Am) and Si. (b) PEELS elemental Fe profile, (c) HREM image at the Am-Si interface.

for the Type (I) OR (~10%), which introduces elastic energy amounting *~19 kJ per molar atoms*, resulting in the β phase being remarkably less stable. This makes other OR variants energetically more competitive than Type (I) for embedded β crystals.

A striking feature of the IBS βFeSi$_2$ crystals, particularly of those buried in the silicon substrate, is the presence of lamellar domains in the silicide grains [2, 4, 5]. These domains were ultra-thin plates of the β phase which are 90° to each other around [100]β. A grow-in mechanism for the formation of the domain boundaries was proposed [2]. They were defined as order domains (ODs), as the [010]β and [001]β directions are geometrically nearly identical and chemically different [2]. Fig. 2 shows a high resolution image showing the atomic-level flat interface between these domains. It is noticed that some of the domains have thickness of only half of the unit cell length in the [100]β direction (indicated by arrows). It is worth mentioning that the β phase demonstrates a layered structure in the [100] direction, which is mechanically the weakest [5]. Growth in the [100]β direction is naturally the slowest, due to the expected low surface energy of the (100)β plane. While such lamellar domains are usually present in IBS βFeSi$_2$, they tend to lose the lamellar feature in the surface β layer fabricated by RDE [11].

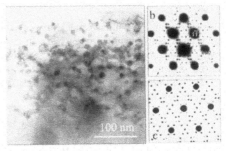

Fig. 4 (a) Bright field TEM image showing Ru$_2$Si$_3$ nanocrystals in Si. (b) SADP from the same region at [110]Si, and four corresponding OR variants of Ru$_2$Si$_3$. (c) Simulated pattern for (b).

It is interesting to notice that the amorphous form of the FeSi$_2$, in particular the relaxed amorphous form, also has a band gap similar to that of the β phase [12]. Such amorphous FeSi$_2$ can be fabricated either by ion beam mixing or by physical vapour deposition and sequential annealing at low temperatures. It is shocking to notice that the amorphous FeSi$_2$ made by ion beam mixing tends to have a constant stoichiometry, as shown in Fig. 3.

3. IBS Ru$_2$Si$_3$

The Ru$_2$Si$_3$ phase (oP40, space group Pbcn, a=1.1057 nm, b=0.8934 nm, c=0.5533 nm) can be in direct equilibrium with Si, making it easy for *in situ* fabrication within the Si wafer using IBS. Similar to the IBS βFeSi$_2$ phase, the IBS Ru$_2$Si$_3$ phase embedded in Si has ORs quite different from the surface layers. The OR between IBS Ru$_2$Si$_3$

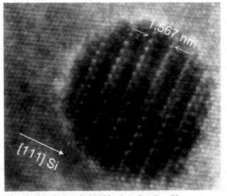

Fig. 5 HREM at [110]Si, showing Ru$_2$Si$_3$ nanocrystal. The indicated More fringe interval is due to (111)Si // (1$\bar{1}$0)Ru$_2$Si$_3$.

nanocrystals and the Si substrate is, [110]Si // [111]Ru_2Si_3, and (1$\bar{1}$1)Si // ($\bar{1}$10)Ru_2Si_3, which contains *24 (i.e. 6×4) crystallographically equivalent OR variants* [6]. Fig. 4 shows a bright field TEM image (Fig. 4a) that reveals IBS Ru_2Si_3 nanocrystals, together with a selected area electron diffraction pattern (SADP) from the same region (Fig. 4b). The pattern was taken at [110]Si which also contains four corresponding Ru_2Si_3 OR variants. The corresponding simulated pattern is shown in Fig. 4c. Fig. 5 shows a typical HREM image of Ru_2Si_3 nanocrystals in Si. The nanocrystalline Ru_2Si_3 phase maintains a considerably large interfacial strain due to coherent lattice constraint by the Si substrate (~11 % [6]), largely due to the high thermodynamic stability of the silicide phase (~60 kJ per molar atoms [13]).

4. Conclusions

IBS semiconducting silicides could have crystallographic relationship remarkably different from epitaxially grown surface layers. In the nano-scale, large lattice mismatch could be maintained coherently, leading to evidently lowered thermodynamic driving force for further growth.

Relaxed amorphous iron disilicide is also a semiconductor that exhibits similar band gap to the crystalline β phase.

References:

1. Leong D., Harry M., Reeson K.J. and Homewood K.P., Nature 1997, 387: 686.
2. Yang Z., Shao G., Homewood K.P., Reeson K.J., Finney M.S. and Harry M., Appl. Phys. Lett. 1995, 67: 667.
3. Yang Z. and Homewood K.P.,Finney M.S., Harry M. and Reeson K.J., J. Appl. Phys. 1996, 69: 4312.
4. Shao G., Yang Z. and Homewood KP, J. Mater. Sci. Lett 1998; 17: 1243.
5. Shao G. and Homewood KP, Intermetallics 2000; 8: 1405.
6. Shao G., Ledain S., Chen Y.L., Sharpe J., Gwilliam R.M., Homewood K.P. Reeson-Kirkby K.J., and Goringe M.J., Appl. Phys. Lett. 2000, 76: 2529.
7. Gerthsen, D., Radermacher, K., Dieker C. and Mantl, S., J. Appl. Phys. 1992, 71: 3788.
8. Geib K.M., Mahan J.E., Long R.G., Nathan M. and Bai G., Appl. Phys. Lett. 1991, 70: 1730.
9. Chierief N., D'Anterroches C., Cinti R.C., Nguyen-Tan T.A., and Derrien J., Appl. Phys. Lett. 1989, 55: 1671.
10. Arakawa T., Shao G., Makiuchi S., Ono T., Tatsuoka H., and Kuwabara H., Journal of Crystal Growth 2002; 237: 249.
11. Gao Y., Wong S.P., Cheung W.Y., Shao G. and Homewood K.P., Appl. Phys. Lett. 2003; 83(1): 42.
12. Milosavljević M., Shao G., Bibic H., McKinty C.N. and Homewood K.P., Appl. Phys. Lett 2001; 79: 1438.
13. Liu Y.Q., Shao G. and Homewood K.P., J. Alloys Compounds 2001; 320: 72.

Inst. Phys. Conf. Ser. No 179: Section 2
Paper presented at Electron Microscopy and Analysis Group Conf. EMAG2003, Oxford, 2003
©2003 IOP Publishing Ltd

TEM analysis and fabrication of magnetic nanoparticles

T J Bromwich, D G Bucknall, B Warot, A K Petford-Long, C A Ross*

Department of Materials, University of Oxford, Parks Road, Oxford, OX2 3BH

*Department of Materials Science and Engineering, Massachusetts Institute of Technology, 77 Massachusetts Avenue, Cambridge, MA 02139-4307

Abstract. Arrays of magnetic pillars 175 nm tall and 75 nm in diameter have been analyzed using Foucault mode Lorentz TEM. This paper describes a sample preparation route for TEM cross sections and shows that even in such small pillars sufficient magnetic contrast is visible to reveal the reversal of magnetisation in the pillars after application of a 2 kOe field. The reversal mechanism is shown to be incoherent by micromagnetic modelling.

1. Introduction

As fabrication techniques have improved, the possibility of exploring the potential of small ferromagnetic particles for data storage media and even random access memory has arisen. This development has occurred at a time where the present technology is approaching its fundamental limit. Current technology for media involves the sputtering of thin films of Co-Cr-Pt alloys, which consist of randomly orientated grains which couple together in groups as the domains for single written bits. The areal density has been increasing rapidly, but is limited by the thermal instability or superparamagnetism of the grains. As the grains get smaller so small thermal fluctuations provide the activation energy for the switching of the magnetisation in the grains, thus resulting in the loss of data at room temperature.

One possible system which could replace the current technology, involves the use of a single domain, ferromagnetic nanodot for each bit. These are then largely independent of each other. They consist either of pillars with easy axis of magnetisation perpendicular to the substrate, or elements with the easy axis in the plane of the substrate. Work has been conducted by Ross et al [1,2,3] using magnetic force microscopy (MFM) and superconducting quantum interference devices (SQUID) to characterise the magnetic properties of arrays of pillars fabricated using interference lithography or nanolithography with block copolymer templates. The magnetic properties have been shown to depend on size and shape, as one would expect. The work has shown that tall pillars, i.e. long axis perpendicular to the substrate show a stable magnetisation direction parallel to the pillar axis.

The work detailed in this paper aims to continue the work started by Ross et al and investigate these relationships using electron microscopy techniques, such as Lorentz microscopy. Little work has been done using Lorentz microscopy of small samples with perpendicular easy axis of magnetisation. Some has been conducted on slightly larger arrays of elements with in-plane magnetisation by Kirk et al [4,5,6]. Those workers show that an increase in coercivity is a result of decreasing dimensions and that the shape of the ends of the elements has an effect on the switching field. Dunin-Borkowski et al have used holography to analyse arrays of particles with sizes of the order of 200nm [7], and concentrated on the magnetisation reversal of the particles. Our future work will concentrate on using electron holography to study the interaction of the pillars and their interactions and will be compared to the micromagnetic modelling results presented in this paper.

2. Experimental Details

Arrays of nanopillars were fabricated using interference lithography followed by electrodepostion and a lift-off process. Cross section TEM samples were then fabricated from these arrays using a FEI focussed ion beam (FIB) system. A 150nm layer of C was sputtered on top of the pillars. This layer was intended to protect them from implantation by Ga+ ions and from the Pt deposition process used to prevent Ga^+ implantation. A schematic and an actual TEM micrograph taken on a JEOL 4000SE 400KV TEM are shown in Figure 1.

Lorentz TEM was conducted using the Foucault mode. The pillars were mounted in a holder with a rotation stage to allow them to be aligned with the field in the microscope. The TEM was carried out using a modified JEOL 4000SE 400 kV microscope fitted with a low-field objective pole piece.

3. Results and Discussion

Figure 2 shows a pair of Foucault mode Lorentz microscopy images taken after the application of a 570 Oe field along the pillar axis. This mode produces contrast by covering part of the split diffraction spot which has been deflected by the Lorentz force resulting from the magnetic field of the specimen. In the pair of images above, opposite parts of the diffraction spot have been removed and the images thus show reverse contrast for oppositely magnetised areas of the sample. Note the second pillar from the left marked with an arrow shows lighter contrast than neighbouring pillars in image (a) and darker contrast in image (b). This shows that the magnetisation of this particular pillar has reversed relative to its neighbours. This has been observed using MFM and is the result of the interaction fields with their neighbours.[8]

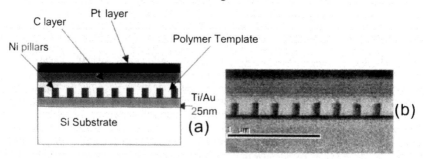

Figure 1 (a) Cross Section schematic and. (b) Image of pillars

(a)

(b)

Figure 2 (a) Foucault image of nano-pillars after application of 570 Oe field along the pillar long axis. (a) and (b) are taken with the objective aperture displaced in opposite directions.

It was not possible to achieve magnetisation reversal and saturation of all of the pillars in situ. Therefore the holder containing the specimen was magnetised ex situ using an electromagnet to +/- 2 kOe. Images were taken of opposite magnetization directions and are shown in Fig 3. This shows the same aperture position but after application of fields in opposite directions. The difference in contrast is clearly visible. Fig 3 (a) show light pillars on a dark background with a light stray field emerging from the top left of the pillars. Fig 3 (b) shows dark pillars on a light background and a reversal in the stray field contrast. This indicates that the magnetization of the pillars has reversed.

A micromagnetic simulation of the reversal of the pillars is shown in Fig 4. The magnetic moments point mainly in the z direction with a small component of magnetisation in the x-y plane. The reversal is inhomogeneous and occurs by the moments at the outer edges of the pillar rotating into the x-y plane and curling to form a vortex at remanence. The moments then flip to point along the opposite direction of the z-axis and the vortex uncurls to complete the reversal. The hysteresis loop shows that the majority of the magnetisation has reversed in an applied field of 400 Oe.

4. Conclusions and future work

The authors have demonstrated the ability to study the magnetisation reversal of magnetic pillars with all dimensions of the order of 100nm using Lorentz microscopy. The micromagnetic modelling has shown that the reversal is inhomogeneous and occurs via the formation of vortices in the magnetisation. Further study of Ni and Co samples is intended as is the fabrication of dots using block copolymer lithography.

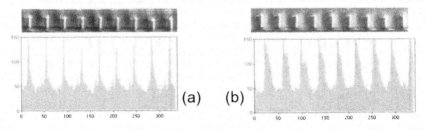

(a)   (b)

98

Figure 3. (a) Foucault mode micrograph of pillars after application of 2KOe Field along pillar axis. (b) As (a) but field applied in opposite direction.

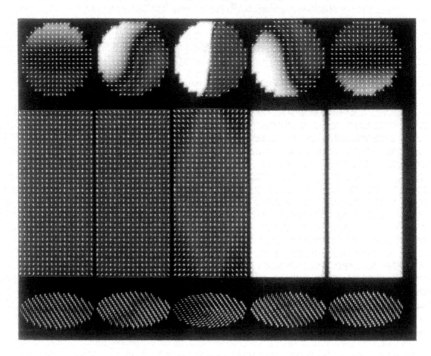

Figure 4 Micromagnetic simulation of the magnetisation reversal of a Ni pillar, 75 nm in diameter and 175 nm tall. Row 1 shows the magnetisation in an x-y slice, row 2 shows the magnetisation in the central x-z slice, and row 3 is a 3-D representation of the magnetisation in an x-y slice.

5. Acknowledgements

T. J. Bromwich would like to thank the EPSRC for their financial support.

References
[1]   M. Hwang et al. J. Appl. Phys., Vol 87, No 2, 2000
[2]   CA Ross et al. J. Vac. Sci. Technol. B, Vol 17, No 6, 1999
[3]   CA Ross et al. Phys. Rev. B, Vol 62, No 21, 2000
[4]   KJ Kirk et al. J. Appl. Phys., Vol85, No. 8, 1999
[5]   KJ Kirk et al. J. Appl. Phys., Vol 85, No 8, 1999
[6]   T Schrefl et al. J. Magn. Magn. Mater., Vol 175, 1997
[7]   RE Dunin-Borkowski et al. J. Appl. Phys., Vol 84, No 1, 1998.
[8]   CA Ross et al. J. Appl. Phys. Vol 89, No 2, 2001

Inst. Phys. Conf. Ser. No 179: Section 2
Paper presented at Electron Microscopy and Analysis Group Conf. EMAG2003, Oxford, 2003
©2003 IOP Publishing Ltd

TEM studies of dislocation-based silicon light emitting devices

M Milosavljević[1], M A Lourenco[1], M S A Siddiqui[1], G Shao[2], R M Gwilliam[1] and K P Homewood[1]

[1] School of Electronics and Physical Sciences, University of Surrey, Guildford, Surrey GU2 7XH, United Kingdom
[2] School of Engineering, University of Surrey, Guildford, Surrey, GU2 7XH, United Kingdom

Abstract. Results of TEM studies of dislocation-based silicon light emitting devices are presented. The basics of these devices are dislocation loops, which were formed by implantation of boron ions in n-type Si (100) wafers, and by subsequent rapid thermal annealing. Dislocation loops were found to be of the interstitial type, with Burgers vectors of a/3 <111>, sitting in the four non-parallel {111} lattice planes. Their appearance, distribution and density were correlated to ion implantation parameters.

1. Introduction

Development of silicon-based light emitters is crucial for further progress in ultra large scale integration (ULSI) technology, in providing a possibility of all silicon electronics and opto-electronics and integration of such components onto a single chip. Numerous methods and techniques were investigated to produce silicon-based light emitters, the disadvantage of silicon being in its inefficient light emission due to the indirect nature of the electronic band gap. Recently it was demonstrated that dislocation engineering, involving ion implantation and subsequent thermal treatments, is a promising route for fabrication of efficient light emitting devices in silicon. [1] This method makes use of controlled introduction of dislocation loops to create a strain field that modifies the band gap, thus preventing non-radiative and enhancing radiative transitions of electrical carriers in silicon. Indeed, light emitting diodes efficient at room temperature were fabricated by implantation of boron into silicon, where the role of boron was both as a dopant to form a p-n junction, as well as a means to introduce dislocation loops. [1,2] Here we present TEM analysis of such structures. The aim was to study the effects of boron ion irradiation parameters on the appearance, distribution and density of dislocation loops formed in silicon under the combined ion implantation and rapid thermal annealing processing. Correlation to electro-luminescence properties of the fabricated light emitting devices is beyond the scope of this paper, and will be published elsewhere.

2. Experimental details

The substrates used in these experiments were (100) n-type silicon wafers. They were implanted at room temperature with boron ions, at energies ranging from 20 – 70 keV.

The ion doses were adjusted to give a peak dopant concentration of 10^{20} cm^{-3} for each energy. For the ion energy of 20 keV the implanted dose was 7.6×10^{14} ions/cm^2, and for 70 keV the dose was 1.6×10^{15} ions/cm^2. After ion implantation all samples were annealed under the same conditions, in a nitrogen ambient for 20 min at 950°C, by rapid thermal annealing (RTA). Fabrication of light emitting diodes was then completed by deposition of front Al contacts and back AuSb contacts. [1,2]

TEM analysis was performed on both cross-sectional and plan-view Si specimens, prepared using low angle argon ion beam thinning at 5 keV. In order to observe the dislocation loops using diffraction contrast, the {220} reflections were avoided for double beam imaging conditions. We chose the [110] zone axis for imaging loops in the cross-sectional specimens, and the [004] zone axis for imaging loops in the plan-view specimens.

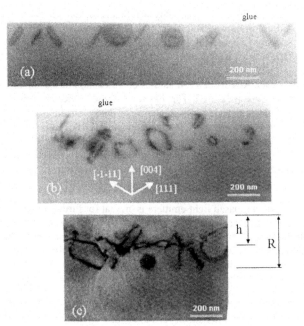

Figure 1. Cross-sectional TEM images taken along [110] Si direction, of samples implanted at 20 keV (a), 35 keV (b) and 50 keV (c). Distance h is from the silicon surface to a half-depth of dislocation loops and R to the far inner edge of the loops.

3. Results and discussion

TEM analysis revealed that the applied ion implantation and RTA processing results in formation of interstitial dislocation loops near the silicon surface. Cross-sectional TEM analysis of samples implanted at 20 keV, 35 keV and 50 keV is given in Fig. 1. Dislocation loops can be seen in all samples, exhibiting a variety of shapes, from nearly circular, or an elongated oval shape, to just a straight line, depending on orientation of their habit planes relative to the zone axis. These loops are interstitial in nature and lie in the four non-parallel {111} lattice planes, i.e. (111), ($\bar{1}$11), ($\bar{1}\bar{1}$1), and ($1\bar{1}$1), with the Burgers vectors of a/3 <111>. The loop diameters could be measured directly from the length of the longer axis or of the straight lines. As can be seen from the figure, their dimensions range from a few tens to a few hundreds of nm, increasing in size and being buried deeper in the substrate with increasing the ion energy.

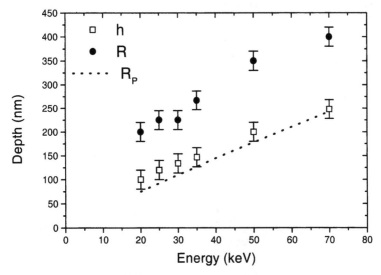

Figure 2. Dependence of h, R and R_P upon implantation energy.

In Fig. 2 we have plotted the measured values of h and R (illustrated in Fig. 1 c) as a function of the ion implantation energy, and with a dashed line we have given the values of the mean projected ion range R_P, calculated by the TRIM code.[3] It is seen that the mean projected ion range is within the error bars of h, but the inner edge of the loops R extends to a much higher depth. Similar behaviour was reported in the literature. For example when dislocation loops were formed by high dose (10^{16} cm^{-2}) implantation of Si ions in silicon and subsequent RTA [4], the loops were found to form below the original amorphous/crystalline interface, and were referred to as the end of range (EOR) dislocation loops. In our case the implanted doses of boron are below the amorphisation threshold. An estimated amount of disorder introduced in the silicon substrate is around 50%, spreading roughly to the mean ion range and then reducing rapidly, as deduced by SUSPRE. [5] However, a small portion of ions can reach and induce disorder at depths denoted by R, which is the probable reason for spreading of dislocation loops to such high depths in our case.

Plan-view images taken from samples implanted with 25 keV and 50 keV boron ions are shown in Fig. 3, as examples. Due to the inclined position of the {111} lattice planes relative to the [004] zone axis, the dislocation loops have an elongated appearance. The loops formed at the lower implantation energy are smaller and more uniform in size and shape, having mainly two mutually perpendicular orientations of their longer axis. Comparing to the image shown in Fig. 1 (a), it is suggested that the loops formed by low energy implants have quiet uniform circular shapes in their habit planes. Dislocation loops formed after boron implantation at higher energies are rather irregular in shape and not so uniform in size and distribution, as compared to those formed after low energy boron implantation. However, in Fig. 3 (b) we still see two preferred orientations of the longer axis, meaning that although irregular in shape the dislocation loops remain in {111} habit planes. From the plan-view images we have determined the density of the loops, from $\sim 1 \times 10^9$ cm^{-2} for the lowest and the highest boron energy, and going through a maximum of $\sim 1.5 \times 10^9$ cm^{-2} for the intermediate energies.

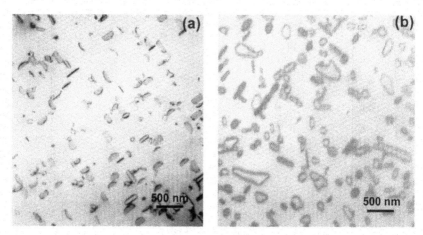

Figure 3. Plan-view images taken in [004] Si direction of samples implanted with 25 keV (a) and 50 keV (b) boron.

4. Summery

We have studied the influence of boron ion irradiation parameters on the formation of dislocation loops in (100) Si wafers by ion implantation and RTA processing. It was found that the depth distribution and size of the loops can be controlled by altering the ion energy. Ion energies of 20-25 keV induce a formation of rather uniform shallow loops, starting from the silicon surface. Higher energies, from 50-70 keV, induce buried loops, increasing in size distribution and being more irregular in shape. Despite the relatively low ion doses, below the amorphisation threshold, the dislocation loops extend much deeper into the substrate than the mean projected ion range.

References

[1]   Wai Lek Ng, Lourenco M A, Gwilliam R M, Ledain S, Shao G and Homewood K P 2001 Nature 410, 192-194

[2]   Lourenco M A, Siddiqui M S A, Shao G, Gwilliam R M and Homewood K P 2003 Towards the First Silicon Lasers, Pavesi L et al. eds. (Netherlands: Kluwer Academic Publishers) 11-20

[3]   Ziegler J F, Biersak J P and Littmark U 1985 The Stopping and Range of Ions in Solids (New York: Pergamon)

[4]   Pan G Z, Tu K N and Prussin A 1997 J. Appl. Phys. 81 78-84

[5]   Webb R P 1991 in Practical Surface Analysis eds. M.Seah and D.Briggs, (New York: John Willey) vol 3 app 3; SUSPRE is available online at: http://www.ee.surrey.ac.uk/SCRIBA/simulations/Suspre/

Inst. Phys. Conf. Ser. No 179: Section 2
Paper presented at Electron Microscopy and Analysis Group Conf. EMAG2003, Oxford, 2003
©2003 IOP Publishing Ltd

Composition modulation and growth-associated defects in GaInNAs/GaAs multi quantum wells

M Herrera, D González, H Y Liu[1], M Gutiérrez[1], M Hopkinson[1] and R García

Departamento de Ciencia de los Materiales e I. M. y Q. I., Universidad de Cádiz, Apartado 40, 11510 Puerto Real, Cádiz, Spain.
[1] Department of Electronic and Electrical Engineering. University of Sheffield. Mappin Street, Sheffield S1 3JD. United Kingdom.

ABSTRACT: Group III-nitride-arsenides are promising materials for communications in the optimum wavelength range for the optical fibre (1.3-1.5 µm). In this work, we analyse the structural properties of GaInNAs multi-quantum wells (MQWs) grown by Molecular Beam Epitaxy (MBE) with increasing In and N contents by means of Cross-Section (X-) and Planar View (PV-) Transmission Electron Microscopy (TEM). Phase separation due to In or N segregation, or both, has been observed in all the samples, causing the undulation of the wells in the N-rich structures. In addition, threading dislocations, stacking faults and dislocation loops are observed. The formation of these structural defects is discussed.

1. Introduction

Since Kondow et al. (Kondow 1996) proposed and created the novel system GaInNAs in 1995, many researchers have focused their work in this promising alloy. In comparison to the conventional system GaInAsP/InP, GaInNAs/GaAs provides better confinement for electrons and thus better characteristic temperature T_0, as well as higher efficiency and higher output power for lasers. The strong negative bowing parameter of this alloy causes a huge decrease of the bandgap by the addition of relatively small amounts of N (<5%) to GaInAs (Kondow 1996). Moreover, $Ga_{1-x}In_xN_yAs_{1-y}$ can be grown lattice-matched to GaAs for compositions with $x \approx 3y$, and therefore, the problems of plastic relaxation observed in the classical alloy GaInAs could be avoided.

However, incorporation of N in very small quantities to produce high quality devices remains a challenge. The difference in the chemical energy between GaN and InN bonds due to the small size and high electronegativity of N atoms (placed in As sites) produces a large miscibility gap in the GaInNAs quaternary alloys and the system tends to phase separate. Increasing N content degrades the luminescence efficiency, decreasing photoluminescence peak intensity and broadening its full width at half maximum (Xin 1998). In order to find the reasons of this behaviour, we report on the structural properties of GaInNAs MQWs with increasing In and N contents, by means of Transmission Electron Microscopy.

2. Experimental

GaInNAs multi-quantum well structures with different In and N contents were grown on (001) on-axis GaAs substrates in a VG V80H Molecular Beam Epitaxy (MBE) system equipped with an Oxford Applied Research HD25 radio-frequency plasma source for N. The N composition was controlled by monitoring the intensity of the 425 nm atomic N plasma emission. All samples consisted of a GaAs buffer layer (200 nm) followed by a $GaN_{0.007}As$ layer (52 nm), five periods of GaInNAs (8 nm)/GaNAs(52 nm) (grown at 460°C) and a cap layer of GaAs (100 nm) plus $Al_{0.25}Ga_{0.75}As$ (100 nm). In Table 1, the In and N contents for each structure are given, inferred by x-ray diffraction and SIMS measurements on bulk and quantum well test structures. The samples were not annealed after growth.

Table 1: N and In contents of the studied samples.

%In MQWs	%N MQWs	% N barrier
20±2	1.5±0.2	0.7±0.2
	1.9±0.2	0.7±0.2
35±2	1.1±0.2	0.7±0.2
	1.4±0.2	0.7±0.2
40±2	2.3±0.2	0.9±0.2

3. Results and discussion

XTEM study of samples with 20% of In showed that neither dislocations nor other structural defects exist and the well interfaces appear reasonably flat. However, the g400BF reflection indicates surface stress contrasts with a periodicity between 20 and 40 nm (see Fig. 1.a), which are more pronounced in the sample of higher N content, despite the lower mismatch in this structure.

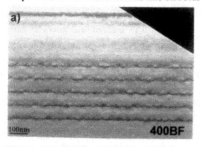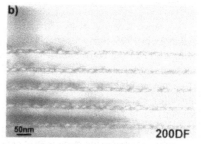

Fig. 1: XTEM images of $In_{0.2}Ga_{0.8}AsN_{0.019}$ (a) and $In_{0.35}Ga_{0.65}AsN_{0.014}$ (b) quantum wells. Flat interfaces with periodic stress contrasts evolve to SK behaviour with increasing In and N.

These contrasts could be due to a phase separation in the wells, because of In or N segregation, or both, and seem to be accentuated when the N incorporation is increased. Due to the large miscibility gap in mixed group V nitride-arsenides, the alloy GaAsN tends to phase separate easily into GaAs and GaN (Kondow 1995). Moreover, In-concentration fluctuations of ±5% on a length scale of 20 nm in GaInNAs QWs have been measured (Albrecht 2002). Therefore, a phase separation with a wavelength of 20 nm due to In and probably N segregation is likely the responsible of the strain contrasts observed in these structures. Because of the reduced size of N atoms, they could be located in In-rich areas in order to minimize the elastic stress in the structure.

With regard to the structures with 35% of In, the lower N concentration sample has shown similar phase separation-associated stress contrasts than the previous samples. Furthermore, very slight undulations of the wells are observed. The stress increase in the structure due to the higher mismatch results in elastic relaxation of the wells through surface corrugation. Nevertheless, when increasing the N content in the MQW (and thus reducing the lattice mismatch), a perfectly defined Stranski-Krastanov (SK) growth mode takes place. As it is shown in Fig. 1.b, undulations with a size about 20-30 nm are present in all the wells. The dimensions of these undulations are similar to the periodicity of the stress contrasts observed in the structures with lower In or N content. This suggests that the change from 2D to 3D growth could be due to the accumulated stress in the wells because of the phase separation. The addition of N to InGaAs has been proved to stimulate the QW interface corrugation (Xin 1999), although it reduces the lattice mismatch with the substrate. The formation of a spontaneous N-rich layer on the surface reduces the (001)-surface energy through anion-anion dimerization (Zhang 1997) which leads to an undulation of the well to stabilize the system.

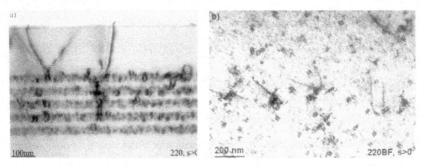

Fig. 2: TEM micrographs of $In_{0.35}Ga_{0.65}AsN_{0.014}$ quantum wells: a) 220BF XTEM, which shows TDs in the upper wells; b) 220BF PVTEM, which shows Frank loops and half-loop dislocations.

Moreover, the sample with $Ga_{0.65}In_{0.35}N_{0.014}As$ MQWs showed an irregular distribution of 60°-type threading dislocations that cross the cap layer (Fig 2.a). Most of them seem to come from the upper QW, although there are some dislocations that come from even the 3rd or 4th well. The location of these dislocations indicates that its formation is not due to overcome the critical layer thickness of plastic relaxation of the whole structure since such defects would have nucleated in the lower quantum well. Instead, these dislocations could have been formed due to local tensions in the structure produced by point defect accumulations and/or phase separation. Furthermore, it is worth mentioning that a large amount of the dislocations appear in pairs, which seem to nucleate at the same region of the quantum well. To clarify the origin of these dislocations, a PVTEM study has been carried out in this sample. We highlight that no evidence of misfit dislocations exists and that the dislocations adopt in many cases a half-loop configuration. Neither precipitates nor any other type of defect associated to these dislocations has been observed, as it is shown in Fig. 2.b. It seems that the structural imperfection that produces the apparition of these dislocations is too small for our resolution level in TEM. Local stresses due to a high accumulation of N interstitials in the InGaAsN alloy (Spruytte 2001, Li 2001) or an N-rich coherent cluster (Mintairov 2001) could be responsible for the formation of these dislocations. In any case, the high concentration of point defects in this structure is corroborated by the observation of a high density of Frank loops with an average size of 20 nm distributed for all the wells. They have been characterized by the inside-outside contrasts as extrinsic. Moreover, extrinsic stacking faults have been observed. This defect structure enlarges when increasing both the In and N contents (40% and 2.3%, respectively) in the MQW. Thus, a high density of threading dislocations that cross all the wells are observed, with a density higher than 10^9cm^{-2} in the cap layer.

The lack of misfit dislocation at the interfaces indicates that the structure does not overcome the critical layer thickness for plastic relaxation. The dislocation nucleation occurs at located strain regions, preferably in the upper wells. As homogeneous nucleation has a large energy activation barrier, dislocations probably nucleate in regions with a loss of coherence due to phase separation joined with a high density of point defects. The defective N incorporation that increases with N content not only affects the planarity of QW interfaces but also it is the source of several crystalline defects that degrade the performance quality of the heterostructure.

4. Conclusions

GaInNAs MQWs have been studied by TEM. Our results have shown that the addition of N to GaInAs stimulates phase separation in the wells. The increase of In and N contents enhances the segregation and produces an undulation of the wells in order to relax the elastic strain in the system. Moreover, structural defects such as threading dislocations, stacking faults and dislocation loops have been observed. The defect generation is due to local stresses by phase separation or a high density of point defects and not to overcome the critical layer thickness of plastic relaxation.

ACKNOWLEDGEMENTS

The present work was supported by CICYT project MAT2001-3362 and the Andalusian government (PAI TEP-0120).

REFERENCES

Albrecht M, Grillo V, Remmele T, Strunk HP, Egorov AYu, Dumitras Gh, Riechert H, Kaschner A, Heitz R and Hoffmann A 2002 Appl. Phys. Lett. **81**(15), 2719

Kondow M, Uomi K, Niwa A, Kitatani T, Watahiki S and Yazawa Y 1996 Jpn. J. Appl. Phys. **35**, 1273

Kondow M, Uomi K, Niwa A, Kitatani T, Watahiki S, Yazawa Y, Hosomi K and Mozume T 1995 Solid-State Electron. **41**, 209

Li W, Pessa M, Ahlgren T and Dekker J 2001 Appl. Phys. Lett. **79**, 1094

Mintairov AM., Kosel TH., Merz JL, Blagnov PA, Vlasov AS, Ustinov VM and Cook RE 2001 Phys. Rev. Lett. **87**(27) 277401

Spruytte SG, Coldren CW, Harris JS, Wampler W, Krispin P, Ploog K and Larson MC 2001 J. Appl. Phys. **89**, 4401

Xin HP and Tu CW 1998 Appl. Phys. Lett. **72**(19), 2442

Zhang SB and Zunger A 1997 Appl. Phys. Lett. **71**(5), 677

Inst. Phys. Conf. Ser. No 179: Section 2
Paper presented at Electron Microscopy and Analysis Group Conf. EMAG2003, Oxford, 2003
©2003 IOP Publishing Ltd

Microstructural study of multilayer YBCO on a curved Ni surface

Y L Cheung, G Passerieux, I P Jones, J S Abell

Metallurgy and Materials, School of Engineering, The University of Birmingham, Edgbaston, Birmingham, B15 2TT, U.K.

Abstract. Pulsed laser ablation has been used to prepare *in situ* multilayer YBCO thin films on buffered curved Ni-based, textured substrates. Standard buffer layer architecture ($CeO_2/YSZ/CeO_2$) was employed to improve the epitaxy of the YBCO layer and STO was used to insulate the YBCO layers . Tc measurement and X-ray θ-2θ scan of curved samples showed similar superconducting and structural properties to those achieved on flat surfaces. The microstructures and interfaces of the multilayers were characterized using electron microscopy particularly cross-sectional transmission electron microscopy (XTEM).

1. Introduction

One of the commercial applications of superconductors is superconducting coils which consume almost no power whilst being able to generate high fields for relatively small size. In order to obtain the highest J_E (Engineering Current Density, equal to the critical current divided by the total cross-sectional area of the superconductor and its supporting matrix) and replace the conventional winding of coils from preformed wires or tapes, an innovative integrated fabrication technology for High Temperature Superconducting (HTS) coils is proposed in this project. Superconducting layers and buffer layers are deposited epitaxially onto a textured substrate with a rotating cylindrical heated former using pulsed laser deposition. A suitably lattice-matched insulating layer is subsequently deposited followed by another YBCO layer. This combination is then repeated to provide a multilayer biaxially textured YBCO coil with each layer electrically isolated. The individual superconducting layers are connected together in series or parallel to produce a coil after the deposition. The end result is a multi-turn, multi-layered helical solenoid prepared *in situ* without the requirement of producing long lengths of superconducting tape [1].

In this paper, coated conductors were produced by rolling-assisted, biaxially textured substrates (RABiTS) to fabricate YBCO multilayer. The detailed structure of multilayer YBCO on a curved Ni surface has been investigated by means of Transmission Electron Microscopy (TEM). The cross-sectional TEM (XTEM) study gives insight into the nature of epitaxial growth and interface structures. The goal of our study is to reveal the possibility of growing thin films on Ni curved surfaces.

2. Experimental

Triple YBCO layers were grown on a buffered curved Ni RABiTS substrate using STO insulating layers (YBCO/STO/YBCO/STO/YBCO/CeO_2/YSZ/CeO_2//Ni). All layers were prepared *in situ* by a Lambda Physik LPX200i Excimer laser running at 248nm (KrF). A special rotating cylindrical heated former (3cm diameter) was used in this research in order to produce *in situ* formation of multi-turn and multi-layered HTS coils. The deposition parameters were chosen to be the same as for optimised flat tapes [2]. The target-substrate distance was 6cm, laser fluence of 1.5 J/cm^2, substrate temperature of 780°C, oxygen pressure of 0.6mbar, cooling down at 10°C/min in 700mbar oxygen after the ablation. To reduce the possibility of substrate oxidation and to remove oxide on the substrate, 0.15 mbar of reducing gas: 4% H_2 in argon, was introduced to the chamber during heating of the substrate.

To prepare the XTEM sample, a Gatan model 691 Precision Ion Polishing System (PIPS) was used to mill the sample. During ion milling, the ion-milling holder was rotated and two ion guns were used. A typical ion milling condition was 10° tilt angle and 5keV ion beam energy. When coloured fringes were observed, the ion milling condition was changed into an ion polishing procedure. The ion polishing condition was 8° tilt angle and 3keV of ion beam energy. The electron micro-images were observed on a transmission electron microscope (Model Tecnai F20, FEI) with a field emission gun (FEG) and energy dispersive spectrometer.

3. Results and Discussion

The results in this paper were obtained on the same curved sample. Figure 1 gives a low-magnification overview of a YBCO/STO/YBCO/STO/YBCO/CeO2/YSZ/CeO2 sample grown on a curved Ni substrate and the diffraction patterns from Ni and NiO are shown in the right top corner. The different layers could be easily identified by their different image contrast. All layers were rather uniform and single crystal like. The thicknesses of YBCO layers were decreasing although the number of pulses and growth parameters were the same. The reason for this decrease may be that the thicker the grown film, the larger the radius of the whole sample. Therefore, the thicknesses of the grown films decreased.

In our sample, the NiO reaction layer was still present at the CeO_2/Ni interface and the plane of NiO is parallel to the (001) plane in Ni, although a reducing gas (4% H_2 in argon) was used to prevent the formation of NiO and to remove NiO before deposition. The thickness of the NiO layer was about 10-70nm; which was not uniform and the interface of the thicker NiO was curved. This reaction layer may be formed by oxygen diffusion after the deposition process at high temperature through the cracks in the CeO_2 layer [3, 4], and it may influence the mechanical and adhesive properties of the films. The YBCO layer on a thin NiO layer grew with a completely c-axis orientation throughout the thickness. However, the YBCO layer on the thicker NiO layer was full of defects, such as misoriented grains, stacking faults and dislocations (Figure 2). The volume of the NiO layer may, therefore, be related directly to the quality of the YBCO layer. On the other hand, the microstructure and surface morphology of YBCO would change with the grown thickness of the YBCO. So, finding a critical thickness of the YBCO layer to prevent the formation of defects is another consideration [5].

Figure 3 shows the interface between the top CeO_2 layer and the YSZ layer. The topCeO2/YSZ interface was atomically sharper than the YSZ/seedCeO2 interface. There were only periodic dislocations ($\approx$6nm) at the topCeO2/YSZ interface. From the TEM micrographs, CeO_2/YSZ/CeO_2 was a good architecture for a buffer layer between the YBCO and the Ni tape. Figure 4 shows the diffraction patterns from the buffer layers CeO_2/YSZ/CeO_2. This shows that both the YSZ and CeO_2 layer had a 45° in-plane rotation because of the lattice mismatch of CeO_2 and Ni.

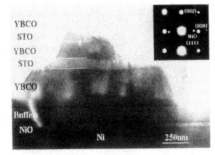

Figure 1. XTEM image of YBCO/STO/YBCO/STO/YBCO/buffer layers grown on curved textured Ni, diffraction patterns from Ni and NiO show that the (111) plane of NiO is parallel to the plane in Ni.

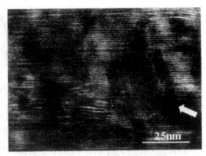

Figure 2. HREM image of 1st YBCO layer on thicker NiO layer. A mis-oriented grain is indicated by a white arrow; this YBCO is full of stacking faults (shorter black arrow) and dislocations (longer black arrow).

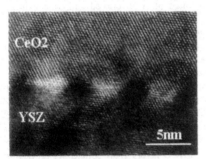

Figure 3. High-resolution image of the top CeO_2 /YSZ interface. Periodic dislocations are indicated by arrows.

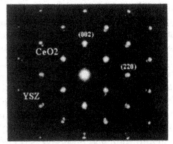

Figure 4. CeO_2 [110] and YSZ [110] zone axis patterns

CeO_2 was used as the last buffer layer because its lattice mismatch with YBCO is small. Figure 5 shows an HREM image of the YBCO/CeO_2 interface and the diffraction pattern of YBCO layer. There was no obvious evidence of reactions between the YBCO and CeO_2 nor the presence of any disordered perovskite layer within the YBCO. And both YBCO layers grew with a completely c-axis orientation throughout the thickness. The interface was atomically sharp and the surface of CeO_2 was flat, CeO_2 is ideal for growing highly c-axis oriented YBCO film.

STO was used in this research as the insulating layer for separating YBCO layers. From Figure 1, both STO, the first and second YBCO layers grew epitaxially on the buffered curved Ni substrate. However, in some areas, STO grew less perfectly due to the roughness of the YBCO surface. Figure 6 shows a high-magnification micrograph of the 2nd YBCO/STO interface. A larger a-axis grain has grown in the 2nd YBCO layer on a rough STO surface, which was full of defects and it should influence directly the properties of the YBCO. A good surface morphology not only depends on the deposition conditions but also on the thickness of the YBCO and on the presence of suitable buffer layers.

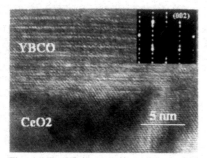

Figure 5. High-resolution image of YBCO/CeO₂ interface and the YBCO diffraction pattern showing its c-axis orientation.

Figure 6. High-resolution image of 2nd YBCO/STO interface. A large a-axis grain is grown on the rough STO surface (indicated by a black arrow).

4. Conclusion

Triple YBCO layers were successfully deposited on a Ni curved substrate using the same deposition conditions as for the optimised flat Ni tapes. Standard ORNL (Oak Ridge National Laboratory) buffer architecture (CeO₂/YSZ/CeO₂) was used to grow on a curved Ni substrate and STO was used to insulate consecutive YBCO layers from each other and to copy the texture to provide a multilayer structure.

From the XTEM results, CeO₂/YSZ/CeO₂ buffer layers grew epitaxially on a curved Ni substrate. A NiO reaction layer was found at the CeO₂/Ni interface. The thickness of the NiO layer was about 10-70nm; it was not uniform and the thicker NiO was curved. The YBCO layer grew with c-axis orientation through the thickness. However, the YBCO layer on the thicker NiO was full of defects, such as stacking faults and dislocations. Both YBCO/CeO₂/YSZ/CeO₂/NiO/Ni interfaces were atomically sharp and no obvious interdiffusion was observed. STO and other YBCO layers grew epitaxially on buffered curved Ni substrate as well. However, in some regions, the STO grew less perfectly due to the roughness of the YBCO surface. Achieving a critical thickness of the YBCO layers to produce a good surface morphology is very important.

Electron backscatter diffraction (EBSD) will be used to further determine the top layer grain orientation and texture information of curved samples copying from the RABiTS. XRD ω-rocking curves and ϕ scans will be used to analyse out-of-plane and in-plane texture of curved samples in the future.

References

[1]   Cheung Y L, Chakalova R I, Abell J S, Button T W and Maher E F 2003 to be published in 6th European Conference on Applied Superconductivity

[2]   Chakalova R I, Cai C, Woodcock T, Button T W, Abell J S and Maher E 2002 Physica C 372-376 846-850

[3]   Sun E Y, Goyal A, Norton D P, Park C, Kroeger D M, Paranthaman M, and Christen D K 1999 Physica C 321 29-38

[4]   Norton D P, Park C, Prouteau C, Christen D K, Chisholm M F, Budai J D, Pennycook S J, Goyal A, Sun E Y, Lee D F, Kroeger D M, Specht E, Paranthaman M, and Browning N D 1998 Materials Science and Engineering B56 86-94

[5]   Leonard K J, Goyal A, Kroeger D M, Jones J W, Kang S, Rutter N, Paranthaman M, Lee D F and Kang B W 2003 J. Mater. Res. 18 1109-1122

Inst. Phys. Conf. Ser. No 179: Section 2
Paper presented at Electron Microscopy and Analysis Group Conf. EMAG2003, Oxford, 2003
©2003 IOP Publishing Ltd

111

The Effects of Oxygen on the Electronic Structure of MgB$_2$

J C Idrobo

Department of Physics, University of California Davis, One Shields Ave,
Davis CA, 95616, USA

S Ögüt

Department of Physics, University of Illinois at Chicago, 845 W. Taylor
(MC 273), Chicago IL, 60607, USA

R Erni

Department of Chemical Engineering and Materials Science, University of
California Davis, One Shields Ave, Davis CA, 95616, USA

N D Browning

Department of Chemical Engineering and Materials Science, University of
California Davis, One Shields Ave, Davis CA, 95616, USA

National Center for Electron Microscopy, MS 72-150, Lawrence Berkeley
National Laboratory, Berkeley CA, 94720, USA

Abstract. The discovery of superconductivity in MgB$_2$, with a transition
temperature of T$_c$=40K [1], has focused scientific attention towards an
understanding of its superconducting properties. Superconductivity in
MgB$_2$ is driven by hole transport through the boron orbitals [2]. The
presence of oxygen as segregates in the grain boundaries and precipitates
in the bulk of polycrystalline materials could have a large effect on the
hole carrier concentration and therefore change the superconducting
properties of MgB$_2$. In this study, we show experimental as well as
theoretical evidence that oxygen segregates in the bulk of MgB$_2$ can have
an effect on the hole carrier concentration. The experiments were
performed using atomic resolution Z-contrast imaging, electron energy
loss spectroscopy (EELS) techniques, and density functional theory (DFT)
calculations.

1. Introduction

MgB$_2$ is a recently discovered superconductor, T$_c$= 40 K [1]. In this material the
boron atoms are responsible for the hole carrier concentration which causes
superconductivity within the BCS mechanism. There have been reports from different
groups showing that MgB$_2$ samples are not a single phase material, but rather a rich

collection of different phases including: MgB$_2$, MgO, B$_y$O$_x$, Mg$_x$B$_y$, Mg$_x$B$_y$O$_z$ [2,3]. Evidently, the presence of these different phases have a considerable effect on the superconductivity properties, and a detailed characterisation will have to be carried out in order to understand as well as to try to improve their overall effect on the transport properties. Oxygen, unlike other doping elements, is present in MgB$_2$ as an unwanted impurity due to its high reactivity with MgB$_2$. Therefore, oxygen acts as a modulator element on the formation of new phases, controlling the presence of Mg-vacancies, reported as an important factor in decreasing the T$_c$, forming precipitates into the bulk and segregating at the grain boundaries. In this work we show experimental and theoretical evidence of MgB$_2$-oxygen precipitates which present changes on the electronic structure of MgB$_2$. These results could explain the variation of the transport properties in different MgB$_2$ samples.

We present in this report a study of oxygen precipitates in the bulk of polycrystalline MgB$_2$ sample using atomic resolution EELS, Z-contrast imaging in a scanning transmission electron microscope (STEM), and DFT calculations. In particular we found that oxygen segregates into the bulk of MgB$_2$ forming different phases. We performed EELS measurements to quantify the concentration of oxygen and boron, and studied in detail the electronic structure of the boron K edge in these precipitates. We observed that the precipitates present clear differences in the boron fine structure, i.e. the pre-peak of the boron K-edge has a sharper feature than in pure MgB$_2$. It has been discussed earlier that the change of intensity and shape of this pre-peak could affect the transport properties (i.e. critical temperature) of MgB$_2$ [4].

2. Instrumentation and Theoretical Calculations

The experimental EEL spectra and Z-contrast images [5] were obtained from a JEOL 2010F field emission STEM/TEM, and the newly developed monochromated STEM/TEM, FEI G2 Tecnai. Both microscopes operate at 200 kV. [6] The lens conditions in the microscopes were defined for imaging with a probe size of 0.13 nm (0.2 nm for spectroscopy). In this experimental condition, the incoherent high-angle annular dark field or "Z-contrast" image allows the structure of the grains to be directly observed, and the image can also be used to position the electron probe for EELS [19]. Core loss EELS probes the unoccupied density of states above the Fermi energy. In the experiment performed here, the EEL spectra are acquired directly from the grain with characteristic core losses used to determine the composition. The EEL spectra were interpreted using DFT calculations. In particular, we used the Vienna ab-initio simulation package (VASP) [8]. This code uses the positions and atomic number of the constituent elements of the crystal structure as the only input to obtain the electronic properties of the system, (i.e. band structures, density of states).

3. Results

Figure 1a shows a low-magnification image of a typical oxygen precipitate observed in these samples (dark spot, low contrast region). The size of the grain is about 2 μm. Figure 1b shows the same region magnified. Here, the hexagonal shape of the precipitate can be distinguished, with a diameter of about 60 nm. EEL spectra of the boron K-edge and oxygen K-edge were taken from the precipitate and the region around it (which also contains oxygen). Both spectra were background subtracted and corrected for multiple-scattering contributions. Figure 2a shows the boron K-edge spectrum for the precipitate (i), the region around the precipitate (ii) and a spectrum obtained for pure MgB$_2$ (iii) displayed for comparison. The fine structure of spectrum ii shows clear differences from the precipitate and the pure MgB$_2$. The pre-peak intensity (labelled a) in spectrum ii shows a sharper feature than the precipitate and the pure MgB$_2$ spectra with a decrease in its intensity at 191 eV. The peaks (labelled b and c) are slightly higher and more pronounced for spectrum ii than the precipitate spectra and both are higher than the pure MgB$_2$ spectrum. The oxygen spectra (figure 2b) for the two regions look completely different, and this can be taken as evidence of two different phases. Additionally, we show spectra of the boron K-edge obtained using monochromated STEM with an energy resolution of 0.3 eV (figure 3a). These spectra were acquired in regions with unknown concentrations of oxygen and without. We observed differences in the fine structure due to oxygen that are consistent with our previous results. We quantify those changes within the energy range (figure 3b) using DFT calculations and taking into account the momentum transfer dependence of crystal orientation and optical conditions [4].

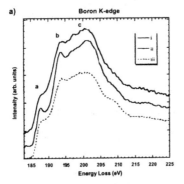
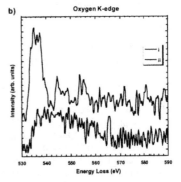

Figure 1. (a) Low magnification of polycrystalline MgB$_2$. (b) High magnification of figure 1a, the change of contrast in the images indicates the differences phases in the material, bulk and precipitate

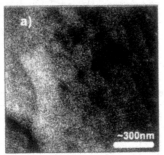
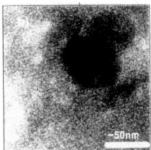

Figure 2. (a) Boron K-edge spectra, where differences can be observed in the fine structure. The changes of shape between the spectra in the pre-peak marked as *a* and peak marked as *b* are due to oxygen segregation in the bulk. (b) Oxygen K-edge spectra. The difference in the spectra for the two regions shows that oxygen has precipitated in MgB$_2$ forming two different phases.

114

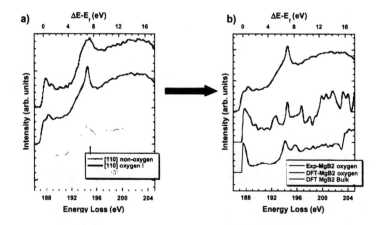

Figure 3. (a) Boron K-edge spectra monochromated with an energy resolution of 0.3eV. Changes in the fine structure can be observed, these are due to the presence of oxygen (b) These changes can be explained within the energy range using DFT calculations.

Conclusions

Different concentrations of oxygen in MgB_2-oxygen nanostructures show changes of the electronic structure with respect to MgB_2, presenting an increase or decrease of the density of states at the Fermi energy. Using a monochromated electron beam in STEM we have shown that the states close to the Fermi level can be resolved, proving that EELS Z-contrast imaging and DFT could help to understand the variation of the transport properties in MgB_2.

References

[1]   Nagamasu J, Nakagawa N, Muranaka T, Zenitani Y and Akimitsu J 2001 Nature 410 63-64

[2]   Klie R, Idrobo J, Browning N, Regan K, Rogado N and Cava R 2001 Appl. Phys. Lett. 79 1837-1839

[3]   Klie R, Idrobo J, Browning N, Serquis A, Zhu Y, Liao X and Mueller F 2002 Appl. Phys. Lett. 80 3970-3972

[4]   Klie R, Su H, Zhu Y, Davenport J, Idrobo J, Browning N and Nellist P 2003 Phys. Rev. B. 67 144508

[5]   Nellist P and Pennycook S 1999 Ultramicroscopy 78 111-124

[6]   James E and Browning N 1999 Ultramicroscopy 78 125-139

[7]   Browning N, Chisholm M and Pennycook S 1993 Nature 366 143-146

[8]   Kresse G and Furthmüller J 1996 Phys. Rev. B 54 11169-11186

Inst. Phys. Conf. Ser. No 179: Section 2
Paper presented at Electron Microscopy and Analysis Group Conf. EMAG2003, Oxford, 2003
©2003 IOP Publishing Ltd

Correlation of electrical and structural properties of Au contacts to KOH treated n-GaN

Grigore Moldovan,[1,2] Ian Harrison,[2] Martin Roe[1] and Paul D Brown[1]

[1]School of Mechanical, Materials and Manufacturing Engineering and Management, [2]School of Electrical and Electronic Engineering, University Park, University of Nottingham, Nottingham, NG7 2RD, UK.

Abstract. A correlated current-voltage (I-V), electron beam induced conductivity (EBIC) and X-ray photoelectron spectroscopy (XPS) study of Au contacts to KOH treated n-type GaN is presented. A strong degradation of I-V characteristics occurs following the KOH treatment, mirrored in a reduction in the magnitude of the EBIC current, even though the EBIC images look visibly unaltered. XPS demonstrates a modification in the surfaces states, e.g. resulting in a –0.3eV shift in the binding energy of Ga_{3d} for MBE GaN following KOH processing.

1. Introduction

One of the major limitations to the performance of the GaN based devices arises from the metallic contacts [1]. It has been demonstrated that improvements in contact performance can be achieved by means of careful GaN surface preparation prior to metal deposition. The etching of n-GaN reduces the ohmic contact resistance [2] whilst producing leaky Schottky contacts [3]. This is illustrated by the case of the KOH surface treatment that has been shown to produce a reduction in ohmic contact resistance by up to one order of magnitude [4]. Further investigations are required to investigate the intimate relationship between the structural and chemical modifications to the GaN surface introduced by KOH treatment and the resulting contact performance. The present study uses an electron beam induced current (EBIC) approach to investigate the influence of KOH treatment on the barrier height of Schottky Au contacts, correlated with X-ray photoelectron spectrometry (XPS). Comparison is made between contacts to etched Ga-polar GaN grown by MBE and MOCVD.

2. Experimental

Au/n-GaN Schottky contacts were deposited onto untreated and KOH treated n-GaN samples. The MOCVD GaN/sapphire wafer was 3µm thick with a 0.9µm thick Si-doped capping layer with an estimated n-type carrier concentration of $\sim 10^{17} cm^{-3}$. The MBE GaN/sapphire wafer was 2.52µm thick and demonstrated an n-type carrier concentration of $6.38 \times 10^{16} cm^{-3}$ and electron mobility of $212 cm^2/Vs$. Cleaved samples were degreased in ultrasonic baths of lotoxane, methanol, acetone, propanol and de-ionised water, each for three minutes. The KOH treatment consisted of dipping in a 6 molar solution of KOH and de-ionised water for 1 minute at 60°C, followed by a 1 minute dip in de-

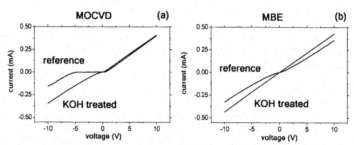

Figure 1. Current-voltage characteristics of reference and KOH treated (a) MOCVD and (b) MBE GaN.

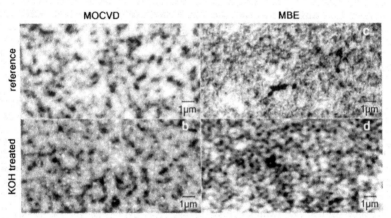

Figure 2. 10kV EBIC images of reference and KOH treated (a, b) MOCVD and (c, d) MBE GaN.

ionised water and finally N_2 blown dry. Prior to each metallisation, the samples were cleaned by dipping into a 37% solution of HCl and de-ionised water for three minutes and again N_2 blown dry. Glass shadow masks were employed for the formation of Au Schottky contacts and ohmic Ti pads for the purpose of ground connection during the EBIC experiments and I-V measurements. Au was deposited simultaneously on all samples at a rate of 3nm/s, to a thickness of 125nm, using a thermal evaporator at a chamber pressure of 4×10^{-6} Torr. Ti was deposited at 0.7nm/s, to a thickness of 200nm, using an e-beam evaporator at a chamber pressure of 2.5×10^{-6} Torr. EBIC measurements were performed using a JEOL 6400 SEM and a Matelect IV5 amplifier. XPS experiments were performed using a VG scientific ESCALAB with an Al K_α cathode.

3. Results and Discussion

The current-voltage (I-V) behaviour of each sample was recorded at room temperature under light conditions, as presented in Figure 1. The I-V traces demonstrate that the Au contacted MOCVD GaN samples have a stronger Schottky response than the MBE samples, while the reference samples exhibit a stronger Schottky behaviour as compared with the KOH treated samples.

The plan-view EBIC images shown in Figure 2 were intentionally recorded at a low acceleration voltage of 10kV to accentuate the contribution of the GaN surface features.

		Ga (%)	N (%)	O (%)	C (%)	Cl (%)
MOCVD	reference	28.0	64.4	0.7	6.2	0.7
	KOH treated	27.2	68.2	0.3	3.4	1.0
MBE	reference	30.0	64.3	0.9	4.0	0.8
	KOH treated	26.6	68.6	1.0	3.3	0.5

Table 1. GaN surface content quantified from the XPS survey spectra, for the reference and the KOH treated samples.

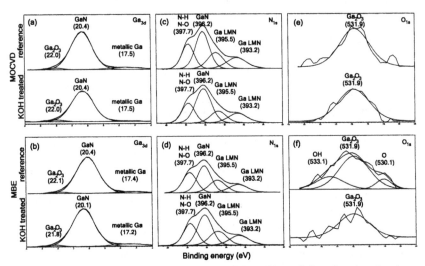

Figure 3. Detailed XPS scans of Ga_{3d}, N_{1s} and O_{1s}, showing the shape and energy (eV) of the de-convoluted peaks.

The MOCVD samples exhibited a smooth surface with large sub-grains of ~1 μm in size and a GaN/sapphire threading dislocation density of ~5x10^6 cm^-2 (presuming that each feature within the EBIC image corresponds to an isolated threading dislocation). The MBE sample displayed a rough surface with a high density of small sub-grains of ~350nm in size. The location of low signals within the MBE sample EBIC images appears to correlate with the position of grain boundaries in the secondary electron images [5]. It is noted the MOCVD samples exhibited a larger induced current as compared with the MBE samples, whilst KOH etching was associated with a reduction in both EBIC signal and contrast (Figs. 2c,d). Overall, however, the EBIC images provide no visible change attributable to the effect of KOH treatment (Figs. 2a,b), in contrast to the induced current and I-V measurements. Hence, the suggestion is that the KOH treatment acts to uniformly change the properties of the GaN surface but does not have a localised effect.

In the absence of evidence for any differences in the surface structures with processing, an investigation was made of the changes in surface chemistry induced by the etching procedure to try and explain this difference in contact performance. The sample-set was replicated from the same wafers and processed under identical condition for the purpose of XPS analysis of the GaN surfaces prior to metallisation. The acquired survey spectra were dominated by N and Ga peaks, with only traces of O, C and Cl (attributed to the cleaning treatment in HCl solution). The relative atomic content of Ga, N, O, C and Cl was determined using the Ga_{3d}, N_{1s}, O_{1s}, C_{1s} and Cl_{2p} peaks, as

118

summarised in Table 1. No K content was identified in the samples even following KOH treatment. The KOH treatment resulted in a small surface Ga content decrease and a small N content increase for both MOCVD and MBE samples. The surface C content significantly decreased following the KOH treatment of the MOCVD sample but only slightly decreased in the case of MBE GaN.

Detailed scans of the Ga, N and O peaks were recorded for more detailed analysis (Fig. 3). The Ga_{3d} peaks were de-convoluted into the contributions of metallic Ga, GaN and Ga_2O_3. The Ga_{3d} binding energies remained unchanged or showed a shift of -0.3eV for the MOCVD and MBE GaN samples, respectively, after KOH treatment, consistent with [6]. A small reduction in the metallic Ga surface content of the MOCVD sample was also recorded due to the KOH treatment. The N_{1s} peaks were de-convoluted into four contributions: a N-O or N-H ionisation peak at 397.7eV, a N-Ga ionisation peak at 396.2eV and two Auger Ga peaks at 395.5eV and 393.2eV, respectively. No changes in the binding energy of these peaks were observed with GaN type or KOH treatment. However, a slight increase of ~ 10% was detected in the N-O or N-H to N-Ga ratio, for both MOCVD and MBE GaN, due to the KOH treatment. The binding energy of the O_{1s} peak similarly remained unchanged with sample type or treatment. However, for the reference MBE sample, the O_{1s} peak could be de-convoluted into three contributions, a chemiabsorbed O peak at 530.1eV, a Ga_2O_3 peak at 531.9eV and an OH peak at 533.1eV, respectively. The KOH treatment was found to eliminate the O and OH peaks, while the overall O content remained constant.

This XPS study indicates that the KOH treatment removes the native C contamination and Ga_2O_3, reducing the surface Ga and resulting in a corresponding increase in surface N. It might be envisaged that atmospheric re-oxidation quickly occurs after the KOH treatment, as the O content remains unchanged, but it is apparent that changes have appeared in the form of the oxygen at the MBE surface. This is again consistent with [6]. These changes in the surface states result in the -0.3eV shift in the binding energy of the Ga_{3d} peak, as demonstrated for the MBE grown GaN, which is believed to be responsible for the degradation of the I-V characteristic of the resulting contact [6]. A corresponding energy shift for KOH etched MOCVD grown GaN was not observed in this work, but shifts of ~ -0.3eV and ~ -0.4eV have been reported for MOCVD GaN following more aggressive treatment in KOH [6, 7]. The implication therefore is that the increased surface roughness associated with the MBE sample assists with the action of the KOH to modify the native surface.

Acknowledgement

This work was supported under EPSRC grant GR/M87078.

References

[1] Pearton S J, Ren F, Zhang A P and Lee K P 2000 Mat. Sci. Eng. R30 55-212
[2] Ping A T, Chen Q, Yang J W, Asif Khan M and Adesida I 1998 J. Electron Mat. 27 4 261-265
[3] Ping A T, Schmitz A C, Adesida I, Asif M, Chen Q and Yang J W 1997 J. Electron Mat. 26 3 266-271
[4] Moldovan G, Harrison I, Humphreys C J, Kappers M and Brown P D 2003 Mat. Sci. Eng., in press
[5] Moldovan G, Harrison I and Brown P D 2003 Inst. Phys. Conf. Ser., in press
[6] Li D, Sumiya M, Fuke S, Yang D, Que D, Suzuki Y and Fukuda Y 2001 J. Appl. Phys. 90 4219-4223
[7] Rickert K, Ellis A B, Himpsel F J, Sun J and Kuech T F 2002 Appl. Phys. Lett. 80 2 204-206

Inst. Phys. Conf. Ser. No 179: Section 2
Paper presented at Electron Microscopy and Analysis Group Conf. EMAG2003, Oxford, 2003
©2003 IOP Publishing Ltd

ELNES of titanate perovskites – a probe of structure and bonding

P Harkins[1], D W McComb[2], M MacKenzie and A J Craven[1]

[1]Dept. of Physics and Astronomy, University of Glasgow, G12 8QQ
[2]Dept. of Materials, Imperial College London, London SW7 2AZ

Abstract. The oxygen K-edge of a range of perovskite (ABO_3) materials has been investigated using electron energy loss spectroscopy (EELS), where A = Ca, Sr, Ba, Pb and B = Ti, Zr. These substitutions strongly affect the ferroelectric properties of the materials as the bonding and symmetry present in each unit cell is altered due to the increasing size of the cations and the stresses that result. These alterations also affect the electronic structure of the materials, which has been modelled using band structure and multiple scattering methods in order to interpret better the EELS data.

1. Introduction

Although perovskite materials have been the subject of much research over the years, the relationship between their atomic structure and the range of ferroelectrics properties displayed is not fully understood [1]. The balance of forces within these materials strongly affect their properties, i.e. short-range repulsive forces within a unit cell favour a cubic, paraelectric phase, while longer-range, electrostatic forces induce a distorted, ferroelectric phase [2]. There is increasing interest in ferroelectric thin films for applications such as non-volatile RAM devices. However, structural and compositional inhomogenieties as well as interfacial phenomena can significantly influence the structure-property relationship.

EELS is a useful technique for investigation of perovskites as the energy loss near edge structure (ELNES) is directly influenced by the local site- and symmetry-projected unoccupied density of states (DOS). Thus, information on the local bonding and co-ordination environments of O atoms within the unit cell can be inferred from the ELNES. However, it is important to understand the electronic structure found in the bulk materials before moving on to more complex, spatially inhomogeneous thin films. Hence, this work concentrates on five polycrystalline perovskites, $CaTiO_3$, $SrTiO_3$, $BaTiO_3$, $PbTiO_3$, and $PbZrO_3$. In the current study, interpretation of the O K-edges was aided with theoretical predictions from the real space, multiple scattering code, FEFF8.2 and partial DOS from the reciprocal space, band structure code, WIEN97 [3,4].

2. Experimental

Polycrystalline powders of $BaTiO_3$, $SrTiO_3$, and $CaTiO_3$ were synthesised via a sol-gel process [5]. A solution of the metal-oxide (M = Ba, Sr, Ca) was slowly added to titanium isopropoxide and methanol to produce a clear gel. This was vacuum dried to obtain a white 'sol' that was calcined at 900°C for 8 hours. Both Pb-containing powders were purchased from Alfa Aesar. The phase purity of all specimens was confirmed using powder x-ray diffraction. To prepare a TEM sample, the powder was ground in a mortar and pestle, dispersed in propanol and dropped onto a grid of holey carbon film. The O K-edges were collected on an FEI Tecnai F20 field emission gun STEM equipped with a Gatan-ENFINA spectrometer for EELS collection. Apertures were selected to give a convergence semi-angle of 10mrad and a collection semi-angle of 9mrad for all EELS data sets. Data was collected from 420eV for compounds containing Ti and from 480eV otherwise, using a dispersion of 0.2eV per channel. The full-width half-maximum (FWHM) of the zero-loss peak (ZLP), was in the range 0.75-0.85eV for all of the data collected.

Analysis of the EELS data was performed using Gatan's Digital Micrograph software package. The Ti $L_{2,3}$-edge threshold is ~80eV below that of the O K-edge. In order to leave only the energy-loss from the O atoms for those samples with Ti in the B-site, a window was placed in the energy range below the Ti $L_{2,3}$-edge and the background contributions subtracted using a simple power-law. This was repeated at the O K-edge to remove the contribution from the tail of the Ti $L_{2,3}$-edge. While not ideal, this is a reasonable approximation. Many sets of EELS data were collected, aligned and averaged in order to reduce random noise present in the spectra. Energy dispersive x-ray (EDX) analysis performed on the samples confirmed that the crystallites studied were compositionally pure.

3. Calculations

WIEN97 uses density functional theory (DFT) to simulate the electronic structure of a crystal in reciprocal space [4]. This approach gives an excellent approximation to the electronic structure and allows accurate comparison of the total energies of a group of atoms with different configurations. However, it requires a periodic structure and the computation time is significantly increased if the effect of a core hole is to be simulated, as this requires the creation of a supercell with lower symmetry. Calculations have been performed on a unit cell using a mesh of ~500 k-points in the Brillouin zone and an RK_{max} value of 8. FEFF8.2 uses ab initio self-consistent real space Green's function for clusters of atoms. The electronic structure of a non-periodic cluster can be calculated and the effects of increasing the number of atoms in the cluster can be determined in a straightforward way, both to check for their effect and also to test for convergence of the calculation. It is also possible to simulate the effect of a core hole without significantly increasing the time required for the calculation.   Simulations were performed for clusters of ~300 atoms. FEFF8.2 automatically includes final state lifetime broadening in its simulation of near-edge structure whereas WIEN97 does not. The effect of the final state lifetime is to increase the broadening of features as the square of their energy above the Fermi level so that fine detail is only present near the threshold. FEFF8.2 calculates the x-ray absorption near edge structure (XANES), but this is equivalent to ELNES in the dipole approximation.

4. Results and discussion

Initially, work focused on the perovskites with group II elements on the A-sites, i.e. $CaTiO_3$, $SrTiO_3$ and $BaTiO_3$. The A-site cations all have a noble gas electronic configuration, $[\ ]ns^0$, and with systematically varying ionic radii, (Ca<Sr<Ba). Ca is smaller than is ideally required for the A-site, causing strain within the unit cell, which distorts to orthorhombic symmetry. The Sr ion sits in the A-site without causing undue

strain and is a good example of the ideal cubic perovskite structure at room temperature. The Ba-cation is too large for the A-site and the unit cell distorts to tetragonal to compensate. These distortions influence directly the bonding, electronic structure and ferroelectric characteristics of each material.

The three panels in Figure 1 show the experimental O K-edges (upper), the lifetime broadened O K-edge XANES from FEFF8.2 (middle) and the site and symmetry projected O p-DOS from WIEN97 (bottom). In each panel the data from the compounds studied are presented in the same order, with each plot vertically offset for clarity. The left-hand edge of all partial DOS plots is the first unoccupied state in the conduction band while it is the ionisation edge threshold for the O K-edges. All partial DOS plots have been normalised.

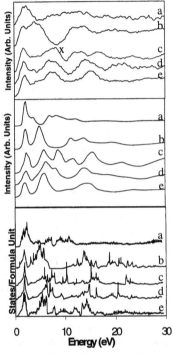

Figure 1. Experimental O K-edge (top), XANES spectra from FEFF8.2 (middle) and oxygen p-DOS from WIEN97 (bottom) after all edge onsets were aligned to each other for (a) PbZrO$_3$, (b) PbTiO$_3$, (c) BaTiO$_3$, (d) SrTiO$_3$, (e) CaTiO$_3$

Both sets of calculations agree quite well with each other. The experimentally collected EELS data for the perovskites containing the two smallest A-cations, i.e. Ca and Sr, show reasonable agreement with the K-edge shapes and the predicted p-DOS. However, for BaTiO$_3$ in the region marked X, the O K-edge does not follow the predicted structure. It is not clear why this should be the case.

In all of these materials, the predicted splitting of the first two peaks is underestimated by both codes. The shapes of these peaks are strongly influenced by the d-states of the nearby Ti cations as the Ti-O bond has significant covalent character. This is essential for ferroelectricity [1]. Figure 2 shows the relationship between the Ti d-DOS and the O p-DOS. The O DOS has been scaled for easy comparison. The definition of these peaks results from the completion of the octahedral coordination of Ti by O, as demonstrated in Figure 3. Here, the cluster size is increased from 16, where the Ti has 5 O neighbours, to 21, corresponding to 6 O neighbours. In the low symmetry environment the ligand field splitting of the Ti d-orbitals is poorly defined. The addition of the next shell of 6 O atoms allows the strong octahedral crystal field symmetry to split the d-states into t$_{2g}$-e$_g$ symmetries. This is reflected on the O-site by the creation of additional p-like states, giving a much more intense peak around 6eV in the O K-edge.

Returning to the discrepancies observed for BaTiO$_3$, there must be some effect that the calculations do not take into account. It was hypothesised that these discrepancies may be the result of unoccupied f-states on the Ba cation. In order to test this theory, PbTiO$_3$ was investigated. PbTiO$_3$ is similar to BaTiO$_3$ in that Pb is a large cation, its unit cell is tetragonal at room temperature and it goes through rhombohedral-

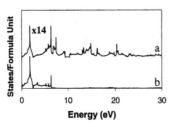

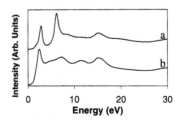

Figure 2. WIEN98 simulation of the (a) O p-DOS and (b) Ti d-DOS for SrTiO₃.

Figure 3. FEFF8.2 simulation of the O K-edge ELNES for SrTiO₃, with (a) 21 and (b) 16 atom

orthorhombic-tetragonal-cubic phases with increasing temperature. However, Pb is a p-block element with fully occupied f-states. Also, the lone pairs associated with Pb allow a degree of covalent mixing with the O p-states, whereas the Ba-O interaction is more ionic in nature [1,6]. There is a marked difference in the ELNES of PbTiO₃ compared to that for materials with group II cations in the A-site. This is reflected by the calculations. This suggests that the f-states are being accurately portrayed by both codes and that other factors must be affecting the ELNES of BaTiO₃. Further investigation of this effect is in progress.

Many solid solutions of perovskites are used in technologically important thin films. Prior to investigating these spatially inhomogeneous systems, it is important to fully understand the properties of the bulk materials. $Pb(Zr_{1-x}Ti_x)O_3$ (PZT) is one such important system. Therefore, the other end member of this system, PbZrO₃ has been investigated. Again, good agreement is found with experiment and theory, seen in Figure 1. As with PbTiO₃, the electronic states are brought closer to the Fermi level by the hybridisation with Pb. Zr is directly below Ti in the periodic table but has a larger radius and denser ion core, resulting in a pronounced effect on the fine structure, agreeing with other investigations [7]. Building on these results, an EELS investigation of PZT thin films is now in progress.

Acknowledgements

The authors would like to thank EPSRC for a doctoral training award for PH.

References

[1]   Cohen R E 1992 Nature 358 136-138
[2]   Zhong W, Vanderbilt D and Rabe K M 1995 Phys. Rev. B 52 6301-6312
[3]   Ankudinov A L, Ravel B, Rehr J J, Conradson S D 1998 Phys. Rev. B 58 7565-7576
[4]   Blaha P, Schwarz K and Luitz J 1999 WIEN97, FLAPW Package for Calculating Crystal Properties (updated version of Blaha P, Schwarz K, Sorantin P and Trickey S B 1990 Comput. Phys. Commun. 59 399-415)
[5]   Day V W, Eberspacher T A, Frey M H, Klemperer W G, Liang S and Payne D A 1996 Chem. Mater. 8 330-332
[6]   Mitchell R H 2002 Perovskites: Modern and Ancient (Ontario: Almaz Press Inc)
[7]   Craven A J, 1995 J. Micros. 180 250-262

Inst. Phys. Conf. Ser. No 179: Section 2
Paper presented at Electron Microscopy and Analysis Group Conf. EMAG2003, Oxford, 2003
©*2003 IOP Publishing Ltd*

Deformation in GaAs under Nanoindentation

F Giuliani, S J Lloyd, L J Vandeperre and W J Clegg

Department of Materials Science and Metallurgy,
University of Cambridge, Pembroke Street, Cambridge, CB2 3QZ

Abstract. Compound semiconductors, such as GaAs, have a similar hardness to Si, even though their flow stresses (as a proportion of their shear modulus) are lower. Previous observations under nanoindents in Si have shown that deformation occurs by a phase transformation while deformation in GaAs is dominated by twinning [1]. In this paper we have studied the influence of temperature on the deformation of GaAs under a Berkovich indenter, using an atomic force microscope and a transmission electron microscope to characterise the deformation behaviour. Electron transparent cross-sections through the deformed region were prepared using a focused ion beam. It is shown that at low temperatures, the hardness impression is accommodated within the material, deformation occurring predominantly by twinning. At higher temperatures (300 °C) pile up around the indent is observed with deformation occurring predominantly by dislocation motion.

1. Introduction

During indentation experiments material displaced by the indenter must be accommodated either within the bulk of the material or by piling up around the indent. It has been proposed by Marsh [2] that these different flow mechanisms alter the relationship between hardness, H, and yield stress, Y, during indentation and that the dominant accommodation mechanism is determined by the ratio of, Y, to elastic modulus, E. Where the ratio of Y/E is low, pile up is generally observed and the value of H/Y is ≈ 3 [3] while when Y/E increases, little pile up is observed, and the ratio of H/Y approaches 1 [2]. Previous measurements by Suzuki *et al* [4] have shown a significant drop in flow stress of GaAs as the temperature is increased from room temperature to 500 °C, while the elastic modulus is expected to change little over this temperature range. This allows a large change in the ratio of Y/E on heating. GaAs has therefore been indented over this range of temperature and any changes in deformation that occurs on the surface and under the indent have been studied.

2. Experimental Methods

Measurements of the hardness of GaAs were made by indenting the (001) surface with loads ranging from 2-100 mN using a Nanotest 600 machine (Micro Materials, Wrexham, UK) with a Berkovich diamond indenter. For measurements at elevated temperature the specimen was mounted on a heated stage, allowing indents to be made

124

at temperatures up to 500 °C. The indenter was heated by holding it against the sample surface for 2 minutes. Indents were made in a line and spaced 20 μm apart to ensure their plastic zones did not overlap. The surface topography around the indents was observed using an AFM (Nanoscope III). To analyse the deformation under the indents cross-sectional TEM samples were produced using a focused ion beam (FEI FIB200) workstation. The liftout technique, as outlined by Langford [5], was used to machine section through the indents. This was preferred to the trench technique as the substrate is left intact and processing time in the FIB is reduced. This technique produced a final thin area of approximately 100-200 nm thick and 20 μm square. The thin window was detached by tilting the substrate by 45° to allow the bottom and sides to be cut. The substrate was then removed from the FIB and the thin area could be lifted out and placed on a fine meshed Cu grid. This was achieved using a micromanipulator and a long working distance microscope. TEM was performed on a Philips CM30 microscope operating at 300 kV.

3.   Results

The results of hardness measurements for a range of temperatures are shown in Fig.1 and are compared with the microindentation results from Gridneva *et al* [6] and flow stress measurements by Suzuki *et al* [4]. Good agreement is seen with the nanoindentation results here and micro indentation experiments [6]. If these hardness results are compared with the flow stress measurements [4], Fig.1, then at room temperature the value of $H / Y \approx 6$, but as the temperature is increased this rises rapidly to ≈ 300. The high temperature value is much greater than predicted and may arise due to the differences in the strain rate between the hardness and flow stress measurements.

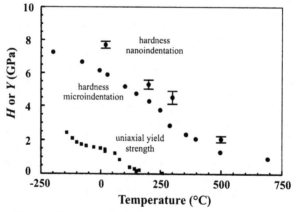

Figure 1 Hardness, *H*, obtained by nanoindentation here compared with microindentation data [6]. Also shown is the uniaxial yield strength [4], *Y*, taken as twice the shear flow stress.

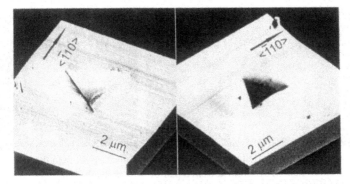

Figure 2 3-dimensional impression of AFM images of the surface displacements around (a) a 30 mN indent made at room temperature, and (b) a 10 mN indent made at 300 °C. The loads were selected to obtain indents of a similar size.

Fig.2a shows an indent formed at room temperature. There is little pile up around the edge of the indent, suggesting that the majority of the material displaced by the indenter has been accommodated within the bulk of the material. This is consistent with observations by Marsh in materials which have a high ratio of Y / E [2]. However at 300 °C, Fig.2b, the volume of material displaced by the indenter is very similar to the volume of material piled up around the indent.

The predominant mechanism of deformation below the indent at room temperature in Fig.3a, is twinning with the twins lying on the (111) planes as shown in the diffraction pattern Fig.3a. Fig.3b is a dark field image formed from a pair of twin spots and shows the twins extended to the full depth of the deformed zone. They can also be seen to extend the complete length of the contact area with the greater number being associated with the steeper side of the indent. An approximation of the amount of vertical displacement accommodated by twins as opposed to dislocation motion can be made by assuming that the twins have an average thickness of 20 nm and shear at an angle of 35.3°. The 25 visible twins account for 290 nm of vertical displacement, which is ≈ 0.8 of the total vertical plastic depth. The depth of the plastic zone is approximately half of the width as predicted by Marsh who assumed the plastic zone is hemispherical [2].

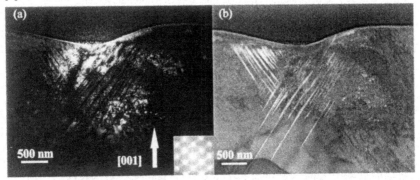

Figure 3 (a) Bright field image and (b) dark field image of a 50 mN indent made at room temperature.

126

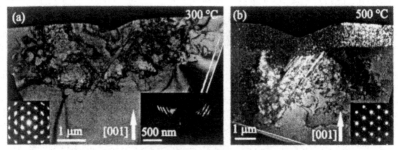

Figure 4 (a) Indent at 300 °C, 16 mN load, bright field the inset is a dark field image formed from a pair of twin spots. (b) Dark field image of indent at 500 °C, 5 mN load, formed from a twin spot.

At 300 °C, the shape of the plastic zone has been elongated horizontally, see Fig.4a, compared to Fig.3. This is consistent with the deformation patterns predicted for soft materials. Some twins are still found close to the indent impression but there are fewer than at room temperature and they do not extend so far into the sample. Twinning is still observed but is not the dominant mechanism. In this case the plastic zone contains many dislocations, which are visible around the periphery of the deformed area. The diffraction pattern in Fig.4a taken from below the point of the indent, shows far weaker twin spots and with significant asterism consistent with a greater proportion of deformation by dislocation flow causing local misorientation of the lattice. Note that the twins, see dark field inset in Fig.4.a, are curved suggesting that they were formed in the early stages of deformation and have later been bent. The twinning is possibly due to initial local cooling by the indenter. Further high temperature indentation experiments on GaAs have been carried out at 500 °C but oxidation obscures a large portion of the deformed area Fig.4.b. At these temperatures some of the defects associated with the deformation are likely to have been annealed out.

4.   Conclusions

The material displaced by the indenter in GaAs is accommodated within the bulk at room temperature but is pushed out of the surface at higher temperatures (and lower Y/E). At room temperature deformation around the indent is primarily by twinning, whereas at 300 °C deformation occurs mainly by dislocation movement.

5.   Acknowledgements

The authors would like to thank the DTI and the Royal Society for financial support.

6.   References

[1] Lloyd S J, Molina J M and Clegg W J 2001 J. Mater. Res. **16**. 3347-3350.
[2] Marsh D M 1963 Proc. R. Soc. A **279** 420-435.
[3] Tabor D, The Hardness of Metals, Oxford University Press, 1951.
[4] Suzuki T, Yasutomi T, Tokuoka T 1999 Phil. Mag. A **79** 2637-2654.
[5] Langford R M and Pettford-Long A K 2001 J. Vac. Sci. Technol. A **19** 2186-2193.
[6] Gridneva I V, Milman Y V and Trefilov VI 1972 Phys. Stat. Sol. (a) **14** 177-182.

Inst. Phys. Conf. Ser. No 179: Section 2
Paper presented at Electron Microscopy and Analysis Group Conf. EMAG2003, Oxford, 2003
©2003 IOP Publishing Ltd

Characterisation of C supported Pt nano-particles using HREM

D Ozkaya, D Thompsett G Goodlet, S Spratt, P Ash, and D Boyd

Johnson Matthey Tech. Centre Blount's Court, Sonning Common,
Reading RG4 9NH, UK

Abstract: C supported Pt nano-particles are studied in order to characterise their structural shape and their interactions with the support as well as size and distributions using high resolution transmission electron microscopy. High-resolution in-situ observations of effects of electron beam damage on particles have also been carried out. The particles were analysed in an edge-on configuration. Most of them have surface structures consistent with three-dimensional cuboctahedron although upon alloying, the structure can change to a disk-like form. Extended exposure to electron beam can cause particles to coalesce in some support/heat treatment configurations. This behaviour requires more investigation to understand particle-support bond and dynamic interactions and their correlation to catalytic activity.

1. Introduction

Platinum group metal (PGM) based catalysts are crucial in the development of environmentally friendly technology. They are used in a number of applications such as proton exchange membrane fuel cells (PEMFC) where continuous improvements in performance have made them a very strong candidate to replace internal combustion engines (Ralph and Hogarth, 2002). The fact that these catalysts are dispersed on supports in the nano-scale means that electron microscopy has a major input to make in the development and property control of these catalysts.

High-resolution electron microscopy so far has been the major contributor to the development and the analysis of these particles and their interaction with a whole host of supports (Bernal et al, 1998, Yacaman et al, 1995, Datye, 2003). Various types of carbon with 20-100nm conglomerated graphite balls are the most popular of PGM catalyst supports (Frenkel et al 2001). The support itself has a very important role in that it helps to disperse the catalyst evenly and also keep them as small nano-particles during activity. Thus, the way these nano-particles bond with the support is crucial to the performance of the catalyst.

The size and the exposed orientation of surfaces of a particle affect the activity and selectivity of a catalyst. In theory, it should be possible to change the selectivity of a catalyst by changing the exposed surfaces of a catalyst. The tailoring of catalyst surfaces in this way could provide another dimension to the more controllable parameters such as size and distribution.

A number of ways of observing particles in the support have been used in the past. Two of the methods are illustrated in figure 1. The most popular one is the plan-view where the particle imaged is superposed on the support. This complicates interpretation of the image as thickness of the support fluctuates in different parts and it is difficult to judge whether a particle is spherical or disk-like on the support. The other method of observation is edge-on observation of a particle that is more difficult due to limited number of the particles in this configuration, especially if also a specific diffracting tilt condition is required for HREM imaging.

In this paper we have used the edge-on technique to investigate Pt and PtCr catalyst particles on carbon black medium subjected to a range of heat treatments. We have investigated the structural shape, size and stability of the particles in relation to the heat treatments they received.

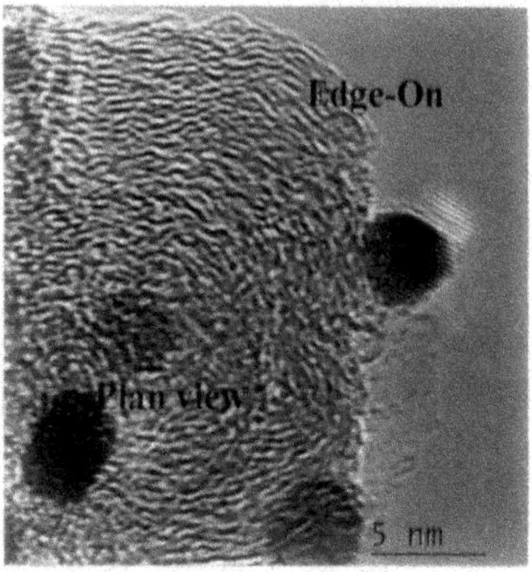

Figure1 Platinum particles on Carbon showing edge-on and plan view particle positions.

2. Experimental

40 wt %Pt and 40 wt % PtCr (4wt %Cr) on C black samples were made using established techniques explained elsewhere (Ralph and Hogarth, 2002)

All samples were in powder form and a small amount was crushed between two glass slides and a holey carbon coated copper TEM grid was dusted with the powder. The samples were examined in a Tecnai F20 Field emission TEM/STEM operated at 200kV with a lattice resolution of 0.14nm. The microscope has an EDX detector attached and is capable of acquiring high angle annular dark field images with a point-to-point resolution of around 0.24nm. The images were analysed using Gatan Digital Micrograph software.

3. Results

Figure 2 shows an HREM image of two Pt particles in different orientations. Both particles are edge-on and they show HREM lattice fringes that can be analysed. The particle shapes although seemingly rounded, show specific surface orientations and they are consistent with a cuboctahedron shape in line with previous observations (Yacaman et al, 1995).

It is possible to suggest some models for the surfaces based on the projection, contrast features and the lattice distances readily available in the HREM image. The lattice spacing for the first particle is d=0.198nm ($d_{Pt\ 001}$ =0.1961nm) and for the second, it is d=0.22 ($d_{Pt\ 111}$ =0.226nm). Note that uncertainty of measurements of lattice distances from nano-particles can be as much as 11% (Malm and O'Keffee, 1997).

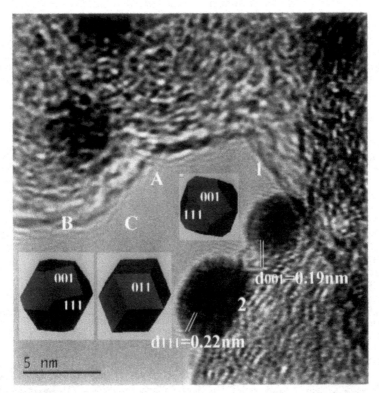

Figure2 HREM image of two platinum particles and possible models based on the lattice fringes and the electron beam directions. The shape A and B are cuboctahedron (001 and 111 faces) in two different orientations and C is formed of 110 faces.

For the second particle two possible models have been included in the image. It could be either a tilted version of a cuboctahedron, which consists of 001 and 111 planes (figure2B), or it could be a prism formed of 110 faces (figure 2C). It is not possible to make a conclusive judgement without HREM simulations but it is essential that we develop methods to make this distinction because this directly affects the selectivity of a catalyst.

Figure 3 shows a PtCr particle heat-treated at 900°C. The <111> type lattice fringes can be seen all through the sample and the intensity distribution indicates that this particle has wet the surface and looks more disk like on the surface of the support.

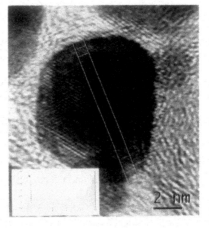

Figure3 A disk-like PtCr particle and the intensity profile (inset) when heat-treated at 900°C.

130

Figure 4 shows two time series images during electron beam exposure taken over a period of 20 min showing the difference in coalescence behaviour depending on the heat-treatment. In figure 4a the particles coalesce and in figure 4b the particles stay intact (although there is some movement due to damage to the carbon support). This implies that particles, having reached a critical mass, do not find coalescing as energetically favourable.

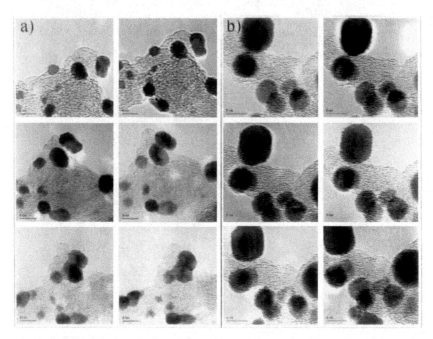

Figure 4 HREM time series images from two samples: Pt particles as received (A) and after heat-treated at 900°C (B) for a period of 20mins. Note how the particles in A are coalescing whereas in B particles are intact (the movement is due to Carbon).

4. Conclusion

An HREM study of Pt particles in C support has been carried out. Specific attention has been drawn to the importance of understanding surface orientations in three dimensions as this forms the basis of a catalysts selectivity and activity. An in-situ study of electron beam exposure highlighted the importance of support-particle interactions in two separate heat treatment conditions and the need for further investigations.

References

S Bernal F J Botana JJ Calvino Clopez-Cartes JA Perez-Omil JM Rodriguez-Izquierdo 1998 Ultramicroscopy **72** 135

A K Datye 2003 Journal of Catalysis **216** 144

A Frenkel CW Hills RG Nuzzo 2001 J. Phys.Chem.B **105 (51)** 689

J O Malm and M A O'Keefe 1997 Ultramicroscopy **13** 23

T R Ralph and M P Hogarth 2002 Platinum metals Rev. **46 (1)** 3

M J Yacaman G Diaz, A Gomez 1995 Catalyst Today **23** 161

Inst. Phys. Conf. Ser. No 179: Section 2
Paper presented at Electron Microscopy and Analysis Group Conf. EMAG2003, Oxford, 2003
©2003 IOP Publishing Ltd

Indium fluctuations analysis inside InGaN quantum wells by Scanning Transmission Electron Microsocopy

A M Sanchez[1], M Gass[1], A J Papworth[1], P Ruterana[2], H K Cho[3], and P J Goodhew[1]

[1] Department of Engineering, Materials Science & Engineering, University of Liverpool, Liverpool L69 3GH, United Kingdom
[2] LERMAT-FRE 2149 CNRS, Institut des Sciences de la Matiere et du Rayonnement, 14050 Caen Cedex, France
[3] Department of Metallurgical Engineering, Dong-A University, Hadan-2-Dong 840, Saha-gu, Busan, 604-714, Korea

ABSTRACT: Energy Dispersive X-Ray and Parallel Electron Energy Loss Spectroscopy analyses have been carried out to determine the indium composition fluctuations in InGaN/GaN quantum wells grown by Metal Organic Chemical Vapor Deposition. Information about the chemical composition and indium segregation in the InGaN/GaN system has been obtained. The compositional distribution is not homogeneous along the quantum wells, and indium rich clusters are observed inside the quantum wells.

1. INTRODUCTION

Over the last few decades III-N semiconductor materials have been subject of intense research activity. The development in the field of nitride based III-V materials is due to the wide range of practical applications of these semiconductors. III-N heterostructures are of great interest for optoelectronic devices, particularly with short wavelength, and in transistors (Nakamura et al. 1997 and Pearton et al. 2000). InGaN/GaN quantum wells constitute the active structure of the light emitting diodes (LEDs) and laser diodes (LDs) fabricated in these materials. However, the miscibility between InN and GaN is quite poor and the presence of In rich nanometer islands has been detected inside the InGaN quantum wells, presenting chemical inhomogeneity (Narukawa et al. 1997).

Detailed analysis to characterize the indium clustering has been carried out using quantitave High Resolution Transmission Electron Microscopy (HRTEM) images. This method allows the measurement of the strain due to the lattice mismatch in the heterostructures (Ruterana et al. 2002). Energy filtered TEM (EFTEM) has also been used to reveal the indium concentration in InGaN/GaN MQWs (Cho et al. 2001).

In this work we analyse the indium composition fluctuations in InGaN/GaN quantum wells in samples grown by Metal Organic Chemical Vapor Deposition (MOCVD). Differences in the compositional distribution inside the quantum well have been observed using energy dispersive x-ray analysis (EDX) and spectrum imaging (SI) parallel electron energy loss spectroscopy (PEELS).

2. EXPERIMENTAL

The growth of the InGaN/GaN multi-quantum well (MQW) LED was carried out by MOCVD on 2 μm n-GaN buffer layer on c-plane sapphire substrates with a nominal 25 nm thick GaN nucleation layer, and a p-type GaN capping layer.

Cross-sectional specimens were thinned down to 100 μm by mechanical grinding and next dimpled to below 10 μm. Electron transparency was achieved using a Precision Ion Polishing System (PIPS) at 5 kV. A final stage at 3kV was used to reduce ion beam damage. The analysis was carried out in a VG HB601 UX scanning transmission electron miroscope (FEG-STEM) working at 100kV. This microscope is equipped with an energy dispersive x-ray (EDX) spectrometer (Oxford Instrument) and a Gatan Enfina system for spectrum imaging (SI) parallel electron energy loss spectroscopy (PEELS).

3. RESULTS AND DISCUSSION

The presence of In rich clusters can be determined by conventional transmission electron microscopy (CTEM). The chemical inhomogeneity is shown through dark contrast corresponding to the In rich islands inside the InGaN quantum wells. HRTEM has been previously used to analyze the structure at nanometer scale. Information about the chemical composition, diffusion and segregation in the InGaN/GaN system can be obtained from these high resolution micrographs. The inhomogeneity inside the InGaN quantum wells can be determined using STEM and associated techniques such as EDX and PEELS.

The sample was aligned with $<11\bar{2}0>$ axes parallel to the electron optic axis. Figure 1(a) shows an annular dark field image. The InGaN/GaN MQW heterostructure with five quantum wells can be observed. EDX spectra, linescans and map analysis were performed to highlight the chemical compositional variations inside the MQWs.

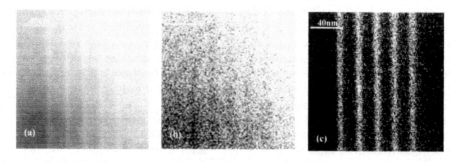

Figure 1. (a) STEM annular dark field image of five InGaN/GaN MQWs. EDX maps of **(b)** Ga **(c)** In

Figures 1(b) and (c) correspond to EDX maps of Ga and In respectively. A detailed analysis of these maps shows a non-homogeneous In distribution along the quantum wells. This inhomogeneous In distribution matched with the Ga depletion inside the wells. A homogeneous distribution of N has been observed in this heterostructure, in agreement with the assumption that the nitrogen composition is the same in GaN and InGaN (Sharma et al. 2000).

EELS measurements were undertaken in the same sample in order to confirm the In segregation inside the InGaN quantum wells. The EELS analyses have been carried out in areas of 43×28 nm^2. These spectra were recorded in an energy range between 40 and 180 eV to obtain compositional information in the heterostructure from the EELS core loss. In this case we analyzed the Ga distribution by EELS, since segregation of In matches the Ga depletion inside the well. The Ga $M_{2,3}$-edge occurs at 103 eV, which can be used to map the compositional fluctuations through the QWs. These spectra were normalized in order to remove the thickness effect. Figure 2 shows the superposition of a STEM bright field image and the normalized Ga core loss mapping. The intensity in the Ga core loss edge changes along the quantum well. A detailed analysis of this EELS mapping shows that inside the InGaN quantum well there are areas where similar intensity values as in GaN can be found. The Ga composition (and the In distribution, in the same way) is not uniform in the quantum well. The mapping clearly shows the minima (dark grey), that can be attributed to indium rich clusters and maxima (light grey), corresponding to areas with a poor In concentration.

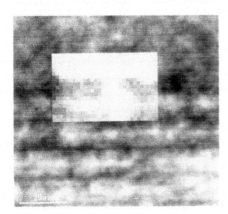

Figure 2. Superposition between STEM bright field and EELS mapping using the Ga $M_{2,3}$-edge .

In order to confirm this observation, EELS measurements with subnanometric resolution have been carried out inside the quantum well along a distance of about 75 nm. Figure 3 shows a bright field image corresponding to one of the quantum wells. EELS spectra were recorded along the line in the same figure. Again, the spectra were normalized to avoid the thickness effect. The graph, superposed onto the bright field image in Figure 3, corresponds to the Ga edge intensity along the spectra line. The different Ga distributions along this quantum well can be determined with higher accuracy than in the EELS mapping, because of better signal statistics. In Figure 3, maxima (e.g. a) and minima (e.g. b) in the Ga intensity can be observed, which can be related to quantum well regions, which are In-poor and In-rich. Further work is in progress to make these observations more quantitative.

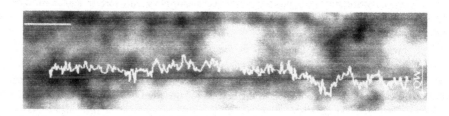

Figure 3. Superposition between STEM bright field and EELS spectra line using the Ga $M_{2,3}$-edge.

4. CONCLUSIONS

We have investigated the indium distribution inside the InGaN quantum wells grown by MOCVD. The EDX analysis has shown chemical compositional variations inside the MQWs. The Ga depletion inside the wells matches with the indium segregation in the analyzed areas while the N distribution is homogeneous. EELS measurements, carried out in areas of 43×28 nm^2 and inside the quantum well, clearly shows the minima, that can be attributed to indium rich clusters and maxima, corresponding to areas with a poor In concentration.

ACKNOWLEDGEMENTS

The authors acknowledge EPSRC for support of the NW STEM. AMS acknowledges the support of "Ministerio de Educacion, Cultura y Deporte", Spain under the postdoctoral grant EX2002-0807.

REFERENCES

Cho HK, Lee JY, Sharma N, Humphreys CJ, Yang GM, Kim CS, Song JH and Yu PW, 2001 Appl. Phys. Lett. **79**, 2594

Nakamura S, Sench M, Nagahama S, Iwasa N, Yamada T, Matsushita T, Sugimoto Y and Kiyoky H, 1997 Appl. Phys. Lett. **70**, 868

Narukawa Y, Kawakami Y, Funato M, Fujita S and Nakamura S, 1997 Appl. Phys. Lett. **70**, 981

Pearton SJ, Ren F, Zhang AP and Lee KP, 2000 Mat. Sci. Eng. R **30**, 55

Ruterana P, Kret S, Vivet A, Maciejewski G and Dluzewski P, 2002 J. Appl. Phys. **91**, 8979

Sharma N, Thomas P, Tricker D and Humphreys C, 2000 Appl. Phys. Lett. **77**, 1274

Inst. Phys. Conf. Ser. No 179: Section 2
Paper presented at Electron Microscopy and Analysis Group Conf. EMAG2003, Oxford, 2003
©2003 IOP Publishing Ltd

Cathodoluminescence spectral mapping of selectively grown III-nitride structures

R W Martin[1], P R Edwards[1], C Liu[2], C J Deatcher[2], H M H Chong[3], R M De La Rue[3], and I M Watson[2]

[1]Department of Physics, University of Strathclyde, 107 Rottenrow, Glasgow G4 0NG, UK

[2]Institute of Photonics, University of Strathclyde, 106 Rottenrow, Glasgow G4 0NW, UK

[3]Department of Electronics & Electrical Engineering, Glasgow University, Glasgow, G12 8LT, UK

Abstract. An array of pyramids containing templated InGaN/GaN quantum wells have been fabricated using selective overgrowth above patterned silica masks and studied using cathodoluminescence hyperspectral imaging. The cathodoluminescence reveals bright luminescence at the peaks of the micropyramids, red shifted from that due to similar, conventionally grown planar quantum wells. The possibility that this emission is due to quantum dots is discussed.

1. Introduction

Lateral overgrowth of GaN-based materials using patterned silica masks has enabled significant advances in material quality, through dramatic reductions in defect density. Selective growth, upon which lateral overgrowth depends, also allows the design of novel structures, including arrays of hexagonal pyramids or prisms, if growth is stopped before features formed in the earliest growth stages have coalesced. Incorporation of InGaN quantum wells in the final stages of pyramid growth can lead to the formation of quantum dot arrays [1]. Selective growth also has important applications in the fabrication of field-emitter arrays and III-N microcavity devices [2,3]. Microring and microdisk devices have previously been fabricated by lithography and etching [4] but the selective growth approach potentially produces more efficient cavities due to smoother facet walls, untainted by process damage.

In this paper we present results from arrays of selectively grown structures studied using an electron probe micro-analyser (EPMA). GaN is grown through an array of micron-scale holes in silica masks, resulting in arrays of mesa structures whose geometry reflects the hexagonal symmetry of wurtzite-phase GaN. Using growth steps similar to those used to produce conventional planar InGaN quantum wells complex low-dimensional structures can be produced through the templated growth of InGaN / GaN quantum wells. The topography and crystallographic faceting of the resulting structures are imaged using secondary electrons in the EPMA. This instrument has been modified to allow full spectral cathodoluminescence (CL) mapping using the optical microscope built into the electron column [5,6]. A spectrograph and cooled CCD array allow the rapid collection of a complete room temperature CL spectrum at each pixel in maps generated either by scanning the electron beam or stepping the sample stage. The best spatial resolution of the resulting CL maps is approximately 100 nm, determined by a combination of the volume probed by the exciting electrons and the scan parameters. The CL spectra have a maximum spectral resolution of 0.2 nm and typical acquisition times are 100's of ms per pixel. CL spectral mapping of regions typically 10 by 20 microns generates large 3-D data sets. These hyperspectral images are then used to generate various 2-D images (e.g. integrated intensity, peak wavelength, intensity within a wavelength band, chromaticity) and compared with the composition or topography maps acquired at the same time. They are also suitable for analysis using numerical methods such as principal component analysis.

The CL spectra show marked variations across the selectively grown structures and the images generated from the CL maps provide details of the modifications to the InGaN quantum wells generated by templated growth.

2. Selectively grown structures

The structures were grown in an Aixtron metalorganic vapour phase epitaxy reactor using (0001)-oriented GaN-on-sapphire seed layers covered with silica mask layers. The 100 nm thick mask layers were patterned into arrays of holes by lithography and dry etching. This paper describes structures grown using circular holes with diameters in the range of 5 - 8 μm and patterned into hexagonal arrays, with a pitch of 6 - 10 μm. Overgrowth of GaN results in arrays of 3D GaN structures as shown by the secondary electron images in Fig. 1. The flat-topped structures on the left result from early termination of the GaN overgrowth. Extending the overgrowth time results in an array of hexagonal pyramids as shown for the sample on the right. The six facets of each pyramid have formed naturally due to the symmetry of the GaN and are extremely smooth. One or more InGaN quantum wells are grown near the end of the overgrowth run using steps similar to those used to produce conventional planar InGaN quantum wells. We describe different structures with one and five InGaN wells with parameters expected to give 440 nm emission when grown conventionally. The quantum wells are clearly modified by the 3D-growth mode and complicated arrays of low-dimensional InGaN structures will result.

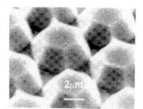 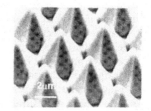

Figure 1. Secondary electron images of selectively grown GaN pyramids terminated at different stages in the growth.

3. Experimental results

The CL map from the flat-topped pyramids shown in Fig. 1 (left) showed that the most intense luminescence was concentrated within an orange band centred at 550nm, strongest on the sloping side-walls. In addition a blue emission band, centred at 430 nm, was also present in the emission from the flat tops. This structure contained a single InGaN/GaN quantum well and the blue emission corresponds to that expected for a conventional planar InGaN/GaN quantum well.

Extending the overgrowth time resulted in a dense array of sharper topped pyramids as shown in Figs. 1 (right) and Fig. 2. The region shown includes pyramids with a variety of apices. A 40 × 40 μm area was mapped using 200 × 200 pixels and a 5 keV electron beam with a current of 4 nA. Data from the room temperature CL hyperspectral image are shown in Fig. 2. The sharper tops can be seen to correspond to localised regions of higher intensity than their surroundings, with a single peak centred at a

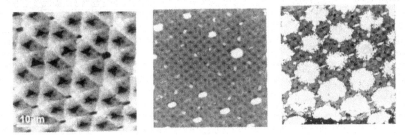

Figure 2. Left: Plan-view secondary electron image of the structures shown on the right of Fig. 1. Middle: Data from the RT CL map are shown by plotting integrated intensity (log scale from 10^3-10^6) and Right: Peak wavelength (black = 350 nm, white = 500 nm)

wavelength in the region 470-480 nm. The flatter topped peaks emit more intense luminescence centred at shorter wavelength (~450 nm), close to that expected for a planar well. This red-shift is most likely related to a thickening of the quantum well and possible quantum dot formation (as discussed later) at the apex of the sharp peaks but could also be due to an increase of the InN fraction. The CL intensity is clearly concentrated at the tips of the structures (and also at ridges between coalesced pyramids) although it should be noted that the CL is collected from above the pyramids, out to an angle of ~23° to the normal and some side-wall light may be lost.

The overgrowth of a similar structure using a new mask and incorporating five InGaN/GaN quantum wells produced sharp-tipped pyramids of improved quality and a further enhancement of the apex luminescence. Fig. 3 shows a plan-view secondary electron image taken with a 5 keV electron beam. The new mask has allowed growth of

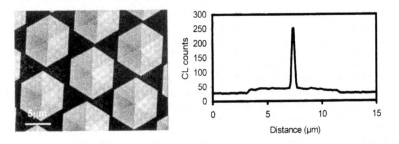

Figure 3. Left: Plan-view secondary electron image of the 5-quantum well structure. Right: CL intensity linescan across pyramid apex.

138

fully developed but separated pyramids (the dark regions are the silica mask). Data from a room temperature CL hyperspectral image are shown in Fig. 4, representing a 15 × 15 μm area mapped using a 5 keV, 500 pA electron beam. Increased brightness compared to the previous sample permits lower current and shorter acquisition times for the CL. The images of the CL intensity within the wavelength bands 410 – 450 nm and 500 – 540 nm are shown along with representative spectra. A single blue luminescence peak is now collected from the pyramid sidewalls whilst an intense green emission is localised at the peak, which shows a reduced "blue" emission. A linescan of the 500 – 540 nm CL

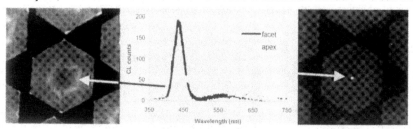

Figure 4. 15 × 15 μm images of the CL intensity within the wavelength bands 410 – 450 nm (left) and 500 – 540 nm (right) for the 5-quantum well structure. Representative CL spectra are shown.

intensity across the apex of the pyramid (Fig. 3) shows that the bright luminescence to be restricted to a region of less than 300 nm. This is clearer evidence of the formation of quantum dots at the tip of the pyramid [1], although we can not rule out contributions due to light guiding effects and/or InN concentration. The suggestion that quantum dots might emit at lower energies than the parent quantum wells may seem surprising. However the critical dimension of the dots will be larger than the width of the wells, leading to the possibility of lower confinement, and in this case there will also be a stronger red-shift due to the intense in-built electric fields within strained InGaN structures.

4. Conclusion

We describe the fabrication of arrays of pyramids containing templated InGaN/GaN quantum wells using selective overgrowth. The resulting structures have been studied using cathodoluminescence hyperspectral imaging. This reveals bright luminescence concentrated at the peaks of the micropyramids, which is red shifted from that due to similar conventionally grown planar quantum wells. The effect of in-built fields within InGaN quantum dots formed at the pyramid apices could account for this luminescence.

References
[1]   Tachibana K, Someya T, Ishida S, Arakawa Y, Appl. Phys. Lett. **76** 3213 (2000)
[2]   Jiang H X, Lin J Y, Zeng K C, Yang W, Appl. Phys. Lett. **75** 763 (1999)
[3]   Pritchard R E, et al., J. Appl. Phys. **90** 475 (2001)
[4]   Choi H W, Jeon C W, Dawson M D, Edwards P R, Martin R W and Tripathy S, J. Appl. Phys. **93** 5978 (2003)
[5]   Martin R W, Edwards P R, O'Donnell K P, Mackay E G, Watson I M, phys. stat. sol. (a) **192** 117 (2002)
[6]   Edwards P R, Martin R W, O'Donnell K P, Watson I M, phys. stat. sol. in press

Inst. Phys. Conf. Ser. No 179: Section 2
Paper presented at Electron Microscopy and Analysis Group Conf. EMAG2003, Oxford, 2003

Structural modelling of nano-amorphous Si-O-N films in silicon nitride ceramics

D Nguyen-Manh [1], D J H Cockayne [1], M Doeblinger [1], C Marsh [1], A P Sutton [1] and A C T van Duin [2]

[1] Department of Materials, University of Oxford, Parks Road, Oxford OX1 3PH, UK.

[2] Beckman Institute, California Institute of Technology, Pasadena, California, 91125, USA.

Abstract. In order to obtain a detailed computational description of the structure and fundamental properties of silicon oxynitride films as thin as 1nm in Si_3N_4 ceramics, a reliable and transferable many-body inter-atomic potential has been developed for Si-N-O systems and used to perform molecular dynamic (MD) simulations. The modelling results of radial distribution function (RDF) are well compared with experimental measured ones for the inter-granular glassy films (IGFs). The predicted variation of bond-order around Si atoms allow to compare with spatially resolved electron energy loss spectroscopy (SR-EELS) studies of compositional and chemical profiles of O and N atoms across IGFs.

1. Introduction

Understanding of the structure and fundamental properties of crystal/glass/crystal interface provides an important knowledge base for developing new materials, especially in the field of ceramics. There has been a great deal of speculation about the structure and composition of IGFs, but there are no firm experimental or theoretical results. The extremely small width of these glassy films will push back the limit of the electron spectroscopy techniques. At the same time, the principal challenge facing the modeling will be to treat the films at *ab-initio* level as an open system, capable of exchanging matter with reservoirs. In this work, we address the detailed structure and composition of Si-O-N films in silicon nitride ceramics from atomistic modeling. We note that $2x+3y \sim 4$ in the SiO_xN_y oxynitride films, well within spatial resolution EELS experimental measurements [1]. An "ionic" interpretation of this observation is that formal charges of +4, -2 and -3 are assigned respectively to the Si, O and N ions. Our density functional theory (DFT) calculations, however, reveal that Mulliken partial charges are much smaller and in fact covalent bonding is dominant in these systems. Therefore, a new many-body inter-atomic potential incorporating bond energy/order relation has been developed for structural modeling of Si-O-N system within reactive force field (RFF) scheme [2]. Computational structural properties and composition profile across the IGFs will be compared with corresponding RDF and SR-EELS experimental studies.

2. Theory and modelling of IGFs structure

The DFT calculations of total energy for various molecules and solids have been carried out within the package of linear combination of atomic-type orbitals (PLATO) [3] in order to create a large data base for developing RFF potentials in Si_3N_4-SiO_2 ceramics. Our main focus is paid for analysing bonding properties and Table 1 shows predicted data of bond energy and bond order for Si-N and Si-O bonds in different molecules and solids. It is found that bond-order values characterizing covalent strength of materials [4] are strong for all considered systems. This covalent character of bonding is unfortunately ignored in previous modelling studies in which the only ionic contribution has been taken into consideration for inter-atomic potential description of Si_3N_4-SiO_2 system (e.g. [5]).

Table 1. DFT data base for various Si-N and Si-O bonding properties

Si-N bond	Bond length ( )	Bond energy (eV)	Bond order
N(SiH3)3	1.741	-12.936	0.705
Si(NH2)4	1.735	-13.132	0.752
Si3N4	1.734	-13.484	0.778
Si2N2O	1.745	-12.493	0.697
Si-O bond	-	-	-
SiO2	1.629	-19.339	0.826
Si2N2O	1.625	-20.034	0.849

Within the RFF scheme [2], the system energy is given by

$$E_{system} = E_{bond} + E_{vdW} + E_{coulb} + E_{val} + E_{tor} + E_{over/under} + E_{pen} + E_{conj} \qquad (1)$$

where the first three terms are the two-body interactions representing bond, van de Walls and Coulomb interaction energies, respectively. The relationship between bond energy and bond order, Θ_{ij} between two atom i and j can be written as

$$E_{bond} = -D_\sigma \cdot \Theta_{ij}^\sigma \cdot \exp\{a[1-(\Theta_{ij}^\sigma)^b]\} - D_\pi \cdot \Theta_{ij}^\pi - D_{\pi\pi} \cdot \Theta_{ij}^{\pi\pi} - D_{\pi\pi\pi}\Theta_{ij}^{\pi\pi\pi} \qquad (2)$$

for σ, single, double and triple π bonds. E_{val} , E_{tor} , $E_{ober/under}$ in equation (1) represent three, four and multi-body interactions and finally E_{pen} ,E_{conj} describe the contributions of penalty and conjugation effects to the molecular energy. All RFF parameters are optimized by fitting with DFT and experimental available data to reproduce heat of formations, bond lengths and bond angles. We expect that the RFF will allow accurate dynamical descriptions of reactions and bond breaking processes in large silicon oxy-nitride systems.

The obtained RFF potential has been used to perform MD simulations for different crystal(Si_3N_4)/glassy(Si-O-N)/crystal(Si_3N_4) configurations. In this paper, our preliminary results for basal/glassy/basal configuration in which the basal orientation of β-Si_3N_4 is perpendicular to the IGFs interface. To form an amorphous layer in contact with the crystalline Si_3N_4, we carried out NVT MD option (constant volume and constant temperature using a Berendsen thermostat [6]) using different temperature regimes as follow. We set the temperature of the Si_3N_4 slab at 300 K with a temperature-damping constant of 10 fs and set the temperature of the IGFs phase at 1500-2500K with a temperature-damping constant of 50 fs. Next we use MD to cool the IGFs region down to 300K in 7.5-10 ps while keeping the β-Si_3N_4 region cooled at about 300K. Figure 1a

a)   b)

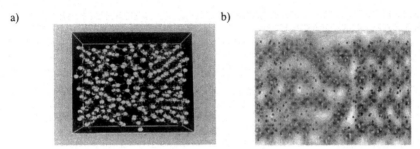

Figure 1. a) Structure of an IGFs model obtained by MD simulation; b) HREM image calculated from the same model.

shows a snapshot of MD simulated 848-atom model with the IGFs width of 1nm and average density of 2.4 g/cm^3 close to those of amorphous cristobalite SiO$_2$. The HREM image obtained from a double atom-unit cell averaged over the last 10 steps of MD simulation is showed in Fig. 1b. The image calculated from EMS software [7] indicates clearly a network of Si atoms (seen by dark field contrasts) within the IGFs of Si-O-N.

3. Radial distribution function of IGFs

In order to understand the role of composition profile within IGFs, different models with various N/(N+O) ratios have been investigated from MD simulations. Figure 2a shows the calculated RDF decomposed into partial contributions for a 496-atom model with the IGFs with of 1.1 nm and average density of 2.8 g/cm^3. The model has an average ratio N/(N+O)=0.312 and it is found as one of stable IGFs phases because there is almost absence of N$_2$ formation due to a very strong triple bond energy of nitrogen dimer obtained from equation (2). Our result is in an excellent agreement with the SR-EELS experimental data indicating that a stable IGF in Si$_3$N$_4$-SiO$_2$ composition corresponds to N/O ~ 0.5±0.1 [1]. The predicted RDFs peak positions are: Si-O (1NN) at 1.6 , Si-N (1NN) at 1.75 , O-O (1NN) at 2.7 , Si-Si (1NN) at 3.1 , Si-O (2NN) at 4.1 and O-O (2NN) at 4.8-5.2 . Our predictions agree well with experimental measurement of

a)   b)

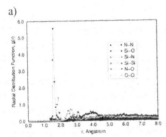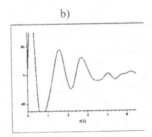

Figure 2. a) Theoretical partial RDFs of nano-amorphous Si-O-N films b) Experimental RDF of IGFs obtained from electron diffraction techniques

total RDF for un-doped IGFs (Fig. 2b) obtained from electron diffraction technique for nanovolume amorphous materials [8]. We note, however, that both Si-N and Si-Si peak position can not be distinguished from the first and second experimental RDFs ones,

142

respectively. Therefore, further theoretical and experimental studies are needed to probe the N concentration within IGFs.

4. Bond-order across IGFs

Within the RFF scheme, it is straightforward to analyse bonding properties in the annealed IGFs structure. Figure 3 shows decomposed bond order around Si atom in the bulk β-Si_3N_4 as well as inside the IGFs. It shows that the total bond order at Si is about 4 at each layer parallel to the IGFs indicating that there are no Si radicals in the system.

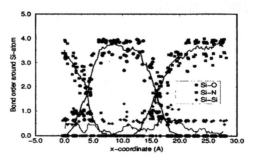

Figure 3. Variation of bond-order around Si atom across the IGFs.

While the average Si-O bond order increases from the bulk to middle of the IGFs, the corresponding value of Si-N bond gradually decreases at Si_3N_4-SiO_2 interfaces. The steep width of bond-order variation is approximately 1nn within our MD simulation. This behaviour is very similar to experimental studies of O and N profiles across the grain boundary films from SR-EELS counts at least in the pocket region of Si_3N_4 ceramics [1].

5. Conclusion

Structural and compositional properties of IGFs have been investigated by MD simulations using a new RFF potential developed for Si-N-O system. Our computational results are well compared with experimental RDFs and SR-EELS measurements. We find that covalent bonding character plays an important role in these systems and the bond-order notion is very useful for understanding interfacial properties of ceramics.

This research is supported by the EU/US collaborative project GRD2-200-30351.

References

[1] Gu H Cannon R M Seifert H J Hoffmann M J and Tanaka I 2002 J. Am. Ceram. Soc. 85, 25-32
[2] van Duin A C T Dasgupta S Lorant F and Goddard III W A 2001 J. Phys. Chem. A, 105, 9396-9409
[3] Nguyen-Manh D Kenny S Cockayne D J H and Pettifor D G 2003 Proceeding of American Physical Society, March 3-7, 48, B26.14, 187 (Austin, Texas)
[4] Pettifor D G 1995 Bonding and structure of molecules and solids (Oxford: Claredon)
[5] Yoshiya M Tanaka I and Adachi H 1998 Acta Mater. 48, 4641
[6] Berendsen H J C Postma J P M and van Gunsteren W F, J. Chem. Phys. 1984 81 3684
[7] Stadelmann P http://cimesg1.epfl.ch/CIOL/ems.html
[8] McBride W Cockayne D J H and D Nguyen-Manh, 2003 Ultramicroscopy 96, 191

Inst. Phys. Conf. Ser. No 179: Section 2
Paper presented at Electron Microscopy and Analysis Group Conf. EMAG2003, Oxford, 2003
©2003 IOP Publishing Ltd

High-resolution scanning electron microscopy of dopants in p-i-n junctions with quantum wells

Z Barkay[*], E Grunbaum[**], Y Shapira[**], P Wilshaw[##], K Barnham[#],
D B Bushnell[#], N J Ekins-Daukes[#], M Mazzer[#]

[*]Wolfson Applied Materials Research Centre, Tel-Aviv University, Israel
[**] Dept. of Physical-Electronics, Faculty of Engineering, Tel-Aviv University, Israel
[##] Dept. of Materials, University of Oxford, UK
[#] Dept. of Physics, Imperial College of Science, Technology and Medicine, London, UK

[**] Enrique Grunbaum's e-mail for correspondence: enrique@eng.tau.ac.il

ABSTRACT. We have studied the electric field distribution in p-i-n structures with multi-quantum wells (MQW) using the method of dopant (ionization potential) contrast in high resolution (with a cold field emission electron gun) scanning electron microscopy (HRSEM). The samples are GaAs-based ternary compound layer structures used for high-efficiency solar cells, consisting of *p-i-n* junctions, in which various numbers of InGaAs quantum wells have been inserted into the intrinsic (undoped) region. The results show an increasing secondary electron signal across the *n, i* and *p* regions while the series of 8 nm wide quantum wells and their corresponding barriers within the *i*-region are clearly distinguished. The field distribution is obtained by differentiating the dopant contrast curves. This study highlights the capability of HRSEM to provide information on active doping and associated electric fields within electronic nanostructures.

1. INTRODUCTION

The knowledge of dopant concentration and distribution in electron devices is of utmost importance. Amongst several methods of dopant detection and analysis, the observation of secondary electron contrast by high resolution scanning electron microscopy (HRSEM) is promising [1-4]. This signal, which depends partly on doping concentration can be detected with nanometer spatial resolution and needs very simple specimen preparation.

The change in internal energy across a p-n junction, or any junction with different dopant concentrations, gives rise to electrostatic "local (patch) fields" between the differently doped materials due to the redistribution of charge around a p-n junction. It is these electrostatic fields, external to the material, that result in the *n-* and *p*-type material having different ionizations energies, and hence give rise to secondary electron dopant contrast in the HRSEM. It has been shown that the n-doped regions produce weaker secondary electron emission (appear darker) than the p-type regions and that the secondary electron emission intensity in the p-regions is logarithmically proportional to the active dopant concentration [2]. However, if there is significant density of surface states the Fermi level can be pinned mid-gap over a range of relatively low doping concentrations and under these conditions the secondary electron signal, which is actually dependant on the ionization energy, will be independent of the actual doping concentration. It is this effect, which explains why, under some conditions, the doping contrast is insensitive to changes in doping for concentrations lower than $\sim 10^{16} \mathrm{cm}^{-3}$ [2].

2. Experimental

A high-resolution scanning electron microscope with a cold field emission electron gun, providing a highly coherent primary electron beam (JEOL 6700F) has been used, an in-lens secondary electron detector provides an efficient collector. The signal is displayed and measured using a line scan. Contamination of the specimen during observation is reduced to a minimum by using an air lock for its introduction into the UHV microscope chamber. The specimens are a 20-quantum well solar cell and a single-quantum well laser, grown by metallo-organic chemical vapor deposition (MOCVD) at Sheffield University. The layout of their layers is given in Table 1. Cross-sections are prepared by cleavage of the sample very shortly before its introduction into the HRSEM. Observations are made with a primary energy of 2 keV, to reduce the effect of specimen charging, and a primary beam current of 100 pA.

Table 1. Layer layout of the studied samples

Sample QT 1410R			Sample 432		
Layer	Thickness nm)	Doping (cm^{-3})	Layer	Thickness (nm)	Doping (cm^{-3})
GaAs	1000	$p\ 3 \times 10^{19}$	GaAs	400	$p\ 2 \times 10^{19}$
$Al_{0.8}GaAs$	43	$p > 5 \times 10^{18}$	$Al_{0.6}GaAs$	1500	$p\ 5 \times 10^{18}$
GaAs	200x2	$p\ 5 \times 10^{18}$			
GaAs	*10*	*i*	$Al_{0.3}GaAs$	*250*	*i*
GaAsP$_{0.06}$	*22.7x20*	*i*	$Al_{0.3}GaAs$	*70*	*i*
In$_{0.17}$GaAs	*8x20*	*i - QW*	*GaAs*	*10*	*i-QW*
GaAsP$_{0.06}$	*22.7x20*	*i*	$Al_{0.3}GaAs$	*70*	*i*
GaAs	*10*	*i*	$Al_{0.3}GaAs$	*250*	*i*
GaAs	3000	$n\ 2 \times 10^{17}$			
GaAs	300	$n\ 1.5 \times 10^{18}$	$Al_{0.6}GaAs$	1500	$p\ 5 \times 10^{18}$
GaAs	substrate	n^{+}	GaAs	substrate	n^{+}

3. RESULTS AND DISCUSSION

Figure 1 shows an increasing secondary electron signal between the n, i and p regions which are clearly distinguished by their different contrast. The secondary electron signal increases at the junction of the n-region with the intrinsic region and extends into the latter (about 5 QW periods). In the remaining part of the intrinsic region the signal appears to only slightly increase. At the $i/p+$ junction there is again a narrow region of a strongly increasing signal.

The band diagram, according to the theory of ionization potential contrast, is drawn in fig. 2. This theory states that the changes in the secondary electron signal correspond to those of the ionization potential, i.e., to changes in the position of the valence band edge. The diagram demonstrates the existence of an electrostatic field due to the built-in potential V_{bi} in the depletion region; the field is strongest at the n/i and i/p junctions.

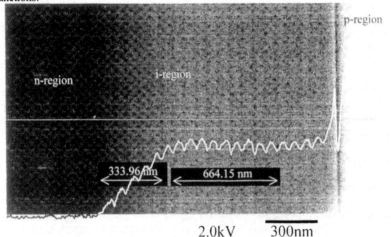

Fig. 1. HRSEM secondary electron line-scan micrograph of sample QT1410R

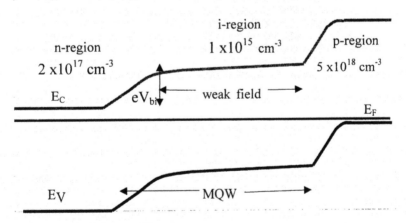

Fig. 2. Proposed band diagram of the *p-i-n* structure

However, for most of the width of the intrinsic region the secondary electron signal is unchanged. This can be explained by a) this being a truly field free region or alternatively b) by there being a small field present in the bulk layer which is absent at the surface where the doping contrast effect arises. The absence of the field at the surface may be due to the effect of surface states, pinning the Fermi level mid-gap in

146

this region. The latter explanation requires that the doping concentration in this region be lower than a few 10^{16} cm^{-3} and that the bulk field in the intrinsic region is small. This is consistent with measurements of the quantum efficiency (QE) for an applied positive voltage at 0.9 V, which show this to be the case.

The residual doping concentration and its type can be obtained by the measured width of the depletion region (1100 nm) and using equations for abrupt n-p junctions [6]: the residual doping is 1×10^{15}cm^{-3} (p type) and the built-in potential is 1.2 V. For comparison the HRSEM micrograph of a single QW laser sample (432) is shown (fig. 3). The continuously increasing contrast in the 650 nm wide intrinsic region shows that a strong field exists in the entire region.

In addition to the changes in contrast due to the relatively gradual bending of the valence band produced by the varying doping concentration, abrupt changes can also be seen due to the QWs themselves. This is because the change in semiconductor composition on the nanometer scale associated with these QWs leads to abrupt discontinuities in the valence band edge (and hence ionization potential) which are then imaged in the secondary electron signal.

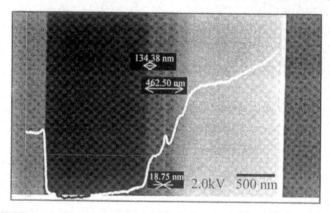

Fig. 3. HRSEM secondary electron line-scan micrograph of sample 432

4. CONCLUSIONS

The results are manifestation of the HRSEM capability to provide useful and easily obtainable information, from a variety of electronic and electro-optical devices, on:

a) The ionization potential and its distribution on a nanometer scale

b) Active dopant concentrations which produce changes in the ionization potential

c) The electric fields due to the built-in potential associated with the dopant distributions.

References

[1] M.R. Castell, D.A. Muller and P.M. Voyles, Nature Materials **2**, 129-131 (2003)

[2] C.P. Sealy, M.R. Castell and P.R. Wilshaw, J. Electron Microsc., **49**, 311 (2000)

[3] S.L. Elliott, R.F. Broom and C.J. Humphreys, J. Appl. Phys., **91**, 9116 (2002)

[4] D. Venables, H. Jain, D.C. Collins, J. Vac. Sci. Tech., **B16**, 262 (1998)

[5] K.W.J. Barnham et al., Physica E**14**, **27** (2002)

[6] S.M. Sze, *Physics of Semiconductor Devices*, 2nd ed. (John Wiley, New York, 1981)

Inst. Phys. Conf. Ser. No 179: Section 2
Paper presented at Electron Microscopy and Analysis Group Conf. EMAG2003, Oxford, 2003
©2003 IOP Publishing Ltd

SEM / EDXS and Polymers – A sociable couple?

P Poelt, F P Wenzl*⁾, E Ernst**⁾ and E Ingolic

Research Institute for Electron Microscopy, Graz University of Technology,
Steyrerg. 17, A-8010 Graz, Austria; *⁾ Institute for Solid State Physics, Graz
University of Technology, Petersgasse 16/II, A-8010 Graz, Austria;
**⁾ Borealis GmbH, St. Peter Str. 25, A-4021 Linz, Austria.

Abstract: Special care has to be taken by performing x-ray analysis of
polymers, because they are extremely prone to radiation damage. Whereas it
is possible to map the ion distribution in conjugated polymers blended with
solid state electrolytes, a sound oxygen analysis seems impossible for the
majority of different types of polymers. By use of EDXS it is additionally
possible to prove the occurrence of radiation damage even in cases, where
no change in the surface topography is visible.

1. Introduction

The SEM, especially the FESEM, is widely used for the investigation of polymers, but
mainly for imaging the surface topography, and rather rarely in connection with energy
dispersive x-ray spectrometry (EDXS). Imaging can be performed both at low electron
energies at around 1 keV and at low probe currents. Thus radiation damage can be kept
at a minimum.

In contrast to imaging, in the case of EDXS measurements the minimum electron
energy is determined by the energy of the x-ray lines used for the analysis, and in
general will be at least 3 – 5 keV. To get a useful x-ray yield, additionally the probe
current has to be rather high, especially if x-ray maps should be recorded. Thus, in many
cases specimen damage is nearly unavoidable. Cooling down bulk specimens to the
temperature of liquid nitrogen, which would reduce specimen damage, is in many cases
impossible.

EDXS analysis of polymers is interesting for many applications: Investigating the
different types of radiation damage and their influence on specimen properties;
Recording the phase distribution of blends, which are characterised by varying oxygen
contents of their phases, without the necessity of staining, which besides will not work
at specimen surfaces; Revealing the distribution of the ions in conjugated polymers,
blended with a solid state electrolyte, as used in light emitting cells.

2. Experimental

All measurements have been performed by use of a ThermoNORAN energy dispersive
x-ray spectrometer (Si(Li), 30 mm²) attached to a Zeiss DSM 982 Gemini SEM. To
avoid surface charging, the specimens have been coated with either a Cr-layer of 1 nm

148

or a C-layer of several nm thickness. The ThermoNORAN Voyager standardless quantification routine has been used in case of quantitative analysis, without any corrections for the coating and for the film thickness in case of thin polymer films.

3. Results and Discussion

3.1 Surface stability of PP compounds

Thermoplastic polyolefins based on polypropylene (PP) have extremely poor adhesion properties for waterborne paints. Therefore, the surface activity is generally enhanced by flaming the base material or treating it in an oxygen plasma. Investigations of failures at painted systems caused by peel forces demonstrated, that these failures could always be located in the polymer itself, in or below the modified surface layer, and not directly at the polymer - paint interface. Thus it is not only the degree of modification of the PP surface, that does directly determine the degree of paint adhesion.

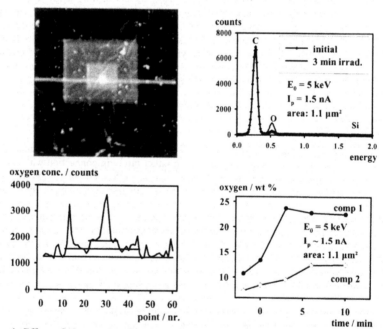

Fig. 1. Effect of electron irradiation on PP compounds; left: BSE-image (E_0 = 5 keV, image width: 114.7 µm) of compound 1 with oxygen linescan along the white line, after irradiation of the sample surface at varying magnifications (visible as areas of different brightness, higher brightness corresponds to longer irradiation times) for several minutes (E_0 = 5 keV, I_p ~ 1.5 nA); upper right: EDX-spectra of compound 1 prior to and after irradiation; lower right: change of the oxygen concentration at the surface as function of irradiation time for two different compounds.

In the present work instead of flaming or oxygen plasma treatment electron irradiation in vacuum was used for the surface modification of two PP compounds with different amounts of oxygen rich additives (high in compound 1, low in compound 2). Fig. 1 demonstrates, that diffusion of oxygen or oxygen containing molecules to the surface does take place. Because of the vacuum (around 10^{-5} Torr), the only source for oxygen is definitely the specimen itself. This is also proven by the dependence of the increase in the oxygen concentration at the surface on the content of oxygen-rich additives.

But diffusion does imply that specimen damage below the surface does take place, most probably because of sample heating. At a certain depth, depending on the composition of the PP-compound, the damage will be maximum. And possibly at that depth the failure at peel tests could occur. Nihlstrand et al. [1] argue, that paint adhesion properties are strongly influenced by the extent of chain scission reactions in the near surface regions of the PP-compound during flaming or plasma treatment. The results obtained by electron irradiation of the specimens seem to confirm their hypothesis. They also measured by XPS oxygen concentrations of up to 20 at% at the specimen surfaces (samples treated with oxygen plasma). Additionally they observed a decrease in the oxygen concentration after irradiation of the samples with x-rays for longer times. A similar decline can be seen in Fig. 1 with increasing electron irradiation time.

The results prove additionally, that for these types of polymers, just because of the strong oxygen diffusion, phase analysis based on a varying oxygen content of the individual phases of blends would not be possible by SEM / EDXS.

3.2 Polymer films for ion mediated organic electronics

In light-emitting electrochemical cells (LECs) an electrically conducting conjugated polymer is blended with a solid state electrolyte, consisting of an ion co-ordinating and transporting component and a salt [2]. In the actual case thin films prepared by spin-casting blends of mLPPP (methyl substituted ladder-type poly (p-phenylene), kindly provided by U. Scherf) and a crown ether derivative, DCH18C6 (Dicyclohexano 18crown6), with a lithium trifluoromethane-sulfonate ($LiCF_3SO_3$) on both glass and Si-wafer substrates were investigated. Of interest were the phase separation of mLPPP and DCH18C6, the dissociation of the salt into ions and the ion distributions.

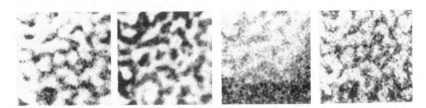

Fig. 2. BSE image and x-ray maps of a blend with mLPPP, DCH18C6 and $LiCF_3SO_3$ (film on a Si-wafer); from left to right: BSE – image, C-map, O_2-map, F-map (image width: 5.6 µm, 128x128 pixel, $E_0 = 3$ keV, $I_p \sim 0.3$ nA, dwell time per pixel: 250 msec; images filtered with mean filter).

SEM revealed the presence of a large-scale lateral variation in the film thickness of the size of several tens of micrometers and an additional fine structure superposed on it [3]. The film thickness varied roughly between eighty and several hundred nm. Fig. 2 shows the fine structure of the film in an area, where the large-scale thickness is small. The bright areas in the BS-image correspond to a lower film thickness, because the Si-substrate has a higher mean atomic number than the film itself, and the electron penetration depth is even at 3 keV much greater than the film thickness. A comparison of the carbon map with the fluorine map proves, that a high fluorine concentration is always correlated with a low carbon concentration and vice versa. Whereas these two maps gave a clear-cut distribution of the respective elements, no reliable information could be gained from the oxygen-map! The upper corner of the map conveys the impression, that the carbon and oxygen concentrations run parallel, the rest of the map looks somehow washed out. This could be an indication of oxygen diffusion as a result of the electron irradiation.

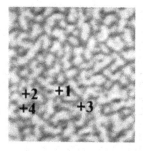

x--ray intensity analysis region	Carbon net counts	Oxygen net counts	Fluorine net counts
spot 1, prior	8503	1550	736
spot 2, prior	12336	1965	462
spot 3, after	14674	1494	515
spot 4, after	17815	2125	289
area, prior	15580	1000	438
area, after	16159	1006	460

Fig. 3. SE-image of a blend with mLPPP, DCH18C6 and LiCF$_3$SO$_3$ (image width: 11.45 μm; E_0 = 3 keV). Image recorded after EDX-analyses at the spots 1 and 2.

Table 1. Net counts of the elements C, O$_2$ and F, measured by EDXS at both the spots marked in and the overall area of Fig. 3 (E_0 = 3 keV; I_p ~ 0.3 nA; 45 sec analysis time); prior / after: prior / after recording of the SE-image in Fig. 3 with a slow scanning time.

The presence of a strong oxygen diffusion is confirmed by Table 1. The oxygen analyses at spots at both the thick and thin areas of the film gave much higher oxygen net counts than measured by an analysis of the overall area of Fig. 3. As a result, it was impossible to ascertain, whether demixing of mLPPP (no oxygen) and the crown ether (contains oxygen) does take place or not. The results for fluorine seem to indicate, that also diffusion of this element does take place, at least at long irradiation times.

A surprising result was the influence of image recording (much longer dwell times per pixel than at normal scanning speed) on the results of the x-ray analyses. Prior to image recording, the carbon signal at spot analysis was lower than that for area analysis. Most probably the beam burns a hole into the thin film. After image recording, this loss in the carbon signal was no longer observable (spots 3 and 4 of Fig. 3). Obviously, the film has become somehow more resistant to radiation damage. Also the fluorine signal from the spots was now, taking the thickness variations into account, comparable to that from the area measurements. But no change in the surface topography could be observed! Thus EDXS measurements definitely seem to be a more sensitive indicator for radiation damage than images of the surface topography.

References

[1] A Nihlstrand, T Hjertberg and K Johansson 1997 Polymer 38 (14) 3591-3599
[2] Q Pei, G Yu,C Zhang, Y Yang, AJ Heeger 1995 Science 269 1086-1088
[3] FP Wenzl, C Suess, A Haase, P Poelt, D Somitsch, P Knoll, U Scherf, G Leising 2003 Thin Solid Films 433 263-268

Inst. Phys. Conf. Ser. No 179: Section 2
Paper presented at Electron Microscopy and Analysis Group Conf. EMAG2003, Oxford, 2003
©2003 IOP Publishing Ltd

Nanostructural studies of biomineralised composites

I M Ross [1] **and P Wyeth** [2]

1. Department of Electronic and Electrical Engineering, University of Sheffield, Sheffield, S1 3JD, UK
2. Department of Chemistry, University of Southampton, Southampton SO17 1BJ, UK

ABSTRACT: Scanning and transmission electron microscopy has been applied to study the microstructural detail in the shell of the rock boring mollusc *Pholas-dactylus*. A cross-lamellar structure was revealed consisting of a hierarchical arrangement of sub-micron sized aragonitic laths. These laths themselves consisted of a series of <110> twinned crystallites 5 to 400nm in width and several micron in length. HREM analysis indicate the twin boundaries within the crystallites to be defect free.

1. Introduction

Many invertebrates have evolved hard shells to protect themselves against predators and their surrounding environment. Over millions of years highly specialised biomineralised structures have evolved to suit a variety of life-styles [Currey and Taylor 1974]. It is known that the special mechanical advantages inherent in these ceramic-reinforced biological materials can be attributed to their micro-composite form, in which a high volume fraction inorganic (ceramic) component is combined within a tenuous organic polymer matrix [Carter 1980, Jackson *et al* 1990]. The underlying structure-function relationship of these remarkable composites is of considerable interest, in particular for the extraction of novel engineering design concepts that may provide the insight into the design of stronger, stiffer and more fracture-resistant materials [Vincent 1990, Katti *et al* 2001].

2. Experimental

In this contribution, electron microscopy has been used to characterise the cross-lamellar layers of the hard wearing shell of the bivalve mollusc, *Pholadidae:Pholas-dactylus*, which has evolved to mechanically bore into stiff mud and solid rock. Previously prepared fractured surfaces were mounted onto SEM stubs with conducting paint and gold coated to prevent charging prior to analysis in a JEOL 6400 scanning electron microscope (SEM). Nano-structural detail was revealed using high-resolution transmission electron microscope (HREM). In this case, specimens were prepared by lightly grinding a small fragment of shell in deionised water followed by dispersion of a few drops of the slurry onto a holey carbon support. Initial analysis was performed on a JEOL 2000EX operating at 200kV using low dose illumination to reduce the effects of electron beam damage. Further HREM studies were performed using a JEOL 2010F field-emission gun TEM equipped with a Gatan Image Filter (GIF200). To further minimise electron beam induced damage, specimens were examined at liquid nitrogen temperatures using a Gatan liquid nitrogen cooled double-tilt stage.

3. Results and Discussion

Figure 1 shows a classic cross-lamellar microstructure which dominats the outer layer of the anterior margin of the adult shell. The structural sub-units consist of alternating blocks of aragonite crystals, 10 to 20μm wide, referred to as "first order lamellae", labelled (**a**) and (**b**). Each of these first-order lamellae consisted of stacks of individual thin, elongated aragonitic platelets termed "second-order lamellae", labelled (**c**).

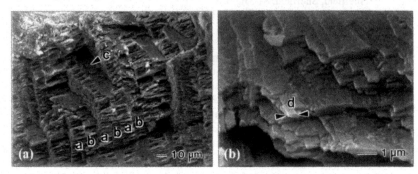

Figure 1: (a) Typical fracture surface displaying a classic cross-lamellar structure, (b) Detail of the 2nd order lamellae labelled "c" in Figure 1a consisting of long lath-like 3rd order crystallites labelled d.

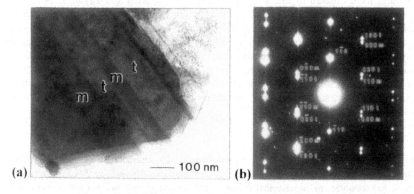

Figure 2: (a) TEM image of a fragment of a third order lamella showing the alternating bands of contrast labelled "m" and "t", (b) An indexed SAED pattern obtained from the fragment shown in Figure 2a illustrating the extensive twinning about the $(1\bar{1}0)$ plane.

Neighbouring first-order lamellae differ only in their relative orientation, such that each stack of second-order lamellae is rotated by an angle of 70 to 90° with respect to the next. Figure 2 shows stacked second-order lamellae with an average thickness of 200 to 300nm. Each second-order lamella was found to consist of an arrangement of narrower 400 to 600nm wide laths or "third-order lamellae" labelled (**d**).

It has been reported that the cross-lamellar structure in some other mollusc shells may contain as little as 0.5% organic matrix [Laraia 1990] and no evidence for any organic matrix was observed during the SEM examinations in this current study. Illumination of higher order structural detail required the use of transmission electron microscopy. This revealed thin laths between 400 and 600nm in width and a minimum of 3μm in length as shown in Figure 2a. These dimensions correspond to the size and shape of the crystallites assigned as third-order lamellae, observed during SEM analysis (Figure 1). Selected area electron diffraction revealed the presence of extensive twinning within the crystallite bound by the aragonite (1$\bar{1}$0) plane (Figure 2b). It is these twinned regions which give rise to the alternating bands of diffraction contrast in the bright-field image (Figure 2a), examples of which are arbitrarily labelled *matrix* (m) and *twin* (t).

An indexed high-resolution lattice image of a typical lath-like crystallite is shown in Figure 3a, corresponding to a near [00$\bar{1}$] aragonite zone axis projection. The image clearly illustrates a twin boundary parallel to the long axis of the crystallite, the twinned regions again being arbitrarily labelled. Twin boundaries were, within the limits of the observations made, found to be free of steps and dislocations.

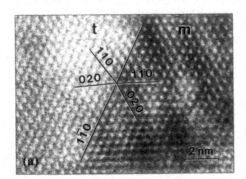 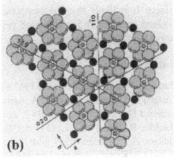

Figure 3: (a) An HREM lattice image of a single twin boundary in the third order lamella showing an [00$\bar{1}$] aragonite zone axis projection, (b) Schematic projection of an {1$\bar{1}$0} twin boundary: the dark spheres correspond to Ca atoms and the large and small grey spheres correspond to O and C respectively.

The widths of the twinned regions ("fourth-order" lamellae) were seen to vary markedly from one crystallite to another, ranging from 5nm to 400nm. A schematic representation of the (1$\bar{1}$0) twin boundary in aragonite is shown in Figure 3b. The existence of twinning in biogenic aragonite crystals was first reported by Marsh and Sass (1980) within the hinge ligaments of bivalve molluscs. These needle-shaped crystals when observed perpendicular to their long axis were found to be twinned on both the (110) and (1$\bar{1}$0) plane. Gauldie and Nelson (1987) also revealed the presence of twinning in aragonite when examining the growth rings in fish otoliths. It was proposed by both these workers that biogenic aragonite forms growth twins in the presence of foreign ions resulting in rapid precipitation and crystal growth.

It is well known that the structure of many materials may be modified by the influence of high-energy electron beams (O'Neill *et al* 2003) Biological materials and minerals such as carbonates are particularly susceptible to such beam induced damage. Examination of the third order lamella in the TEM, even at a relatively low current density of around $150pA/cm^2$, exhibited the development of a blotchy contrast across the crystallites within minutes of exposure to the electron beam. Alteration of the specimen was confirmed by analysis of the corresponding SAED pattern. The additional diffraction rings observed were conclusively assigned to those originating from polycrystalline CaO. It has been proposed by Walls and Tence (1990) that the transformation to CaO is most likely a result of direct beam damage effects, rather than beam heating induced thermal decomposition as previously suggested by Burrage and Pitkethly (1962).

4. Conclusions

A microstructural study of the shell of the mollusc *Pholas-dactylus* has been performed. The cross-lamellar structure revealed, forms a hierarchical arrangement of aragonite laths each consisting of a series of nano-scale <110> twins. HREM observations indicate that the twin boundaries are smooth and defect free. The mineral component was found to be particularly beam sensitive, decomposing to polycrystalline CaO within several minutes. Therefore, it was necessary to image in the TEM using low dose conditions ($<150pA/cm^2$) with the specimen at liquid N_2 temperatures to minimise such effects. It is likely that the complex hierarchical structure observed in this study has evolved specifically for increased resistance to abrasion in the context of increasingly active burrowing life habits.

5. References

Burrage B J and Pitkethly D R 1962 Phys. Stat. Sol. 32 399-405
Carter J G Skeletal Growth of Aquatic Organisms 1980 Plenum Press 69-168
Currey J D and Taylor J D 1974 J. Zool. (London) 173 395-406
Gauldie, R W and Nelson, D G A 1988 Comp. Biochem. Physiol. 90A (3) 501-509
Jackson A P, Vincent J F V and Turner R M 1990 J.Mat.Sci. 25 3173-3178
Katti D R, Katti K S, Sopp J M and Sarikaya M 2001 Comput. Theor. Polym. Sci. 11 397-404
Laraia V J, Aindow M and Heuer A H 1990 Mat.Res.Soc.Symp.Proc. 174
Marsh M E and Sass R L 1980 Science 208 1262-1263
O'Neill J P, Ross I M, Cullis A G, Wang T and Parbrook P J, 2003 Appl.Phys.Lett. 83 (10) 1965-1967
Vincent J F V 1990 Metals and Materials (1) 7-11
Walls M G and Tence M 1990 Inst.Phys.Conf.Ser. 98 255-258

6. Acknowledgements

The authors would like to thank the Engineering & Physical Science Research Council for their financial assistance and the University of Liverpool (Department of Materials Science) and University of Sheffield (Department of Electronic and Electrical Engineering) for their logistic support.

Inst. Phys. Conf. Ser. No 179: Section 2
Paper presented at Electron Microscopy and Analysis Group Conf. EMAG2003, Oxford, 2003
©2003 IOP Publishing Ltd

RHEED characterisation of the near surface microstructure of Ti-O based biocompatible coatings

S Marlafeka, D M Grant and P D Brown

School of Mechanical, Materials, Manufacturing Engineering and Management, University of Nottingham, University Park, Nottingham NG7 2RD, UK.

Abstract. Mechanically polished, annealed, nitric acid treated and aged in boiling water after nitriding, commercially pure Ti substrates have been characterised using reflection high-energy electron diffraction (RHEED) and secondary electron imaging, in terms of their naturally formed or 'accelerated' oxide layers. Annealing induced crystallisation and transformation of anatase to the rutile phase and led to increased roughness, with localised fracture and balling up of the surface oxide layer as the time and temperature of annealing were increased. Nitric acid modification produced no influence on the anatase to rutile transformation, whilst further aging in boiling water induced an acceleration of this transformation. RHEED data acquired at differing accelerating voltages have indicated a Ti-O phase gradation within annealed sol-gel derived V modified TiO_2 layers deposited by spin coating onto Ti substrates.

1. Introduction

The success of Ti and Ti-based alloys for load bearing applications such as artificial implant biomaterials is due to the low elastic modulus combined with the natural formation of a surface oxide passivating layer. The oxide provides protection against metal ion release [1] and hence, there is interest in engineering thicker TiO_2 coatings using techniques such as the sol-gel method with the aim of further improving the material biocompatibility [2]. Recent work indicates that the biological cell response varies with the surface composition of modified Ti surfaces [3]. Nevertheless, it is not clear why the coating crystallinity influences this response, and in this context, the development and application of the surface sensitive technique of reflection high energy electron diffraction (RHEED) for the rapid characterisation of biomaterials could be beneficial [4]. RHEED provides a non-destructive, rapid, dynamical method of obtaining crystallographic information, taking the form of a half diffraction pattern produced following the interaction of an electron beam at glancing angle with the near surface microstructure of a material.

2. Experimental

The surfaces of mechanically polished, commercially pure Ti (cp-Ti) discs of 10 and 6 mm diameter and 1 mm thickness were annealed in air under conditions of 400°C for 30 min up to 700°C for 18 h; were treated in a 30% nitric acid bath for 10 min; and were aged in boiling

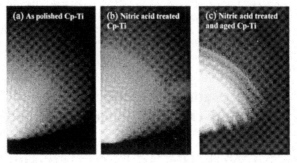

Figure 1. 80 kV RHEED diffraction patterns from cp-Ti specimens illustrating the surface microstructure following: (a) Mechanical polishing (amorphous); (b) nitric acid treatment (amorphous); and (c) immersion in boiling water following nitric acid treatment (indicative of the development of intermediate grain sized polycrystalline anatase-rutile TiO_2 (tetragonal)).

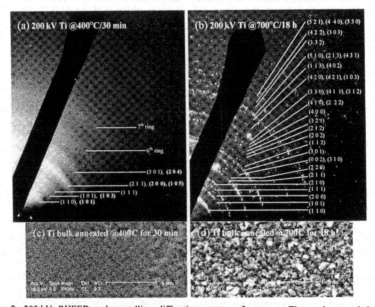

Figure 2: 200 kV RHEED polycrystalline diffraction patterns from a cp-Ti sample annealed at (a) 400°C in air for 30 min and (b) 700°C in air for 18 h, with the assignment of d_{hkl} Miller indices. (c,d) show the corresponding SE images from these samples acquired at 10 kV. Fig. 2a is consistent with the development of fairly small grains of polycrystalline anatase-rutile TiO_2 (tetragonal), with anatase being more dominant, and there is also the possibility of some TiO (hexagonal); Fig. 2b is consistent with the development of larger grained polycrystalline rutile TiO_2 (tetragonal). (Assignments in normal font: rutile TiO_2 (tetragonal); bold font: anatase TiO_2 (tetragonal).)

double distilled water for 24 h subsequent to a nitric acid treatment. Vanadium modified titania sol-gels were produced by the evaporation of an aqueous colloidal sol and deposited onto 10 mm diameter cp-Ti discs by spin coating, using a custom-built spin coater controlled by an EMC TOP-5200 syringe pump. The 16wt.%V modified titania sol-gel sample reported on here was heat-treated in air at 300°C for 18 h. All the sample heat treatments were performed using a Vecstar 91e tube furnace. Topographical and morphological information on the samples was

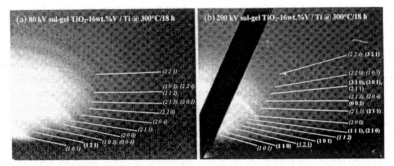

Figure 3. (a) 80 kV and (b) 200 kV RHEED patterns from a sol-gel TiO_2-16wt.%V / Ti sample annealed at 300°C in air for 18 h, illustrating some differences in the surface microstructure. Fig. 3a is consistent with the development of fine grains of polycrystalline TiO_2-anatase (tetragonal) with the possibility of some TiO_2-brookite (orthorhombic). Fig. 3b is consistent with grains of polycrystalline TiO_2-anatase (tetragonal), some TiO_2-rutile (tetragonal) with the possibility of some TiO_2-brookite (orthorhombic), TiO (hexagonal) and V (cubic). (Assignments in italic font: anatase TiO_2 (tetragonal); bold font: rutile TiO_2 (tetragonal); normal font: TiO (hexagonal); underlined font: brookite TiO_2 (orthorhombic).)

obtained using a Philips XL30 FEG-ESEM operated in high vacuum mode at 10 kV. Sub-projector RHEED stages inserted within either a Jeol 2000fx or a Philips 410 TEM operated at 200 and 80 kV, respectively, were used for near surface study of the specimens positioned vertically in the microscope column. The RHEED patterns were calibrated using a particulate Au sample that produced polycrystalline rings and a crystalline GaN specimen.

3. Results and discussion

The RHEED patterns from the as-polished cp-Ti sample presented in Figure 1a and the annealed cp-Ti samples of Figures 2a,b indicate a transformation of the outer surface layer with processing. The initial amorphous microstructure, evident after mechanical polishing (Fig. 1a), is replaced by nanocrystalline anatase TiO_2 particles (Fig. 2a), which in turn provide the nucleation sites for the formation of a few, initially small rutile TiO_2 grains (Fig. 2b), as the temperature and time of annealing are increased. Full rutilisation is apparent at 700°C annealing for 18 h, with the development of relatively large grains. The observation of some slight preferred orientation in a direction normal to the surface of the heat treated samples is characteristic of alignment of some of the grain growth within the film following the anatase to rutile transition. The related secondary electron (SE) images of the annealed specimens (Figures 2c,d) indicate uniform layer contrast at low annealing temperatures, which is consistent with the RHEED observations and the concept of uniform coverage of nanocrystalline TiO_2 anatase particles at this stage. SE imaging also reveals a rough, porous, large-grained layer for the sample annealed at 700°C. This layer seems to have balled-up and become broken due to the process of densification. The additional presence of flake-like polygonal shaped features is commensurate with RHEED supporting the observation of rutile grain growth.

The amorphous surface structure evident from the RHEED pattern of the nitric acid treated cp-Ti sample (Figure 1b) suggests that this modification has had no influence on the surface microstructure, being similar to that of the mechanically polished sample (Figure 1a). The early formation of polycrystalline anatase, however, at the surface of the cp-Ti sample aged in boiling water after nitric acid treatment (Figure 1c) is considered to be due to the presence of the

158

hydrogen-anodic environment that must have initiated the anatase to rutile transformation at such a low temperature of 100°C.

In the attempt to investigate the extent of applicability of the RHEED technique at differing accelerating voltages, diffraction patterns from the surface of a sol-gel TiO_2-16wt.%V / Ti sample annealed at 300°C in air for 18 h were compared at 80 and 200 kV, respectively (Figures 3a,b). Both RHEED patterns are dominated by the anatase TiO_2 phase, nevertheless, it is only at 200 kV (Figure 3b) where there is an additional indication of the formation of some rutile TiO_2 phase. The sequence of intensities and the location of the polycrystalline rings for different RHEED voltages allow these different assignments for Figures 3a,b, being clearly dissimilar, in addition to the poorer definition of the 80 kV rings as compared with those obtained at 200 kV. Hence, it is suggested that a variable voltage RHEED technique might be able to provide tentative insight into the depth distributions of certain microstructures. Thus, a sub-surface oxide phase gradation is indicated from these RHEED patterns, and this is consistent with cross-sectional TEM observations reported on previously [5]. The development of a graded microstructure is probably related to the initial oxidation / amorphisation of the Ti substrate when exposed to the atmosphere prior to the sol-gel deposition, combined with further oxidation during sol-gel processing.

4. Summary

The development of the surface oxide layer of bulk Ti samples under different surface treatments has been investigated using the combined characterisation techniques of RHEED and SE imaging. RHEED results from annealed cp-Ti specimens demonstrate the transformation of the anatase to the rutile phase of TiO_2, as the annealing time and temperature are increased. SE observations reveal fracture and balling up of the oxide layer surfaces and void development, due to densification of the surface. Nitric acid treatments and mechanical polishing introduce amorphous surface oxide layers. Specimens aged in boiling water after nitric acid treatment exhibit much lower anatase to rutile transformation temperatures compared to the annealed samples. 80 and 200 kV RHEED results indicate several significant differences in the near surface microstructure of a 16wt.%V modified TiO_2 sample deposited by spin coating onto a Ti substrate, implying that depth crystallographic profiling of the surface becomes possible to investigate graded oxide layers at the near sample surface.

5. Acknowledgments

This work was supported under EPSRC contract GR/M74603 and GR/L55209.

6. References

[1] M Long and H J Rack, 1998 Biomaterials **19** 1621-1639

[2] M P Casaletto, G M Ingo, S Kaciulis, G Mattogno, L Pandolfi and G Scavia, 2001 Appl. Surface Science **172** 167-177

[3] J Ryhanen, E Niemi, W Serlo, E Niemela, P Sandvik, H Pernu and T Salo, 1997 Journal of Biomedical Materials Research **35**(4) 451-457

[4] B Kasemo and J Lausmaa, 1988 'Biomaterials from a surface science perspective' in Surface characterization of Biomaterials (ed. B.D. Ratner; Elsevier Science)

[5] S Marlafeka, F Allison, D M Grant, P G Harrison and P D Brown 2001 Inst. Phys. Conf. Ser. No. **168** 381 – 384

Inst. Phys. Conf. Ser. No 179: Section 3
Paper presented at Electron Microscopy and Analysis Group Conf. EMAG2003, Oxford, 2003

Towards sub-0.5 angstrom beams through aberration corrected STEM

P D Nellist[1], N Dellby[1], O L Krivanek[1], M F Murfitt[1], Z Szilagyi[1], A R Lupini[2] and S J Pennycook[2]

[1]Nion Co., 1102 8th St., Kirkland, WA 98033, USA.
[2]Oak Ridge National Laboratory, Condensed Matter Division, Oak Ridge, TN 37831-6030.

Abstract. Correction of spherical aberration (C_S) in the scanning transmission electron microscope (STEM) has enabled routine sub-angstrom resolution imaging and increased the current available in an atom-sized probe by a factor of 10 or more. Both high-angle annular dark field (HAADF) imaging and EELS spectrum imaging (SI) results are shown from instruments fitted with Nion aberration correctors.

1. Introduction

After some 40 years of not wholly successful attempts, improving the performance of an electron microscope by aberration correction has recently become a commercially available reality. A C_S-corrected dedicated STEM operating at 120 kV has demonstrated sub-angstrom spatial resolution [1], and a corrected 300 kV STEM is routinely imaging at resolutions better than 0.8 Å [2]. In a high-resolution transmission electron microscope (HRTEM), the ability to adjust the phase contrast transfer function through control of C_S has allowed O atoms in an oxide material to be observed for the first time, with the machine operating at a spatial resolution of 1.3 Å [3].

The first theoretical design for a C_S corrector was proposed in 1947 [4], so it is relevant to ask why the above development took nearly half a century. One of the reasons is that, contrary to popular perception, the lenses in electron microscopes are already of high optical quality making it hard to improve upon them. In a previous paper [5] we have shown how even an uncorrected high resolution electron microscope reaches a higher optical figure of merit than the "corrected" Hubble space telescope. Indeed, a significant fraction of the C_S of a modern high resolution objective lens can be explained by geometry and deviations from the small angle approximation. Consider a so-called Gaussian lens that deflects a ray by an angle proportional to the radius of the beam in the lens. In the small angle approximation, such a lens will focus the beam (Fig. 1), with

$$\theta = x/f \tag{1}$$

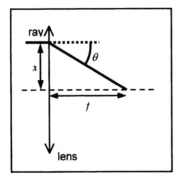

Figure 1. A schematic diagram of the focusing of a ray at radius x in a parallel beam by a lens to form a crossover a distance f (the focal length) from the lens.

A more exact form of Eq. (1) for the geometry in Fig. 1 is

$$\theta = \arctan(x/f) = x/f - x^3/3f^3 + \ldots \tag{2},$$

where we have expanded to the third order term. In practice we are dealing with beams with a maximum convergence angle of 25 mrad, for which the third-order term in Eq. (2) will be equal to 5 μrad and we might reasonably expect to neglect it. In terms of the spherical aberration, however, this third order deviation in the angle corresponds to a coefficient of spherical aberration, C_S, equal to $f/3$, which is similar to the value that might be expected in a state-of-the-art objective lens. This analysis is valid for the infinite demagnification condition shown in Fig. 1; conditions of finite magnification require a slightly different third order term. This example shows how the C_S of a modern TEM objective lenses is commensurate with the small deviations from a Gaussian lens required by geometry, and is not the major optical defect it is often perceived to be.

Given that electron lenses are already of high optical quality, it is now clear that the corrections being made to the wavefront by a spherical aberration corrector are relatively small and need to be extremely precise. In order to function effectively, an aberration corrector also needs to compensate for the additional "parasitic" aberrations that arise from imprecise machining and assembly, and magnetic inhomogeneities. Practical aberration correction has only now become practical through the development of techniques for measuring and correcting the parasitic aberrations, which has itself required the development of computers fast enough to make such measurements in a practical length of time.

2. The performance of installed Nion correctors

There are now five Nion C_S correctors installed in VG dedicated STEM instruments operating in the field. Four of these machines are installed in 100 kV instruments (one of which is operated at 120 kV) and one in a 300 kV instrument. The design and operation of the corrector has been described in previous papers [5,6]. As mentioned above, the key to the performance of these correctors is the ability to measure and correct for the parasitic aberrations that arise once the C_S has been corrected. Our software measures the aberrations using Ronchigrams [6]. Tuning the microscope at the start of a session typically takes less than 10 minutes and is an automated procedure.

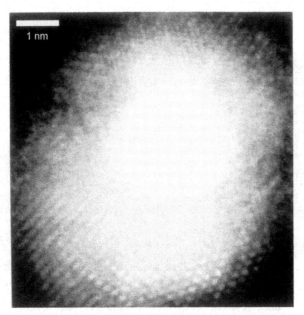

Figure 2. An HAADF image of a Au nanoparticle supported on amorphous carbon taken using the Nion corrected 100 kV VG STEM. The near horizontal fringes at the top of the particle have a spacing of 1.2 Å.

All of the 100 kV instruments have attained a resolution approaching or beyond 1 Å in HAADF images. Figure 2 illustrates how the resolving power of the microscope is improved. Prior to correction, the 2.3 Å {111} fringes in the <110> oriented domain at the bottom of the particle would just have been resolved. With correction, much smaller spacings can be observed in a domain at the top of the particle and {113} fringes with a spacing of 1.2 Å are observed with good contrast.

The corrector installed in the 300 kV HB603 STEM at Oak Ridge National Laboratory has allowed that machine to routinely image at a resolution of 0.8 Å [2]. Figure 3 shows a similar Au nanoparticle with the contrast enhanced to show the many single atoms near the edge of, and between, the nanoparticles. The ability to observe single atoms so close to a nanoparticle indicates that the probe is free of the long range tails that can occur if the imaging conditions are not optimized. An intensity profile of a pair of Au atoms shows that they have a FWHM of 0.7 Å, indicating that the probe diameter is smaller than this (since the atom potential has a finite width). Note also that the profiles of the atoms both show a similar slight asymmetry in the probe, perhaps due to some small residual coma. The ability to image routinely at this resolution is a remarkable development, but measurement of the aberrations suggests that the aberration correction is sufficient for imaging at a resolution better than 0.6 Å. Such performance has yet to be achieved, and is probably being limited by instabilities of the microscope that have yet to be cured. The corrector overcomes spherical aberration but not instabilities, and it is remarkable that the resolutions described above have been

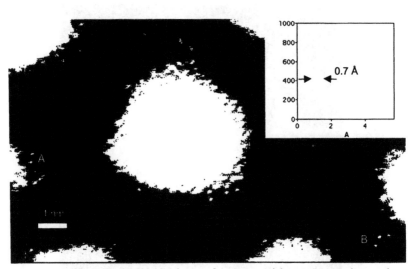

Figure. 3. An HAADF image of Au nanoparticles on an amorphous carbon film taken using the corrected 300 kV VG HB603 STEM at Oak Ridge. Inset: A profile through the pair of atoms indicated at A. The image of the atoms has a FWHM of 0.7 Å, and the atoms are separated by 2.8 Å. The trimer of atoms at B have similar dimensions.

achieved in columns that were not originally designed and specified for such performance.

3. EELS and spectrum imaging in a corrected STEM

In addition to improving the spatial resolution, a further major benefit of aberration correction is the increased beam current that becomes available. Correction of the C_S allows larger objective apertures to be used which, for no change in demagnification of the source, will lead to a significant increase in the probe current. In a 100 kV VG STEM with a cold field emission source, it is now possible to focus more than 200 pA into a probe smaller than 1.4 Å. Previous to correction, 200 pA would only have been available in a probe of about 5 Å. The increased probe current makes it likely that elemental mapping by electron energy-loss spectroscopy at 1-2 Å resolution will become possible in a minute or so for a 256 by 256 pixel elemental map.

Making use of the additional current for spectroscopy requires that the increased beam angles in the probe can be coupled into the spectrometer, which means that the spectrometer collection semiangle needs to be at least 20 mrad. Most VG instruments are not fitted with a post-specimen lens, and the additional optics can only easily be added after the lid of the machine, about 240 mm after the sample. Even assuming that the objective lens magnetic field provides a typical post-specimen compression of 3 times, a spectrometer at this position would require an entrance aperture of greater than 3 mm to collect the whole beam. At this size, the highest energy-resolution will not be available due to the aberrations of the spectrometer.

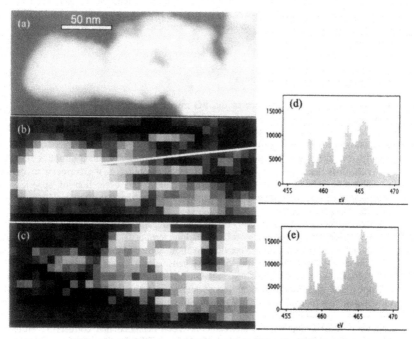

Figure 4. Data from a spectrum image (SI) of a TiO_2 catalyst support particle. Spectra were recorded on a 32 by 13 array of probe positions, with 0.5 s exposure per spectrum, 0.3 eV per channel dispersion. (a) The HAADF reference image of the particle. (b) and (c) The resulting images from a singular value decomposition (SVD) fit of (b) rutile and (c) anatase reference Ti $L_{2,3}$ spectra to the SI data. (d) and (e) Spectra from the pixels marked with the arrows showing the obvious changed to the fine structure.

At Nion we have developed a post-specimen EELS coupling module that uses 4 quadrupoles in a configuration similar to a regular C_S corrector. It provides increased compression of the beam prior to entry into the spectrometer while also ensuring that the virtual cross-over of the beam is at the height expected by the spectrometer to avoid injecting second order aberrations. The similarity of the coupling module to the aberration corrector goes further in that both a- and b-type (skew) octupoles can be placed at positions where the beam is either elliptical or round. This means that the third order aberrations of the spectrometer (and of the couping lenses) can be corrected, allowing larger physical entrance apertures to be used.

To demonstrate the ability to record spectra quickly in a spectrum image, Fig. 4 shows data recorded from a sample of TiO_2 [7] that contains both the rutile and anatase phases. A 2 mm entrance aperture was used, which allowed a 20 mrad collection semiangle with the additional compression of the coupling module. Individual spectra from the data cube can be easily identified as being either predominantly rutile or anatase simply by inspection of the fine structure. For a more automated map, reference spectra can be used in a singular value decomposition fit to map the regions of anatase and rutile. Such data can be used to "colour in" an ADF image of the region of TiO_2 identifying the phases, but for monochrome reproduction we display two separate images here.

4. Discussion and future directions

The improvement in performance of the VG microscopes that have been fitted with Nion aberration correctors has been substantial. Our experiences with these projects have, however, indicated a number of limiting factors that will prevent further significant improvement [5]. The most important of these are:

1. The construction of the VG objective lens prevents the placement of optics close enough to the objective lens to allow optical coupling of the corrector to the objective lens. Without this optical coupling, the correction of spherical aberration generates increased 5th order aberrations that limit the resolution to about 0.8 Å at 100 kV.
2. The condenser lenses of the VG are mechanically instable, particularly after excitation changes when they suffer from long term drift. In addition there are not enough electrical alignments provided to allow correct alignment.
3. The post-sample field of the objective lens is not strong enough to provide sufficient angular compression to allow large angle electrons to be collected into the EELS spectrometer without loss of energy resolution.

It is apparent that to make optimal use of the ability to correct the spherical aberration, a new column is required that addresses these issues. At Nion we are presently constructing such a column. A corrector has been designed that corrects 5th order aberrations in addition to the spherical aberration. The limiting aberration of this corrector is 8-fold astigmatism, which is a 7th order aberration. In our design this is kept to about 1 cm, which would limit resolution at 100 kV to 0.4 Å. Of course, chromatic aberration will become increasingly significant at these resolutions. In our new column we have minimized C_C through the design of the objective lens and by minimizing the number of round lenses present in the probe forming optics. We anticipate that the C_C of the probe forming optics will be about 1.1 mm at 100 kV. Even so, achieving resolution better than 0.5 Å will probably require a 200 kV gun. The column is being designed for beams of up to 200 kV. Although the first column is to be mounted on an existing 100 kV VG gun, we are also engaged in developing our own 200 kV electron source.

References

[1] Batson P E, Dellby N and Krivanek O L 2002 *Nature* **418** 617-620.
[2] Pennycook S J et al. in *Spatially Resolved Characterization of Local Phenomena in Materials and Nanostructures,* ed. D. A. Bonnell, A. J. Piqueras, P. Shreve, F. Zypman, (Materials Research Society, Warrendale, Pennsylvania, 2003) p. G1.1.
[3] Jia C L, Lentzen M and Urban K 2003 *Science* **299** 870-873.
[4] Scherzer O 1947 *Optik* **2** 114-132.
[5] Krivanek O L, Nellist P D, Dellby N, Murfitt M F and Szilagyi S Z 2003 *Ultramicroscopy* **96** 229-237.
[6] Dellby N, Krivanek O L, Nellist P D, Batson P E and Lupini A R 2001 *J. Electron. Microsc.* **50** 177-185.
[7] R Klie and N Browning are gratefully acknowledged for providing the TiO_2 sample.

Inst. Phys. Conf. Ser. No 179: Section 3
Paper presented at Electron Microscopy and Analysis Group Conf. EMAG2003, Oxford, 2003
©2003 IOP Publishing Ltd

Development of a C_S corrector for a Tecnai 20 FEG STEM/TEM

S A M Mentink, M J van der Zande, C Kok and T L van Rooy
Philips Research, 5656 AA Eindhoven, The Netherlands

Abstract: The resolution of scanning transmission electron microscopes can be improved by the correction of the aberrations of the objective lens. We discuss the development of a quadrupole-octupole type C_S corrector for a FEI Tecnai 20 STEM/TEM with a S-TWIN lens. Experiments demonstrate that C_S correction has been achieved. High-resolution TEM can be performed in combination with corrected STEM.

1. Introduction

The electron probe size of scanning transmission electron microscopes (STEM) is limited by the spherical aberration of the objective lens. This aberration can be eliminated by the introduction of an aberration corrector, as demonstrated by Krivanek (Krivanek 1999). This corrector is of the quadrupole-octupole type, where two octupole elements are placed in or close to the two line foci created by a system of quadrupole elements. A third octupole element is placed in the round beam. An alternative based on two hexapole elements has been presented by Haider et al. (Haider 1998).

At Philips Research we have developed a quadrupole-octupole type corrector to correct for third-order spherical aberration of a FEI Tecnai 20 STEM/TEM with a S-TWIN lens. The design of the corrected STEM column has been improved to minimize the sensitivity to changes in temperature and magnetic fields, which occur upon switching TEM operating modes. This has led to uncompromized TEM functionality. Here we discuss the design of the corrector and modified TEM column, demonstrate C_S correction and show our experimental HAADF STEM images. It is also shown that TEM imaging is still possible while the probe corrector is operating.

2. Design of corrector and TEM integration

Our corrector consists of seven multipole layers for C_S correction and two additional octupole layers for alignment of the optical axis of the corrector to the condensor system and condensor stigmation. See Fig. 1. The resolution is limited by fifth order spherical aberration, determined mainly by the distance d=120 mm between the exit plane of the corrector and objective lens. This aberration coefficient can be estimated from:

$$C_5 = 3d \left(\frac{C_S}{f} \right)^2 \tag{1}$$

With lens parameters $f=2.0$ mm, $C_S=1.3$ mm, $C_5=152$ mm. Full numerical analysis yields $C_5=150$ mm.

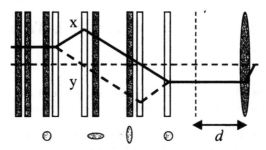

Figure 1. Schematic view of our C_S corrector. The four quadrupole elements are indicated by open boxes, the correcting octupoles by shaded boxes. Ray paths in the *(x,z)* and *(y,z)* planes and the beam shape are also shown. The distance between the (dashed) exit plane of the corrector and the objective lens is indicated by *d*.

The spot size versus current can be evaluated using formulas given in the literature (Barth and Kruit 1996, Batson 2003). Important input parameters besides the accurate values of high-order geometric aberrations are the reduced brightness and energy spread of the Schottky field emitter source. We have used a reduced brightness of $B_r=2 \cdot 10^7$ $Am^{-2}sr^{-1}V^{-1}$ and an energy spread $\Delta U=0.6$ eV. The resulting current in a given spot size with and without the use of our C_S corrector for a 200 kV system with S-TWIN lens is depicted in Fig. 2. The spot size is defined as the diameter of the spot that contains 50% of the total current in the probe.

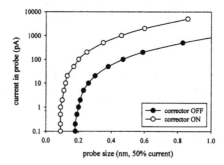

Figure 2. Theoretical probe current (in pA) versus probe size (d_{50} value in nm) for our corrected STEM system.

The theoretical probe size in the zero current limit with corrector improves from 0.2 to 0.09 nm, while the current in a 0.20 nm probe increases from 2 to 100 pA with respect to the uncorrected case. Further improvements on these numbers can be achieved by correction of the fifth order geometric aberrations, reduction of the energy spread and/or chromatic aberration of the objective lens, and an increase of the gun brightness.

Apart from the improved resolution and probe current, other important design objectives have been realized. The demands on the electronic stability have been reduced by mechanically accurate multipole design. The C_S corrector behaves almost linearly in the excitation thanks to excellent magnetic material and software normalization procedures. Minimal cross-talk within the corrector and between the corrector and the condensor and objective lens modules was obtained by dedicated magnetic shielding. Special emphasis was put on the magnetic material, design and construction of the multipoles. The final geometry of the multipoles has been defined by spark erosion. Each individual multipole has accurate reference planes, resulting in a <2 µm positioning error of a multipole layer. The quality of the multipole elements has been checked before introduction to the microscopes, using a rotation table with optical, mechanical and magnetic probes. This way it was found that the mechanical and magnetic centers of the multipoles coincide within several micrometers. We have used non-magnetic material as a stiff mechanical frame for the multipole layers. The thermal expansion of the frame material is close to that of Permalloy, which reduces the sensitivity to temperature variations.

The corrected system has been installed in a low-vibration environment at the Philips Research Laboratories, as shown in Fig. 3. The original Tecnai F20 microscope has been adapted to incorporate the corrector module. Accurate alignment of the condenser, objective lens and corrector module has been achieved by mechanical design. Software has been developed to simultaneously control the corrector and the TEM and to numerically analyze the aberrations.

Figure 3. The Tecnai F20 with probe corrector at Philips Research.

Initial alignment of the corrector has been done manually. Coarse alignment is facilitated by the presence of the TEM projection system, which is used to analyze the probe and its caustic. Fine alignment is done with the Ronchigram as is common practice in high-resolution STEM. The coarse alignment of the corrector is stable for many days. Fine alignment is required every few hours.

3. Experimental results and discussion

Figure 4 shows experimental Ronchigrams for the uncorrected and corrected system, with C_S=1.3 mm and C_S=0.0 mm. The area of constant phase has increased from 24 to 48 mR (cf. Batson 2003), giving an optimum half-opening angle of 24 mR for our corrected system. The hexagonal shape of the Ronchigram of the corrected system is due to fifth order geometric aberrations. Our first STEM HAADF images on gold islands have resolved the 0.204 nm lattice spacing (Fig. 4), while 0.16 nm fringes have been observed on Si [110] (not shown).

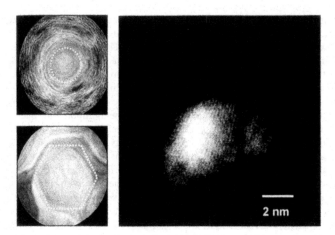

Figure 4. Left: Ronchigrams of the uncorrected (top) and corrected (bottom) probe, with optimum opening angles (dotted lines) of 12 mR and 24 mR, respectively. Right: C_S=0 HAADF STEM image of a gold island on amorphous carbon.

For many applications, such as navigation on the sample, it is advantageous to use regular TEM imaging before switching to STEM imaging or analysis. Figure 5 demonstrates this possibility while the probe corrector is switched on.

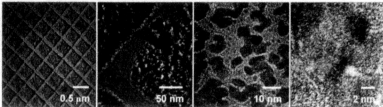

Figure 5. TEM images of a specimen of gold particles on carbon at various magnifications. The probe C_S corrector is switched on.

In summary, we have presented initial results of a probe-corrected Tecnai TF20 STEM. Further research will concentrate on the correction of fifth order geometric aberrations, automation and software integration to enable sub-0.1 nm STEM imaging in a state-of-the-art TEM instrument.

References

Barth J E, Kruit, P 1996, Optik **101**, 101

Batson P E 2003, Ultramicroscopy **96**, 239

Haider M, Rose H, Uhlemann S, Schwan E, Kabius B, Urban K 1998, Ultramicroscopy **75**, 53

Krivanek O L, Dellby N, Lupini A R 1999, Ultramicroscopy **78**, 1

Inst. Phys. Conf. Ser. No 179: Section 3
Paper presented at Electron Microscopy and Analysis Group Conf. EMAG2003, Oxford, 2003
©*2003 IOP Publishing Ltd*

Accurate FOFEM computations and ray tracing in particle optics

B Lencová

Institute of Scientific Instruments, Academy of Sciences of the Czech Republic, Královopolská 147, 61264 Brno, Czech Republic

Abstract. It is shown that accurate 2D FOFEM developed in our institute is available for computations of lenses, deflection systems and correcting multipoles. With some limitations we calculate also spectrometers or detectors. High accuracy of computed potential allows the evaluation of aberrations from accurate ray tracing. Application of the software will be shown on the design of a low voltage electrostatic SEM with cathode lens and Wien filter, intended for detection of angular and energy distribution of low energy electrons.

1. Introduction

Electron microscopes become more complicated whenever we increase the demands on the detected signals. Low energy scanning electron microscopes and integrated circuit testers require a small probe at low energy 10-1000 eV. The beam is usually decelerated in front of the specimen so that the signal electrons follow similar or almost the same paths as the primary beam and thus a suitable means of separating of the secondary beam must be incorporated. Moreover, the secondaries contain information on the energy and angular distribution, and it may be desirable to extract it. This necessarily makes new demands on the CAD software for the computation of electromagnetic fields and beam properties, some of which are not fully covered by aberration theory (misalignments, Wien filters, detectors and spectrometers). Fortunately the software development keeps pace with these requirements; first, some improvements of programs we developed are described, followed by application to elements of a low energy SEM.

2. The software

The first order FEM introduced over 30 years ago [1] is capable of providing accurate 2D distributions of potential for rotationally symmetric lenses [2,3] as well as for multipole elements (deflectors, quadrupoles, etc) [4]. The papers provide a survey of improved features we introduced in order to make the software more accurate (by proper evaluation of coefficients of FEM equations and by the use of a mesh with graded mesh step) and faster (the use of the preconditioned conjugate gradient method). The newest addition implemented recently [5] for rotationally symmetric lenses provides an estimate of computational accuracy as well as the usual potential at mesh points and on the axis. This is made possible by using up to one million points in the mesh.

    From the user's point of view it is a substantial advantage that the computation programs are equipped with an interface that allows easy editing of input data (generation of the coarse mesh defining geometry or fine mesh, input of material and coil regions and editing their properties) and the display of results, including saturation of magnetic materials in any part of the circuit. Our old DOS-based interfaces are slowly being replaced with Windows programs that integrate the user interface, FEM computation and ray tracing, including electron and ion guns and space charge effects, into a single program [6].

Optical properties of the lens and deflector fields can be derived from the results of accurately computed particle trajectories (ray tracing). In principle it is possible to evaluate in this way geometrical aberrations up to an arbitrary order or chromatic aberrations of any desired rank in this way, even for such systems as electrostatic mirrors or Wien filters where the aberration theory is not easy to implement or the order of the aberration is not known in advance as is the case for cathode lenses where the beam starts at low energy. Ray tracing also allows the design of non-standard elements in the electron optical column, like filters, spectrometers, and detectors, or evaluation of misaligned systems.

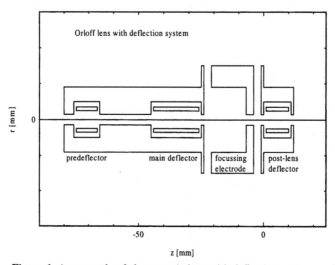

Figure 1. An example of electrostatic lens with deflection system analyzed.

3.   The application

We shall illustrate the flexibility of our programs by considering the design of elements of the electron optical system of a low voltage electrostatic SEM equipped with a cathode lens, a Wien filter to separate the primary and signal electrons, and a deflection-spectrometer unit needed for the use for angle- and energy-resolved observation of signals detected with a special CCD sensor [7]. Electrostatic systems may be more suitable for use in an ultra-high vacuum environment to enable surface sensitive studies.

We consider a well-known lens studied by Orloff and Swanson [8]. We selected the beam energy to be 5 keV, which is close to the extraction voltage of the Schottky electron gun. For simplicity we do not at present consider the second lens close to the gun and concentrate only on the properties of the probe-forming lens. This lens can act as a cathode lens if we put the sample holder at high voltage close to cathode potential.

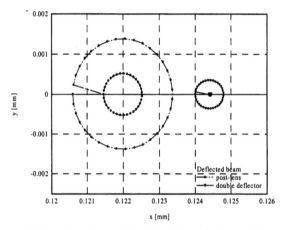

Figure 2. Aberration figure after deflection by both possible systems.

Figure 1 shows the geometry of the Orloff lens with deflectors. Two deflection concepts are possible, post-lens deflection, which consumes some of the space below the lens and forces us to use longer working distance, and the use of standard double deflection system allowing the deflected beam to go through the lens centre. We have examined both. With the selected length 8.5 mm of the predeflector we get smaller deflection aberration with 40 % less voltage – see Figure 2 for x-deflection to about 0.12 mm for a parallel beam with diameter 0.2 mm entering the lens. Obviously, the field curvature of the deflection field is larger for the post-lens system.

We are interested in slowing down the primary beam from 5 keV to a landing energy of just 100 eV on the specimen. Notice that the direction of the optical axis is reversed. We have to increase slightly the voltage on the focusing electrode (from 2030 V for 5 keV beam to 2153 V with the cathode lens on). Figure 3 shows 10 eV secondary electrons starting on the axis; only the particles with angles of less than 30 degrees pass through the lens. At this beam energy the electron wavelength is around 0.1 nm, and we would like to know what happens with the diffracted low energy electrons. However, for such a large working distance and field on sample below 200 V/mm only the beam with 10 degrees to the axis gets through (the 100 eV beam is started 0.2 mm off-axis). The ray tracing in Figure 3 was done for a two-stage deflection system switched on to the same voltage as in Figure 2. Obviously the central beam, expected to be deflected to 0.12 mm off-axis, lands at 0.2 mm because of the action of the cathode lens. Moreover, the secondary electrons do not focus to about the same point as in case of combined electrostatic-magnetic lens studied by Kienle and Plies [9], and so the beam quality for further use of Wien filter at the position around z=100 mm is not sufficient. The Wien filter [10] will be followed by a system of lenses and by a spherical sector deflector (see [9] for a similar concept). Detailed analysis of these elements is not given here because of lack of time and space.

172

4. Conclusions

Obvious problems in the initial design of an electrostatic SEM were shown. They show only the function of lenses and deflectors and illustrate the approach to computations based on accurate ray tracing instead of aberration theory. They also indicate that the cathode lens must be made stronger by having shorter working distance below 5 mm instead of 25 mm used before; the short working distance does not allow the use of post-lens deflection. The design of the beam separator (Wien filter) and detection system after the filter (sector analyser, transfer lenses) is still a subject of further research.

Acknowledgement. Supported by grant 202/03/1575 of the Grant agency of the Czech Republic.

References

[1] Munro E 1997 in Handbook of Charged Particle Optics (ed. J. Orloff), CRC Press.
[2] Lencová B 1999 Nucl. Instr. Meth. **A427,** 329-337
[3] Lencová B 1995 Nucl. Instr. Meth. **A363,** 190-7
[4] Lenc M and Lencová B 1998 Rev. Sci. Instrum **68,** 4409-14
[5] Lencová B 2002 Ultramicroscopy **93** 263-270
[6] Lencová B and Zlámal J 2000 Proc. EUREM 2000, Brno, Vol. I, 101-102
[7] Horáček M 2003 Rev. Sci. Instrum **74,** 3379-84
[8] Orloff J and Swanson LW 1979 J. Appl. Phys. **50,** 2494-2501
[9] Kienle M and Plies E 2003 Proc. CPO6 (Nucl. Instr. Meth. **A**), in print.
[10] Lencová B and Vlček I 2000 Proc. EUREM 2000, Brno, Vol. I, 87-88

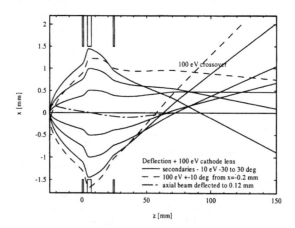

Figure 3. Ray tracing of 10 eV and 100 eV electrons going through system from Figure 1 in cathode lens arrangement (the optical axis direction is reversed). The figure shows clearly the large beam size above deflector, the effect of cathode lens on deflection of the primary beam and the angular extent of rays that can pass through the lens.

Inst. Phys. Conf. Ser. No 179: Section 3
Paper presented at Electron Microscopy and Analysis Group Conf. EMAG2003, Oxford, 2003
©*2003 IOP Publishing Ltd*

Remote Microscopy using the Grid

R R Meyer[1], A I Kirkland[1], M Dovey[2], D J H Cockayne[1] and P Jeffreys[2]
[1] Department of Materials Science, Parks Road, Oxford, OX1 3PH
[2] Oxford University Computing Services, 13 Banbury Road, Oxford, OX2 6NN

Abstract: As part of the UK e-science initiative (within the Oxford e-science centre) we are developing an implementation of a remote JEOL SEM operating persistently and pervasively in a shared and secure multi user environment. It is anticipated that "GridSEM" will be operated as part of the computing grid by a wide range of user groups including schools, museums and small industry and in particular will facilitate remote collaborative investigations

1. Introduction

Electron beam instrumentation, such as electron microscopes, electron probes and electron beam lithography systems are becoming increasingly expensive, increasingly sophisticated, abut at the same time more essential to the research activities of a widening group of users both in industry and in Universities. Because of the rising capital cost, the need for expert operators and maintenance support, and the fact that many users need only occasional access, it is inevitable that shared instrumentation with remote access will become necessary. To address this as part of the UK e-science initiative (within the Oxford e-science centre) we are developing an implementation of a remote JEOL SEM operating persistently and pervasively in a shared and secure environment.

A computational grid is the next generation computing infrastructure to support the growing need for computational based science. This involves utilization of widely distributed computing resources, storage facilities and networks owned by different organisations but used by individuals who are not necessarily a member of that organization [1,2]. A descriptive explanation of computational grids draws on an analogy with the electric power grid. The latter provides us with instant access to power which we use in many different ways without any thought as to the source of that power. A computational grid is expected to function in a similar manner [2]; the end user will have no knowledge of what resource they used to process their data and, in some cases, will not know where the data itself came from and their only interest is in the results they can obtain by using the resource. Today computational grids are being created to provide accessible, dependable, consistent and affordable access to high performance computers, databases and even people across the world. It is anticipated that in the near future these new grids will become as influential and pervasive as their electrical counterpart. Of particular relevance to electron microscopy is the role of shared instrumentation within the grid and the possible use of the Access Grid™(or similar) environments. The latter comprises an ensemble of resources including multimedia large-format displays, presentation and interactive environments and interfaces to Grid middleware and to visualization environments. These resources are used to support group-to-group interactions across the Grid including large-scale distributed meetings, collaborative work sessions, seminars, lectures, tutorials, and training. The Access Grid thus differs from desktop-to-desktop tools that focus on individual communication.

In addition to the Access Grid the basic Grid environment also offers a number of additional benefits which may of use in remote microscopy. In this regard it provides a well defined set of security protocols, contains provision for multiple user conferencing and enables users to seamlessly access other grid based resources e.g. for high performance adaptive video compression, real time image processing etc.

2. System Overview.

The successful implementation of remote instrumentation requires solution of a number of technical challenges as follows:

- Fully duplexed real time instrument control (optics, sample positioning, ancillary devices).
- High Bandwith persistent connection for image feeds (> 1MB / sec).
- Multi-level user interface.
- Multiple remote users with appropriate authentication and security.
- Robust Operation / Remote Diagnostics.

To address the above Oxford GridSEM will incorporate a number of specific hardware and software features. The basic instrument is based around the JEOL JSM 5510LV SEM together with the JEOL WebSEM interface providing operation of the microscope via a remote PC client and addressing the first technical challenge via a SCSI connection between the server PC and the internal microscope hardware. In order to address the second issue it will almost certainly be necessary to implement adaptive video compression in order to cope with the widely varying bandwidths available to different groups of users. For the final challenge the use of the Access Grid environment which provides several streams of multicast video together with audio links will provide a robust remote diagnostic environment. Finally the grid will enable facile remote storage and retrieval of the large amounts of image data that may be generated from a remote instrument.

One promising candidate for the latter is the use of the embedded zerotree wavelet algorithm [3]. The particular advantage of this algorithm is that it is an efficient (lossy) compression algorithm with the particular property that the encoded data stream can have any desired bit rate. Of equal importance is the fact that this algorithm can be readily implemented in software without a need for dedicated hardware, thus increasing its portability.

For an instrument which can be operated b y a wide rage of users spanning other research organisations through to schools it is important to provide a variety of access levels. We propose to do this through the use of different user interfaces which will be coded as standard Web service applications to ensure maximum portability. These Web service applications will run on the local client and communicate directly with the microscope server using Web services. Thus, under this arrangement no modifications to the microscope server hardware or software are required and the client software to control the SEM can be run with only a standard Web browser.

The provision of access for multiple users (using secure digital certificates to determine which user has control) with robust security and authentication is provided for within various standards embedded within the Open Grid Services Architecture (OGSA) [4].

Overall therefore the architecture for the Oxford GridSEM containing these features is summarised in Figure 1. This comprises a series of key functions passed through appropriate grid middleware conforming to the above OGSA standard to provide the necessary microscope controls in a form that is accessible by all users across the grid.

In order to achieve this we will utilize the network topology shown in Figure 2 in which the SEM is connected to the Oxford University 10GBit backbone via a microwave link from the Begbroke site. In the initial phase we plan to install a grid cluster to enable dynamic video compression at the Begbroke site together with an Access Grid room at the microscope. This will be linked to a variety of clients and to other Access Grid rooms thus providing multi user "collaboratories" [2] centered around one instrument.

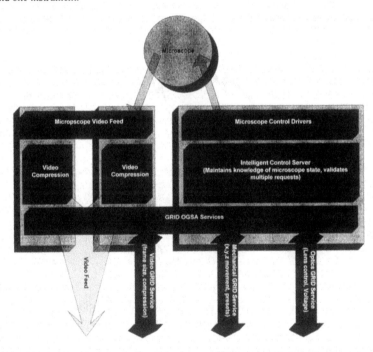

Figure 1. Overview of the proposed service architecture showing the necessary microscope control services. The basic microscope control functions are passed through appropriate grid middleware to provide a grid based instrument. The video streams from the microscope and Access Grid environment are exposed to appropriate grid based resources for dynamic data compression and processing.

176

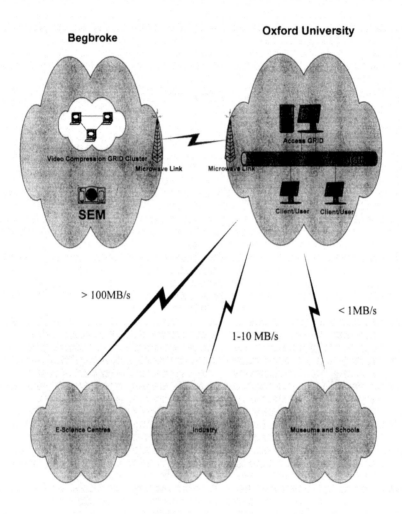

Figure 2. Network topology linking the GridSEM within the Oxford University sites and typical currently available bandwidths to external users.

References.
[1] Foster I and Kesselman C 1999 The Grid: Blueprint for a new computing infrastructure, Morgan Kaufmann ISBN 1-55860-475-8.
[2] Hey T and Trefethen A 2003 Phil. Trans. Roy. Soc (Lond) **361**, 1809.
[3] Shapiro J.M 1993 EEE Trans. on Signal Processing **41(12)**, 3445.
[4] Welch V, Siebenlist F, Foster I, Bresnahan J, Czajkowski K, Gawor J, Kesselman C, Meder S, Pearlman L and Tuecke S 2003 IEEE Press In press.

Inst. Phys. Conf. Ser. No 179: Section 3
Paper presented at Electron Microscopy and Analysis Group Conf. EMAG2003, Oxford, 2003

Miniature SEM with permanent-magnet lenses

B Buijsse

Philips Research, Eindhoven, The Netherlands

Abstract. A very compact low-voltage SEM has been designed, fabricated, and tested. The system is a two-lens column equipped with a Schottky emitter electron source. The lenses are energized by permanent magnets. The low aberration coefficients of magnetic lenses and the short column length allow high performance of the column. We have investigated imaging results obtained at electron beam currents of 18 nA and 220 nA at landing energies close to 1 keV. Predicted spot sizes were 13 nm and 90 nm, respectively. The measured spot sizes were 20 nm and 100 nm.

1. Introduction

A SEM equipped with a high brightness electron source, like a Schottky emitter, can be miniaturized to dimensions in the centimetre range. Several low-voltage minicolumns have been built during the last 15 years, most of them based on electrostatic lenses (e.g. Krans et al 1999, Winkler et al 1997, Chang et al 2001). Miniaturization allows parallel positioning of multiple electron beam columns which can improve throughput for electron beam lithography or for large area electron beam inspection, important for semiconductor industry. Miniaturized SEM columns can also be applied in UHV systems where only limited space is available for SEM functionality.

In electron microscopy magnetic lenses are generally preferred above electrostatic lenses. They have smaller aberration coefficients, which results in higher resolution images. However, in miniature design current carrying coils cannot offer sufficient ampere turns to generate strong enough lens fields. These can only be realized with permanent magnets. Permanent-magnet lenses have the additional advantage that no electrical connections or connections for water cooling are needed. This reduces the complexity of the SEM column. A disadvantage of permanent-magnet lenses is that the lens fields cannot be tuned in strength. This limits the use of a magnetic version of a miniature SEM column to applications where a fixed landing energy is used.

Adamec et al (Adamec et al 1995) have investigated the use of permanent magnet materials based on rare earth metals for the design of small magnetic lenses. Using these lenses they successfully built a low-voltage TEM of compact dimensions. Khursheed et al (Khursheed et al 1998) were the first to build a miniaturized SEM employing permanent magnet lenses. In this work an alternative design is proposed of even smaller dimensions.

2. The column design

A difficulty encountered in column design with permanent magnet lenses is the fact that the integral over the magnetic field along the optical axis is always equal to zero, obeying Ampères law. It is not possible to design a single lens without a surrounding stray field. This problem can be avoided by spatially confining the stray field and forming an additional lens from this confined field. Such a two-lens system can serve as the condenser and objective lens of a two-lens miniature SEM. Obviously, the strength of these lenses is related to each other as a result of Ampères law.

There are several ways to form a condenser-objective unit using one piece of permanent magnetic material, examples have been given by Adamec et al. Here an alternative way of forming a two-lens unit is suggested, using two small magnets made of Sm_2Co_{17} in the shape of hollow cylinders (outer diameter: 16 mm, inner diameter: 10 mm, thickness: 2 mm). Figure 1 shows our prototype SEM column equipped with these magnets. The magnets have their magnetization vector in opposite directions. The total height of the column is 5 cm.

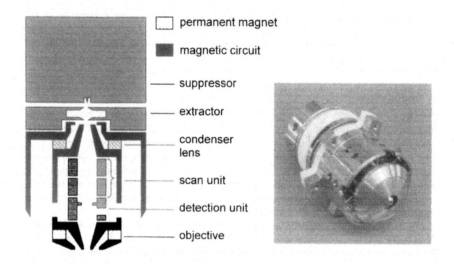

Figure 1. Miniature scanning electron microscope. (a) Schematic drawing (b) Photograph of the unwired column. The total column length is 50 mm, the diameter is 28 mm.

A magnetic circuit of cobalt-iron and carefully designed pole pieces guide the magnetic flu lines to the lens regions. Saturation of the pole tips is thereby avoided. Also flux leaks to th environment (apart from the sample region) and to the optical axis are avoided. The objectiv lens is of immersion type and has a bore diameter of 1 mm. The magnetic field reaches a valu of 0.3 T on the optical axis. The usual electrostatic condenser lens of the Schottky emitte source module is replaced by a magnetic condenser lens.

The beam current is defined by the extractor aperture. Inside the column there are no additiona apertures. The column was successively tested with extractor apertures of 25 μm and 72 μr diameter, resulting in respective currents of 18 nA and 220 nA. High voltages are restricted t the source module, the remainder of the column is operated at ground potential. Thus the beam energy in the column is equal to the landing energy.

Scanning is performed electrostatically using a double layer of octupole electrodes. Below the deflection layers a detection unit is placed, that can act as mirror for SE travelling upwards. The retarded electrons can be deflected towards a detector mounted just behind a hole in the detection chamber sidewall (not shown in Figure 1). In the present prototype we have not tested this detector but used the sample current as detection signal.

A typical working distance is between 1 and 1.5 mm. Some fine focussing is possible by a slight variation of the beam energy. There is a limited range in landing energy where optimal focus conditions are present and the smallest spot size can be achieved.

3. Simulation and imaging results

Calculations were performed on the magnetic and electron optical properties of the configuration shown in Figure 1. Vector Fields OPERA-2d and 3d software was used for the calculation of the magnetic flux distribution of the column. The axial magnetic field served as input for Munro's electron optical software (CMECH). The objective lens has a coefficient of spherical aberration of 2.3 mm at a working distance of 1.5 mm and a landing energy of 1.2 keV; the coefficient of chromatic aberration is 1.1 mm. Due to the high beam current that is used and the low beam voltage, also statistical Coulomb interactions will contribute to the final spot size. Thanks to the short column length this contribution is of the same order of magnitude as the other contributions. The Coulomb blurring was analytically calculated (Jansen 1990) using in-house developed software. The total spot size was calculated according to the formula of Barth and Kruit (Barth et al 1996).

In Figure 2 the spot size as a function of landing energy has been plotted, operating with a beam current of 18 nA. According to the calculations, the column provides optimal resolution in the range from 1.2 keV to 1.3 keV with a spot size of 13 to 14 nm. At a beam current of 220 nA the contribution of Coulomb blurring and spherical aberration drastically increases and a smallest spot size of 90 nm is reached at 1.1 keV.

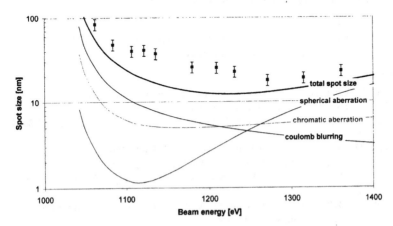

Figure 2. Calculated and measured spot size ($d_{12\%-88\%}$) as a function of landing energy for a current of 18 nA. The various contributions to the total spot size have been indicated.

180

The performance of the column was tested by making micrographs for various values of the beam landing energy, both in 18 nA and 220 nA beam current mode. A carbon TEM foil was employed as sample. In Figure 3 it can be seen that good quality micrographs can be obtained at high spatial resolution.

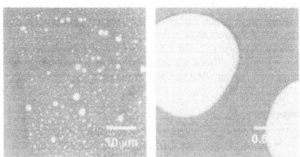

Figure 3   SEM micrographs of holey carbon foil obtained with the column shown in Figure 1 (U=1.3 keV, I=18 nA). The sample current is used as contrast signal.

The $d_{12\%-88\%}$ edge resolution was determined for these images, the result of which is displayed in Figure 2. For the 18 nA beam there is some discrepancy. At a beam current of 220 nA the agreement is good. The discrepancy at 18 nA can be attributed to small misalignment of some parts of the column and associated aberrations. Moreover, mechanical vibrations deteriorated the performance.

Conclusions

A prototype miniature SEM equipped with permanent-magnet lenses has successfully been built. High-resolution micrographs have been obtained. Analysis of these micrographs shows close-to-predicted performance of the column.

Acknowledgements

The excellent support of M. van Wely, M.T. Meuwese and C.A.N. van der Vleuten in the design and assembly of the column is acknowledged.

References

Adamec P, Delong A and Lencova B 1995 J. Microscopy **179** 129-132

Barth J E and Kruit P 1996 Optik **101**, 101-109

Chang T H P, Mankos M, Lee K Y and Muray L P 2001 Microelectron. Eng. **57-58**, 117-135

Jansen G H 1990 Nucl. Instr. Meth. Phys. Res. A, **A298** 496-504

Khursheed A, Phang J C and Thong J T L 1998 Scanning **20**, 87-91

Krans J M and van Rooy T L 1999 Microscopy and Microanalysis 5, Microscopy Society of America, 322-323

Winkler D, Bubeck C D, Fleischmann A, Knell G, Lutsch Y and Plies E 1998 J.Vac.Sci.Technol. **B16**, 3181-3184

Inst. Phys. Conf. Ser. No 179: Section 3
Paper presented at Electron Microscopy and Analysis Group Conf. EMAG2003, Oxford, 2003
©2003 IOP Publishing Ltd

Backscattered SEM imaging of high-temperature samples for grain growth studies in metals

I M Fielden, J Cawley*, J M Rodenburg**

Materials Research Institute, Sheffield Hallam University, Howard Street, Sheffield S1 1WB, U.K. **Mappin Building, Mappin St. Sheffield S1 3DJ, U.K. *Solutio (Sheffield) Ltd, 22 Carterknowle Road, Sheffield S7 2DX, U.K

Abstract. A novel technique is presented for SEM backscattered electron imaging of hot specimens, in-situ, in real time. The technique has been applied to recrystallisation, grain growth and phase transformation studies in metals, principally steel. Temperatures attained were in the order of 850°C and were limited by the capability of the specimen heater, not the detector. Representative results are presented as stills and video.

1. Introduction

1.1. The materials problem

Grain size and (where they occur) phase transformations are key determinants of microstructural evolution and hence materials properties. Large gains in strength and toughness are available by engineering a smaller finished grain size. There are many methods for characterising static or room-temperature microstructures. However, in almost all real-world thermo-mechanical processing (e.g. hot-rolling of steel) it would be highly advantageous to have some knowledge of the kinetics of these phenomena. Many methods exist for modelling the dynamics of recrystallisation, growth and/or phase transformation, but these have proved very hard to confirm experimentally; kinetics are generally inferred from "post-mortem" studies.

1.2. The electron microscopy problem

The most attractive ways of generating grain-to-grain contrast are crystal orientation sensitive techniques relying on backscattered electrons. Established backscatter detectors are intrinsically sensitive to the IR and light photons emitted by a hot specimen. Imaging with diode detectors rapidly becomes problematic with increasing temperature, and is impossible beyond temperatures of about 450°C. Scintillator/photomultiplier detectors are similarly affected. Topographic techniques (e.g. SE imaging) show grain boundaries via secondary phenomena (e.g. thermally-etched boundary grooves). These features do not form rapidly (or even at all) under all conditions. Thus high-temperature SE imaging via the ESEM is unhelpful.

2. Technique development

Initial trials used room-temperature specimens and conventional diode detectors. These covered several configurations: zero-tilt and high-tilt (EBSD-like "forward scatter") geometries, large and small detector areas, single/summed/difference signals from two detector elements and differing distances between two detector elements. An electron-channelling contrast (ECCI) geometry [1] – of a single detector subtending a large solid angle at a specimen of zero tilt – proved to be superior.

Diode detectors overlaid with thin polymer films were trialled at room temperature. The intention was, if successful, to thinly coat the polymer with a light metal to exclude/reflect photons. Results were variable, but at best somewhat disappointing.

The micro-channel plate detector was considered and discarded, due to cost and risk of damage from the contaminants anticipated from a heated specimen.

2.1. Modified converter plate detector

The converter plate appears to be little-used in current SEM practice. It appears in the literature, notably the more comprehensive textbooks [2], but the authors have found only one reference to it on the Internet and no indication that the technique is in regular use, or was ever applied to hot specimens or orientation contrast.

The converter plate emits secondary electrons (designated SE3) when struck by energetic backscattered electrons. This allows images to be generated from BSE information via an SE detector. The conversion gives little or no amplification, and SE1+SE2 signal from the specimen must be suppressed if a "pure" BSE image is desired; hence its lack of popularity. However, the converter plate is inherently immune to photon radiation, can be made of thermally robust materials and is easily made in any desired size and shape. It is equally applicable to SEM and ESEM, with an SE detector shielded from the specimen, or one immune to photon effects.

Converter plates were trialled at room temperature, with encouraging results. The plate designed for the zero-tilt ECCI geometry was not only an effective detector, but could also function as a heat shield. An early prototype plate was coated with diamond-like carbon (DLC), but there was no obvious advantage over the established MgO coating, which allowed coating in-house and rapid prototyping of plates.

A Philips/FEI XL30 ESEM-FEG was fitted with a Philips/FEI hot stage (nominal capacity 1000°C) and an optimised converter plate. A low-deformation polished specimen of cold-deformed 1050 aluminium (commercially pure) was heated, and images captured as digital video. The instrument was operated in "hi-vac" conventional SEM mode, but with differential pumping activated and a "hot-stage" ESD/PLA assembly in place to protect the column.

Grains were clearly seen and in accelerated video playback the growth process could be followed. However, acceptable image quality dictated a scan time of 20-30 seconds per frame, so highly dynamic events could not be followed. 660°C was attained.

The experiment was repeated with a heavily cold-deformed, nominally-eutectoid carbon steel wire, cut in transverse section. Image quality was greatly improved, due to the higher back-scattering coefficient of the steel. Maximum specimen temperature was ≈860°C, limited by the specimen heater. Improved image quality allowed a faster scan speed: ≈5s per frame. A subsequent experiment gave acceptable images with a ≈1s per-frame scan time, meeting the "in-situ, in-real-time" goal and allowing highly dynamic events to be followed. Increasing temperature did not degrade image quality. Gold has since been investigated, as a model FCC material.

3. Results

To appreciate the kinetics, it is best to watch a video sequence. These have been made available in highly-compressed form as supporting media files, and will be placed on or linked from: http://extra.shu.ac.uk/fielden-emag2003 and http://www.srama.demon.co.uk/iain/publications/emag2003. No address can be indefinitely guaranteed not to change. In case of difficulty, try this paper's title as a search term, or contact the author for information or a CD containing uncompressed video.

Space dictates that only the first steel results are considered here. Full-size figures show the same specimen area, others a selection from it. The full field is ≈150μm wide; all scale marks are 20μm; specimen temperatures are approximately calibrated (estimated ±15°C), as the thermocouple monitors the heater, not the specimen.

3.1. Transformation to Austenite

The initial highly-deformed pearlite structure is irresolvable at this magnification. Above the transformation temperature the structure becomes "unlocked", presumably by the dissolution of carbide in the newly-transformed Austenite, and a plethora of fresh γ crystals rapidly appear (fig1).

3.2. Grain growth

After an initial flurry of grain growth activity (figs 1-3), grain growth occurs in a localised manner, often with only one boundary in the field moving (fig 4).

Fig 1. Austenite grains at 765°C.

Fig 2. As Fig 1, +70 seconds, 785°C.

Fig 3. As Fig 1, +140 seconds, 795°C.

Fig 4. The only growing grain in the field. 860°C, 60-second interval.

Movement is a mixture of smooth gliding with some stop/start behaviour and speed/direction changes. In one instance it is clear that a growing grain's boundary has reversed direction and retreated immediately after having annihilated a neighbour. While the average velocity of boundaries is slow enough to be easily followed, some boundaries (e.g. those in the act of annihilating a neighbour grain) move very rapidly indeed (fig 5).

Fig 5. The last 19 seconds in the life of a previously stable grain, 845°C.

Later in the growth process, thermally-etched boundary grooves appear (fig 6, 7; upper part of fig 4). The grain boundaries do not show any strong tendency to become pinned by these (this becomes particularly evident when the specimen is re-heated).

3.3. Transformation to pearlite

Transformation from Austenite on cooling appears to nucleate mostly from grain boundaries and possibly twin boundaries. The first evidence of transformation visible in the microstructure are small, relatively slow-growing islands that appear dark in the ECCI/BSE-Z image (fig 6).

184

These features do not go on to play a part in the main "wavefront" growth of pearlite originating from grain boundaries. The most obvious/active growth front in this sequence appears to originate from a triple point (fig 6c, 7).

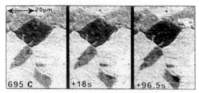

Fig 6abc. At 695°C. High-temperature microstructure and first signs of transformation (at +18s & 96.5s).

Fig 7. (right) At +97s. Shows context of fig 6 & rapid motion of wavefront (see 6c).

Fig 8. At +200s. A transformation front, originating from A, has crossed X-X but not X-Y. More "dark island" features.

Fig 9. At +232s & 660°C. Transformation fronts from several origins sweep rapidly across the specimen.

The "wavefront" is briefly checked, but seems to cross austenite grain boundaries with ease. However, where the direction of the front is approximately parallel to the boundary (e.g. X-Y in fig 8), crossing to the neighbouring grain occurs less readily. The motion of the transformation fronts is as if they had a physical momentum in the direction of their movement, which does not transfer across boundaries hit at a shallow angle.

The black & white mottled appearance (fig 9) of the transformed structure is unexplained, but the ECCI set-up is extremely sensitive to any Z contrast, if present.

4. Conclusions

The converter plate is a practical method of imaging high-temperature grain structures and would be easily capable of giving atomic number contrast at high temperature.

References

[1]   Simkin B A Ben Simkin's Electron Channelling Contrast Imaging (ECCI) How-To [online] http://www.egr.msu.edu/~simkin/ECCI.html (last accessed 26/08/03)

[2]   Reimer L 1998 Scanning Electron Microscopy: Physics of Image Formation and Microanalysis, 2nd ed. (Heidelberg: Springer-Verlag)

Inst. Phys. Conf. Ser. No 179: Section 4
Paper presented at Electron Microscopy and Analysis Group Conf. EMAG2003, Oxford, 2003
©2003 IOP Publishing Ltd

Can Ronchigrams provide a route to sub-angstrom tomographic reconstruction?

J M Rodenburg
Department of Electronic and Electrical Engineering, University of Sheffield, Mappin Street, Sheffield S1 3JD UK

Abstract: High-angle aberration-corrected Ronchigrams can be processed as a function of probe position to pick out intensity fringes arising from many pairs of Bragg conditions. We can obtain an estimate of several planes through reciprocal space, which correspond to projections of the object function in certain directions: the first step of a tomographic reconstruction at sub-atomic resolution. A principal set of beams lie at or near the achromatic ring, thus maximising instrument transfer to very high resolution.

1. Introduction

Ronchigrams observed in an aberration-corrected STEM exhibit elaborate interference fringes (Figure 1). The angular width and clarity of the region where inference fringes can be seen is much larger than in uncorrected machines. What can we do with these patterns; do they hold any information which we can't obtain from the ADF or conventional TEM image? If we observe a thick region of amorphous material and vary defocus, we see contracting and expanding speckle. Regions near the edge of the pattern change in appearance dramatically. We can also discern different rates of movement (expansion and contraction) as a function of a smooth change of defocus: this is because each level in the specimen is at a different defocus, so that a variety of different apparent local magnifications occur simultaneously. Even when the probe is focussed accurately on the centre of an amorphous specimen, the central disc has elaborate structure, because specimen layers above and below the plane of focus are significantly out of focus. In other words, when their angular range is expanded, Ronchigrams are rich in z-direction data. Indeed, in his original paper on holography, Gabor modelled this exact scattering geometry with light and was surprised to observe 'It is a striking property of these diagrams that they constitute records of three-dimensional as well as plane objects. One plane after another of extended objects can be observed in the microscope...' [1].

In practice, Gabor holography relies on a major part of the object plane being transparent, in order to provide the in-line reference wave. This is generally not easy to achieve if the specimen is self-supporting. In what follows, we are concerned with processing many such patterns collected from many probe positions. This extended data set has several advantages over any one single Ronchigram (or Gabor hologram).

186

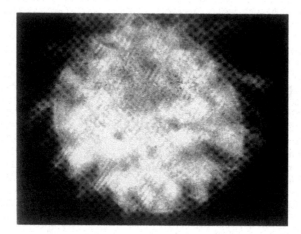

Figure 1: An electron Ronchigram from silicon, collected on the aberraction-corrected superSTEM in Daresbury, recorded with large defocus. Note the extensive interference fringes: it is these that can be processed to give a projection image of the atomic potential.

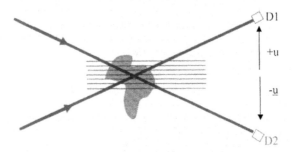

Figure 2: Two incident beams within a STEM probe. The detector pixels lie in the line of sight of the beams. The specimen appears illuminated by sheets of intensity.

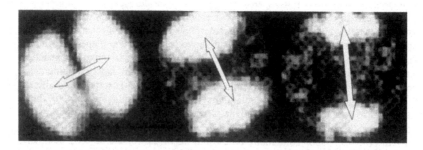

Figure 3: The result of Fourier-filtering many Ronchigrams as function of probe position obtained from thin amorphous carbon. The arc-shaped features represent those pairs of pixels (detectors D1 and D2 in Fig 2) which are separated by the vector of the relevant Fourier component marked (of length 2<u>u</u>).

It should be noted that this data is simultaneously available during any conventional annular dark field (ADF) scan. For ultimate imaging microscopy, it would be ideal to collect all the dark- and bright-field information at once. The only experimental obstacle lies in developing a two-dimensional detector that can collect (with low noise) and save (via a wide band-pass data bus) the enormous quantities of data which are produced.

2. The double-resolution projected image

Let us start by considering a pair of beams lying within the STEM probe which reside at equal distances (in angle space) either side of the optic axis. Because both these beams lie on the same diameter of the probe forming lens (aberration-corrected or not) they are less affected by chromatic spread in the beam or instability in the lens supply. Consider two detector pixels in the far-field (Ronchigram plane) which lie in direct line of sight of these two beams, as illustrate in Figure 2. As we move the probe, that is, alter the relative phase of the two incident beams in the STEM, the interference condition at the detectors will alter periodically. One way to think about this is as follows. The two incident beams interfere at the specimen to produce a series of flat planes of intensity which lie parallel with the optic axis. As the phase of the incident two beams is varied, these planes shift across the specimen. We would expect more or less scattering depending on whether the planes of intensity coincide with well-populated atomic planes or not. The easiest way to calculate the intensity at either D1 or D2 is via reciprocity. If we place a source at D1, this is equivalent to illuminating the specimen in a tilted-beam condition in TEM. The variation we expect as function of probe position is then given by the two-beam fringe interference pattern in the image plane of a TEM.

Of course, a STEM probe consists of a range of incident k-vectors, and so any one detector pixel in the Ronchigram can also have amplitude scattered to it from all of the other beams. This is where processing the data as a function of probe position is so useful. We can separate the pairs of interfering beams as follows. Let us designate the position of a pixel in the Ronchigram by a reciprocal position vector $\underline{u}$, measured from the optic axis, so that our detectors D1 and D2 lie at $+\underline{u}$ and $-\underline{u}$ respectively. As a function of probe position, the Fourier components of intensity at either of these detectors due to interference from the equal and opposite beam must have a spatial frequency of $\pm 2\underline{u}$, the separation of the two beams. Figure 3 shows a set of data which has been processed in this way. A large number of Ronchigrams have been collected from an amorphous material. In the original data, the intensity at each pixel appears to be random. However, each of the pictures in Figure 3 shows just one Fourier component of the intensity measured at each pixel, as resolved as a function of probe movement. The vector component of that periodicity is shown drawn on each pattern. It is remarkable that even though the object is amorphous, the areas which have strong amplitude are shaped like the overlap region of three apertures: the disc of the central beam (the Gabor hologram), and two diffracted discs, of the same diameter, shifted by $\pm 2\underline{u}$. The shapes represent all pixels from which we can draw a vector $\pm 2\underline{u}$ which fits inside the objective aperture. Of course, no diffracted disc is visible in the original Ronchigram itself because the specimen is amorphous: this is simply the region where two-beam interference can occur. It is important to note that the value of amplitude shown here is a complex quantity: it has both phase and modulus because, in general, the interference fringes measured, as the probe moves, will have arbitrary position.

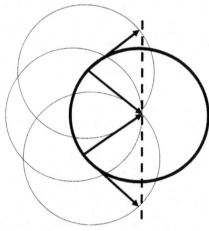

Figure 4a: Greatly exaggerated Ewald sphere construction. The heavy circle is the locus of points which lie at the origin of the k-vectors of the incident waves in STEM. All these point towards the origin at the centre of the heavy circle. Scattered beams subtend from a point on the heavy circle and explore reciprocal space, on Ewald spheres shown as thin lines. Pairs of symmetric beams, as shown in Fig 2, which also lie on the achromatic ring of the objective lens, explore a plane in reciprocal space shown as a dotted line. This is a projection in real space.

Figure 4b: In any one full set of Ronchigrams, collected as a function of all probe positions, scattering data pertaining to the grey shaded regions of reciprocal space can be reached. Any two pixels, separated by 2$\underline{u}$, as in Fig 2, but which are not symmetric to the optic axis, explore other planes (one is shown dotted) in reciprocal space. Very high spatial frequencies are not available on these tilted beams, because we still require an incident beam to lie parallel with any one particular scattered beam, so that the two can interfere in the far-field.

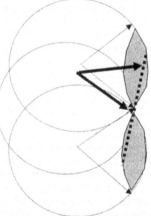

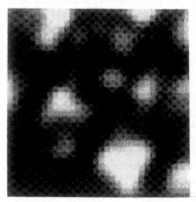

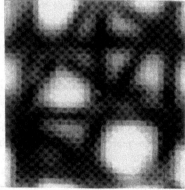

Figure 5: Optical demonstration of projection imaging. The left figure is a bright-field image of two TEM grids, separated in the z-direction: neither grid is clearly visible. The right image is a double-resolution projection image.

By Fourier processing the data in this way, we have separated one component of the set of all Bragg conditions shown in Figure 4a, which is a representation of reciprocal space drawn through the Ewald sphere. The locus of all such points which represent the scattering vectors described by this geometry is a plane through reciprocal space, perpendicular to the optic axis. Interference between pairs of pixels that do not lie exactly at $-\underline{u}$ and $+\underline{u}$ (i.e., are not on the achromatic ring) but are still separated by $\pm 2\underline{u}$, describe scattering vectors that lie on other planes in k-space, tilted slightly. The total region of reciprocal space we can access in this way is shown in cross-section in Figure 4b. For low frequencies ($\underline{u}$ less than 1/3 the objective aperture radius), and for low angle average orientations, more than two beams can interfere at once. This transfer function is identical to that which is obtained in a scanning optical microscope (SOM). The principal benefit of SOM is its ability to focus on one layer within a three-dimensional specimen in a way that filters out multiple scattering. We might hope that we could employ this similar method in STEM.

What we need is some algorithm for processing Ronchigrams to give us a quantity which is related to the three dimensional structure of the specimen. Apart from all the usual problems of specimen damage, contamination and (especially in the case of STEM) drift, there are fearsome complications associated with dynamical scattering. However, this Fourier-resolved processing is an excellent first step.

When the specimen is relatively weak, we can combine the complex amplitudes of Fourier components shown in Figure 3 in a cunning way [2,3] to obtain an estimate of the projected complex transmission function of the specimen: that is to say, an estimate of

$$T(x,y) = \exp(i\sigma \int V(x,y,z)dz)$$

where x and y are coordinates in the projected image plane, V(x,y,z) is the three-dimensional atomic potential, and σ is the scattering cross-section, dependent on electron incident energy. If we could concertina the specimen into a truly two-dimensional plane (that is, 'switch off' the Fresnel propagation that occurs from one layer of specimen to the next), this equation would hold true provided the total z-integrated inner potential was a small fraction of the incident energy.

An optical test of this method is shown in figure 5. Figure 5a is a conventional bright-field optical STEM image of two TEM gratings separated in the z-direction by several millimetres. Neither grating can be easily resolved. In Figure 5b, synthesised from all symmetric two-beam conditions, we see a double-resolution image of the object in perfect projection. The gain in resolution, which is a characteristic of this method, results from the fact that we are not interfering an off-axis beam at $\underline{u}$ with the central beam at $\underline{u}=0$ (as in bright-field imaging) but with an opposite beam at $-\underline{u}$, leading to a doubling of the effective scattering angle and hence resolution. Why do we obtain a projection? Firstly, because beams on the achromatic ring are by definition insensitive to defocus. Secondly, because a plane in reciprocal space (not the curved surface of the Ewald sphere in conventional imaging) described by the vectors in Figure 4a, is the Fourier transform of a projection of the object function. This is a well-known result in 3D tomographic imaging. In fact, from any one set of position-resolved Ronchigrams, we can synthesise a number of projections of the object, but then not all the beams involved have the benefit of lying exactly on the achromatic ring, which will always have maximum information transfer.

In the electron case, dynamical scattering complicates the phase and amplitude of the two-beam interference fringes we need to use in this method. If the specimen is crystalline, then at certain orientations (excitation errors combined with thickness) the top and bottom atoms of adjacent columns will be seen just to overlap in projection, which means that no scattered beam is excited. At this orientation, the reconstruction method described above will give zero Fourier component in the estimate of V(x,y): that is to say, the correct result. This is in contrast to the ADF image, where at small specimen tilt, channelling still gives rise to apparent atomic columns. When the zone axis is perfectly aligned with the optic axis, the achromatic signal consists of pure Bragg conditions, each with zero excitation error. The two beam solution for this case gives image fringes which are π/2 out of phase with the atomic columns: when combined with the method described in [3], this still gives the correct estimate of V(x,y) (i.e. as expressed in the phase of T(x,y)) up to about a quarter of the extinction distance. In practice, multiple beam effects also leads to the attenuation of the transmitted beam. However, because in any one data set we have variety of excitation errors to explore for pairs of beams not lying on the perfect Bragg condition, we might hope in future to use a more sophisticated inversion process to obtain a better estimate of V(x,y,z).

3. Experimental issues

The benefit of the data set described is that it is insensitive to chromatic aberration, which is an outstanding limitation to increased resolution. Unfortunately, it is particularly vulnerable to an alternative source of instability – any lateral motion or drift of the electron source or, equivalently, drift in the absolute position of the specimen. For this reason, it would good to use rather fewer Ronchigrams, taken from well-spaced probe positions. To span a greater field of view of the specimen, one could employ a large amount of defocus, rather like in a Gabor hologram, which would have the effect of putting much more intensity structure into the detector pixels. Unfortunately, though, this data cannot allow unique separation of all the Fourier components described above. An alternative strategy, given that the ultimate intention would be to construct a three-dimensional tomographic reconstruction, would be to rotate a rod-like specimen within the defocused STEM probe, and use the angular stepping as an alternative set of measurements, resolved as a function of angle, instead of probe position.

4. Conclusions

We can use the wide transfer function of an aberration-corrected STEM to improve resolution and achromatic transfer. The ADF image indirectly exploits all the coherent interference by detecting those electrons which are scattered out from the transmitted beams to high angle, via destructive interference. But in ADF we only measure one data point (intensity) per image pixel. In the Ronchigram, we have a very rich set of interference data, collectable from each real-space probe position. The challenge is to make a detector which can capture all this vast wealth of information quickly enough and with low enough read-out noise.

Thanks are due to Andrew Bleloch for helping me obtain Figure 1 on the Daresbury SuperSTEM, and to Archie Howie for helpful email discussion.

5. References

[1] Gabor D Nature 161 (1948) 777
[2] Rodenburg J M, McCallum, B C and Nellist, P D Ultramicroscopy 48 (1993) 304
[3] Plamann T and Rodenburg J M Optik 96 (1994) 31

Inst. Phys. Conf. Ser. No 179: Section 4
Paper presented at Electron Microscopy and Analysis Group Conf. EMAG2003, Oxford, 2003
©2003 IOP Publishing Ltd

Chromatic aberrations of electrostatic lens doublets

B Lencová and P W Hawkes[(*)]
Institute of Scientific Instruments, Czech Academy of Sciences,
Královopolská 147, 61264 BRNO, Czech Republic
[(*)] CEMES-CNRS, B.P. 4347, 31055 TOULOUSE cedex 4, France

Abstract. The asymptotic aberration coefficients of the primary aberrations of electron lenses can be written as polynomials in reciprocal magnification (or object position). Expressions are available for the polynomial coefficients of a doublet in terms of those of the individual members of the doublet for both geometrical and chromatic aberrations. Here, these expressions are used to search for favourable lens doublet configurations when the chromatic aberration is the most important.

1. Introduction

That the asymptotic aberration coefficients of electron lenses can be written as polynomials in reciprocal magnification m ($m = 1/M$, where M denotes the magnification) can be seen immediately from the integral expressions for these coefficients. For the chromatic aberration coefficients the regular axial chromatic aberration coefficient is quadratic in m and the chromatic aberration of distortion is linear in m. The anisotropic aberration of magnetic lenses is independent of m.

Expressions for the polynomial coefficients of the chromatic aberration coefficients of electrostatic and chromatic lens doublets have recently been established in terms of those of the individual members of the doublet [1]. Here, we exploit these relations to search for favourable lens doublets and to shed light on the behaviour of the chromatic aberrations of doublets in different working conditions.

2. Electrostatic doublets

For a single electrostatic lens, imaging in the presence of chromatic aberration can be conveniently expressed in terms of a transfer matrix T [2] (the matrix elements can be extracted from [3]):

$$\begin{pmatrix} u_i + \Delta u_i \\ u_i' + \Delta u_i' \end{pmatrix} = T \begin{pmatrix} u_o \\ u_o' \end{pmatrix} \quad ,$$

$$T = \begin{pmatrix} M - M\left(C_D + C_e\right)\dfrac{\Delta\phi}{\hat{\phi}_o} & -M C_c \dfrac{\Delta\phi}{\hat{\phi}_o} \\ c + \left[-c\left(C_D + C_e\right) + r\,m\,C_\alpha\right]\dfrac{\Delta\phi}{\hat{\phi}_o} & \left[-c C_c + r\,m\left(C_D - C_e\right)\right]\dfrac{\Delta\phi}{\hat{\phi}_o} \end{pmatrix} \quad ,$$

where u_o and u_o' represent the position and slope of the incoming asymptotic paraxial ray in the object plane, u_i and u_i' in the image plane, and Δu_i and $\Delta u_i'$ the deviation from the paraxial trajectory arising from a difference in potential $\Delta \phi$ from the relativistically corrected beam voltage at the object side $\hat{\phi}_o = \phi_o (1 + \varepsilon \phi_o)$, $\varepsilon = e/(2m_o c^2)$, $r = (\hat{\phi}_o / \hat{\phi}_i)^{1/2}$, $\gamma = 1 + 2\varepsilon \phi$, $c = -1/f_i$ is the convergence. The integrals in the chromatic aberration coefficients take the form $\int f(\phi) G^p H^{2-p} dz$, where H and G are solutions of the paraxial equations satisfying the boundary conditions $H(z \to -\infty) \to z - z_o$, $G(z \to -\infty) \to 1$. By using another solution of the paraxial equation independent of the object position $\bar{G}(z \to \infty) \to 1$, the ray H can be replaced by $H = f_o(\bar{G} - mG)$, where f_o denotes the object focal length ($f_o = r f_i$) and it is then clear that the integrals can be written as polynomials in m:

$$\begin{pmatrix} C_c \\ C_D \\ C_\alpha \end{pmatrix} = \begin{pmatrix} f_o^2 e_{20} & -2 f_o^2 e_{11} & f_o^2 e_{02} \\ 0 & -f_o e_{20} & f_o e_{11} \\ 0 & 0 & e_{20} \end{pmatrix} \begin{pmatrix} m^2 \\ m \\ 1 \end{pmatrix} ,$$

in which

$$e_{ij} = \hat{\phi}_o^{1/2} \int \frac{\gamma(3 + 2\varepsilon \hat{\phi})}{8 \hat{\phi}^{5/2}} \phi'^2 G^i \bar{G}^j \, dz$$

and

$$C_e = \frac{1}{4}\left(r^2 \gamma_i - \gamma_o\right) .$$

We consider a doublet consisting of two electrostatic lenses, the first lens described by T_m, the second one by T_n. The transfer matrix of the doublet is then $T_p = T_n T_m$. Comparison of the individual terms in the matrix elements gives trivial results for the magnification of the doublet $M_p = M_n M_m$, the inverse magnification $m_p = m_n m_m$, the convergence $c_p = M_m c_n + r_n m_n c_m$ and the 'refractive index' $r_p = r_n r_m$. The terms in the first power of $\Delta \phi$ express the effect of the chromatic aberration of the doublet in terms of the coefficients for the individual lenses

$$e_{20}^p = \frac{r_m f_n^2}{f_p^2} e_{20}^n - 2\frac{r_p f_n^2}{f_m f_p} e_{11}^n + \frac{R}{r_m f_m^2} = \frac{r_m^3 f_{on}^2}{f_{op}^2} e_{20}^n - 2\frac{r_m^3 f_{on}^2}{f_{om} f_{op}} e_{11}^n + \frac{R_o}{f_{om}^2} ,$$

$$e_{11}^p = \frac{r_m f_n}{f_p} e_{11}^n + \frac{f_m}{r_n f_p} e_{11}^m - \frac{R}{r_p f_m f_n} = \frac{r_m^2 f_{on}}{f_{op}} e_{11}^n + \frac{f_{om}}{f_{op}} e_{11}^m - \frac{R_o}{r_m f_{om} f_{on}} ,$$

$$e_{02}^p = \frac{f_m^2}{r_n^2 f_p^2} e_{02}^m - 2\frac{f_m^2}{r_n^2 f_n f_p} e_{11}^m + \frac{R}{r_m r_n^2 f_n^2} = \frac{f_{om}^2}{f_{op}^2} e_{02}^m - 2\frac{f_{om}^2}{r_m f_{on} f_{op}} e_{11}^m + \frac{R_o}{r_m^2 f_{on}^2} ,$$

where

$$R = r_m f_m^2 e_{20}^m + r_p^2 f_n^2 e_{02}^n , \quad R_o = f_{om}^2 e_{20}^m + r_m^3 f_{on}^2 e_{02}^n .$$

This set of coefficients e_{ij}^p characterizes the chromatic aberration of the doublet. The coefficients are functions of the geometry of the lenses composing the doublet and of their excitations but are independent of the object and image positions, or of the magnification. It is therefore interesting to examine these coefficients for configurations of lenses that have been shown to possess useful properties.

The quantities e_{20} and e_{02} are positive definite as are R and R_o. The smallest values of e_{20} and e_{02} for the doublet can therefore be found by exploring the behaviour of certain ratios. We can give each of these ratios in two forms, in the first of which image-side quantities are employed (notably f_i which is written without a suffix), while in the second form, object-side quantities are used. For e_{20} the ratio is

$$E_{20} = \frac{r_m e_{20}^n + R/\left(r_m D_p^2\right)}{2r_p\left(f_p/f_m\right)e_{11}^n} = \frac{R_o + D_p^2 r_m^3 e_{20}^n}{2D_p r_m^3 f_{on} e_{11}^n}$$

and for e_{02}

$$E_{02} = \frac{e_{02}^m + \left(R/r_m D_p^2\right)}{2\left(f_p/f_n\right)e_{11}^m} = \frac{R_o + D_p^2 r_m^2 e_{02}^m}{2D_p r_m f_{om} e_{11}^m}.$$

For e_{11}, we examine

$$E_{11} = -\frac{R/\left(r_m D_p\right)}{r_p f_n e_{11}^n + f_m e_{11}^m} = -\frac{R_o}{D_p r_m\left(r_m^2 f_{on} e_{11}^n + f_{om} e_{11}^m\right)}.$$

The distance between the image-side focus of the first lens and the object-side focus of the second lens is denoted by D_p: $D_p = f_m f_n/f_p = f_{om} f_{on}/f_{op}$.

3. Preliminary calculations

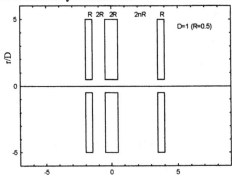

The minimum value of the chromatic aberration for the whole range of magnification is obtained, if all the ratios E_{ij} are close to 1.

This idea has been explored using the lens proposed by Shimizu [4] as a model (Fig. 1). The results are presented for $n=5$ and the second lens shifted by 30 mm. Fig. 2 shows the dependence of the coefficients e_{ij} on strength of the lens (i.e. the inverse of the ratio of the voltages applied to the inner and the outer electrodes V_2/V_1). The thin lens approximation gives $e_{ij} \approx 2f_o$. The doublet is formed with the second lens mirrored in the middle plane of the inner electrode. The focal length of the doublet and ratios E_{ij} are shown on Figs. 3 to 6 in dependence on the strength of the second lens for several values of the strength of the first lens. The results clearly show, that in our explored case the ideal solution does not exist.

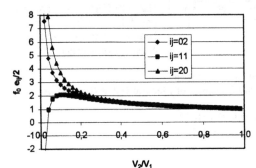

Fig. 2. Normalized e_{ij}.

194

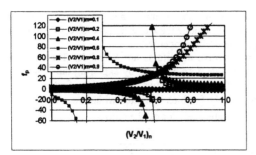

Fig. 3. The focal length of the doublet.

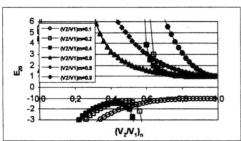

Fig. 4. The E_{20} ratio of the doublet.

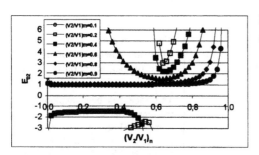

Fig. 5. The E_{02} ratio of the doublet.

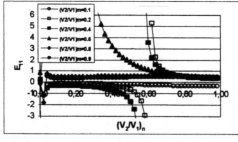

Fig. 6. The E_{11} ratio of the doublet.

Acknowledgement: Work supported by EU grant G5RD CT-2000-00344 NanoFIB and NanoFIB-NAS. We are most grateful to Michal Lenc for helpful comments and much constructive criticism.

References
[1] Hawkes P W 2003 submitted to Ultramicroscopy
[2] Hawkes P W and Kasper E 1989 Principles of Electron Optics (London: Academic Press)
[3] Lencová B and Lenc M 1994 Optik 97 121–126
[4] Shimizu R 1983 Jpn J Appl Phys 22 1623–1626

Inst. Phys. Conf. Ser. No 179: Section 5
Paper presented at Electron Microscopy and Analysis Group Conf. EMAG2003, Oxford, 2003
©2003 IOP Publishing Ltd

Interpretation of Diffraction Patterns from Long-Period out-of-phase Superlattice Structures

G Shao
School of Engineering (H6), University of Surrey,
Guildford, Surrey GU2 7XH, UK

Abstract. The paper presents a generalised analytical formalism for the structure factors and diffraction intensities for long-period out-of-phase superlattices of both constant modulating periods and incommensurate arrangements, showing excellent agreement with experimentally observed diffraction patterns.

1. Introduction

Occurrence of periodically arranged antiphase domain boundaries (APBs) in chemically ordered alloys leads to long-period out-of-phase superlattice structures, i.e. *superstructures*. These superstructures have been observed in both fcc-type base cells such as the Cu_3Au (L1$_2$) [1, 2] and CuAu I (L1$_0$) [3] structures, and bcc-type base cells such as B2 [4] (A base cell is defined as the basic structural unit).

APBs are formed by shifts of the base lattice along some lattice vectors linking different atomic species [4-6]. In the fcc-based superstructures, these shift vectors are the four variants which are equivalent to [0 ½ ½] [5] and the superstructures are characterised by having shift-vectors ($\vec{R}$) that are parallel to the APBs. On the other hand, bcc-based superstructures have $\vec{R}$ equivalent to [½ ½ ½], containing normal component to the APBs. While shift vectors that are parallel with the APBs do not cause changes of the phase composition, vectors that have normal components to the APBs do [4].

This work aims at a generalised mathematical formalism of structure factors for superstructures of both constant periods (fully-ordered) and of incommensurate arrangements of commensurate structural units.

2. Superstructures of Constant Periods

The structure factors of a lattice plane $(h\ k\ l)$ of the base cell with lattice parameter a_0 is defined as F_0 (To simplify descriptions in this paper, all crystallographic indices of lattice planes *refer to the base cell*, unless defined otherwise). The structure factor of a one-dimensional superlattice with the constant period $2Na_0$ in the [100] direction can be derived by the summation of the contributions of all of the $2N$ base cells [4]:

$$F_{hkl}^{(1)} = F_0 \sum_{n=0}^{N-1} \exp(2\pi inh) \ [1+\exp[2\pi i(Nh+\beta)]] \tag{1}$$

where $\beta = \vec{R}\cdot\vec{g}$ denotes the phase-shift of the electron wave at an APB, due to the shift-vector $\vec{R}$ (where $\vec{g}$ is a reciprocal space vector). Therefore, for bcc-based out-of-phase superstructures $\beta=(h+k+l)/2$, and for fcc-based out-of-phase superstructures with long-period array of APBs occurring in the [100] direction, $\beta=(k+l)/2$. The kinematical intensity of $(h\ k\ l)$ is:

$$I_{hkl} = |F_{hkl}^{(1)}|^2 \cdot I_l = \underbrace{|F_0|^2}_{(i)} \cdot \underbrace{\frac{\sin^2 \pi Nh}{\sin^2 \pi h}}_{(ii)} \cdot \underbrace{4\cos^2(Nh+\beta)\pi}_{(iii)} \cdot I_l \tag{2}$$

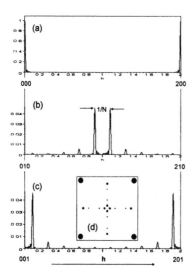

000   200

010   210

001   h   201

Fig. 1 (a-c) Simulation of intensities due to long-range modulation in the [100] direction. (d) <001> electron diffraction pattern containing all orientation variants, with zone axes normal to the modulating direction.

where the Laue function term I_l reflects the size effects of the analysed volume (i.e. number of superstructure cells in the volume). For detailed analysis of the effects of the terms on the right of the eq. (2), one is referred to ref. [5]. Fig. 1 shows the application of eq. (2) to CuAu II structure, which is formed by alteration in one dimension of five unit cells of each of the two variants of the CuAu I structure ($L1_0$). For the $L1_0$ base cell, fundamental reflections are these with h, k, and l being all even or all odd. Lattice planes of mixed odd/even indices are forbidden, unless h and k are both even or both odd to give allowed superlattice reflections due to chemical ordering in the $L1_0$ base cell. According to Eq. (2), for any fundamental reflections of the $L1_0$ structure, $\beta = $ n (n is an integer) and thus term (iii) reaches maximum at $(h\ k\ l)$, and there will be no splitting in the reflection. For any $L1_0$ superlattice reflections such as (1 1 0) and (0 0 1) of the $L1_0$ structure, $\beta = (n\pm 1/2)$ and splitting occurs in the reflections in the [1 0 0] direction, giving superstructure reflections with interval of an integral subdivision of the base cell spot separation, 1/N. Superimposition of all orientation variants of long-period superlattice domains containing the long-period direction [100] is shown in Fig. 1(d). This is in excellent agreement with experimentally observed electron diffraction patterns [6,7].

For superstructures based on the $L1_2$ base cell, simulation shows that both (010) and (110) splits in the [100] direction. The (100) spots, however, does not split (Fig. 2).

It is straightforward to extend the formalism to two-dimensional superstructures. For a superstructure modulated in both the [100] and [010] directions of the base cell ($a_1 = 2N_1 a_0$, $a_2 = 2N_2 a_0$), for example,

$$F_{hkl}^{(2)} = F_{hkl}^{(1)} \sum_{n=0}^{N_2-1} \exp(2\pi i n k) \; [1+\exp[2\pi i(N_2 k + \beta_2)]] \qquad (3)$$

3. Incommensurately Modulated Out-of-Phase Superstructure

Usually, the experimentally observed superstructure reflections such as shown in Fig. 1 do not correspond to regular subdivision of base cell spot separations [2,5,6]. According to Eq. (2), this corresponds to experimentally observed long-period repeat distances that are not integral multiple of the base cell size, i.e. N is not an integer. It is not to be concluded that the APBs occur regularly at intervals of, say, 4.7 rather than 5 base cells. Rather, it is considered that for N=4.7, the APBs maintain the same form and occur at intervals of either 4 or 5 base cells, with a random distribution of the 4 or 5 cell spacings occurring with relative frequencies such that the average spacing is N=4.7 [6]. The one-dimensional modulated structure such as this may be regarded as a locally disordered sequence of commensurate superstructure units with a statistically long-range ordered, incommensurate super-structure [5,6].

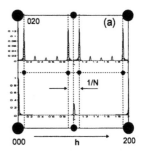

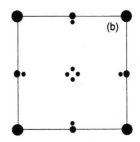

Fig. 2 (a) [001] pattern with modulation in the [100] direction, being superimposed with simulated intensities. (b) Schematic <001> pattern with all orientation variants.

Assuming that such a modulated superstructure is made of one-dimensional commensurate units containing N_1 and N_2 base cells respectively, the probabilities with which the two structural units occur are x_1 and x_2, with $x_1+x_2=1$. The random occurrences of any one of the two structural units at a sequence number (m) on a one-dimensional array are independent events. An APB occurs between any two neighbouring units. The scattering potential, due to the time averaged electron distribution density, of the two structural units are defined as ϕ_1 and ϕ_2 respectively. For the first structural unit (m=0) along the modulating direction, the probabilities for the two structural units to occur at position $r_0 = 0$ are x_1 and x_2 respectively, giving real space scattering potential as the weighted scattering potentials convoluted to the delta function centred at $r_0 = 0$, i.e. $(x_1 \phi_1 + x_2 \phi_2)*\delta(\vec{r}-0)$. For the second position in the modulated direction, there is the probability x_1^2 for the presence of two consecutive unit 1, with the contribution to the overall scattering potential $(x_1^2\phi_1)*\delta(\vec{r}-N_1\vec{a_0}-\vec{R})$. Similarly, there is

$(x_1 x_2 \phi_2) * \delta(\vec{r} - N_1 \vec{a_0} - \vec{R})$ due to the consecutive arrangement of unit 1 and unit 2, $(x_2 x_1 \phi_1) * \delta(r - N_2 a_0 - \vec{R})$ due to the consecutive arrangement of unit 2 and 1, etc. The overall scattering potential is obtained by summation [5],

$$\phi_a(\vec{r}) = (x_1 \phi_1 + x_2 \phi_2) \sum_{m=0}^{M-1} \sum_{k=0}^{m} \frac{m!}{(m-k)! \, k!} x_1^{m-k} x_2^{k}$$
$$* \delta[\vec{r} - (m-k) N_1 \vec{a_0} - k N_2 \vec{a_0} - m \vec{R}] \tag{4}$$

The structure factor, which is the Fourier transform of Eq. (4), is,

$$F(\vec{u}) = (x_1 f_1 + x_2 f_2) \cdot \frac{1 - [x_1 e^{2\pi i h(N_1 + \beta)} + x_2 e^{2\pi i h(N_2 + \beta)}]^M}{1 - [x_1 e^{2\pi i h(N_1 + \beta)} + x_2 e^{2\pi i h(N_2 + \beta)}]} \tag{5}$$

where f_1 and f_2 are the Fourier transforms of ϕ_1 and ϕ_2 respectively. For the case of $M=2$, $N_1 = N_2 = N$, and hence $f_1 = f_2 = f_N$, Eq. (5) reduces to the form of $f_N \cdot \{1 + \exp[2\pi i u(N + \beta)]\}$, the same as Eq. (1). The diffraction intensity is proportional to $|F(\vec{u})|^2$. For further details about the application of eq. (5), one is referred to ref. [5].

4. Conclusions

The structure factors for out-of-phase superstructures of both constant periods and incommensurate arrangements are formalised on the basis of diffraction physics. The results are applicable to all long-period out-of-phase superlattice structures. The interval between superstructure reflections corresponds to a fraction of the base cell spot separation, $1/N$. A non-integral N is attributed to incommensurate arrangement of commensurate structural units.

References

[1] Sato H and Toth RS, in *Alloying Behaviour and Effects in Concentrated Solid Solutions*, Ch. 4, Massalski TB ed., Gordon and Beech, New York 1963.
[2] Cowley JM, in *Advances in High-temperature Chemistry*, Vol. 3, Ch. 17, Academic Press, New York 1971.
[3] Glossop AB and Pashley DW, Proc. Roy. Soc. 1959; A250 : 132.
[4] Shao G, Appl. Phys. Lett., 1999; 74 : 2643.
[5] Shao G, Intermetallics 2002; 10 (5): 493.
[6] Cowley JM, *Diffraction Physics, 3rd Revised Edition*, Elsevier Sciences (1995).
[7] Oshima K and Watanabe D, Act Cryst. 1973; A29: 520.

Inst. Phys. Conf. Ser. No 179: Section 5
Paper presented at Electron Microscopy and Analysis Group Conf. EMAG2003, Oxford, 2003
©2003 IOP Publishing Ltd

Measuring Isoplanaticity in High-Resolution Electron Microscopy

R R Meyer and A I Kirkland

University of Oxford, Department of Materials, Parks Road, Oxford, OX1 3PH, UK

Abstract. We recently reported on a new method for the accurate determination of the symmetric aberration coefficients (defocus and two-fold astigmatism) from a focal series of high resolution images based on an analysis of the image Fourier transform phases. This can be extended to also cover the antisymmetric aberrations (axial coma and three-fold astigmatism) by recording a combined tilt-focal series where at each beam tilt a short focal series of three images is recorded. In this paper we apply this method to investigate the variation of the wave aberration function coefficients across the field of view by an aberration measurement for an array of 7x7 independent subregions. For some datasets, a very significant variation of the axial coma was found, consistent with a convergence of the incident illumination that leads to a mistilt of up to 1 mrad at the borders of the field of view. This shows that the setting up of truly parallel illumination is important and difficult. No significant variations were found in the other aberration parameters. The non-parallelity of the illumination can also be assessed by measuring the focus-induced change in the magnification, and a good agreement was found between the two methods.

1. Introduction

In general, the wave aberration function W depends on scattering direction (u,v) and position (x,y). However, because the area of the field of view is very small, the position dependence of W is often neglected in HREM (isoplanatic approximation) and W can be written as expansion by the complex scattering angle $w=u+iv$ as

$$W(\omega) = \Re\left(\frac{1}{2}A_1\omega^{*2} + \frac{1}{2}C_1\omega^*\omega + \frac{1}{3}A_2\omega^{*3} + \frac{1}{3}B_2\omega^{*2}\omega + \frac{1}{4}C_3\omega^{*2}\omega^2 \right)$$

Where C_1 and C_3 are defocus and spherical aberration and A_1, A_2 and B_2 are two- and three fold astigmatism and axial coma. The degree to which the isoplanantic approximation holds can be investigated by measuring variations in these expansion coefficients across the field of view.

A prominent cause for such variations are deviations from parallel illumination conditions (Fig1) [1]. When the illumination is tilted by a complex angle τ, W can be expanded about the origin given by the new beam direction with new expansion coefficients given by:

$$A_1^{'} = A_1 + 2A_2\tau * + \frac{2}{3}B_2\tau + C_3\tau^2, \qquad C_1^{'} = C_1 + \Re\left(\frac{4}{3}B_2\tau *\right) + 2C_3\tau * \tau$$

$$B_2^{'} = B_2 + 3C_3\tau, \qquad A_2^{'} = A_2, \qquad C_3^{'} = C_3$$

It is immediately apparent that the axial coma can be compensated with a beam tilt by τ=-$B_2/(3C_3)$; this is known as coma-free alignment. In the following, the aberration coefficients are quoted with respect to the coma free axis, which means that instead of $B2$, the mistilt from the coma-free axis is reported.

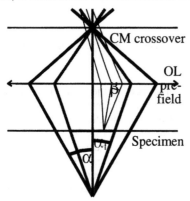

CM crossover

OL pre-field

Specimen

Fig 1: Parallel illumination is only achieved when the Condensor Minilens (CM) crossover coincides with the Objective Lens (OL) pre-field focus plane. If this is not the case, the mean beam direction varies across the specimen. The variation over the illuminated patch is denoted with a and the variation across the field of view with a_f. These are not to be confused with the beam divergence b, given by the spread of illumination directions at fixed position due to imperfect lateral coherence.

2. Measuring the local aberrations from tilt/focus series

The traditional way of measuring aberration coefficients involves recording a Zemlin tableau of diffractograms recorded under different known injected beam tilts. From each diffractogram, the apparent defocus and astigmatism can be measured, and the other aberrations can be deduced from the tilt-induced change in these values.

However, diffractograms from small image subregions have a poor signal to noise ratio, making measurements difficult and laborious if measurements have to be taken for many subregions (Fig 2). When instead of a single image a short focus series of three images is recorded at each beam tilt, focus and astigmatism can be automatically determined very accurately using the PCF/PCI method [2], and then all other aberrations determined [3]. The standard 'tiltf6' dataset acquired comprises 27 images: Three-membered focus series at axial condition and 6 different tilt azimuths, plus additional axial images between the tilted sets to allow a correction for focus drift.

3. Measuring non-parallelity from focus-induced shifts

A simpler method can be applied if only the local variation of the beam tilt needs to be measured, because non-parallelity also leads to a change in magnification. If the illumination converges as in Fig. 1, the magnification increases as the focus is changed towards underfocus moving the imaged plane upwards. When the shift vectors between images acquired at a focus difference D is measured for two subregions separated by r, the results differ by $Dr=rDa'+rD/f_{obj}$, where a' is the rate of change of the beam direction. The second term relates to the change in the objective lens focal length when the focus is changed, however, for a typical value of f_{obj}=1mm, this term corresponds to an α' of only 1mrad/μm and can be neglected. Using phase-compensated phase correlation functions PCF, the shift vectors can be measured accurately even for large focus differences, leading to accurate results even when the deviation from parallelity is small.

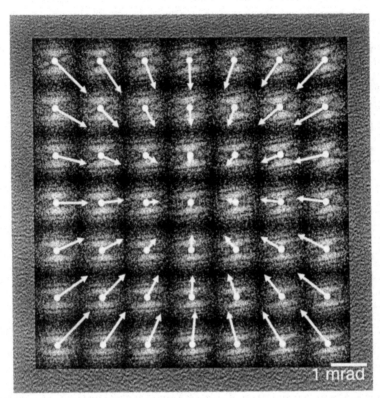

One image with tilted illumination of the tiltf6 dataset of amorphous germanium. The shape and orientation of the diffractograms calculated from 128 pixel subregions varies across the field of view. The arrows show the local beam tilt direction relative to the beam tilt at the image centre determined from the whole dataset.

4. Experimental results

In order to assess the isoplanaticity for the Oxford JEM 3000F microscope, tilt/focus datasets of an amorphous germanium foil were acquired under typical conditions for high resolution electron microscopy. It was found that parallel illumination was not often achieved. In the example in Fig 2, the beam is converging with a'=80mrad/μm. No systematic changes in the other parameters across the field of view were found. The statistical variations were typically 1nm for $C1$ and $A1$ and 60nm for $A2$, indicating the high accuracy of the PCF/PCI method even for small subregions.

Fig 3 shows results of both methods for with a sample of the layered perovskite $Nd_4SrTi_5O_{17}$. In addition to the tilt series, a focus series with a range of $D=380$nm was acquired, so that both methods could be applied, consistently indicating a small beam divergence. (14 and 17mrad/μm, respectively).

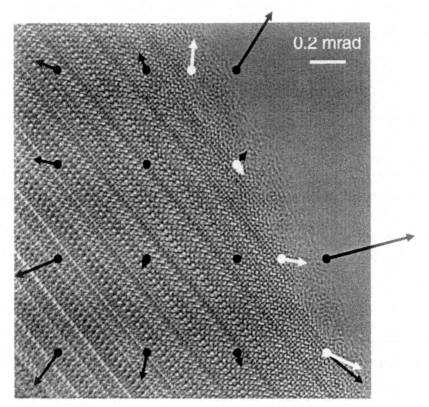

Fig 3: Axial image of the layered perovskite $Nd_4SrTi_5O_{17}$. The arrows indicate the local deviation of the beam tilt from that in the image centre. Green arrows: measured from the induced coma from a tiltf6 series. Since the sample thickness increases rapidly, only subregions close to the specimen edge were considered. Red arrows: measured from the focus-induced shift in a 24 member focus series. Note that the non-parallelity is much smaller than in Fig 2.

5. Discussion

It was shown that the PCF/PCI method applied to a combined focus series provides an accurate way to investigate variations of imaging parameters over the field of view. It was found that while the beam tilt sometimes varies significantly due to non-parallelity of the illumination, the other aberration parameters are constant. A simpler method to investigate the illumination utilising just two images with a large focus difference is also presented and satisfactory agreement is found in an experimental example.

References

[1]   Christenson K K and Eades J A 1988 Ultramicroscopy 26 113-132
[2]   Meyer R R, Kirkland A I and Saxton W O 2002 Ultramicroscopy 92, 89-109
[3]   Meyer R R, Kirkland A I and Saxton W O 2002 Ultramicroscopy in press

Inst. Phys. Conf. Ser. No 179: Section 5
Paper presented at Electron Microscopy and Analysis Group Conf. EMAG2003, Oxford, 2003
©2003 IOP Publishing Ltd

A comparison of the column approximation and the Howie-Basinski approach in simulations of TEM images under weak-beam diffraction conditions

Z Zhou†, S L Dudarev‡, M L Jenkins†and A P Sutton†§

† Department of Materials, University of Oxford, Parks Road, Oxford OX1 3PH, UK

‡ EURATOM/UKAEA Fusion Association, Culham Science Centre, Oxfordshire OX14 3DB, UK

§ Laboratory of Computational Engineering, Helsinki University of Technology, PO Box 9203, FIN 02015 HUT, Finland

E-mail: zhongfu.zhou@materials.ox.ac.uk

Abstract. Weak-beam diffraction contrast images of dislocation loops have been simulated by solving numerically the Howie-Basinski equations, which avoid the column approximation. Quantitative comparisons are made with similar simulations which do make use of the column approximation. The images predicted by the Howie-Basinski equations were displaced relative to images calculated using the column approximation, and were generally broader and somewhat lower in intensity. The differences between the images were most marked for very small loops.

1. Introduction

The usual formulation of the dynamical theory of electron diffraction developed by Howie and Whelan [1] makes use of the so-called column approximation (CA). This approximation may fail in weak-beam diffraction contrast imaging if the thickness of the foil is relatively large and we are interested in reproducing fine details of the observed experimental images [2, 3]. The approach of Howie and Basinski [4] avoids the CA. In this paper we describe a new numerical algorithm for solving the Howie-Basinski (HB) equations. WB images of small dislocation loops have been simulated using a code based on this algorithm, enabling comparison of predictions made with and without the column approximation.

2. Theoretical Basis

The HB equations give an approximate solution of the time-independent Schrödinger equation describing high-energy electrons propagating through a crystalline foil containing defects. The wave function of the electron is assumed to have the form $\psi(r) = \sum_g \phi_g(r)e^{2\pi i(k+g+s_g)\cdot r}$ and the crystal potential is evaluated using the deformable ion approximation $V(r) =$

$\sum_g V_g e^{2\pi i g \cdot (r - R(r))}$. Here $R(r)$ describes the (continuous) field of atomic displacements around the defect [4]. By neglecting the second derivatives we arrive at the HB equations:

$$(k+g)_x \frac{\partial \phi_g}{\partial x} + (k+g)_y \frac{\partial \phi_g}{\partial y} + (k+g+s_g)_z \frac{\partial \phi_g}{\partial z}$$

$$= -\pi i U_0 \phi_g - \sum_{g'} (1-\delta_{gg'}) \pi i U_{g-g'} e^{2\pi i (g'-g) \cdot R(r)} e^{2\pi i (s_{g'} - s_g)z} \phi_{g'} \qquad (1)$$

Here it is assumed that the reciprocal lattice vectors involved in the diffraction process lie in the (x,y) plane, and that the zone axis z is perpendicular to the (x,y) plane. $s_g = |s_g|$ is the magnitude of the excitation error: the vector s_g is generally assumed to be parallel to z.

Equations (1) can be simplified using a gauge transformation [5, 6] defined by the equation

$$\Phi_g(r) = \phi_g(r) e^{2\pi i g \cdot R(r)} e^{2\pi i s_g z} e^{\pi i \frac{U_0}{(k+g+s_g)_z} z} \qquad (2)$$

Since ϕ_g and Φ_g differ only by phase factors, the gauge transformation (2) does not affect the intensities of the transmitted beam and the diffracted beams. As a result of the gauge transformation the HB equations acquire the form

$$(k+g)_x \frac{\partial \Phi_g}{\partial x} + (k+g)_y \frac{\partial \Phi_g}{\partial y} + (k+g+s_g)_z \frac{\partial \Phi_g}{\partial z}$$

$$= -\sum_{g'} (1-\delta_{gg'}) \pi i U_{g-g'} \phi_{g'} + 2\pi i (k+g+s_g)_z s_g^{(R)} \Phi_g \qquad (3)$$

where $s_g^{(R)} = s_g + \frac{(k+g)_x}{(k+g+s_g)_z} \frac{\partial (g \cdot R)}{\partial x} + \frac{(k+g)_y}{(k+g+s_g)_z} \frac{\partial (g \cdot R)}{\partial y} + \frac{\partial (g \cdot R)}{\partial z}$.

In order to solve the HB equations numerically we divide the crystal foil into a set of small cells, where the strain field is assumed to be uniform throughout each cell. The HB equations can then be solved by treating each cell as a small perfect crystal. In a perfect crystal, the amplitudes of the transmitted beam and the diffracted beams are independent of x and y, and vary with z according to

$$\frac{\partial \Phi_g}{\partial z} = -\sum_{g'} (1-\delta_{gg'}) \frac{\pi i}{\beta_g} U_{g-g'} \Phi_{g'} + 2\pi i s_g^{(R)} \Phi_g, \qquad (4)$$

where $\beta_g = (k+g+s_g)_z$. To take into account the inclined propagation of the diffraction beams with respect to the zone axis (i.e. the effects associated with the non-zero projection of the vector $\mathbf{k}+\mathbf{g}$ on the (x,y) plane) we use an interpolation procedure. In this procedure the amplitudes of diffracted beams calculated using equations (4) for neighbouring columns parallel to the zone axis were combined with appropriate weights proportional to the projections of the $\mathbf{k}+\mathbf{g}$ vector on the x and y axes. It can be shown that this interpolation procedure results in an exact solution of the HB equations in the limit of an infinitely small cell size [7]. The column approximation can be made by writing $(k+g)_x = (k+g)_y = 0$ in equations (1) or (3), which recovers the Howie-Whelan formalism. This is equivalent to the assumption that the transmitted beam and the diffracted beams propagate through the foil parallel to the zone axis z.

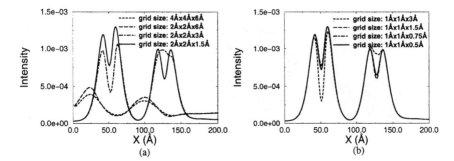

Figure 1. Computed image profiles for $g = [\bar{2}20]$ from two segments of a 10 nm flat-on vacancy loop with Burgers vector $\frac{1}{3}[111]$ in silicon. In (b), profiles for simulations with slice thickness $0.75\mathring{A}$ and $0.5\mathring{A}$ are superimposed.

Simulations were carried out for small hexagonal Frank dislocation loops located at a depth of 20 nm in a silicon foil of thickness 60 nm. The elastic strain fields of the loops were calculated from linear elasticity theory using expressions for the strain fields of angular dislocations given by Yoffe [8] as done previously by Saldin and Whelan [9]. The electron energy was taken as 100 keV, with a beam direction close to [111] and using $g = \pm(\bar{2}20)$ with $s_g \approx 0.2 nm^{-1}$ which corresponds to experimental conditions in the literature.

3. Results and Discussion

Figure 1 shows a test of the convergence of the numerical procedure described above. Intensity profiles of a flat-on hexagonal dislocation loop of size 10 nm are shown for different cell sizes. The most critical parameter appears to be the size of the cell in the z-direction, which has to be of the order of 0.1 nm or smaller in order to achieve convergence in this particular case.

A particular feature of images simulated without the CA is a sideways displacement or translation [3, 10]. In figure 2 the image simulated by the HB approach is displaced by about 1.1 nm in a direction antiparallel to g for the conditions shown. In general the magnitude of the displacement depends both on the diffraction conditions and on the distance from the defect to the bottom surface of the foil. Note also that images simulated by the HB approach are broader and of lower maximum intensity than images obtained using the CA.

The effects associated with a breakdown of the CA are particularly serious for very small dislocation loops, as seen in figure 3. For loops smaller than about 3 nm there are significant differences between images simulated by the HB and CA approaches. For loops larger than about 5 nm the differences are small. A similar conclusion was reached by Howie and Sworn in the case of partial dislocation dissociation [10].

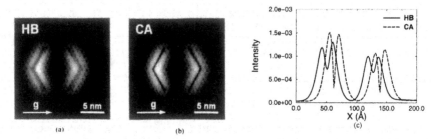

Figure 2. Simulated WB images of a 10 nm flat-on vacancy dislocation loop with Burgers vector $\frac{1}{3}[111]$ in silicon for $g = [\bar{2}20]$. (a) Image simulated using the HB approach, (b) image simulated using the CA; (c) intensity profiles taken from (a) and (b) along a horizontal line 8 nm below the top of images.

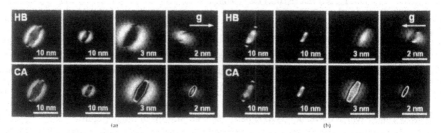

Figure 3. Simulated WB images of inclined vacancy dislocation loops with $b = \frac{1}{3}[1\bar{1}\bar{1}]$ in silicon. The sizes of the loops are 10nm, 5nm, 3nm and 1nm from left to right in both image panels (a) for $g = [\bar{2}20]$ and (b) for $g = [2\bar{2}0]$.

4. Acknowledgments

The authors are grateful to Prof. M J Whelan, Prof. C B Carter, Dr. M A Kirk and Dr. R Schäublin for stimulating discussions. Z Zhou is grateful to the EURATOM/UKAEA Fusion Association for the provision of a studentship. He also acknowledges the Weilun scholarship from St Hughs College, Oxford.

References

[1] Howie A and Whelan M J 1961 *Proc. Roy. Soc.* **A 263** 217
[2] Jouffrey B and Taupin D 1967 *Phil. Mag.* **16** 703
[3] Wiezorek J M K, Perston A R and Humphreys C J 1995 *Inst. Phys. Conf. Ser.* **147** 455
[4] Howie A and Basinski Z S 1968 *Phil. Mag.* **17** 1039
[5] Dudarev S L, Ahmed J, Hirsch P B and Wilkinson A J 1999 *Acta Cryst.* **A55** 234
[6] Williams D B and Carter C B 1996 *Transmission Electron Microscopy* (New York: Plenum Press) p 404
[7] Zhou Z, Dudarev S L, Jenkins M L and Sutton A P *unpublished research*
[8] Yoffe E H 1960 *Phil. Mag.* **5** 161
[9] Saldin D K and Whelan M J 1978 *Proc. Roy. Soc.* **A 292** 513
[10] Howie A and Sworn C H 1970 *Phil. Mag.* **22** 861

Inst. Phys. Conf. Ser. No 179: Section 5
Paper presented at Electron Microscopy and Analysis Group Conf. EMAG2003, Oxford, 2003
©2003 IOP Publishing Ltd

Regression methods for image distortion of 3-D scanning electron microscopy

H Noro [1] and K Yanagi [2]

[1] Steel Research Laboratory, JFE Steel Corporation, Fukuyama 721-8510, Japan
[2] Department of Mechanical Engineering, Nagaoka University of Technology, Nagaoka 940-2188, Japan

ABSTRACT: Regression methods to reduce image distortions of a three-dimensional scanning electron microscope equipped with two pairs of secondary electron detectors have been studied. Such distortions originated in the electron beam scanning cause errors in the measured heights of the surface topographies of specimens. These errors, which decrease rapidly at higher magnifications, are quantitatively evaluated and in order to remove them effectively, some regression methods are examined from the viewpoints of the amplitude transmission characteristics and the processing time. Parabolic regression followed by Spline filtering is proposed as the most effective method. Notice for applying this method is also discussed.

1. Introduction

The surface topographies of materials that influence their characteristics have been measured by various techniques. One of such techniques, three-dimensional scanning electron microscopy (3D-SEM) can offer high resolutions, whole field analysis, high-speed measurements and non-destructive examination of the topographies. The main drawback of this technique is that the topographic data measured at low magnifications contain unnecessary distortions originated in the electron beam scanning. This paper reports data-processing methods to remove such distortions and to extract appropriate features of specimen surfaces from their topographic data.

2. Experimental details

A three-dimensional scanning electron microscope Elionix ERA-8800FE was used for the measurements. The principle of this technique is described in the literature [1]. The topographic measurements were carried out with the accelerating voltage of 5kV and the beam current of pA order. Several tens of nanometres of gold were sputter-coated on all specimens in case the secondary electron yields of the constituents influenced the measurements. Surface topographies of mirror polished silicon wafers were measured to clarify the distortions at low magnifications. The levels of the distortions were evaluated by the maximum errors, i.e. the differences between the highest and the lowest heights in the topographic data. Since the distortions can be treated as a kind of unnecessary

208

Table 1. Prospective regression methods and their characteristics

Regression method	Flexibility of reference surface	Whole field analysis	Processing time	End effect
Parabolic regression	*(worse)*	*possible*	*short*	*advantageous*
Gaussian filtering	*base*	*impossible*	*short*	*NA*
Gaussian regression filtering of 0th order	*base*	*possible*	*short*	*disadvantageous*
Gaussian regression filtering of 2nd order	*best*	*possible*	*very long*	*advantageous*
Spline filtering	*almost best*	*possible*	*short*	*somewhat disadvantageous*

long wavelength components of the data, it is expected that some regression methods can effectively remove such components. Several prospective methods were applied to the data and the residual maximum errors were compared to find out the most effective one. Roughness standards for stylus profilometry were measured to confirm the actual advantage of this method. The regression processing was carried out using three-dimensional surface analysis software 'SUMMIT' developed by Yanagi laboratory in Nagaoka University of Technology.

3. Results and discussion

3.1 Optimisation of regression method

Five regression methods, which can be used to remove unnecessary long wavelength components, and their characteristics are summarised in Table 1 [2, 3]. These methods include parabolic regression, Gaussian filtering, Gaussian regression filtering and Spline filtering [4]. The Gaussian regression filtering of the second order shows the best amplitude transmission characteristic, whereas its processing time is much longer than

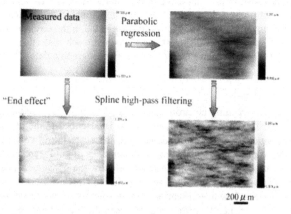

Fig. 1. Topographic data measured from mirror polished silicon wafer and processed by the regression method shown.

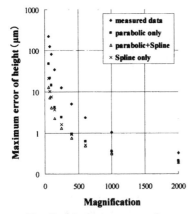

Fig. 2. Maximum errors due to image distortions as a function of magnifications.

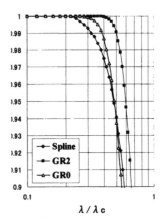

Fig. 3. Amplitude transmission characteristics of the filters under consideration. GR2 and GR0 mean Gaussian regression filters of the second and 0th order, respectively.

the others. The Spline filtering shows the excellent balance of the processing time and the characteristic. The parabolic regression also appears promising because the shape of the distortions in 3D-SEM is relatively close to the one of the parabolic surface.

Figure 1 shows the data measured from a mirror polished silicon wafer and processed by the parabolic regression, the Spline filtering and their combination. The upper and lower limits of the grey scales show the highest and the lowest heights of the topographic data with reference to the average plane. The cut-off wavelength of the Spline high-pass filter was set to the half of the evaluation length along the x-direction. The Spline filtering reduces the distortions more effectively than the parabolic regression, but causes artefacts, so-called 'end effect', at the edges of the processed data. On the other hand, the combination of the parabolic regression and the Spline filtering in this order does not cause the end effect and reduces the distortion more effectively than the independent Spline filtering.

The levels of the distortions as a function of the magnifications are shown in Fig. 2. As can be seen in the figure, the maximum errors of the measured data are more than 100, 10, 1 and 0.1-micron at the magnifications of less than 80, 300, 1,000 and 3,400, respectively, and decrease rapidly at higher magnifications. In addition, the combined regression method reduces the errors to 1/10 or less of the originals at the magnifications up to 300 and the independent Spline filtering reduces the errors to the same extent at the magnifications more than 500.

In order to confirm the advantage of the combined regression method, two kinds of roughness standards for stylus profilometry (Tokyo Seimitsu Co.Ltd.E-MC-S24 B) were measured at the minimum magnification and their arithmetic mean deviations were evaluated according to the ISO International Standards. The results and the specifications are listed in Table 2. The evaluations almost coincide with their specifications and the tenth digits can be considered as the numbers of their significant figures.

Table 2. Calculated arithmetic mean deviations of roughness standards.

Number of times	Specifications	
	0.40μm	3.17μm
1	0.42	3.23
2	0.43	3.30
3	0.43	3.29
Average	0.43	3.27

3.2   Influence of cut-off wavelength

Parts of amplitude transmission characteristics of above-mentioned filters are shown in Fig. 3. If we assume that the wavelength with the transmission characteristic more than 99 % is not affected by the filtering, the threshold wavelength λ for the Spline filtering satisfies the equation $\lambda \leq 0.3\ \lambda c$, where λc shows the corresponding cut-off wavelength. In other words, if the cut-off is set to the half of the evaluation length, this equation means that the wavelength shorter than 15% (=1/2 x 0.3) of the evaluation length is not affected by the filtering. Therefore, in order not to create artefacts by the filtering, it is important to keep this relation in mind and try to choose proper magnifications and fields of view at the topographic measurements.

4.   Conclusion

(1) The errors in the measured heights caused by the image distortions of 3D-SEM are more than 100, 10, 1 and 0.1-micron at most at the magnifications of less than 80, 300, 1,000 and 3,400, respectively, and decrease rapidly at higher magnifications.
(2) Parabolic regression followed by Spline high pass filtering can remove these errors most effectively.
(3) This combined regression method reduces the errors to 1/10 or less of the originals at the magnifications up to 300 if the cut-off wavelength of the Spline filter is set to the half of the evaluation length.
(4) The amplitude transmission characteristic of the Spline filter shows that the filtering causes almost no effect on the wavelength shorter than 30% of the cut-off wavelength. Therefore, if the cut-off is set to the half of the evaluation length, we can consider that there is almost no influence on the shape of interest (e.g. pits) as long as its lateral size is less than 15% of the evaluation length. This rule should be kept in mind when the fields of view and the magnifications are selected.

References
[1]   Suganuma T 1985 J. Electron Microsc. **34**, 328-337.
[2]   Kato M, Hara S and Yanagi K 2001 Journal of the Japan Society for Precision Engineering **67**, 1281-1283 (Japanese)
[3]   Kato M 2000 "A study on surface regression methods for three-dimensional surface topographic data", master's thesis, Nagaoka University of Technology (Japanese)
[4]   Krystek M 1996 Measurement **18**, 9-15

Inst. Phys. Conf. Ser. No 179: Section 5
Paper presented at Electron Microscopy and Analysis Group Conf. EMAG2003, Oxford, 2003
©2003 IOP Publishing Ltd

Advances in aberration corrected STEM at ORNL

A R Lupini, M Varela, A Y Borisevich, S M Travaglini and S J Pennycook
Condensed Matter Sciences Division, Oak Ridge National Laboratory, Oak Ridge, TN 37831-6031

Abstract. Aberration correction has recently made the transition from being merely a technically interesting result to finally becoming a practical tool for extremely high resolution electron microscopy. In this paper we discuss some of the progress that is being made and highlight some of the more unexpected advantages that aberration correction will bring.

1. Introduction

For many years, the optical aberrations of the round lenses have determined the primary resolution limit in most high-performance transmission electron microscopes (TEMs). The proof in 1936 by Scherzer that the spherical (C_s) and chromatic aberration (C_c) will always have positive values for conventional round lenses seemed to set a limit on the performance that can be achieved with any realistic pole-piece design [1]. Attempts to minimize these aberrations have contributed significantly to the present state-of-the-art in TEM and result in limited space available for insertion, in-situ treatment and tilt of the sample. It is therefore not surprising that there have been many attempts to correct the spherical aberration of the objective lens [2]. However, it is only in recent years that spherical aberration correctors have actually improved the resolution of the microscope on which they are fitted. Reasons for the lack of success include stringent stability requirements, but are largely due to the difficulty of aligning such a system. It is no coincidence that all of the successful aberration correctors rely on sophisticated computer control to automate this procedure [3].

One sign that C_s-correction has come of age is that plans for more modern correctors are already going beyond just considering spherical aberration and are tackling higher order aberrations [4,5]. It is hoped that fifth order correctors will provide a further reduction in probe size over the current generation of C_s-corrected systems.

With the geometric lens aberrations corrected, the next limiting factor is likely to be chromatic aberration. It is normally assumed that the defocus spread, Δf, from chromatic aberration depends on the instabilities and energy spread added in quadrature:

$$\Delta f = C_c \left(\left[2\frac{\Delta I}{I} \right]^2 + \left[\frac{\Delta V}{V} \right]^2 \right)^{\frac{1}{2}}$$

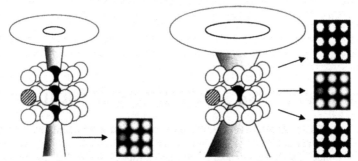

Figure 1. Schematic of the effect of changing aperture size in a model system without channeling. A single dopant atom is shaded. For a small aperture (left), a whole column is illuminated with similar intensities, and so the single image is a projection of the structure. As the aperture size increases (right), the depth of field is decreased, so a series of images is generated at different focal planes.

Where ΔI represents the fluctuations in the lens current, I, and ΔV represents the energy spread of electrons due to both the range of energies from the tip and instabilities, particularly in the high voltage supply, V. Assuming that the fractional instabilities in the lens currents can be kept below 1 part per million (ppm), and the energy spread is of the order of 1-2 ppm from a cold field emitter at 100-300 keV, the range of focus values due to chromatic effects will be approximately 3 nm. This result is a major factor in the information limit of a TEM, and suggests that the high angle annular dark field (HAADF) imaging mode of a scanning transmission electron microscope (STEM) with its reduced sensitivity to C_c [6] provides an attractive site for a pure C_s-corrector. Possible solutions to this problem would include either monochromation or C_c-correction.

2. Three-Dimensional Atomic Resolution STEM

Another extremely interesting prospect for aberration correction is three-dimensional atomic resolution. Harnessing this capability strongly relies on selecting the appropriate objective aperture. The choice of objective aperture size in TEM or STEM is a compromise between the diffraction limit, favoring a larger aperture, and the lens aberrations which necessitate a smaller beam-defining aperture. The resolution increase obtained through aberration correction arises because correcting the geometric aberrations allows the objective aperture size to be increased, resulting in a corresponding improvement in the diffraction limit. Increasing the objective aperture size will result in a \decrease in the depth of field in a STEM. This has been shown by many authors, for example [7], who demonstrate that the increased probe intensity on nearby columns will result in some signal being detected from atoms in neighboring columns for a particular probe position. However, we suggest that the reduced depth of field could really be a significant advantage because it may allow the 3-dimensional location of a single dopant atom or vacancy. This is schematically illustrated in figure 1 for the model case of a system without channeling or aberrations.

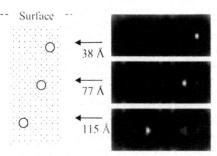

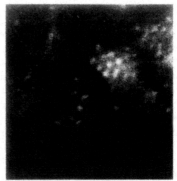

Figure 2. Simulated HAADF images of Bi atoms embedded in a 160 Å thick Si crystal viewed down the [110] axis. The Bi atoms are situated at roughly 38 Å depth steps within the crystal (left). The HAADF images generated for the probe focused at the depth of each dopant in turn are shown (right) for a 35 mrad objective aperture.

Figure 3. Experimental HAADF image of Au atoms on titania taken at 300 kV. The light titania appears as a grey background. The heavy Au atoms show up as bright spots. Both single Au atoms and clusters are visible.

We can construct a very simple estimate for the (incoherent) depth of field by considering the probe propagating in the absence of channeling and aberrations. If we assume that the diffraction limit, r_d, and the geometric spreading, r_g. given by:

$$r_d = 0.61\lambda/\theta \qquad \text{and} \qquad r_g = z\theta$$

add approximately in quadrature, then the total probe size, r_t, is given by:

$$r_t^2 \approx r_d^2 + r_g^2 = 0.61^2 \lambda^2/\theta^2 + z^2\theta^2 .$$

where λ is the electron wavelength and θ is the objective aperture half-angle. If the current density is proportional to the illuminated area, then we might use the points at which the area is twice the diffraction limited value, to estimate the point at which the probe intensity is halved, giving the depth of field as: $\Delta z = 1.22\lambda/\theta^2$.

Thus at 100kV, for the uncorrected case, with an objective aperture semi-angle of approximately 10 mrad, we have a depth of field of around 45 nm, which is rather too large to be useful in most cases. However, if we consider the C_s-corrected case at 300 kV, and 25 mrad, then the depth of field by this estimate is reduced to around 4 nm. This has already been shown to allow La atoms on the top and bottom of a catalyst support to be differentiated [8]. Another doubling of the aperture angle, for instance by C_5-correction, would allow sub-nm resolution for a suitable sample. Simulations indicate that extremely large aperture angles (approaching 100 mrad) should allow the detection of single dopant atoms with atomic depth resolution, depending on the material. At present, it appears this would be limited by the chromatic effects described above, and would present formidable challenges in terms of stability.

Figure 2 shows the output from a multislice calculation, performed with phonons, but with zero chromatic aberration. This shows a silicon sample doped with several bismuth atoms, viewed down the [110] axis. For suitably thin, aligned specimens, the channeling generally enhances the electron density on atomic columns, with a maximum that can (perhaps surprisingly) exceed the intensity in the free-space probe. As the crystal becomes thicker, this intensity will oscillate, as has been described by several authors [7]. Thus there is a strong depth dependence of the intensity in the image, but this is dominated by the channeling effects rather than defocus.

However, at large enough aperture angles, the geometrical probe convergence becomes more significant and it is possible to focus the probe at variable depths within the crystal. A simple explanation is that rays at larger angles to the optic axis are, by definition, also a long way from the strong channeling condition of an aligned crystal. In the example shown in Figure 2, by focusing the probe at different depths it is possible to determine the dopant depth directly. The out of focus dopants still give a significant contribution to the image, but there is a very clear increase in intensity when the probe is focused at the depth of a particular dopant atom. We therefore also suggest that interstitial dopants will become more visible with increasing aperture size.

3. Discussion and future directions

Other work presently underway involves more detailed image simulation [9], accurate structure calculations [10], statistically justified reconstruction of experimental images [11], and further work on improving the spectrometer resolution and efficiency [5]. Single atom detection and even spectroscopy is becoming a regular occurrence, both in the bulk of a material [12] [13] and on the surface of catalyst supports [8] (or figure 3). In this paper, we have shown that with a large enough convergence angle, under suitable conditions, the 3-dimensional location of single dopant atoms appears to be theoretically feasible, although further instrumental development remains necessary.

Research sponsored by the Laboratory Directed Research and Development Program of Oak Ridge National Laboratory managed by UT-Battelle, LLC for the U.S. Department of Energy under contract No. DE-AC05-00OR22725, and by appointment to the ORNL Postdoctoral Research Program administered jointly by ORNL and ORISE.

References

[1]   Scherzer O 1936 *Zeit. Phys.* **101** 593-603.
[2]   Summarized by: Hawkes P W and Kasper E 1989 *Principles of Electron Optics* (Academic Press, London).
[3]   Dellby N, et al. 2001 *J. Electron. Microsc.* **50** 177-185.
[4]   Krivanek O L, et al 2003 *Ultramicroscopy* **96** 229-237.
[5]   Nellist P D, et al 2003 *these proceedings.*
[6]   Nellist P D, Pennycook S J 1998 *Phys Rev Lett* **81** (19): 4156-4159.
[7]   Dwyer C, Etheridge J 2003 *Ultramicroscopy* **96** (3-4) 343-360.
[8]   Borisevich A Y, et al 2003 *Proceedings Microscopy and Microanalysis 2003* (Cambridge University Press, New York).
[9]   Allen L J, et al 2003, *Phys Rev Lett, to be published.*
[10]   Pennycook S J, et al. *Encyclopedia of Materials: Science and Technology* (Elsevier Science Ltd) 1-14.
[11]   See http://www.pixon.com.
[12]   Varela M, et al 2003 *in preparation.*
[13]   Lupini A R, Pennycook S J 2003 *Ultramicroscopy* **96** (3-4) 313-322.

Inst. Phys. Conf. Ser. No 179: Section 5
Paper presented at Electron Microscopy and Analysis Group Conf. EMAG2003, Oxford, 2003
©2003 IOP Publishing Ltd

Direct visualization of electromagnetic microfields by new double-exposure electron holography

Akinori Ohshita, Hiroki Sugi, Masaaki Okuhara, Yohei Yamakawa and Koichi Hata

Dept. of Electrical and Electronic Eng., Fac. of Eng., Mie Univ., Tsu 514-8507, Japan

Abstract. A new double-exposure electron holographic method for direct visualization of pure phase objects such as electromagnetic microfields is proposed and discussed.

1. Introduction

Electron holography is a useful method for observing electromagnetic microfields, because they are displayed as equal-phase lines of object waves in interference micrographs [1]. Since holography is a two-step imaging method, however, it is impossible to observe them in real time. Double-exposure electron holography [2,3] and three-electron-wave interference method [4,5] were developed for direct visualization of pure phase objects such as electromagnetic microfields.

In the three-electron-wave interference method, two electron biprisms should be adjusted to be completely parallel and the voltage applied to each biprism should be controlled separately so that the fringe spacings of both biprisms become identical [6]. However, these operations are not easy tasks.

In this paper, we describe a new double-exposure electron holographic method and present an experimental result of electric-field observation.

2. Theory

Figure 1 shows the schematic arrangements for hologram formation with an electron biprism. Let $\mathbf{k}$ be the wave vector perpendicular to the biprism axis and $\mathbf{r}(x,y,z)$ the position vector. When the incident electron wave, as shown in Fig.1(a), passes through a specimen having the phase distribution $\phi(\mathbf{r})$, the object and reference waves can be described by $\exp[i\mathbf{k}\cdot\mathbf{r} + i\phi(\mathbf{r})]$ and $\exp(-i\mathbf{k}\cdot\mathbf{r})$, respectively. The intensity of a conventional off-axis hologram, I_1, is given by $I_1 = 2 + 2\cos[2\mathbf{k}\cdot\mathbf{r} + \phi(\mathbf{r})]$. Under the same biprism condition, let the specimen be introduced into the interference fringes from the opposite side, as illustrated in Fig.1(b). In this case, the object and reference waves can be represented by $\exp[-i\mathbf{k}\cdot\mathbf{r} + i\phi(\mathbf{r})]$ and $\exp(i\mathbf{k}\cdot\mathbf{r})$, respectively, and the intensity of an off-axis hologram, I_2, is expressed as $I_2 = 2 + 2\cos[2\mathbf{k}\cdot\mathbf{r} - \phi(\mathbf{r})]$. When these two holograms are superposed, the resultant intensity I is given by

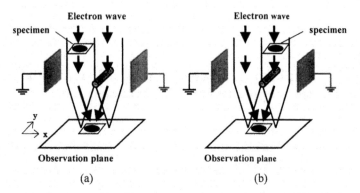

Figure 1. Schematic arrangements for hologram formation with an electron biprism.

$$I = I_1 + I_2 = 4 + 4\cos(2\,\mathbf{k}\cdot\mathbf{r})\times\cos\phi. \tag{1}$$

In this equation, $\cos\phi$ indicates that equal-phase lines of the object wave can be visualized as the amplitude modulation of periodical interference fringes expressed as $\cos(2\mathbf{k}\cdot\mathbf{r})$. On the other hand, the intensity of a conventional double-exposure hologram is written as

$$I = 4 + 4\cos(2\mathbf{k}\cdot\mathbf{r} + 0.5\,\phi)\times\cos(0.5\,\phi). \tag{2}$$

Comparison of Eq. (1) with Eq. (2) shows that the new double-exposure electron holographic method is twice as sensitive as the conventional one. The sensitivity of the new method is the same as that of the three-electron-wave interference method [4,5]. Although the control of the lateral coherence length of electron beam is indispensable in the three-electron-wave interference method, it is unnecessary in the new method.

3. Experimental results

3.1 Observation of an electric field around a latex particle

Our method was experimentally investigated using an electron microscope Hitachi HF-2000 equipped with an electron biprism. The accelerating voltage was 200kV. The holograms were recorded using a Gatan slow scan CCD camera.

Latex is a kind of electrically insulating organic material and its spherical particles are often used for calibrating the magnification of electron microscopes. Figures 2(a) and 2(b) are off-axis image holograms of a spherical latex particle about $0.5\,\mu$ m in diameter on a thin carbon film. For these two holograms, the directions of the introduction of the latex sphere into the interference fringes were the opposite. Figure 2(c) shows the superposition of two holograms. In this double-exposure hologram, two almost concentric equipotential lines of the electric field induced by charging up of the latex particle are observed.

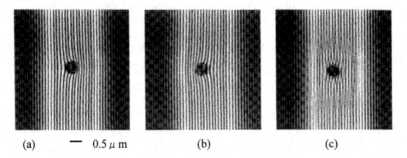

(a)   — 0.5 μm   (b)   (c)

Figure 2. Spherical latex particle on a thin carbon film. The particle is
charged up by electron irradiation and an electric field is produced
around the particle: (a) off-axis electron hologram, (b) off-axis electron
hologram and (c) double-exposure electron hologram.

3.2 Estimation of electric charge of the latex particle

To estimate the amount of electric charge, the simulation of electron holograms was
performed for the electric field generated by the charged latex sphere on a thin carbon
film. In the simulation, we assumed that the particle is uniformly charged and further
the carbon film has an infinite area and is grounded. Under these assumptions, the
electric field can be modeled by that of an equally charge valued point sphere located in
front of an infinite conducting plane at a distance equal to the radius of the sphere. As
the electrostatic potential $V(x,y,z)$ of this field can be simply calculated by the method of
images, we obtain the phase shift

$$\phi = \frac{\pi}{\lambda E} \int V(x,y,z)\, dz = \frac{q}{2\varepsilon_0 \lambda E} \operatorname{arc\,sinh}\left(\frac{a}{\sqrt{x^2 + y^2}} \right), \qquad (3)$$

where E is the accelerating voltage, λ the electron wavelength, q the charge value, ε_0
the dielectric constant of vacuum and a the radius of the sphere [6]. Substituting Eq. (3)
into Eq. (1), we can simulate the interference patterns for various values of q. Figure 3
shows the simulated double-exposure hologram, for E=200kV, q=6.41×10^{-17}C, a=

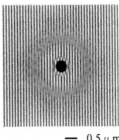

— 0.5 μm

Figure 3. Simulated double-exposure electron hologram.

$0.25\mu m$ and $|\mathbf{k}|=2.617\times10^7 m^{-1}$. Two concentric equipotential lines of the electric field are observed as the intensity modulation of the fringes. This simulated pattern and the experimental pattern shown in Fig. 2(c) resemble. Therefore the electric charge of the latex particle is estimated to be approximately $6.41\times10^{-17}C$. This charge is equal to about 400 electrons.

4. Conclusion

We have presented a new double-exposure electron holography to visualize pure phase objects such as electromagnetic microfields, which was achieved by superposition of two kinds of electron holograms. The new method is twice as sensitive as the conventional double-exposure one. Using this technique, an electric field around a latex sphere induced by electron beam irradiation was successfully observed. The amount of electric charge of the particle was estimated by comparing the experimental pattern with the simulated one to be $6.41\times10^{-17}C$, which is equal to about 400 electrons. This method is very simple and useful for observing electromagnetic fields.

Acknowledgements

We would like to thank Professor Takayoshi Tanji of Nagoya University and Tukasa Hirayama of Japan Fine Ceramic Center for their valuable discussions. We are also grateful to Dr. Kazuo Yamamoto of Japan Fine Ceramic Center for his kind technical support in our experiments. This work was partially supported by Japan Society for the Promotion of Science (Grant-in-Aid for Scientific Research (C) No. 14550024).

References

[1]   Tonomura A 1992 Adv. Phys 41 59-103
[2]   Fu S, Chen J, Wang Z and Cao H 1987 Optik 76 45-47
[3]   Matteucci G, Missiroli G F, Chen J W and Pozzi G 1988 Appl. Phys.Lett. 52 176
       -178
[4]   Hirayama T, Tanji T and Tonomura A 1995 Appl. Phys. Lett. 67 1185-1187
[5]   Hirayama T, Lai G, Tanji T, Tanaka N and Tonomura A 1997 J. Appl. Phys. 82
       522-527
[6]   Ohshita A, Ito S, Kuwata R, Sugi H and Iida K 2002 Proc.15th Int. Cong. on
       Electron Microsc. Durban vol 3 315-316

Inst. Phys. Conf. Ser. No 179: Section 6
Paper presented at Electron Microscopy and Analysis Group Conf. EMAG2003, Oxford, 2003
©2003 IOP Publishing Ltd

Strategies for One Ångström Resolution

C J D Hetherington

Department of Materials, Oxford University, Department of Materials,
Parks Road, Oxford OX1 3PH

Abstract. Various types of instrument have emerged that offer one ångström or close-to-one ångström resolution. The different challenges for each of the approaches – high voltage, image restoration and aberration correction in TEM and STEM – are outlined, as are the applications to materials characterisation. The projection problem is considered for higher index zones and defects in crystals and for amorphous specimens.

1. Introduction

It is proving to be an interesting time for high resolution electron microscopists. The remarkable developments of individual components of a microscope - the field emission gun, aberration correctors for the objective lens, CCD cameras - are leading to further dramatic improvements in the image resolution. Other new instrumentation, including the biprism, the energy filter, the high angle dark field aperture, is allowing new imaging techniques. Furthermore, since many of the microscopes which boast the higher resolutions also carry excellent analysis facilities, the high resolution microscopist finds him or herself delving into Xray and electron energy loss spectroscopy.

In this paper we will give an overview of the different approaches that offer one ångström or close-to-one ångström resolution picking out some relevant points for each technique. Sample preparation, beam damage or contamination may all affect the quality of results. We will start however by examining the structural information which may be revealed in materials at a resolutions approaching one ångström

2 What is there at 1 Ångström?

It would be nice to think that among the considerations of the Republican Government in France in 1793, the year they decided on the metre as the basic unit of length (Jerrard and McNeill, 1986), was that time in the future when microscopes would start resolving details of precisely one ten thousand millionth of that distance. But even if the size of the ångström unit is arbitrary when it comes to the atomic structure of materials, that spacing has been used as a goal for several microscope projects. We should therefore consider the atomic structures of materials and which spacings approach one ångström, all the while keeping in mind the possibility of potential HREM test specimens.

In crystal structures of the elements we find that most atom-atom distances lie between 2 and 4Å (Kelly and Groves 1970). A notable exception is carbon in which the atom-atom distance is 1.54Å in diamond and 1.42Å in the basal plane of graphite; note that this latter case occurs in nanotube structures when the graphene plane is viewed face-on. In the general case of HREM imaging, it is of course the *projected* atom-atom distances and the interplanar spacings that are important. Thus for gold, with bond length of 2.88Å, the relevant spacing is 2.35Å when viewed down the [110] direction, or 1.44Å when viewed down [111]. Silicon, with the {111} spacing of 3.1Å at the [110] zone, was a specimen well-suited to HREM tests on earlier microscopes. The same specimen remains useful as the dumb-bells have a separation of 1.36Å – ideal for testing the newer machines. In general, higher-index zones give projected structures with the smallest spacings. The [125] and [356] zones in Ge have pairs of {113} planes (spacing 1.70Å) and {133} planes (spacing 1.29 Å) respectively, and have proved to be useful test specimens (Hetherington et al. 1989).

Disordered structures, such as amorphous materials or the atomic rearrangement at interfaces and defects, also present small spacings in projection. Figure 1 illustrates this for a twin boundary in gold in which one model (unrelaxed) shows gold columns separated by 1.0Å, less than the spacings in the projection of the perfect lattice.

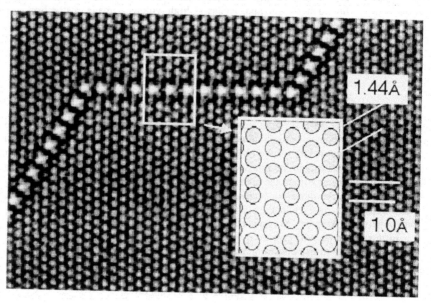

Figure 1 An experimental image taken at 1250kV of a 112 twin boundary in gold with, inset, an unrelaxed model of the boundary atomic structure: note the reduced spacing.

As we consider smaller projected atom distances, we have also to be consider the finite size of the atom and its projected potential. As the atoms approach in projection, the potentials overlap, and the "contrast" (which we loosely describe here as maximum minus minimum) in the projected potential decreases. An example is shown for Si at the [110] zone in Figure 2. Profiles of the projected potential are plotted for lines through the pair of atoms separated by 2.35Å and the pair separated by 1.36Å. Even at 1.36Å, the potentials start to overlap. Resolution of the two atoms is now not only about the smaller distance between points but also a question of imaging the reduced "contrast" of the projected potential and hence signal-to-noise ratio in the image.

In addition to this effect, there is also a reduced scattering at greater angles – visible in any diffraction pattern - and this too will lead to reduced contrast of the small spacings in HREM images.

Si 110 projected potential

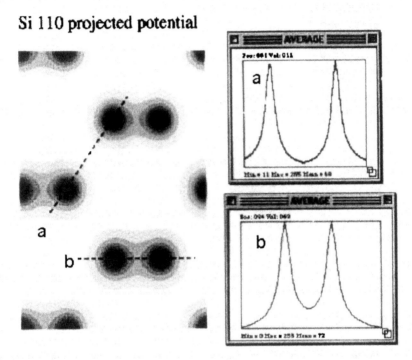

Figure 2 The projected potential of Si viewed down [110], with profiles plotted through pairs of atom columns of spacing 2.35Å (a)and 1.36Å (b). Increasing overlap of potentials leads to a decreased "contrast" in the object.

3 Different types of high resolution electron microscopes

A pre-requisite for resolutions approaching 1Å is the extension of the information limit. In the 1990's, there were two relevant developments for transmission microscopes, the high voltage microscope with a reduced HT ripple and the field emission gun, replacing LaB_6 as the electron source. Instabilities introduced by the environment, or present in the lens supplies had also to be minimised.

More recently the aberrations, principally spherical aberration, of the objective lens in the lower voltage microscope could be removed. An image series (*e.g.* through focal) can be processed to restore the exit wavefunction – the software solution – or an aberration corrector can be fitted to the microscope column - the hardware solution!.

The resolution of the scanning transmission electron microscope can also be improved by correcting the spherical and other aberrations of the probe forming lens, so we have the four principal types of microscope that can now offer resolutions at or approaching one ångström, as shown schematically in figure 3. (A TEM/STEM version of a STEM is illustrated here)

222

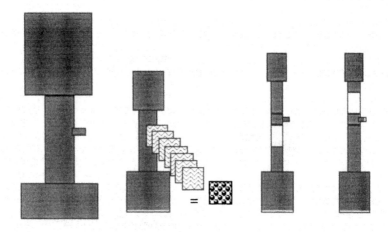

Figure 3 Schematic of high resolution electron microscopes, (left to right) high voltage, FEG with image series/restoration, TEM with C_s-corrector, STEM with C_s-corrector

3.1. High voltage electron microscopes

The point-to-point resolution (for a lens with C_s) is given by $d_{pt} = 0.66C_s^{1/4}\lambda^{3/4}$. Increasing the accelerating voltage reduces the wavelength λ and therefore d_{pt}. An early version of a high voltage HREM, the JEOL ARM-1000 in Berkeley (Hetherington et al. 1989) had HT instabilities that limited the information limit to 1.6-1.7Å even though the first zero in the CTF was around 1.3Å. On the JEOL ARM-1250 in Stuttgart, much attention was paid to the HT stability and the theoretical point resolution of ~1.2Å was attained in practice. (Phillipp et al 1994)

The expense of the equipment unfortunately limits the number of such microscopes. A further problem is radiation damage, and the basic strategy here is to minimise exposure to the beam of the important region in the specimen. A particular advantage of the high voltage microscope is the high specimen tilt angles (but the ideal of multiple views of the specimen through a wide range of angles is often undone by the effects of beam damage). Another advantage is that the weak phase approximations etc. hold for slightly thicker specimens compared to the case for lower voltages.

3.2. FEGTEM with image restoration

In order to overcome the many zeros in the phase contrast transfer function before the information cut-off, a series of images at different defoci can be taken and the results combined (Kirkland et al. 1999). Typically there will be higher order aberrations of the lens such as three-fold astigmatism, that have to be accounted for in the restoration in addition to the defocus and C_s. The end result though is the complex wavefunction corresponding to the exit surface of the specimen. There are clearly advantages in having both phase and amplitude mapped, and the method has been usefully applied also to data from high voltage HREM.

Given the high number of the FEG machines that have been delivered in the past few years, this strategy is perhaps the one that will become most prevalent. The lower accelerating voltage means that beam damage during the taking of multiple images is less of a problem.

Alternatives to the through-focal series for high resolution imaging on the FEGTEM. are through-beam tilt series and holography through the use of an in-column biprism.

3.3. FEGTEM with C_s corrector in imaging lens

Clearly one of the more dramatic developments in electron microscopy has been the C_s-corrector. Multipole lenses placed after the objective lens remove C_s and other higher order aberrations (figure 4). The point-to-point resolution stretches towards the information limit – although the calculation and display of the contrast transfer function is less straightforward than for the case of a lens dominated by C_s.

Alignment of the corrector is done *via* beam tilt tableau and initial impressions are that, once higher orders are aligned, only focus and two-fold astigmatism will need to be adjusted regularly to compensate for the introduction or movement of specimen and holder.

Since C_s can be corrected, might it not be possible to use larger pole-piece gaps (which suffer higher C_s values) and enjoy the benefits of increased tilts? The answer is complicated by the simultaneous increase of C_c, the coefficient of chromatic aberration, and the risk that the information resolution limit will cut into the transfer.

Figure 4. JEOL JEM-2200FS with two C_s-correctors in the JEOL factory before shipment to Oxford. The specimen holder is at a height of 2m.

3.4. STEM with C_s corrector in probe-forming lens

Some STEMs (dedicated STEM or TEM/STEM) are now fitted with correctors (see figure 4 also for an example of the latter). The reduced probe size afforded by the aberration-corrected probe-forming lens allows HAADF imaging at resolutions around 1Å. Alignment of the corrector may be *via* Ronchigrams or analysis of a beam-tilt tableau of images and the correlation of those tilted images with an on-axis image.

One major advantage of this microscopy mode is the ability to stop the beam and record EELS or EDX spectra form the very small area of the specimen.

On any TEM/STEM microscope, there is the possibility of imaging directly the probe. Preliminary calculations show however that if the probe is very small – on the same scale as the resolution limit of the TEM imaging system – then the size of the probe *image* may be rather larger than the size of the probe itself: the double C_s-corrector microscope may therefore aid this measurement. Other interesting applications of the microscope might be observing the shape of the beam as it exits the specimen while focussing the probe onto different heights of the specimen. On this latter subject, there may be an advantage if, in the future, it becomes possible to move the specimen in the z-direction by very small distances, in the same way that the specimen stage can at present (on *e.g.* the JEOL 3000F in Oxford) be moved by piezo-drives in the x and y-directions. Through focal series could then be generated without the need to change lens currents.

4. Conclusion

There are now several microscope types that can form images at or close to 1Å resolution. The field emission gun and stable power supplies play important roles in this development, and the C_s-correctors exploit the improved information transfer. Methods to restore the exit wavefunctions are vital for an uncorrected FEGTEM but also beneficial for the high voltage and corrected machines.

As the spacing of projected atom-atom distances decreases in high-index zones, amorphous structures and defect regions, the projected potential will have smaller variations compared to the projected potential of a low index zone.

The last words in this paper concern the specimen itself. FEG illumination is renowned for generating contamination rings and spots. In TEM, higher magnification images (to record the higher resolution information) will require a more focussed illumination (to maintain the exposure time or counts per pixel) and the contamination problem will increase. Special precautions will need to be followed in sample preparation, such as plasma cleaning. Similarly, specially adapted ion milling – low angles or low voltages – will be needed to reduce damage layer thickness on the sample surfaces. Amorphous support films degrade the quality of images of the particles they are supporting, and methods for reliably reducing the support film thickness are emerging.

Acknowledgements

Thanks go to colleagues in Oxford: D J H Cockayne, R C Doole, J L Hutchison, A I Kirkland, J M Titchmarsh, to G Möbus in Sheffield for useful discussions, and to A Robins of Fischione for help with specimen preparation issues.

References

Hetherington C J D, Nelson E C, Westmacott K H, Gronsky R and Thomas G, 1989, Mat. Res. Soc. Symp. Proc. 139, 277-282
Jerrard H G and McNeill D B, 1986, A Dictionary of Scientific Units (5th ed. Chapman and Hall) pp 11, 82 ("...the unit of length would be 10^{-7} of the earth's quadrant passing through Paris, and that the unit would be called the metre.")
Kelly A and Groves G W 1970 Crystallography and Crystal Defects (1st ed Longman) Table A5.1, p411
Kirkland A I, Meyer R R, Saxton W O, Hutchison J L and Dunin-Borkowksi R. 1999 Inst. Phys. Conf. Ser. 161 291.
Phillipp F, Höschen R, Osaki M, Möbus G,, Rühle M, 1994 Ultramicroscopy 56 1-10

Inst. Phys. Conf. Ser. No 179: Section 6
Paper presented at Electron Microscopy and Analysis Group Conf. EMAG2003, Oxford, 2003
©2003 IOP Publishing Ltd

Crystal Structures of the Trigonal U1-MgAl$_2$Si$_2$ and Orthorhombic U2-Mg$_4$Al$_4$Si$_4$ Precipitates in the Al-Mg-Si Alloy System

S J Andersen[1], C D Marioara[1], A Frøseth[2], R Vissers[3] and H W Zandbergen[3]

[1]SINTEF Materials Technology, Dept. of Applied Physics, 7465 Trondheim, Norway

[2]Norwegian University of Science and Technology (NTNU), Dept. of Physics, 7491 Trondheim, Norway

[3]National Centre for HREM, Laboratory of Materials Science, Delft University of Technology, Rotterdamseweg 137, 2628 AL Delft, The Netherlands

Abstract. The atomic structures of two important metastable phases that co-exist with β' in the Al-Mg-Si system have been solved. U1 has a trigonal unit cell with a = b = 4.05 Å, c = 6.74 Å. It contains five atoms in the space group P$_{-3m1}$ (164), with composition MgAl$_2$Si$_2$. U2 is orthorhombic having unit cell parameters a = 6.75 Å, b = 4.05 Å and c = 7.94 Å. It consists of twelve atoms packed in the space group P$_{nma}$ (62) with composition Mg$_4$Al$_4$Si$_4$. Initial models were extracted by analysing electron nano-diffraction (NDP) data. Refinements by *ab initio* quantum mechanical (QM) structural relaxation calculations and by quantitative electron diffraction using a multi slice least square (MSLS) method converged to very similar results. The intensities of the simulated diffraction patterns fit the experiment with overall R-values below 5%.

1. Introduction

The Al-Mg-Si alloy system has been subjected to extensive studies during the years due to its great practical importance, especially for the automotive industry. The main characteristic of these materials is a significant increase in hardness obtained during heat treatment as result of formation of AlMgSi metastable phases inside the Al matrix. Different type of precipitates and microstructures (phase morphology, number density, size) are created at different temperatures that influence the macroscopic behaviour of the material. Knowing the atomic structure and composition of these phases is therefore a necessity in the struggle for designing materials with improved mechanical properties.

In the middle of the 1990's Matsuda & al discovered that at temperatures above 200°C β' is not the only phase to form in Si-rich alloys as previously thought, but it nucleates in combination with three other precipitates called A, B and C [1]. The names U1, U2 and B' respectively are instead used by our group. X-ray (EDS) analysis show that, unlike β" and β', the U-phases contain aluminium in addition to Mg and Si in their composition [1]. In light of the new discoveries the precipitation sequence during heating becomes: ssss -> atomic clusters -> GP zones -> β" -> β'/U1/U2/B' -> β, where ssss refers to the initial super saturated solid solution and GP to Guinier-Preston zones.

The present paper presents the atomic structure of the most common phases found to co-exist with β', namely U1 and U2.

2. Experimental

The material used in this work is a commercial 6082 alloy with composition Al - 0.9Si -0.6Mg -0.5Mn -0.2Fe (wt%). Cubes of 1 cm³ were cut from the direct cast billets. These samples were solution heat treated in a salt bath at 550°C for 55 min and subsequently quenched in water at room temperature (RT). After two days storage at RT the samples were annealed 2 hours at 260°C. Transmission electron microscope (TEM) samples were prepared by conventional electropolishing for imaging, and by ion milling for the specimens investigated by nano-diffraction. The TEM used was a Philips CM30UT/FEG operated at 300 kV, with 1.7 Å point resolution. A slow scan CCD camera (1024x1024 Photometrix with 12 bit dynamical range) attached to the microscope enabled a linear recording of the high resolution (HR) images and diffraction data. A WIEN2k package [2] was employed for the QM *ab initio* atomic relaxation calculations. The WIEN2k is an implementation of Density Functional Theory (DFT) based on the (L)APW+(lo) basis set (*Linear Augmented Plane Wave + local orbital basis set*) [3]. Quantitative electron diffraction refinement was performed with the use of the MSLS software [4,5].

3. Results

TEM pictures were taken with the Al matrix oriented in two different zone axes, <100> and <310>. The following coherency relationships between U1, U2 and matrix have been found: (001)Al//(-120)U1, [310]Al//[001]U1, [-130]Al//[120]U1 and (001)Al//(010)U2, [310]Al//[001]U2, [-130]Al//[100]U2.

NDPs from a total number of 58 particles were recorded. Among them 31 could be identified as β' precipitates, 16 as U1 and 9 as U2. Two patterns could not be identified. For both U1 and U2 phases the unrefined atomic models were based on information given by NDPs taken in specimen areas with thicknesses lower than 10 nm for avoiding dynamical effects. The Al, Mg and Si atoms are assumed to be hard spheres, and the minimum inter-atomic distances were taken from known structures. In the case of U1 a total number of six NDPs were used for the MSLS refinement. The structure was confirmed with an overall R-value of 3.85%. For U2 seven NDPs refined to an overall R-value of 5%.

The atomic coordinates of U1 and U2 are given in Tables 1 and 2. Both figures 3 and 4 present a set of three experimental patterns each from three different particle zone axes together with the corresponding simulated diffraction patterns. The simulations were performed with the computer program *MacTempas*.

Table 1 Atomic coordinates of the U1 phase.

	Unrefined Model			QM Refined Model			MSLS Refined Model		
	Mg	Al	Si	Mg	Al	Si	Mg	Al	Si
x	0	1/3	1/3	0	1/3	1/3	0	1/3	1/3
y	0	2/3	2/3	0	2/3	2/3	0	2/3	2/3
z	0	.621	.22	0	.632(8)	.243(8)	0	.6334(6)	.2463(6)

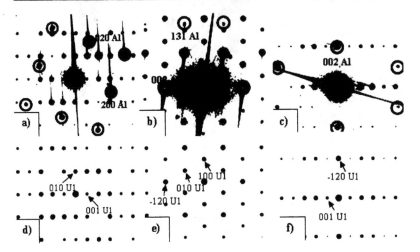

Figure 1 a) NDP, <100>Al, <100>U1. b) NDP, <310>Al, <001>U1. c) NDP, <310>Al, <210>U1. d) Simulation of a). e) Simulation of b). f) Simulation of c). Simulations used the QM refined coordinates and a thickness of 7.5 nm was assumed. For a), b) and c) the Al reflections are indicated by grey circles.

Table 2 Atomic coordinates of the U2 phase.

	Unrefined Model			QM Refined Model			MSLS Refined Model		
	Mg	Al	Si	Mg	Al	Si	Mg	Al	Si
x	.056	.37	.25	.034(9)	.361(4)	.239(3)	.0326(8)	.3567(5)	.2400(6)
y	3/4	1/4	1/4	3/4	1/4	1/4	3/4	1/4	1/4
z	.311	.436	.125	.327(4)	.432(5)	.120(9)	.3307(5)	.4350(5)	.1224(4)

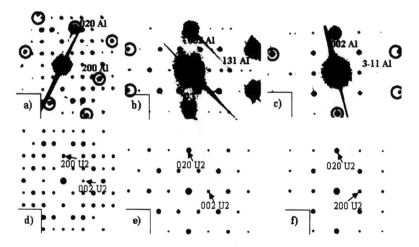

Figure 2 a) NDP, <100>Al, <010>U2. b) NDP, <310>Al, <100>U2. c) NDP, <310>Al, <001>U2. d) Simulation of a). e) Simulation of b). f) Simulation of c). Simulations used the QM refined coordinates and a thickness of 7.5 nm was assumed. For a), b) and c) the Al reflections are indicated by grey circles.

References

[1] Matsuda K Sakaguchi Y Miyata Y Uetani Y Sato T Kamio A and Ikeno S 2000 J. Mat. Sci. 35 179-189
[2] Blaha P Schwarz K Madsen G Kvasnicka D and Luitz J 1999 An Augmented Plane Wave + Local Orbitals Program for Calculating Crystal Properties (Karlhinz Schwarz, Tech. Universitat Wien, Austria) ISBN 3-9591931-1-2
[3] Frøseth A G Høier R Derlet P M Andersen S J and Marioara C D 2003 Phys. Rev. B 67 224106
[4] Jansen J Tang D Zandbergen H W and Schenk H 1998 Acta Cryst. A54 91-101
[5] Andersen S J Zandbergen H W Jansen J Træholt C Tundal U and Reiso O 1998 Acta Mater. 46 3283-3298

Inst. Phys. Conf. Ser. No 179: Section 6
Paper presented at Electron Microscopy and Analysis Group Conf. EMAG2003, Oxford, 2003
©2003 IOP Publishing Ltd

Structural studies of a modulated quaternary layered perovskite

J Sloan[*1,2], **K L Langley**[1], **A I Kirkland**[1], **R R Meyer**[1], **M J Sayagués**[3], **R J D Tilley**[4] **and J L Hutchison**[2]

[1]University of Oxford, Department of Materials, Parks Road, Oxford OX1 3PH
[2]University of Oxford, Inorganic Chemistry Laboratory, South Parks Road, Oxford, OX1 3QR
[3]Division of Materials and Minerals, School of Engineering, University of Wales, Cardiff, CF2 1YH
[4]Instituto de Cienca de Materiales de Sevilla, CSIC c/ Americo Vespucio, s/n 41092, Sevilla

ABSTRACT: The quaternary compound, $Nd_4SrTi_5O_{17}$, is a layered perovskite derived from ternary monoclinic $Nd_5Ti_5O_{17}$. The quaternary version exhibits an usual quasi-sinusoidal streaking in electron diffraction patterns obtained from [010] presumed to be due to an incommensurate modulation caused by periodic compositional substitution. Exit-plane wave image restoration from a focal series of HRTEM images allowed small structural distortions at the interface of the perovskite slabs – presumed to be the origin of the structural modulation – to be studied in detail. Further supporting evidence for this interpretation was provided via nano-beam diffraction experiments. In addition, specimen exit-plane waves and diffraction patterns were recorded from quaternary samples that were subjected to annealing at 1200°C for periods of 48 hours and 7 days. The distinctive streaking was found to be entirely absent in patterns obtained from the specimen annealed for 7 days.

1. INTRODUCTION

Perovskites of the general formula $A_nB_nO_{3n+2}$ have the capacity to accommodate cations of mixed valency resulting in materials that can exhibit potentially useful charge carrying properties, such as $La_5Ti_5O_{17}$ which displays metallic behaviour at low temperatures (Lichtenberg et al., 1991). The relationship between the structure and the electronic properties of such compounds is therefore of interest. Within the homologous series, perovskite $(A_{n-1}B_nO_{3n+2})_\infty$ slabs are produced parallel to the generalised [110] perovskite axis and are bounded by layers of A cations coordinated by oxygen atoms in a rocksalt-type structure (Figure 1(a) and (b); Tilley, 1980). We recently described the refinement of the structure of ternary $Nd_5Ti_5O_{17}$ (Connolly et al., 1996; Sayagués et al., 2003) based on this structure type from a reconstruction of the specimen exit-plane restored from a focal series of high-resolution transmission electron microscope (HRTEM) images (Schiske, 1983; Coene et al 1992; Kirkland et al., 1999). This was a well-ordered phase with a structure similar to the reported phase $La_5Ti_5O_{17}$ (Williams et al., 1991) although in the version based on neodymium, a more pronounced skew angle φ of 3.0° corresponding to the Nd atom rows relative to the expected bulk perovskite layers was observed relative to that reported for the former phase (i.e. 1.0°, corresponding to the La atoms rows) when both materials are viewed along [100] (Figure 1(a); Sayagués et al., 2003; Williams et al., 1991). In 1994, quaternary $Nd_4SrTi_5O_{17}$ was reported which was produced by the technique of arc melting (Sloan and Tilley, 1994). In electron diffraction patterns obtained from the [010] zone of this phase, alternating rows of diffraction maxima were replaced by diffuse streaks which displayed 'quasi-continuous sinusoidal streaking' thought to be due to an incommensurability in the structure. In the present work, we present a detailed microstrutural analysis of this phase and also investigate the stability of the modulation behaviour following extended annealing of the sample at elevated temperature.

2. EXPERIMENTAL

$Nd_4SrTi_5O_{17}$ was synthesised by arc melting a stoichiometric mixture of Nd_2O_3, TiO_2 and $SrTiO_3$. $SrTiO_3$ was prepared by arc melting together SrO (calcined from $SrCO_3$) and TiO_2. All of the starting materials used were Johnson Matthey 'Specpure' grade. The arc melted beads were initially crushed in a percussion mortar and then finely crushed in an agate mortar and pestle under acetone. Portions of the as-prepared quaternary sample were also subjected to annealing at 1200°C in a Carbolite tube furnace for periods of 48 hr and 7 days. Suspensions of each preparation were pipetted onto lacey carbon coated copper grids (Agar, 300 mesh). Focal series were obtained from various crystallographic zones using a JEM 3000F FEGTEM (C_s = 0.6 mm at 300 kV) from crystals oriented using selected area diffraction patterns. The images were recorded digitally using a 1024 × 1024 pixel CCD camera mounted axially and primary microscope magnifications of 400kX and 600kX.

The microscope was manually aligned to the coma-free axis and the two-fold astigmatism corrected using on-line diffractograms of the amorphous carbon support film. Focal series of images were recorded with the microscope under automatic control using scripts running under the Gatan Digital Micrograph software. For each orientation a series of 30 images was recorded of a thin crystal edge beginning underfocus with a nominal focal increment of 10 nm between images. A final image at the starting defocus was recorded to assess the focal drift. The image processing techniques subsequently employed to recover the modulus and exit-plane wave function are described elsewhere (Kirkland et al., 1995, 1997, Meyer et al., 2002).

3. RESULTS AND DISCUSSION

Figs. 1(a) and (b) show schematic structural representations of the generalised $A_5B_5O_{17}$ structure common to $La_5Ti_5O_{17}$, $Nd_5Ti_5O_{17}$ and $Nd_4SrTi_5O_{17}$ viewed along the [100] and [010] directions, respectively.

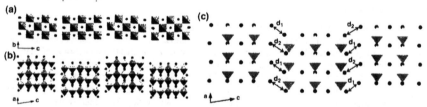

Fig. 1. (a) Generalised [100] and [010] projections common to the $Nd_5Ti_5O_{17}$, $La_5Ti_5O_{17}$ and $Nd_4SrTi_5O_{17}$ structures. The A columns are indicated by black circles and the BO_6 octahedra are indicated by the shaded squares or diamonds. The skew angle φ is given by the displacement of the A atom rows relative to <001> in the [100] projection. (c) Schematic structure model indicated differential displacement of alternating (Nd,Sr) columns (i.e. d_1 and d_2) observed in the quaternary version of the phase.

Figs. 2(a) and (b) show electron diffraction patterns obtained from the equivalent [010] zone of both the ternary $Nd_5Ti_5O_{17}$ and quaternary $Nd_4SrTi_5O_{17}$ specimens. The former was indexed on the basis of a $P2_1/c$ space group in which the reflections $l = 2n + 1$ are systematically absent (Sayagués et al., 2003). With respect to the pattern obtained from the same zone for $Nd_4SrTi_5O_{17}$ (Fig. 2(b)) it is immediately apparent that alternating $(h0l)$, $h = 2n + 1$ rows of reflections parallel to c^* are replaced by the complex diffuse streaking described above. It was expected that a diffraction pattern from a smaller field would not contain this streaking as no disorder should be present over the short range. A nano-beam diffraction pattern was therefore collected with a probe of only 0.4 nm FWHM and, as expected, a pattern which more closely resembles the equivalent pattern from the unmodulated ternary phase (cf. Fig. 2(a)) was obtained. Similarly, it was anticipated that the disorder giving rise to the diffuse scattering observed in Fig. 2(b) was probably a kinetic phenomenon given that the samples were prepared by the rapid quenching arc melting technique. Consequently we have annealed portions of the arc melted material at 1200°C for periods of 12 h and 7 days. Electron diffraction patterns obtained from the sample following annealing for 12 h (not shown) still exhibited

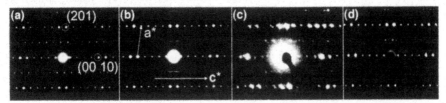

Fig. 2. (a) Electron diffraction pattern obtained from the [010] zone of the ternary $Nd_5Ti_5O_{17}$ phase indexed according to S.G. $P2_1/c$. (b) Electron diffraction pattern obtained from the [010] zone of the quaternary $Nd_4SrTi_5O_{17}$ phase. (c) Nano-beam diffraction pattern obtained from [010] of the unannealed quaternary phase. (d) Electron diffraction pattern obtained from [010] of the quaternary phase following annealing of the sample at 1200°C for 7 days.

the diffuse scattering behaviour observed in Fig. 2(b) but, following annealing for 7 days, this streaking is almost entirely absent. The 'quasi-sinusoidal streaks' may now be seen to be replaced by rows of reflections in which are now doubled in frequency relative to the corresponding rows of reflections observed for the ternary form (i.e. Fig. 2(a)). Some stacking disorder is still evident in this pattern although it is apparent that this pattern represents a more thermodynamically ordered form of the structure.

In the previously reported structural refinement of ternary $Nd_5Ti_5O_{17}$, the individual cation positions were extracted from the more strongly scattering Nd and Ti cation columns clearly visible in restored modulus images obtained from the respective [010] and [100] zones (Sayagués et al., 2003). The more weekly scattering anion lattice was then extracted from the corresponding phase images produced from the same focal series from the two respective zones. In the present work, a source of the modulation apparent in the [010] zone of the quaternary phase has been extracted by direct comparison of restored modulus images obtained from [010] zones of the ternary and quaternary phases, respectively (Figure 3(a)-(d)).

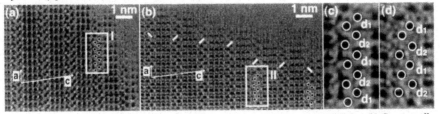

Fig. 3. (a) Restored modulus image obtained from the [010] zone of $Nd_5Ti_5O_{17}$. (b) Corresponding modulus image restored from the same zone of $Nd_4SrTi_5O_{17}$. (c) and (d) details from the indicated regions I and II in (a) and (b) respectively. The relative displacements of successive Nd or (Nd,Sr) columns in interfacial slab regions are indicated by d_1 and d_2 (see also Fig. 1(c)). For the ternary phase $d_1 \sim d_2$ for the quaternary phase $d_1 < d_2$.

In Fig. 3(a) and (b) phase images of the unmodulated ternary phase and modulated quaternary phase are presented, respectively. In Fig. 3(c) a detail shows the configuration of Nd cation columns at the slab interface. Comparison of the separations of these columns in successive layers (i.e. d_1 and d_2, see also Fig. 1(c)) revels no significant differences in terms of their relative displacements. In Fig. 3(d) we see a similar detail but this time from the quaternary phase. In this instance, the separations of the (Nd,Sr) columns (i.e. d_1 and d_2, see also Fig. 1(c)) is much more marked. A further feature evident in the main phase image (i.e. Fig. 3(b)) is that the position of the cation column shift changes thoughout the image as indicated by the short white lines. Presumably with annealing these sites order to give an effectively twinned phase (i.e. Fig. 2(d)).

4. CONCLUSIONS

The technique of phase image restoration has been used to refine the microstructure of the modulated quaternary phase $Nd_4SrTi_5O_{17}$. Modulus images obtained from the [010] crystallographic zone provide clear evidence of systematic cation column displacements at the perovskite slab interfaces of this material. Nanobeam electron diffraction experiments reveal the long range nature of this modulation while annealing of the sample forces the crystal into a more ordered twin-type arrangement.

ACKNOWLEDGEMENTS

J.S. is indebted to the Royal Society for a University Research Fellowship.

REFERENCES

Coene W, Janssen G, Op de Beeck M and van Dyck D 1992 Phys. Rev. Lett. **69**, 3743.
Connolly E, Sloan J and Tilley R J D 1996 Eur. J. Solid State Inorg. Chem., **33**, 371.
Kirkland A I, Saxton W O, Chau K-L, Tsuno K and Kawasaki M 1995 Ultramicroscopy **57**, 355.
Kirkland A I, Saxton W O and Chand G 1997 J. Electron Microscopy **1**, 11.
Kirkland A I, Meyer R R, Saxton W O, Hutchison J and Dunin-Borkowski R 1999 Inst. Phys. Conf. Ser. No 161: Section 6.
Meyer R R, Kirkland A I and Saxton W O 2002 Ultramicroscopy **92**, 89.
Lichtenberg F, Williams T B, Reller A, Widmer D and Bednorz J G 1991 Z. Phys. Sect. B **84**, 369.
Saxton W O 1988 Scanning Microscopy Supplement 2, 213.
Sayagués M J, Langley K, Meyer R R, Kirkland A I, Sloan J, Hutchison J L and Tilley R J D 2003 Acta Cryst. **B59** 449.
Schiske P 1983, in *Image Processing and Computer-Aided Design in Electron Optics*, ed. Hawkes P W (Academic Press, London, 1973) pp 82-90.
Sloan J, Tilley R J D 1994 Eur. J. Solid State Inorg. Chem. **31**, 673.
Tilley R J D 1980 *Chemical Physics of Solids and their Surfaces*, (London: Royal Society of Chemistry) p. 151.
Williams T, Schmalle H W, Reller A, Lichtenberg F D W 1991 J. Solid State Chem. **93**, 534.

Inst. Phys. Conf. Ser. No 179: Section 6
Paper presented at Electron Microscopy and Analysis Group Conf. EMAG2003, Oxford, 2003
©2003 IOP Publishing Ltd

TEM Characterization of Stress Corrosion Cracks in 304SS

S Lozano-Perez, J M Titchmarsh and M L Jenkins

Department of Materials, University of Oxford, Oxford OX1 3PH, UK.

ABSTRACT: This paper reports the analysis of a sample of type 304 stainless steel tested under constant load (CLT) in a simulated pressurized water reactor environment in which had developed a mixture of both long (of the order of mm) and short (a few microns) cracks. A duplex oxide layer structure was observed along the flanks of all cracks. The chemical composition of the two layers remained constant and independent of crack depth. However, differences were observed as a function of crack depth between the thicknesses and sizes of the spinel crystallites that composed the two layers. The implication of these observations for SCC mechanisms is described.

1. Introduction

Stress corrosion cracking (SCC) and intergranular corrosion have a major impact on the structural integrity of austenitic stainless steels during service. New sample preparation techniques have recently allowed the characterization of stress corrosion cracks (SCC) using TEM. Imaging and chemical analysis with nanometer resolution make possible the observation of previously unknown features at the tips of advancing cracks which require re-assessment of SCC mechanisms [1, 2]. Characterization of corrosion products at different stages of growth, i.e. along the flanks of short and long cracks, is very important for understanding SCC mechanisms. The specific choice of preparation method, using either focused (FIB) or broad ion beams (PIPS), or a combination, depends on the depth and geometry of cracks [1, 3].

2. Experimental

Type 304 Stainless Steel (SS) specimens with composition (weight %) **C**: 0.054, **Si**: 0.52, **Mn**: 1.48, **P**: 0.024, **S**: 0.001, **Ni**: 9.74, **Cr**: 18.44, **Fe**: Bal. and subjected to a heat treatment at 1050°C for 30 min + 750°C for 100 min + 615°C for 5h, were tested under simulated pressurized water reactor (PWR) conditions using an autoclave in which the chemistry of the circulating aqueous environment (500ppm B + 2ppm Li + 0.5ppm DO_2) was continuously monitored and controlled. Testing was performed at 240°C and a load of 410MPa using samples of 15mm^2 rectangular cross-section. The test was stopped after 1000h without breaking the sample.

TEM samples containing deep cracks up to several millimeters long were prepared using Ion Beam Milling (Gatan PIPS691) while a FEI FIB200 was used for preparing specimens with short (~10-30μm) cracks. FIB was also used to thin membranes containing the tips of the secondary cracks emanating from the long cracks. Details of the sample preparation techniques can be found in [1, 3]. TEM characterization was performed with a Philips CM20 (LaB$_6$ TEM) and a VG HB501 (FEG-STEM) equipped with an LINK EDX detector.

3. Results and Discussion

Long cracks were filled with corrosion products along the flanks (Fig. 1a). Although preferential PIPS milling of the corrosion products occurred at widely separated flanks, the narrower tips were immune from such edge attack and it was not necessary to protect such cracks by filling with epoxy resin. The cracked grain boundary in a FIB sample was always a line of potential structural weakness that necessitated used of the modified "lift-out" technique [1] and the original sample surface containing the crack mouth also required protection by deposition of a Pt layer prior to milling. The FIB-milled sample in Fig. 1b shows the Pt layer, corrosion products down the crack and a Cr-rich carbide, labelled *a*, close to the boundary that has also been attacked.

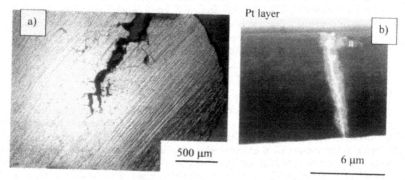

Figure 1. (a) Optical micrograph showing crack tip at centre of a 3mm disk after PIPS milling; (b) STEM BF image of short crack in a FIB cross-sectioned sample.

Fig. 2 compares the tips of deep and shallow cracks. While the deep propagating crack is sharp, it is open even though it lies a few mm below the original surface of the sample (Fig. 2a). The FIB-thinned shallow crack tip, only a few microns below the surface, is in the first stages of growth and the separation between the flanks is almost non-existent (Fig. 2b). More detailed examination revealed that the corrosion product along the deep crack flanks was composed of a nanocrystalline inner layer and an outer polycrystalline layer. Microdiffraction revealed that both these layers had the fcc spinel crystal structure, with similar lattice parameters of ~0.84nm. However, compositional line profiles revealed a distinct compositional difference (Fig. 3a), the inner layer being Cr-rich and the outer layer being Fe-Ni-rich. Consequently, the inner and outer layers were identified as $FeCr_2O_4$ and $NiFe_2O_4$, respectively. Fig. 3b shows the compositional uniformity of the layers.

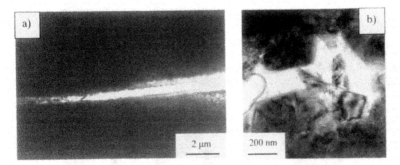

Figure 2. BF micrographs: (a) deep crack (PIPS); (b) shallow crack (FIB).

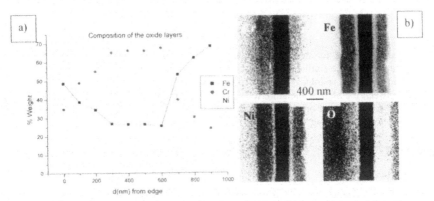

Figure 3. EDX analysis of duplex oxide layers: (a) line profiles; (b) EDX elemental maps.

The area containing the Cr-carbide, '*a*' in Fig. 1b, is enlarged in Fig. 4a. Comparison of the Cr and O EDX elemental maps from this area (Fig. 4b) reveals that the region between the carbide and the grain boundary has been preferentially oxidized. The maps also show mostly Ni-Fe rich particles with rectangular shape along the flanks, which were identified as spinel by microdiffraction. They are, therefore, similar to the $NiFe_2O_4$ layer identified along the flanks of the deep cracks. A Cr-rich inner nanocrystalline spinel layer, too thin to be directly revealed in the Cr elemental map, was seen by colour-overlay of the elemental maps and also by using microdiffraction. This layer was considered to be the same Cr-rich inner layer found along the deep crack flanks, but at an earlier stage of growth.

The individual $NiFe_2O_4$ crystals in the outer layer were almost always epitaxially orientated with one of the two adjacent matrix grains, suggesting that they were once in direct contact. This was supported by studying a shallow crack where the flank was not completely covered by the very thin Cr-rich oxide layer. Similar observations of epitaxy

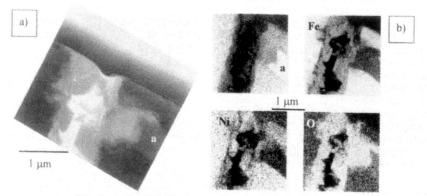

Figure 4. (a) STEM BF image showing details of the crack and the attacked Cr-carbide (labelled as **a**); (b) EDX maps from the same region.

have been reported for deep SCC cracks in 316SS for different test environment and temperature [4]. The similar observations suggest a common mechanism for intergranular attack in 300-series stainless steels that requires oxygen transport through an existing spinel layer between the high temperature water and the grain boundary. Crack advance is likely to be by frequent local brittle fracture through the spinel layers. Duplex oxide layers are reported to grow during aqueous corrosion of stainless steel surfaces [5] where, once formed, the inner Cr-rich layer inhibits attack on the matrix. Rapid corrosion of the grain boundary could occur because the Cr-rich layer is not continuous at the crack tip.

4. Conclusions

TEM characterization of both deep and shallow SCC in type 304 stainless steel was facilitated by PIPS and FIB specimen preparation methods, respectively. The flanks of all cracks were covered by a Cr-rich nanocrystalline spinel layer adjacent to the matrix and an outer microcrystalline Ni-Fe-rich spinel. The latter was oriented epitaxially with the matrix grains, even though separated by the nanocrystalline layer. The results mirror those found earlier for type 316 stainless steel and suggest a common mechanism for crack advance [4].

Acknowledgements

The authors are grateful to Dr K. Fujii (INSS) for providing the samples and for financial support. Support from the RAE, AEA Technology and EPSRC is also acknowledged.

References

1. Huang Y Z et al 2002 J Microscopy 207 129-136.
2. Thomas L E and Bruemmer S M 2000 Corrosion 56 (6) 572-587
3. Lozano-Perez S et al 2001 Inst. Phys Conf. Ser. 168 (5) 191-194
4. Huang Y et al 2001 Inst. Phys Conf. Ser. 168 (5) 203-206
5. Stellwag B 1998 Corrosion Science 40 (2-3) 337-370

Inst. Phys. Conf. Ser. No 179: Section 6
Paper presented at Electron Microscopy and Analysis Group Conf. EMAG2003, Oxford, 2003
©2003 IOP Publishing Ltd

EBSD Study of the Hot Deformation Microstructure Characteristics of a Type 316L Austenitic Stainless Steel

P Cizek, J A Whiteman, W M Rainforth and J H Beynon

IMMPETUS, Department of Engineering Materials, University of Sheffield, Mappin Street, Sheffield S1 3JD, UK

Abstract. Characteristics of the crystallographic texture and deformation microstructure were studied in a type 316L austenitic stainless steel, deformed in rolling at 900°C to true strains of about 0.3 and 0.7, using electron backscatter diffraction (EBSD). The texture was mainly characterised by rotations towards the α fibre orientations with increasing strain. At the lower strain level, there was considerable evidence of a rotation of the pre-existing twin boundaries from their original orientation relationship, as well as the formation of highly distorted grain boundary regions and deformation bands. The subgrains were predominantly arranged in elongated bands, the boundaries of which frequently approximated to traces of the {111} slip planes. The corresponding misorientations were generally small and largely displayed a non-cumulative character across the band widths, while displaying a tendency to cumulate strongly along the band lengths. Misorientation axis vectors appeared non-crystallographic and were largely clustered around the macroscopic transverse direction. At the higher strain level, general characteristics of the deformation microstructure remained qualitatively similar to those observed at the lower strain. However, the subgrain dimensions became finer, the corresponding misorientation angles increased and both these characteristics became less dependent on a particular grain orientation. The extended sub-boundaries largely appeared to maintain an approximately constant inclination towards the rolling plane within the strain interval used. The obtained statistically representative data will assist in the development of physically-based models of microstructural evolution during hot deformation of austenitic stainless steels. .

1. Introduction

Fully automated electron backscatter diffraction (EBSD), in particular when used in conjunction with a high-resolution field emission gun scanning electron microscope, has recently emerged as a powerful tool for microstructural analysis, allowing rapid acquisition of large quantities of orientation data [1]. Orientation averaging techniques, such as the Kuwahara filter [2], can improve the angular resolution of EBSD to less than 0.5°, thus making it suitable for a quantitative investigation of the hot deformation microstructures [3]. The aim of the present work was to undertake a detailed EBSD

238

study of the evolution of crystallographic texture and dislocation substructure during hot rolling of a type 316L austenitic stainless steel. A statistically representative set of data obtained would then assist in the development of a physically-based model of microstructural evolution during hot forming of austenitic stainless steels.

2. Experimental Procedures

The type 316L austenitic stainless steel used in the investigation had a chemical composition of 0.02wt.% C, 1.27% Mn, 0.43% Si, 0.029% P, 0.003% S, 17.0% Cr, 11.3% Ni, 1.98% Mo and the balance Fe. Specimens were rolled at 900°C to reductions of 25 and 50% (true strains of 0.29 and 0.69) at nominal strain rates of 4.8 and 8.1 s^{-1}, respectively, and quenched. An EBSD study was undertaken in the central area of rolling specimens on the RD-ND sections, using both a W-filament JSM 6400 and FEG LEO 1530 scanning electron microscopes operated at 20 kV. Both instruments were equipped with the fully automatic HKL Technology EBSD attachment. The data processing was carried out using the HKL Channel 5 and the VMAP software, kindly provided by Prof. Humphreys of UMIST, which includes a Kuwahara filter routine [2].

3. Results and Discussion

3.1. A Strain of 0.29

At a true strain of 0.29, the crystallographic texture (Fig. 1a) was largely composed of the α and β fibre orientations [1] and its overall strength was about 4.5 times random.

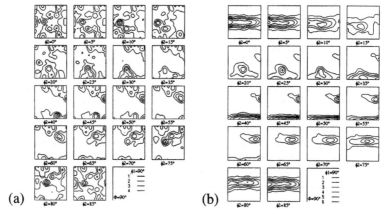

Fig. 1. Crystallographic texture (an area of 5×1 mm, a step size of 2 μm) expressed by the orientation distribution function [1]: (a) a strain of 0.29; (b) a strain of 0.69.

The EBSD study showed that grain boundary regions [4] frequently displayed larger distortions than the grain interiors, which was manifest by a higher density of sub-boundaries (Fig. 2a). The observed cumulative misorientation profiles (Figs. 2c, 2d) showed a pronounced accumulation of misorientation angles towards the boundary. The study also revealed extended large-angle dislocation walls with misorientations exceeding 10°, indicating a tendency of some grains to split into deformation bands.

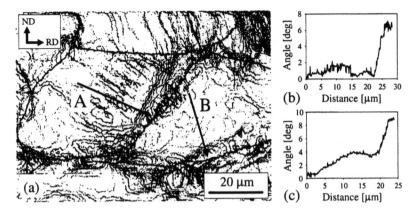

Fig. 2. EBSD analysis of grain boundary regions formed at a strain of 0.29: (a) boundary map (a step size of 0.1 μm); (b),(c) cumulative misorientation distributions for the lines A and B indicated in (a) respectively.

There was considerable evidence of rotations of the pre-existing twin regions away from their original Σ3 coincidence site lattice orientation relationship [1]. The quantification of substructure by EBSD [3] revealed a noticeable dependence of the substructural characteristics on grain orientation (Table 1). Extended sub-boundaries were often aligned close to the {111} crystallographic slip planes and predominantly inclined to the rolling direction at angles between 30° and 50° [5]. Misorientation axis vectors across sub-boundaries had a tendency to cluster around the sample transverse direction, however, their distribution in the crystal lattice coordinates was essentially random.

Table 1. Substructural characteristics estimated by EBSD (d, α and E denote the mean subgrain diameter, misorientation angle and stored energy values, respectively).

Texture Component	A Strain of 0.29			A Strain of 0.69		
	d (μm)	α (°)	E (J·m^{-3})	d (μm)	α (°)	E (J·m^{-3})
Brass	1.4	1.0	$4.29 \cdot 10^5$	0.7	1.1	$1.09 \cdot 10^6$
Goss	2.1	0.8	$3.24 \cdot 10^5$	0.7	1.4	$1.31 \cdot 10^6$
Copper	2.8	1.2	$3.22 \cdot 10^5$	0.6	1.5	$1.69 \cdot 10^6$
Random	2.0	0.8	$3.23 \cdot 10^5$	0.7	1.2	$1.26 \cdot 10^6$

3.2. A Strain of 0.69

When increasing a strain level to 0.69, the texture (Fig. 1b) became dominated by orientations situated along the α fibre [1] and its overall strength increased to about 6.9 times random. General characteristics of the deformation microstructure remained qualitatively similar to those observed at the lower strain. However, the subgrain dimensions became finer, the corresponding misorientation angles increased and both these characteristics became less dependent on a particular grain orientation (Table 1). The formation of deformation bands and grain boundary regions, as well as distortions

240

of the pre-existing twin boundaries, generally became more pronounced. In contrast to the lower strain level, extended straight sub-boundaries delineating elongated "microbands" became more dominant. Figure 3a shows an example of a region of well-developed microbands formed within a Goss-oriented grain [1]. Figures 3b and 3c present the cumulative distributions of misorientation angle values, corresponding to line scans performed along the lines A and B in Fig. 3a, respectively. These distributions illustrate that the misorientation vectors displayed a tendency to cancel each other across the consecutive large-angle extended microband walls (across the band widths), but misorientation angles were largely observed to accumulate across the small-angle transverse short walls (along the band lengths) [5]. Sub-boundaries appeared to maintain their frequent alignment close to the {111} crystallographic slip planes and they remained largely inclined to the rolling direction at angles ranging from 20° to 40°, similar to the lower strain level. This indicates that these sub-boundaries might possibly undergo continuous dynamic reorganisation during plastic straining [5].

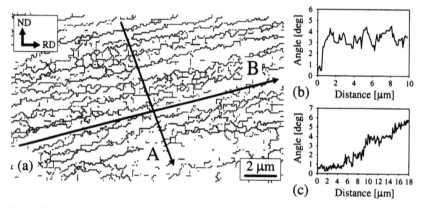

Fig. 3. EBSD analysis of the microbands formed within a Goss-oriented grain at a strain of 0.69: (a) boundary map (a step size of 0.05 μm); (b),(c) cumulative misorientation distributions for the lines A and B indicated in (a) respectively.

4. Conclusions

The evolution of crystallographic texture and deformation substructure was studied in a type 316L austenitic stainless steel, deformed in rolling at 900°C to true strain levels of about 0.3 and 0.7, using the EBSD technique. The obtained statistically representative data will assist in the development of physically-based models of microstructural evolution during hot forming of similarly alloyed austenitic stainless steels.

References

[1] Randle V and Engler O 2000 Introduction to Texture Analysis: Macrotexture, Microtexture and Orientation Mapping (Amsterdam: Gordon and Breach)
[2] Humphreys F J, Bate P S and Hurley P J 2001 J. Microscopy 201 50-58
[3] Humphreys F J 2001 J. Mater. Sci. 36 3833-3854
[4] Randle V, Hansen N and Juul Jensen D 1996 Phil. Mag. A 73 265-282
[5] Hurley P J and Humphreys F J 2003 Acta Mater. 51 1087-1102

Inst. Phys. Conf. Ser. No 179: Section 6
Paper presented at Electron Microscopy and Analysis Group Conf. EMAG2003, Oxford, 2003
©2003 IOP Publishing Ltd

Energy-Filtered Imaging of Cu-nanoprecipitates

S Lozano-Perez, J M Titchmarsh and M L Jenkins

Department of Materials, University of Oxford, Oxford OX1 3PH

ABSTRACT: Energy-Filtered (EF) TEM imaging has been used to reveal and measure the size of individual Cu-precipitates as small as 2nm embedded in ferritic steels. Optimisation of the acquisition and analysis procedures is described. A procedure was developed to identify the contributions to spatial resolution limit from fundamental factors, drift and instrument instabilities. A long data acquisition time is necessary during which drift occurs that ultimately limits detection. Nevertheless, EFTEM provides a viable method for revealing the precipitation.

1. Introduction

The determination of the size, number density and composition of Cu-rich nanoprecipitates in ferritic steels is important for assessing the potential for brittle fracture in nuclear reactor pressure vessels [1]. The smallest particles are known to be coherent with the bcc matrix but normally cannot be imaged in the TEM, while particles larger than a few nanometres transform to the 9R structure [2] and can then be seen using HREM. The relative importance of the two particle types on mechanical properties is presently unknown. 3D Atom Probe (PoSAP) images all the particles without discrimination, but has poor sampling statistics [3]. Small Angle Neutron Scattering (SANS) averages the combined particle characteristics in bulk samples but requires extra information to allow assessment of composition [3]. In principle, EF-TEM offers the means of revealing both types of particle, so direct comparison of precipitates revealed by HREM and in a chemical map of the same area would enable discrimination. However, spatial resolution, drift and signal-to-noise ratio (SNR) in EFTEM become limiting as the precipitate size falls. In this paper we explore the use of EFTEM to determine the size of the particles when, in addition to the effect of aberrations, diffraction and delocalisation on resolution, the influence of stage drift becomes more important as the data acquisition time increases.

2. Experimental Procedures

A Fe-Cu-Ni-Mn-Si alloy was heat-treated at 365°C for 10000h to induce the homogeneous precipitation of Cu-rich clusters with diameters in the range ~2-10nm and examined in a JEOL 3000F TEM ($Cc = 1.4$ mm, $Cs = 0.6$ mm) equipped with a Gatan GIF for energy-filtered (EF) imaging. Due to the difference in inelastic partial differential cross-sections, the decrease in signal in the Fe-L_{23} edge due to the reduced Fe in the volume occupied by the Cu precipitate was approximately ten times larger than the corresponding increase in the signal at the Cu-L_{23} edge. For this reason, it was found to be advantageous to visualize the precipitates as "lack" of iron, using an elemental Fe-L_{23} map, instead of Cu enrichment in a Cu-L_{23} map.

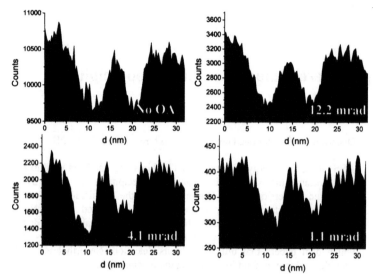

Figure 1. Line profiles across the same two precipitates in a 730 eV EF image acquired with a 10 eV slit and different OA.

The procedure for EFTEM imaging was as follows. A suitable area of interest was selected and oriented before the final alignment of the microscope and GIF, in order to minimise the effect of astigmatism on resolution from the strong magnetic field of the specimen. A thickness map was then acquired to ensure that $t < \lambda$ (normally $t \approx 0.5\lambda$). The sample was focused using inelastically scattered electrons with ~100eV energy loss, the beam centred on the GIF CCD and then defocused until the CCD was completely illuminated. The series of 15 EF images was acquired using a 10eV slit over a 150eV energy-loss range. Images were aligned to compensate for any drift before background subtraction and generation of the Fe-L_{23} elemental map. Similar series of EF images of the same area were acquired using different objective aperture (OA) sizes. The acquisition time was 20s for every image.

The background fitting was performed using a power law, smoothed over 6 pre-edge windows, using a self-written code in Digital Micrograph Script Language. A typical Fe-L_{23} map from 9 added post-edge images (710-800eV) is shown (Fig. 5). The position of the first post-edge image and the width of the window were chosen in order to maximize the SNR [4]. A volume of the order of 1 million cubic nanometres can be typically analysed with each series of EF images. Hence, a suitable compromise between image magnification and pixel resolution was found using the 2k CCD camera with a hardware binning of two, giving a pixel size of 0.27 nm at a magnification of x10k, equivalent to ~10 pixels for a 3nm precipitate image.

3. Results and Discussion

Line-profiles along diameters through the same two Cu-precipitates are shown for different OA in Fig.1. A nominal value of 100mrad has been used when no OA was inserted, (Fig.1a). The noise in each image was estimated from the standard deviation of the counts in an adjacent area of the precipitate-free matrix with homogeneous thickness.

The signal-to-noise ratio (SNR) was then measured by defining the signal as the depth of the depleted region in the centre of the precipitate profile (Fig. 2a). By summing 9 post-edge background-subtracted images, the SNR increased considerably while maintaining the image resolution. The number of counts in the image increased with OA diameter (Fig.2b) but the highest SNR was obtained with an OA of 4.1mrad (Fig.2a), consistent with the profiles in Fig.1.

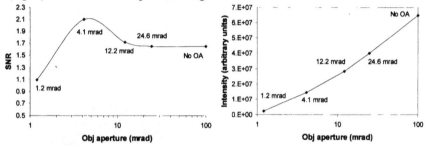

Figure 2. a) SNR vs. OA diameter: b) Intensity (Number of counts) vs. OA diameter for the same illumination conditions.

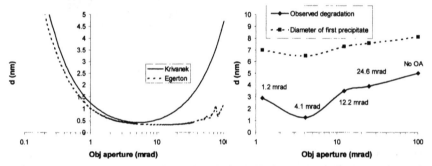

Figure 3. a) Degradation of the spatial resolution with the OA according to Egerton and Krivanek; b) Observed degradation and diameter of the first precipitate vs. OA.

Theoretical estimates of the spatial resolution of a point object in a thin sample due to inherent aberrations, diffraction and delocalisation using the methods of Egerton (point spread function calculations for 50% current) [5] and Krivanek (quadrature summation) [6] are shown in Fig.3a for the current microscope. Predictions from both theories are similar for OA semi-angles <5mrad but rapidly diverge for larger OA (Fig. 3a).

For each OA series, the quadratic sum: $d_m^2 = d_r^2 + d_{deg}^2$, was used to relate the measured precipitate size, d_m, to the real (constant) size, d_r, and the degradation, d_{deg}, due both to inherent aberrations (Fig.3a) and to additional factors such as stage drift and any power supply instabilities. Assuming that d_{deg} (OA=1.2mrad) is ~2.3 times d_{deg}(OA=4.1mrad), as suggested by Fig.3a, solution of the pair of quadrature equations for these two OA series gave values of d_{deg} (Fig.3b) that were slightly worse than the values predicted by Krivanek in Fig.3a. Corresponding analysis for the other OA image series yielded similar high values. Hence, either the theoretical estimates in Fig.3a were both too low or drift and instability were present. The latter was much more likely because theoretical estimates of resolution have been confirmed by others [6].

244

The distance by which successive images in the EF series were laterally shifted to align with the first image provided a direct measure of drift. The plots in Fig. 4 show the drift vectors for the area in Fig.1 during three sequential OA series acquisitions between which the stage piezoelectric controls were used for re-centring the area of interest. The total analysis time was > 2000s. For example, for acquisition series 1 in Fig.4, a drift contribution of ~1.2nm, the average drift between consecutive images, should be then added to the values in Fig.3a. Adding the drift, the measured degradation in Fig.3b gets closer to the predicted values from Krivanek, rather than those of Egerton. Experiments using the same experimental parameters and analysis procedure showed that particles as small as ~2nm could be revealed (Fig.5), consistent with the combined drift and aberration contributions to resolution. EFTEM imaging, therefore, promises to become a very useful method, complementing PoSAP for characterising Cu precipitation.

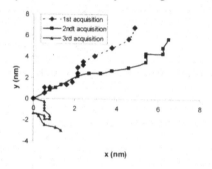
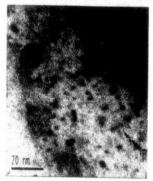

Figure 4. Plot showing the stage drift for consecutive series of acquisitions.

Figure 5. Fe elemental map (OA = 4.1mrad).

4. Conclusions

The imaging of Cu-rich nanoparticles as small as 2nm in diameter in a ferritic matrix was demonstrated by optimising experimental parameters to maximise the intensity reduction in the Fe-L_{23} EF image. A 6-window fitting procedure was developed to improve the background subtraction from the aggregated signal from 9 post-edge images while retaining resolution. After allowing for drift during the long acquisition times, the resolution limit predicted by Krivanek was found to be more consistent than that of Egerton. The achievable spatial resolution shows that EFTEM imaging of Cu-rich particles in ferritic steel is a viable complementary method to PoSAP.

5. References

1. Othen P J et al. 1991 Phil. Mag. Lett. 64 p. 383
2. Nicol A C et al. 1999 Mat. Res. Soc. Symp. Proc. 540 p. 409
3. Miller M K et al. 2003 Mat. Sci and Eng. A353 p. 133
4. Kothleitner G and Hofer F 1998 Micron 29(5) p. 349
5. Egerton R F and Crozier P A 1997 Micron 28(2) p. 117
6. Krivanek O L et al. 1995 J of Microscopy 180(3) p. 277

Acknowledgements

This work was funded by the EPSRC. JMT thanks the Royal Academy of Engineering, AEA Technology and the Institute of Nuclear Safety System, Japan, for support.

Inst. Phys. Conf. Ser. No 179: Section 6
Paper presented at Electron Microscopy and Analysis Group Conf. EMAG2003, Oxford, 2003
©2003 IOP Publishing Ltd

Characterisation of dispersions within annealed HVOLF thermally sprayed AlSnCu coatings

Chang-Jing Kong, Grigore Moldovan, Mike W Fay, D Graham McCartney and Paul D Brown
School of Mechanical, Materials, Manufacturing Engineering & Management, University of Nottingham, University Park, Nottingham, NG7 2RD, UK

Abstract: High velocity oxy-liquid fuel (HVOLF) AlSnCu coatings are characterised following annealing for up to 5 hours at 300°C. A combination of statistical analysis of BSE images and TEM observations demonstrate the decrease in the number of sub-micron and nanoscale Sn particles with annealing, commensurate with a decrease in the coating microhardness. TEM evidence further suggests the coarsening of nanoscale Sn through a mechanism of a liquid phase migration within the Al matrix. EELS and EFTEM additionally allow the identification of the precipitation of θ'.

1. Introduction

The Al-12wt.%Sn-1wt.%Cu (Al12Sn1Cu) alloy deposited by the high velocity oxy-liquid fuel (HVOLF) thermally spray technique may be used for the manufacture of automotive bearings. The soft Sn phase acts as a self-lubricant that introduces good anti-friction characteristics. However, the microhardness of the as-sprayed coating is too high for the bearing material applications and heat treatments are applied to reduce the coating microhardness. In this paper, the microstructure of Al12Sn1Cu as-sprayed and annealed coatings is characterised, with particular regard to the dispersion and evolution of Sn particles and the development of Al-Cu compound precipitates. The factors leading to the decrease in microhardness with increasing annealing time are discussed.

2. Experimental

The HVOLF thermal spraying parameters for the Al12Sn1Cu coating have been described previously (Kong *et al* 2001). The coatings were heat treated at 300°C for between 15 minute and 5 hours. Metallographically sectioned samples were examined using backscattered electron (BSE) imaging in a JEOL 6400 SEM. Three BSE images representing each processing condition were selected for analysis. A grey-level based threshold was applied to each and the resulting binary maps were used to measure the projected areas of the sub-micron Sn particles. A criterion of a minimum of 9 pixels was used to determine the smallest quantifiable area. Histograms of particle number against particle area, averaged across each set of images, were constructed. A Jeol 2000fx TEM was used to examine the development of the nanoscale Sn dispersion within the annealed coatings, whilst a Jeol 4000fx was used to characterise the fine details of the precipitation, using the EFTEM and EELS techniques.

246

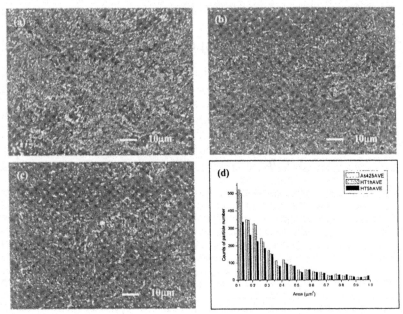

Figure 1. BSE images of HVOLF thermal sprayed coatings and statistical analysis of the sub-micron Sn particles. (a) As-sprayed coating. (b) Coating heat treated at 300°C for 1 hour. (c) Coating heat treated at 300°C for 5 hours. (d) The relationship between number of Sn particles against area as a function of time of annealing.

3. Results and discussion

Figures 1a-c show BSE images of Al12Sn1Cu coatings for the as-sprayed condition and annealed at 300°C for 1 hour and 5 hours, respectively. The light regions within these images correspond to large Sn particles, the dark regions correspond to the Al rich matrix, whilst the intermediate grey contrast regions are attributed to dispersions of nanoscale Sn within the coatings (Kong *et al* 2001). It is apparent that the area fraction of the grey contrast regions within these BSE images decreases as the annealing time increases. However, changes in the light contrast regions due to the effects of annealing is less easy to distinguish. Accordingly, statistical analysis of these images was performed using the ImageJ software package to produce histograms of the number of large Sn particles against project area.

Inspection of Figure 1d representing the Sn particle distributions within the as-sprayed coating and those annealed at 300°C for 1 hour and 5 hours initially shows that the Sn particle number decreases with increasing particle area, tending to zero for an area value of greater than $1.0\mu m^2$. The histogram shows that the number of sub-micron Sn particles with area between $0.1 \sim 0.3\mu m^2$ within the same volume decrease slightly after 1 hour annealing at 300°C, as compared with the as-sprayed coating, but decreases significantly after 5 hours of annealing. For Sn particles of area between $0.3 \sim 0.55\mu m^2$, there is little difference in the decrease of the number of Sn particles within these annealed coatings. For the Sn particles with area greater than $0.55\mu m^2$, the number of Sn particles slightly increases after 5 hours of annealing. These results indicate that the sub-micron Sn particles tend to coalescence and merge with the larger Sn particles as the annealing time is increased.

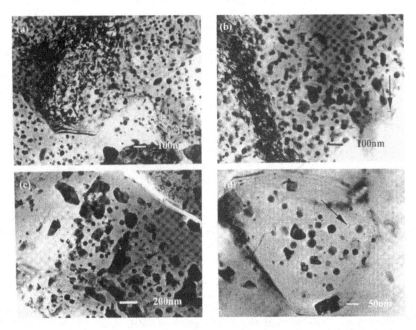

Figure 2 TEM images showing coarsening of nanoscale Sn particles with annealing temperature and increasing time. (a) As-sprayed coating. (b) The coating heat treated at 300°C for 1 hour. (c, d) Coating annealed at 300°C for 5 hours.

The TEM images of Figure 2 are used to illustrate the development of the nanoscale Sn particles with annealing and increasing annealing time. The nanoscale Sn particle dispersoid in the as-sprayed coating (Fig. 2a) coarsens slightly after 1 hour annealing at 300°C (Fig. 2b) while these particles further coarsen after 5 hours of annealing (Fig. 2c,d). It is noted that many nanoscale Sn particles are no longer spherical with annealing, but become elongated and irregular in shape within the Al matrix due to the merging of two or more particles. Larger Sn particles located at the grain boundaries similarly develop irregular shapes with annealing. In addition, annealing at 300°C induced the formation of very fine rod-like precipitates (e.g. Figs. 2 b,d, arrowed) which required EFTEM characterisation as follows.

The EEL spectrum from a rod-like precipitate and the matrix region of a sample annealed at 300°C for 1 hour shows that there are two Cu energy peaks (L_3 with energy 930eV and L_2 with energy 951eV) from the region of the precipitate (Fig. 3a). EFTEM maps for Sn, Cu and Al were acquired from this feature at energies of Sn_M (485eV), Cu_L (931eV) and Al_K (1560eV), respectively (Figs. 3b-d). The maps delineated a precipitate of approximately 10nm by 30nm in size, rich in Cu and depleted in Al, while no Sn was found to be present. Slight distortion of the Al map has occurred due to sample drift during the long exposure time. These results indicate that these fine rod-like precipitates are a Cu-Al compound. This is consistent with Porter et al (1981) who use phase diagrams to predict that 1 wt.% Cu within the Al may form θ' precipitates after annealing at 300°C. No evidence of θ' was found in the as-sprayed coatings, but it was identified in the coatings annealed at 300°C for 5 hour.

Returning to the phenomenon of Sn particle coarsening within the Al matrix and its relationship with the material hardness. It is noted that the melting points of pure Sn and

248

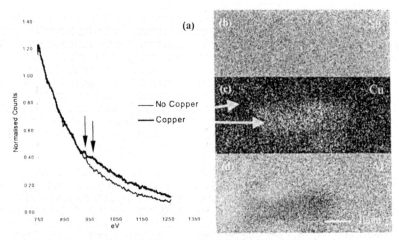

Figure 3 EELS spectrum and EFTEM result from an elongated precipitation. (a) EEL spectrum of matrix and precipitates; (b-d) EFTEM maps of Sn, Cu and Al distribution.

Al are 232°C and 660°C, respectively. It is thus expected that the sub-micron and nanoscale Sn particles adopt a liquid state in the solid Al matrix during annealing at 300°C. The liquid elemental phase is predicted to move as a coherent droplet within the solid phase, driven by local temperature gradients (McLean et al, 1974). During the process of movement the liquid Sn droplet dissolves solid Al at the hot end of the particle and condenses Al at the trailing, cold end. When two or more liquid Sn droplets approach, they will start to merge to form a larger Sn particle. Surface energy of the coalescing Sn droplets will tend to make the irregular shapes more spherical given sufficient time at a high enough temperature. With increasing time of annealing, the nanoscale Sn particles tend to be eliminated, either by the coarsening of particles within the grains or by the coalescence of larger particles at the grain boundaries due to the additional effect of local strain fields.

The reason the microhardness decreases with increasing annealing time (Kong et al, 2001) is therefore attribute to this coarsening of the nanoscale and sub-micron Sn, combined with the local release of strain within the Al grains. The precipitation of θ′ might be expected to increase the microhardness of the annealed coatings by dispersion strengthening. However, the volume fraction of θ′ precipitates is not high enough to significantly effect the microhardness as compared with the effect of Sn.

4. Conclusion

The microhardness decrease with increasing time of annealing is attributed to the coarsening of sub-micron and nanoscale Sn particles. The precipitation of θ′ is additionally identified within the annealed coatings but this is not considered to significantly affect the microhardness. A liquid migration mechanism is invoked to explain the coarsening of Sn particles.

References

Kong CJ, Brown PD, Horlock AJ, Harris SJ and McCartney DG 2001 Inst. Phys. Conf. Ser., **227-230,** 168

McLean M and Loveday MS 1974 Journal of Materials Science **1104-1114,** 9

Porter DA and Eastering KE 1981 Phase Transformations in Metals and Alloys (Van Nostrand Reinhold (UK) Co. Ltd.) **291**

Inst. Phys. Conf. Ser. No 179: Section 6
Paper presented at Electron Microscopy and Analysis Group Conf. EMAG2003, Oxford, 2003
©2003 IOP Publishing Ltd

Characterisation of polluted clay-zeolite landfill liner materials using scanning electron microscopy

Y Güney, H Koyuncu

Department of Civil Engineering, Anadolu University Eskisehir, Turkey

A Kara

Department of Materials Science and Engineering, Anadolu University Eskisehir, Turkey

Abstract. A landfill liner act as a barrier to minimize the migration of leachates in waste containment systems. Thus, proper management of leachates is important to minimize the risk of contamination for groundwater and subsoil. Structure of a liner is affected by different salt and metal solutions. Therefore, microstructural and physico-chemical properties of landfill liners are important. This paper, part of an extended study, reports the morphological characteristics of clay-zeolite liner material mixtures before and after contamination with salt and heavy metals solutions.

1. Introduction

Landfilling is the most common method for disposing of solid wastes. One of the major problems associated with landfilling is the generation of large amounts of heavily polluted leachate. Municipal landfill leachate is identified as a potential source of ground and surface water contamination. Landfill liners act as a barrier to minimize the migration of leachates. [1] The main requirements of liners are minimization of pollutant migration, low swelling, and shrinkage and resistance to erosion. Clay liners (especially bentonite) are widely preferred due to their cost effectiveness, large attenuative capacity and resistance to damage and puncture. Further reasons are their low permeability and high cation exchange capacity. However, such liners crack on drying due to its swelling potential and as a result its hydraulic conductivity increases. In addition, some leachates increase its permeability due to liner material-pollutant interaction. Clay liners also possess high shrinkage potential and hence can crack under unsaturated conditions causing instability and increase in leakage rates. [2,3] Leachates may contain large amounts of organic matter, ammonia-nitrogen, heavy metals and chlorinated organic and inorganic salts. Amongst these pollutants, heavy metal species, including Cu, Zn, Cd, and Pb can be particularly dangerous to humans, animals, and aquatic lives. Heavy metals are introduced into the environment through agricultural practices, transport, industrial activities, and waste disposal. [4] Considering all this, investigation of physical and chemical characteristics of different liner materials, their competetive interactions with the pollutants and possible consequent structural changes

250

to the liner is necessary. In a previous study achieved by the same authors, dynamically compacted kaolinite-zeolite mixtures (UCLM) were prepared for possible usage as a landfill liner at various ratios (K/Z: 0.1, 0.2, and 0.3). Then, these mixtures were contaminated with both salt (NaCl, $CaCl_2$) and heavy metal ($CuCl_2$, $PbCl_2$) solutions at optimum moisture content of 32.5 % (CLM) in order to simulate the contamination of the liner. Geotechnical tests showed that a K/Z ratio of 0.2 was ideal regarding to hydraulic conductivity and unconfined compression strength for the UCLM samples. A certain amount of increase in permeability was observed for the representative CLM samples [5]. The objective of this paper is to investigate the possible morphological changes of the kaolinite/zeolite mixture (K/Z ratio: 0.2) after being polluted experimentally with the above-mentioned solutions.

2. Experimental

A kaolinitic clay was provided from a local tile producer. Natural zeolite was obtained from the Gördes region of Balıkesir/Turkey. The details regarding to the preparation of the uncontaminated (UCLM) and contaminated (CLM) kaolinite/zeolite mixtures are given previously. [5]

The mineral compositions of the as-received raw materials were established by using XRD (Rigaku, Rint 2200, Japan). The representative samples were firstly dried at 105°C, then a Cam Scan S4 analytical SEM was employed to investigate their fracture surfaces under different magnifications in secondary electron (SE) imaging mode. Prior to SEM investigation, the samples were sputtered with gold-palladium for long enough to prevent charging.

Table 1 Salts and heavy metals and their ratios used in the study

ratio	formula	Molecular weight (g/mol)
0.5 N	NaCl	58.44
0.5 N	$CaCl_2.2H_2O$	147.02
2000 ppm	$PbCl_2$	278.1
2000 ppm	$CuCl_2.2H_2O$	170.48

3. Results and discussion

According to the XRD results, clay consists mainly of kaolinite with small amount of free quartz. The main mineral phase in natural zeolite was found to be clinoptilolite. Typical SEM images obtained from the fracture surface of the UCLM sample under two different magnifications are given in Figure 1.

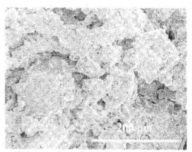

Figure 1. SEM images of the UCLM sample

A typical high magnification image of a group of zeolite crystals embedded in the UCLM sample is given in Figure 2. The typical morphology of the natural zeolite crystals is quite evident. Figures 3, 4, 5 and 6 are the typical SEM images taken under different magnifications from the fracture surfaces of the K/Z mixture contaminated with NaCl, CaCl₂, PbCl₂ and CuCl₂, respectively. As can be seen, these images show a very different pore size distribution than that of the uncontaminated samples.

Figure 2. SEM image of a group of zeolite crystals

As stated earlier, the permeability, or in other words, hydraulic conductivity of a landfill liner is the primary concern. Its capacity for removing toxic pollutants is an extra quality. It is apparent from the images below that when wetted with the salt and the metal solutions employed in the study, the particles in the K/Z mixture were settled and clumped together to form larger clusters or flocs by the process called coagulation-flocculation or agglomeration. As a result, deterioration in the permeability of the mixture is expected. Naturally, this process depends mainly on the physico-chemical forces acting on the liner-pollutant system.

Figure 3. SEM images of the CLM samples (contaminated with NaCl solution)

Figure 4. SEM images of the CLM samples (contaminated with CaCl₂ solution)

Figure 5. SEM images of the CLM samples (contaminated with PbCl₂ solution)

Figure 6. SEM images of the CLM samples (contaminated with CuCl₂ solution)

Regarding to the degree of flocculation effect of the salt and the metal solutions on the morphology of the UCLM sample on the basis of the representative SEM images, the differences were small. However, it is possible to say that the salts appear to be more effective than the metals.

4. Conclusions

The microstructural characteristics of a kaolinite/zeolite mixture (K/Z ratio: 0.2) as a potential liner material were investigated using SEM. The results demonstrated that the presence of salts and heavy metals employed in the study was directly related to the alteration of microstructural orientation such as flocculation and/or agglomeration. Thus, it would not be surprising to obtain an increase in the hydraulic conductivity and a decrease in the strength values of the previously investigated K/Z mixtures.

References

[1] Allen A 2001 Engineering Geology 60 3-19
[2] Kayabalı K 1997 Engineering Geology 46 105-114
[3] Tuncan A Tuncan M Koyuncu H Güney Y 2002 Waste Management & Research 21 54-61
[4] Hani H 1990 In. J. Environ. Anal. Chem. 39 (2) 197-208
[5] Güney Y Koyuncu H and Yılmaz G 2003 Proc. Int. Conf. On New Developments in Soil Mechanics & Geotechnical Eng. N. Cyprus 1 137-144

Inst. Phys. Conf. Ser. No 179: Section 6
Paper presented at Electron Microscopy and Analysis Group Conf. EMAG2003, Oxford, 2003
©2003 IOP Publishing Ltd

Electron Microscopy of Defects in Oxide-Dispersion Strengthened Ferritic Alloys

Y L Chen
School of Engineering, The University of Surrey, Guildford, Surrey GU2 7XH

A R Jones
Department of Engineering, The University of Liverpool, Liverpool, L69 3GH

Abstract. In this work, development of porosity in an iron based oxide dispersion strengthened (ODS) alloy, PM2000, has been examined by scanning electron microscopy (SEM) and electron probe microanalysis (EPMA). Pores about 10μm in size were found in a sample that had undergone recrystallisation annealing at 1380°C for 1hour. However, the size of the pores increased to about 100μm when the sample was annealed at 1380°C for 2200 hours. EPMA results showed that the Al content in the 1 hour annealed sample was 4.50% but this fell to 3.35% after the long anneal. Based on these results, it has been suggested that some of the pores observed in the ODS alloy may be Kirkendall pores due to the loss of Al. In addition, SEM results have identified that two types of pore are present, one of which is part-filled.

1. Introduction

Oxide dispersion strengthened (ODS) alloys produced by mechanical alloying (MA) are materials with important engineering potential because of the combination of excellent high temperature creep strength and oxidation resistance [1]. However, defects such as porosity are often found in ODS alloys after secondary recrystallisation annealing[2-6]. Since porosity will be detrimental to their mechanical strength, a lot of work has been done to understand the occurrence of porosity in ODS alloys [2-6]. Chen, Jones and Miller [3] attributed the origin of porosity in ODS alloy to gas trapped in the alloy during mechanical alloying (MA) process. The rapid kinetics of secondary recrystallisation observed in ODS alloys may leave no time for gas trapped in the material to dissipate to free surfaces via rapid grain boundary diffusion and porosity is formed as a result [2]. Based on this understanding, a double annealing method has been proposed to reduced pores in ODS alloys and the technique has been demonstrated to work very well in an iron based ODS alloy [2].

However, previous work on pores in ODS alloy was mainly concentrated on samples soon after recrystallisation annealing. Since ODS alloy will serve at high temperature, it is important to examine the effect of long time high temperature service on the porosity. In this work, the development of porosity in an iron-based ODS alloy during long time annealing has been examined by electron microscopy.

2. Experimental Methods

254

A model ferritic ODS alloy was used in the current study. It is compositionally similar to commercially available grades such as PM2000 produced by Plansee GmbH and MA956 produced by Special Metals Corporation. The as-received material was a fine grained 2mm thick sheet material. Two kinds of sample have been examined. One was isothermally annealed at 1380°C for 1h to obtain a fully recrystallised, coarse grained structure and the other was annealed at 1380°C for 2200h. All samples were polished to 2400 grit SiC paper and then were further polished using 0.25μm colloidal silica suspension in order to attain a damage-free surface. The polished samples were examined by Scanning Electron Microscopy (SEM) and Electron Probe Microanalysis (EPMA). SEM was performed with a CAMSCAN x500 crystal probe field emission gun SEM and EMPA was performed in a JEOL JXA-8600 superprobe.

3. Experimental Results

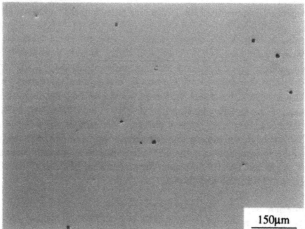

150μm

Fig.1 SEM image from 1380°C/1h annealed samples showing pores.

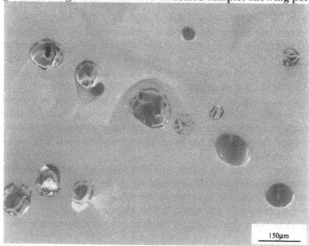

150μm

Fig.2 SEM micrograph from 1380°C/2200h annealed sample showing pores.

Fig.1 is a representative SEM image from the 1380°C/1h annealed sample showing pores with a size from 5μm to 15μm.

Fig.2 is a typical SEM micrograph from the 1380°C/2200h annealed sample. Compared with Fig.1, the pore size is larger in Fig.2. It is from 50μm to 150μm. As a matter of fact, not only the pore size but also the pore volume fraction increased during the annealing. The pore volume in this alloy soon after the recrystallisation annealing has been reported to be 1.5vol%[3] and the pore volume fraction in the 1380°C/2200h annealed sample is 4.7vol% [7].

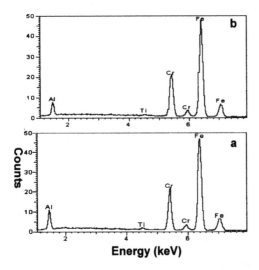

Fig.3 EDX spectra from 1380°C/1h annealed sample (a) and 1380°C/2200h annealed sample, respectively.

From Fig.1 and Fig.2, it has been observed that there are two types of pores; one is empty sphere whilst the other appears to be part filled pores. The latter has been previously referred to as intrusion [5].

The Al content in the two samples were analysed by EMPA. Fig.3 shows two X-ray energy dispersive spectra from the two samples. The two spectra were taken under identical conditions and it can be seen that the Al peak is lower in the long term annealed sample while no difference could be observed for other peaks. Quantitative analysis from the two samples showed that the Al content was 4.50wt% for the sample annealed for 1380°C/1h and 3.35wt% for the sample annealed 1380°C/2200h.

4. Discussion

The reasons why pores occur in ODS alloys during recrystallisation annealing and why ODS alloys are so prone to porosity have been discussed previously [2-3]. However, it will be difficult to explain why more pores will develop and/or the pores will grow larger at high temperature after recrystallisation annealing based on the previous understanding. For this reason the Al content in the two samples were examined by EMPA. The results showed that more than 1wt% of Al was lost during the long time high temperature annealing. The high Al content is one important reason for the excellent oxidation resistance for this FeCrAl ODS alloy. At high temperatures, an adherent alumina layer will be formed that offers good protection against further isothermal oxidation. Thus it is understandable that the Al-level in solution will decrease gradually with prolonged exposure at high temperatures.

There are two possible results due to the loss of Al. The material could shrink or Kirkendall pores will be formed as a result. For the present ODS alloy, because of its high strength, shrinkage is less likely [8]. An alternative mechanism to accommodate the loss of Al is through formation of Kirkendall pores. Hence, the increase of pore size and volume fraction during long term annealing at high temperatures could be attributed to Kirkendall pores.

The above discussion suggested that porosity may be inevitable in ODS alloys after protracted exposure at high temperatures. Further work is needed to understand the implication of this discovery.

Another interesting observation in this work is the two forms of porosity. The part filled pores have been referred to previously as intrusions, and it is believed that these defects can originate during the mechanical alloying process, either as a results of occlusion of fragments of MA powder within other, larger MA powders during processing, or as powder particles from the finest size of MA powder that resist full densification during consolidation processing [5]. The present results identify larger filled pores than seen previously (Fig.2); and there is an absence of smaller intrusions. This may suggest that the so called intrusion may just be another form of pores. However, further work is necessary to establish the nature of these artefacts, their role as incipient pores and defect sinks and the size range observed after different annealing durations.

5. Conclusions

Based on the above results and discussion, the following conclusions could be drawn:

1. The pores in Fe-based ODS alloys increase both in size and volume fraction during high temperature annealing and this has been attributed to the formation of Kirkendall pores due to the loss Al.
2. It has been clarified that there are two forms of pores. One is empty and one is part filled.

6. Acknowledgement

The authors are very grateful to Miss Gill Gibbs at MicroStructural Studies Unit for doing the EPMA and for Mr. U. Miller at Plansee GmbH Germany for providing the samples. Special thanks to Dr. Justin Ritherdon at The University of Liverpool for his interesting and full support for this work.

7. References

1. Czyrska-Filemonowicz A, Dubiel B J, 1997, Mater. Proc. Tech. 64, 53-64
2. Chen Y L, Jones A R, 2001, Metall. Mater. Trans. A. 32A, 2077-85
3. Chen Y L, Jones A R and Miller U, 2002, Metall. Mater. Trans. A, 33A 2713-18
4. Lengauer W, Ettmayer P, Korb G, Sporer D, 1990, Powder Metall. Inter., 22, 19-22
5. Jones A R, Jeager D M, 1998, 12th Ann. Conf. On Fossil Energy Mater., P2.11
6. Heine B, Kirchheim, Stolz U, 1992, Powder Metall. Inter., 24, 158-163
7. Chen Y L, Jones A R, 2002 unpublished work
8. Tatlock G, private communication

Inst. Phys. Conf. Ser. No 179: Section 6
Paper presented at Electron Microscopy and Analysis Group Conf. EMAG2003, Oxford, 2003
©2003 IOP Publishing Ltd

Microstructural characterisation of a rapidly solidified NiMo steel

M Briceno-Gomez, P M Brown*, J M Titchmarsh, S Ahmed and P Schumacher**

Department of Materials, Oxford University, Parks Road, Oxford, OX1 3PH, UK

* QinetiQ ltd, Ively Road, Farnborough, GU14 0LX, UK

** Institut fuer Giessereikunde, Franz-Josef-Strasse 18, A-8700 Leoben, Austria

Abstract. The microstructure in a high strength steel containing impurity elements, prepared by arc melting and melt spinning, has been studied by analytical TEM, electron and ion probe methods. Melt spun alloys have a more refined structure than arc melt alloys, leading to greater hardness. Additions of La resulted in the gettering of residual impurities. However, attack of the crucible during melt spinning caused the unexpected formation of potentially harmful large carbo-borides.

1. Introduction

Compared with conventional processing, improvements in strength, toughness and wear resistance of steels can be achieved by rapid solidification processing (RSP) through refinement of the microstructure, increased solubility of alloying elements and impurities, reduced levels of segregation and, in some cases, the formation of metastable phases [1, 2]. Here, we compare the microstructure of a NiMo steel that has been rapidly solidified by two methods, arc melting (AM) at a cooling rate of $\sim 10^4$ K/s and melt spinning (MS), cooled at up to $\sim 10^6$ K/s. In particular, the removal of P, a potent embrittling element in high strength Ni-Mo ferritic steels, was investigated by deliberate additions of La and P.

2. Experimental procedure

Table 1 lists the compositions of the samples. Appropriate amounts of La, in the form of LaNi$_5$, and P, as FeP, were mixed to a base NiMo steel of low P and S concentration, to make 10gm AM ingots in a Cu boat. The additions of La and P were made to induce the formation, and facilitate the study, of phases that might form during the gettering of impurities. The MS samples were prepared by melting and superheating AM ingots in a BN crucible to $\sim 1600°C$ under He gas and ejecting by high pressure Ar gas onto a rotating Cu wheel to produce ribbons typically 30-50 μm thick. Samples were examined using optical microscopy, SEM (JSM6300), EPMA (JXA8800), TEM (JEM3000F, CM20, HB501) and a Cameca NanoSIMS50 using an 8KeV Cs ion source. TEM foils were prepared by electro-polishing in a Struers Tenupol 5 with a 10% perchloric acid/methanol electrolyte at -30°C, or using a Gatan Precision Ion Polishing system. Specific inclusions were thinned using a FEI FIB200 system using the 'liftout' technique [3].

Table 1. Compositions of alloys

	C	Ni	Mo	La	P	Fe
Base Material	0.4	2	1.5	-	Max 0.005	Balance
Alloy A	0.4	2	1.5	-	0.1%	Balance
Alloy B	0.4	2	1.5	0.1%	-	Balance
Alloy C	0.4	2	1.5	0.5%	0.2%	Balance

3. Results

The matrix structure of all the AM samples was martensitic, of hardness 550H$_V$, with highly dislocated laths, 100-300nm wide, and a dispersion of fine carbides (Fig.1). Elemental maps revealed Mo segregation of ~50µm spacing (Fig.2), consistent with dendritic coring. In addition, alloys B and C contained complex particles, ~1µm in size, rich in La-O-P, La-O or La-O-S that were not present in the base material or in alloy A (Fig.3). TEM diffraction confirmed the presence of both La$_2$O$_3$ and LaPO$_4$. In alloy C, particles rich in La and P were also found and confirmed by diffraction to be LaP. The structure of the S-containing particles has not yet been established but is likely to be La$_2$O$_2$S.

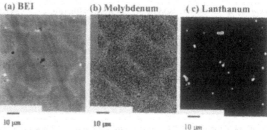

Figure 1. BF TEM micrographs of AM sample alloy B showing (a) laths martensite and (b) high concentration of fine carbides inside martesite plate

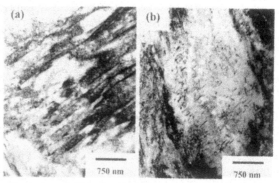

Figure 2. AM sample alloy B: (a) EPMA back-scattered electron image (b) Mo and (c) La elemental maps

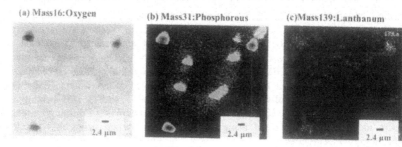

Figure 3. NanoSIMS elemental maps of AM sample alloy C showing LaPO$_4$ particles

There was no apparent Mo segregation but inhomogeneous B segregation was revealed by elemental mapping (Fig.5). EELS and EDX showed ~100nm-sized BN particles in all MS alloys and which had a La-rich core in alloys B and C (Fig.6). The La-P-O particles found in AM alloys were not present in any of the MS samples. However, all MS alloys contained large (1-10µm) rounded particles (Fig.7) of a crystalline phase identified as a carbo-boride, with the orthorhombic Fe₃X structure reported by others [4, 5].

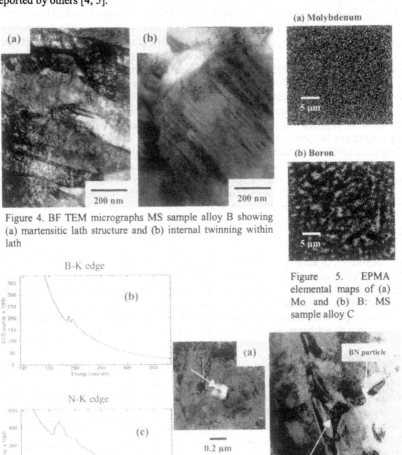

Figure 4. BF TEM micrographs MS sample alloy B showing (a) martensitic lath structure and (b) internal twinning within lath

Figure 5. EPMA elemental maps of (a) Mo and (b) B: MS sample alloy C

Figure 6. (a) BF TEM micrograph of MS base material showing typical BN particle (arrowed) and its corresponding EELS spectra for (b) B and (c) N

Figure 7. BF TEM micrograph of MS sample alloy C, showing (Fe,Mo)₃(B,C) and BN particles

4. Discussion

Clear differences between the AM and MS alloys were observed that were consistent with the difference in cooling rates. The higher cooling rate in MS produces very hard material by refinement of grains, laths and twins during the high undercooling achieved in solidification. The slower cooling of AM alloys resulted in a coarser microstructure, permitting segregation coring of Mo and the nucleation of very small needle precipitates, possibly Fe_3C, as sometimes observed in martensitic steels produced by more conventional methods. The lath width of the MS alloys, ~3 times smaller than for AM material, produces the large increase in hardness. The present results confirm previous reports that stable La_2O_3, La_2O_2S, $LaPO_4$ and LaP can spontaneously nucleate in the melt [6] even though $LaPO_4$ is believed to require significant supercooling to nucleate.

The presence of BN and the carbo-boride in the MS alloys is believed to be caused by the extreme reactivity of La, even in the form of $LaNi_5$, causing dissociation of the BN crucible and the incorporation of BN into the melt. The morphology of the carbo-boride suggests that it solidified at a lower temperature than the matrix. Although high hardness is achieved, such large precipitates are likely to reduce the fracture toughness of the MS samples and, clearly, the BN crucible attack must be minimised before such material can be utilised. The lack of large La-rich particles shows that La has been lost from the high temperature liquid during the MS process, preventing formation of La-P-O particles and consequently it presently remains uncertain whether P and S are removed by the MS process.

5. Conclusions

A combination of analytical techniques has been successfully employed to confirm that RSP methods can be used to generate refined ferritic steel microstructures with very high hardness. Additions of La induced the precipitation of residual P and S during arc melting, which offers the potential for improved fracture toughness. Alternative crucible material is needed to eliminate La attack and the formation of potentially harmful carbo-borides in melt spinning.

Acknowledgments

The work was financed by the UK Department of Trade and Industry (Aerospace Research Programme). The authors are grateful for the support of QinetiQ (MBG); Corus and EPSRC (SA); Royal Academy of Engineering, AEA Technology and the Institute of Nuclear Safety System, Inc, Japan (JMT). Mr Zaaijer (Thermphos International B.V) kindly donated the FeP.

References

[1]   Mawella K J A 1982 J. Mat. Science 17 2850-2854
[2]   Cantor B 1991 J. Mat. Science 26 1266-1276
[3]   Langford R M 2001 J. Vac. Sci. Technol. A19 2186-2193
[4]   Keown S R 1977 Metal Science 7 225-234
[5]   Jimenez J A 1992 Mat. Sci. and Eng. A159 103-109
[6]   Ghosh G 1993 Proc. 51st Annual Meeting of MSA, Evanston, USA 746-747

Inst. Phys. Conf. Ser. No 179: Section 6
Paper presented at Electron Microscopy and Analysis Group Conf. EMAG2003, Oxford, 2003
©2003 IOP Publishing Ltd

Electron back scatter diffraction and x-ray pole figure analysis of c-axis textured α-alumina fabricated by gel-casting

M Wei, D Zhi and D. Brandon[1]

Department of Materials Science and Metallurgy, University of Cambridge, Cambridge, CB2 3QZ, UK

[1]Department of Materials Engineering, Technion-Israel Institute of Technology, Haifa, 32000, Israel

Abstract. In the present research, c-axis textured α-alumina ceramics have been prepared by an alginate-based, gel-cast, doctor-blade process. The macrotexture of the α-alumina ceramics was analysed by x-ray pole figure measurements and the c-axis texture nature was revealed using 00.12, 10.10 and 113 pole figures. In order to examine the regional grain orientation relations (microtexture), electron back scatter diffraction (EBSD) was applied in this work. Using EBSD, the boundary mis-orientation distribution was obtained and analysed. Apart from the well-developed low angle grain boundaries, a large fraction of boundaries were found misoriented in the range of 50-60°, implying the high possibility of forming Σ3 boundary in this c-axis textured α-alumina ceramic. Pole figures and inverse pole figures were also measured by EBSD patterns taken from the grains in the SEM scanned area, which confirms the c-axis texture symmetry determined by x-ray diffraction.

1. Introduction

Texturing a material by controlling the orientation distribution of crystallites with respect to the coordinate system of a component has become a subject of renewed interest, which represents a significant departure from traditional materials processing and could be used to tailor structural properties and obtain some of the single crystal anisotropy (Seabaugh et al 2000). Such texture could be used to reduce thermal expansion mismatch, and hence the risk of microcracking, while texture control offers a useful alternative to aligned second-phase reinforcement (Carisey et al 1995a). Recently, the lowering of residual stress in textured α-alumina ceramics has been predicted (Vedula et al 2001). The macrotexture in such textured α-alumina ceramics can be routinely analysed by x-ray diffraction method. However, in traditional texture determination by x-ray diffraction, large numbers of grains are sampled simultaneously and no information is available regarding the crystal location responsible for individual x-ray reflections. Local grain orientation cannot be identified by the x-ray method. TEM is capable of providing diffraction data from grains down to 10 nm diameter with an angular precision in convergent beam diffraction, but TEM specimens must be electron transparent, typically less than 100nm thick, and the area of the sample thin enough for electron transmission is only of the order of large textured alumina grains, which makes TEM unsuitable for comparing grain orientations in these materials.

In the present work EBSD was used to examine the regional grain orientation relations (microtexture) in the c-axis textured α-alumina ceramics produced by a gel-casting method and the results were compared with the marcotexture analysis by traditional x-ray diffraction method, i.e. pole figure measurements.

2. Experimental

C-axis textured α-alumina ceramics were prepared by a gel-casting process (Carisey et al 1995b). The texture and microstructure of the α-alumina ceramics were optimised by controlling the volume fraction and alignment of high-aspect ratio alumina platelets in a precursor tape. After the lay-up of the tapes and subsequently cold isostatical pressing (CIP) at 250-300 MPa for 30 minutes, textured alumina ceramics were sintered at 1500°C for two hours. Sintered pellet sample sections were ground and polished to 1/4 μm grade diamond paste, followed by thermal etching at 1400 to 1500°C for 5 to 60 minutes in order to delineate the grain and phase boundaries. The polished samples were examined by high-resolution scanning electron microscopy (HRSEM) using a field emission electron gun (LEO 982-FEG-SEM). Grain boundaries of aligned alumina grains were also examined by transmission electron microscopy (TEM-JEOL 2000-FX) at 200 kV. TEM specimens were prepared by conventional method, including mechanical grinding, dimpling, and ion milling.

X-ray pole figures were measured to confirm the C-axis texture nature of α-alumina, using a x-ray diffractometer with a parallel-beam assembly (PW1710 based). Microtexture was investigated using electron back-scatter diffraction (EBSD) in the LEO 982-FEG-SEM. To perform EBSD, surface preparation was critical since the diffraction information originates in a 20 nm layer at the surface, corresponding to the penetration depth for back-scattered electrons. The sample surface was polished, etched, and ultrasonically cleaned before EBSD examination. The EBSD data were collected sequentially by positioning the focussed electron beam on each grain individually. The normal to the sample surface was tilted 70° to the incident beam and the EBSD pattern was captured at a beam voltage of 20 kV. The individual crystal orientations were analysed using commercial software to generate discrete pole figures for each localised, microtexture data set.

3. Results and discussions

A typical microstructure of the cross-section of a sintered alumina sample with 9.1 vol% initial platelet content is given in Figure 1. The microstructural anisotropy of the monolithic alumina is evident in the micrograph. By comparison with the size (15-25 μm) and aspect ratio (~1:10) of the initial platelets, it can be concluded that the Al_2O_3 platelets grew preferentially perpendicular to the gel-cast tapes or the platelet basal direction. The observed anisotropic platelet growth could be a result of several factors. Firstly, the Al_2O_3 platelets prefer to grow in the directions of prismatic planes to produce a lower energy basal surface (Roedel and Glaeser 1990). Microstructural constraint argument would also suggest that the grain growth is greater at the ends of a platelet, where the local microstructure undergoes a compressive strain (Sudre and Lange 1992). Because the platelets are initially aligned along their basal planes by the gel casting process and most of the surrounding random grains are consumed by the platelet growth during subsequent sintering, basal plane grain boundary can be easily formed in final microstructure. Figure 1(b) shows a TEM micrograph taken from a grain boundary between two textured alumina grains for which the boundary is accurately parallel to both basal planes (see SAD pattern in Figure 1(c)) with no sign of any glassy phase or second-phase precipitation.

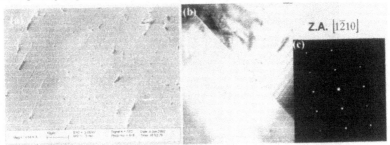

Figure 1. (a) A SEM image of a textured alumina at the cross-section direction (polished and thermally etched at 1450°C for 30 minutes), (b) Bright field TEM micrograph of a two-platelet boundary in a textured alumina, and (c) diffraction pattern was taken from one of the grains in (b).

X-ray 00.12, 10.10, and 113 pole figures were measured from the polished surfaces of the sintered samples initially with different amount of initial platelets (0-20vol%). The pole figure results do confirm the strong C-axis texture nature in the samples. Strongest texture has been measured from the sample with 9.1vol% initial seed platelet content, of which the pole figures are shown in Figure 2. In effect, the degree of crystallographic alignment initially increases with increasing platelet content but eventually reaches a maximum due to increasing interference between adjacent platelets as they rotate into the plane of the tape during tape-casting. Interference between neighbouring platelets is also the probable cause of the lower densification at high platelet contents.

Figure 2. 00.12, 10.10, and 113 (left to right) x-ray pole figures from the textured alumina with 9.1vol% initial seed platelet.

A standard α-alumina EBSD pattern was first recorded and indexed from a polished basal-plane sapphire substrate, figure 3(a), for which the sample normal is close to the 0001 zone axis, and this pattern was used to calibrate the orientation of patterns taken from other grains and index their EBSD patterns. EBSD patterns were measured from plan-view of a c-axis textured alumina ceramic sample with 9.1 vol% initial platelets. Figures 3(b) shows one of the EBSD patterns from the plan-view specimen. The EBSD results demonstrate that the local orientation of each grain relative to the surface normal deviates only slightly from the 0001 zone axis. These results were plotted in the orientation of the plane of the sample as discrete and inverse pole figures, as shown in Figures 3(c) and 3(d). The present EBSD results correlate the grain morphology with the grain orientation, and confirm the results of the X-ray diffraction determination of macrotexture. From the EBSD measurements, misorientation between adjacent grains

264

can be determined. The result shows that large fractions of boundaries are misoriented in the range 50-60°. which suggests a high probability of forming basal twins in the textured alumina.

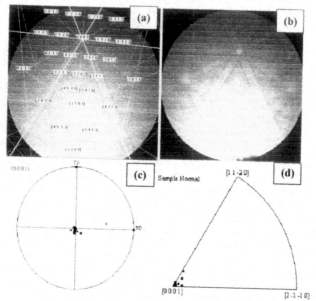

Figure 3. (a) Standard EBSD pattern from a polished basal sapphire substrate, (b) EBSD patterns from one of the grains in a plan-view sample of a textured alumina with 9.1 vol% initial platelets, (c) Discrete 0001 pole figure, and (d) Inverse pole figure.

4. Conclusions

A unique texture structure in alumina ceramics was produced and controlled by aligning alumina seed platelets by water-based gel-casting. The XRD pole figure results show that the texture has c-axis symmetry with the strongest texture developed in samples initially with 9.1 vol% basal plane alumina platelets. The marcotexture studied by X-ray pole figure measurements was confirmed by the local grain orientation measurement by EBSD. The boundary misorientation distribution shows that there is a large fraction of boundaries misoriented in the range 50-60°, implying large probability of the formation of $\Sigma3$ twin boundary.

References

Carisey T, Levin I and Brandon D G 1995a J. Eur. Ceram. Soc. **15** 283
Carisey T, Laugier-Werth A and Brandon D G 1995b J. Eur. Ceram. Soc. **15** 1
Roedel J and Glaeser A M 1990 J. Am. Ceram. Soc. **73** 3292
Seabaugh M M and Messing G L 2000 J. Am. Ceram. Soc. **83** 3109
Sudre O and Lange F F 1992 J. Am. Ceram. Soc. **75** 519
Vedula V R et al 2001 J. Am. Ceram. Soc. **84**, 2947

Inst. Phys. Conf. Ser. No 179: Section 6
Paper presented at Electron Microscopy and Analysis Group Conf. EMAG2003, Oxford, 2003
©2003 IOP Publishing Ltd

A preliminary study into microstructural characterisation of the river Porsuk sediments, Turkey

H Koyuncu, Y Güney, G Yılmaz

Department of Civil Engineering, Anadolu University Eskisehir, Turkey

A Kara

Department of Materials Science and Engineering, Anadolu University Eskisehir, Turkey

Abstract. This study reports the preliminarily scientific data on some of the morphology and surface features of the polluted, soft organics, bottom river sediments collected from depths of 0-40 cm at representative sites along the main course of the river Porsuk, Turkey. The representative SEM images revealed the presence of different structures within the sediments.

1. Introduction

Nowadays, it is essential to know about the dimensions of the damage given by humankind to the environment and to determine the effects of the resultant changes. As more than half of the world is covered with water, wastes from industrial and agricultural activities and also municipal development are discharged into these waters easily. River sediments consist of clay minerals, soil organics, carbonate, silt fraction, amorphous materials, municipal sludge and many industrial wastes that accumulate on the bottom of the water body. Such activities often result in the introduction of nutrients and potentially hazardous levels of trace metals and compounds into the riverine ecosystem. Contaminated sediments can have an impact on aquatic life by making areas uninhabitable. Once discharged into the environment, the behavior and the fate of polluting substances will be determined by the combined effect of different variables such as the compound's physicochemical properties, river hydrology, and possibly interactions with the biota. In particular, many contaminants tend to concentrate in river bed sediments. Assessment of sediment quality is therefore recognized as a critical step in estimating the risks associated with man-made pollution in river systems. By identifying current characteristics of composition and pollution level of sediments, some of the problems associated with sediment contamination may be solved. It is also important to understand how contaminants are held within the sediment structure, since this greatly influences the ease with which they may be removed and the processes that are most effective in doing so. Among the properties of sediments, microstructure is a crucial fundamental property that largely determines the physical and mechanical properties of sediments and also their behavior under static and dynamic loading [1,2].

The river Porsuk, which extends for 436 km within the boundaries of Kütahya, Eskisehir and Ankara provinces, is between 30-220 longitudes and 39-400 latitudes, and has a basin of 11326 km^2 (Figure 1). It is subject to intensive exploitation by both industrial and agricultural activities. The river sediment is highly dense and contaminated. It is often accumulated on the riverbanks and at depths and from time to time it causes shrinkage of river flow and results in an increase in the height of the river basin. The sediment load of the river Porsuk is estimated to be around 5 million tonnes every year.

Both optic and electron optic techniques can be used in order to study the structural arrangements of the constituents particles and associated voids present in sediments. For detailed examination, a range of magnification can be used to cover the range of particles. As a part of an extended project, this study makes an attempt to study qualitatively some of the characteristics of the river Porsuk sediments using SEM.

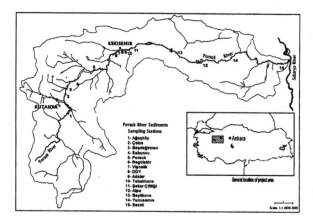

Figure 1. Locations of sampling sites in the river Porsuk

2. Experimental

Bed sediment samples were collected from depths of 0-40 cm at representative sites evenly distributed along the main course of the river Porsuk (Figure 1) in 2002. Firstly, a range of samples were randomly chosen and then were dried at 105°C overnight. Following drying, they were coated with gold-palladium to render the surface conducting during observation. A CamScan S4 analytical SEM was employed for microstructural characterization in the secondary electron (SE) imaging mode.

3. Results and discussion

As stated earlier, most industrial and domestic wastes enter the river without any threatment and as a result most of the discharges are toxic and nutrient. Nutrients are also carried away with sediment soil particles. The presence of nutrients at elevated levels can cause unwanted growth of algae, and can results in the lowering of the amount of oxygen in the water when the algae die and decay. Figure 2 gives an image of an algal filament positioned on sediment particles. Figure 3 is believed to depict aggregations of planktonic foraminifers (marked with arrow). Well-preserved planktonic foraminifers aggregate and contact each other, resulting in a large surrounded by each foraminifer's surface [3]. Larger particles or other matter such as diatoms are also incorporated into the structural arrangement without any general realignment of the surrounding matrix. Figures 4 and 5 are the typical examples of the skeletal remains of both diatom and algae observed in the river Porsuk sediments.

Figure 2. Algal filament on sediment particles   **Figure 3.** Planktonic foraminifers

Figure 4. Silt grains and a diatom   **Figure 5.** Diatom and algae

With increasing thickness of sediment, gradual self–weight consolidation may occur with accompanying general realignment of particles. Such processes will be dictated by overall porosity and permeability, and the open fabric may persist to depths of several meters. If sedimentation rate is high and the permeability is low, then excess pore-water pressures may develop and preserve the open fabric until sufficient water has migrated to allow volumetric changes [3]. The microfabric of Porsuk river sediments can also be characterized by open fabric approximating to a honeycomb and also an open card–house structure consisting of plate-shaped particles with limited preferred orientation (Figures 6 and 7).

Another feature observed in the sediments of the river Porsuk is illustrated in Figure 8. The marked agglomerate possibly indicates organic and inorganic matter bonded together by different forces such as surface tension, organic cohesion, and adhesion.

Figure 6. Honeycomb structure

Figure 7. Open card–house structure

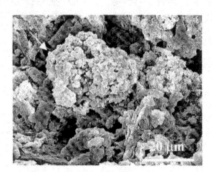

Figure 8. Agglomeration

4.   Conclusions

The qualitative SEM investigation showed that the river Porsuk sediments examined in this study were found to consist of randomly oriented clay-silt-organic particles in varying particle sizes. Biogenic materials such as planktonic foraminifers, diatoms and fine algal filaments were also observed. It is expected that the presence of biogenic materials and flocculated structures (honeycomb and card-house) be related to contamination. However, a more detailed examination of the sediments from Porsuk River involving energy dispersive X-ray microanalysis (EDS) has yet to be carried out for a better understanding of the characteristics of the river Porsuk sediments.

Acknowledgements
We are grateful to Anadolu University Research Fund for financial support thorough the research project number: 030224.

References

[1] Bennett R H O'Brien N R Hulbert M H 1990 Microstructure of Fine-grained Sediments, from Mud to Shale (Springer-Verlag)
[2] Noorany I 1989 J.Geot. Eng., 115/123-37
[3] Hirano S Ogawa Y Kawamura K 2001 Proc. Ocean Drilling Program, Scientific Results 174B Web publication

Inst. Phys. Conf. Ser. No 179: Section 6
Paper presented at Electron Microscopy and Analysis Group Conf. EMAG2003, Oxford, 2003
©2003 IOP Publishing Ltd

Phosphorus segregation and intergranular embrittlement in 2.25Cr1Mo steel

Rengen Ding, Tiesheng Rong and John Knott

Department of Metallurgy and Materials, The University of Birmingham, Birmingham B15 2TT

ABSTRACT: Phosphorus segregation at prior austenite grain boundaries (PAGBs) in 2.25Cr1Mo steel with different degrees of embrittlement has been investigated using X-ray spectroscopy in a field emission gun scanning transmission electron microscope (FEGSTEM). The results indicate that P concentration at PAGBs increases with ageing time, which agrees well with the prediction of McLean's equilibrium segregation model.

1. INTRODUCTION

Localised changes in chemical composition at or close to grain boundaries and interfaces can influence strongly the fracture and corrosion properties of many alloy steels. It is well established that the impurity element P diffuses to PAGBs in ferritic steels to induce temper embrittlement [1, 2]. Such embrittlement can take place both during heat treatment and in components in service at elevated temperatures. The detection and measurement of grain boundary chemistry is therefore of considerable importance in expanding the fundamental knowledge of fracture mechanisms for elucidating the causes of unexpected failures in components.

One of the most widely used techniques for the study of grain boundary segregation in steels is Auger electron spectroscopy (AES) [3], but it requires the small Auger samples to fracture in an intergranular manner inside the high vacuum apparatus: this is not always easily achievable for small testpieces. Another technique used to investigate P segregation is transmission electron microscopy. However, early work on the measurement of segregation of impurities at grain boundaries using conventional LaB6 gun transmission electron microscopes (TEMs) was not entirely successful, mainly because of inadequate spatial resolution. Following the development of field emission gun transmission electron microscopes (FEGTEMs), probes less than 1nm in diameter with very high current can be produced. FEGTEMs have been successfully used to detect P segregation at grain boundaries in structural alloys [4, 5]. In this paper, we describe measurements of thermally induced P segregation at PAGBs as a function of heat treatment conditions and compare the results with McLean's equilibrium segregation model.

2. EXPERIMENTAL

A commercial 2.25Cr1Mo steel with 0.013 wt.% (0.023 at.%) P was austenitised for 2 hr at 1100°C oil quenched to room temperature (RT) and then tempered for 2 hr at 650°C followed by oil-quenching to RT or ageing for different periods at 520°C before oil-quenching. Samples for TEM examinations were prepared by twin-jet electropolishing to perforation with a solution of 5 % perchloric acid in ethanol at 30V and -20°C. Analysis of P segregation at PAGBs was performed in a Philips Tecnai F20 FEG transmission electron microscope equipped with a Link ISIS 300 EDS system.

3.   RESULTS AND ANALYSIS

A TEM micrograph obtained from the sample after austenitising at 1100°C for 2 hr oil quenching and then tempering at 650°C for 2 hr is shown in Fig.1. This indicates a typical tempered martensite structure with carbides at grain boundaries and lath boundaries and in the matrix. Using selected area electron diffraction and EDS, these carbides were identified as M_3C, M_7C_3 and M_2C.

In order to identify P concentration at PAGBs, TEM foils were tilted to an orientation where the PAGBs were edge-on, as shown in Fig. 2. An electron beam with a beam size about 1 nm full width at half maximum (FWHM) was used for EDS analysis. Fig. 3 shows typical spot EDS spectra taken from a PAGB and from the matrix for a specimen aged for 210 hr, indicating P segregation at the PAGB. To examine the variation in the P concentration across the boundary, an electron beam with a beam size about 1 nm (FWHM) was scanned across the boundary with a step length of 1 nm and a dwell time 150 s. EDS data on the P concentration from the scan are presented in Fig 4.

Segregation of phosphorus depends on heat treatment conditions and may also vary from boundary to boundary. To obtain an indicative concentration of phosphorus for each different thermal condition, 5 separate PAGBs were carried-out for each condition. The average values of P concentration at grain boundaries are presented in Table 1. It is clear that the P concentration increases with ageing time.

Table 1. The average P concentration at grain boundaries for different ageing time

Ageing time at 520°C (hour)	0	24	96	210
P concentration (at.%)	1.15 ± 0.39	1.9 ± 0.42	2.35 ± 0.58	3.2 ± 0.54

Because of the spatial resolution of EDS, the 'actual' *grain boundary* concentration of P is likely to be higher than the measured concentration. In order to compare the measured P concentration with that predicted by McLean's equilibrium segregation model [6], we convolute the measured concentration to an 'actual' grain boundary concentration. Following Hall's analysis [7], the 'actual' grain boundary concentration can be written as

$$C_{actual} = \frac{C_m \cdot t}{F(x)}$$
    1

where C_m is the measured concentration, t is the foil thickness, and $F(x)$ is a function related to the profile of segregant, beam size, electron energy, atomic number, atomic weight, density, t, and beam position from grain boundary (x). If the beam is centred on the grain boundary, $x = 0$.

For a grain boundary thickness of 1 nm and a TEM foil thickness of 100 nm, the predicted concentration profile is shown by the solid line in Fig. 4, which indicates that the convoluted prediction agrees with the measurement so that assumption of those parameters is reasonable. The 'actual' grain boundary concentrations of P are presented in Table 2.

Fig. 1 typical tempered martensite structure with carbides at grain boundaries, lath boundaries and in the matrix.

Fig. 2 showing a typical section of grain boundary used to detect segregation.

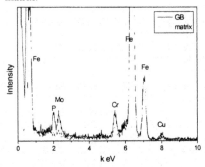

Fig. 3 EDS spectra from a PAGB and the matrix, respectively, showing P segregation at the boundary.

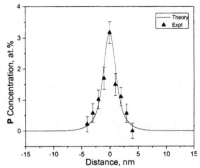

Fig. 4 Variation in P concentration across a grain boundary (210 h).

Table 2. The average corrected P concentration at grain boundaries for different ageing time

Ageing time at 520°C (hour)	0	24	96	210
P concentration (at.%)	2.95±1.02	4.75±1.08	5.85±1.46	7.70±1.35

McLean's equation for equilibrium segregation is [6]:

$$\frac{C(t)-C_g}{C_T(\infty)-C_g} = 1 - \exp(\frac{4Dt}{\alpha^2 d^2}) \cdot erfc(\frac{4Dt}{\alpha^2 d^2})^{1/2}$$

where $C(t)$ is the boundary solute concentration after time t, C_g is the boundary solute concentration at time $t = 0$, D is the solute bulk diffusivity, d is the thickness of the boundary, α is given by $\alpha = \dfrac{C_T(\infty)}{C_0}$, where C_0 is the bulk concentration (0.023 at.%)

and $C_T(\infty)$ represents the equilibrium boundary concentration of P determining to the

Langmuir-McLean equation: $\dfrac{C_T(\infty)}{1 - C_T(\infty)} = C_0 \exp(\dfrac{-\Delta G}{RT})$ [6], where ΔG is the Gibbs free

energy of P grain boundary segregation. In the kinetic calculations, the following parameters were used: $\Delta G = -43.5$ kJ/mol [8], D = 2.4×10^{-20} m^2/s.

Fig. 5 shows that the equilibrium segregation predication is consistent with the experimental values, which indicates that, for quenched and tempered 2.25Cr1Mo steel, P grain boundary segregation during ageing at 520°C is caused by equilibrium segregation mechanism. Fig. 6 shows that the grain boundary P concentrations exhibit a linear variation with the square root of time, suggesting that the diffusion process does not reach equilibrium within 210 hr at 520°C.

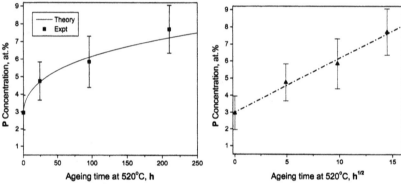

Fig. 5 a comparison of experimental values and the predictions of equilibrium segregation model.

Fig. 6 the grain boundary P concentrations exhibit a linear variation with the square root of ageing time.

4.   CONCLUSIONS

FEGTEM microchemical analysis shows P segregation at grain boundaries. The P concentration is a function of ageing time and is consistent with the predictions of McLean's equilibrium segregation model.

REFERENCES

[1]   Balajiva K, Cook R M, and Worn D K, *Nature*, Vol.178, 1956, P433
[2]   McMahon C J, *"Temper Embrittlement in Steels"*, ASTM, 1968, P127
[3]   Stein D F, Joshi A, and Laforce R P, *Trans. ASM*, Vol.62, 1969, P776
[4]   Titchmarsh J M, in *Proc. 11th ICXOM*, edited by J.D.Brown and R.H.Packwood, University Western Ontario, Canada, 1987, P337
[5]   Vorlicek V and Flewitt P E J, *Acta Metall.*, Vol.42, 1994, P3309
[6]   McLean D, *"Grain Boundaries in Metals"*, Oxford, 1957
[7]   Hall E L, Imeson D, and Sande J B V, *Philo. Mag. A*, 1981, Vol.43, P1569
[8]   Perhacova J, Grman D, Svoboda M, Patscheider J, Vyrostkova A, and Janovec J, *Mater. Lett.*, Vol.47, 2001, P44

Inst. Phys. Conf. Ser. No 179: Section 6
Paper presented at Electron Microscopy and Analysis Group Conf. EMAG2003, Oxford, 2003
©2003 IOP Publishing Ltd

Investigation of grain boundaries in LPS aluminas by TEM

Peter Svancarek[a,b], Dusan Galusek[a] ,Fiona Loughran[b], Clair Calvert[b], Andy Brown[b], Rik Brydson[b] and Frank Riley[b]
[a] Institute of Inorganic Chemistry, Slovak Academy of Sciences, Dúbravská cesta 9, SK-845 36 Bratislava, Slovak Republic
[b]Institute for Materials Research, University of Leeds, LS2 9JT Leeds, UK

Abstract. Hot pressed alumina samples containing 5wt% calcia and silica additions in 2:1 and 10:1 molar ratios respectively were studied in detail using both scanning (SEM) and transmission (TEM) electron microscopy. Grain size, secondary crystalline phases and the compositions of amorphous films at grain boundaries and triple pockets have been characterised and related to mechanical properties.

1. Introduction

Liquid forming additives allow sintering to be carried out at faster rates and at lower temperatures, as well as affecting mechanical properties in ceramic materials. Silica, when used as a sintering additive, can significantly improve the wear resistance of alumina [1]. Magnesia [2,3] and calcia [1,4] are also used to modify the mechanical properties of polycrystalline aluminas. Wear resistance appears to be controlled by median grain size [3,5]. Small median grain sizes ($<1\mu m$) wear by a tribochemical mechanism whereas larger median grain sizes exhibit intergranular fracture and grain detachment as the wear mechanism in wet environments. It appears that the presence of intergranular glassy films, found at both 2 grain junctions (grain boundaries) and 3 grain junctions (triple pockets), can strengthen the grain boundaries, and therefore reduce grain pull out. This is probably caused by the differences in the thermal expansion coefficients of the crystalline and amorphous phases, which result in grain boundary strengthening compressive hoop stresses being formed [2]. This stress can be calculated from Cr^{3+} photoluminescence spectra produced by Raman optical microscopy [2,6], and can be observed with TEM [2,6,7].

Molecular dynamic simulations of silicate intergranular grain boundary films suggest that there are cage like structures between the alumina grain and the glassy phase which can, in principle, accommodate metal cations such as calcium or magnesium [8]. Calcium cations (≤ 12 mol %) prefer to segregate to these cage-like structures and lower the glass/grain interfacial energy [9]. When the calcia content is less than 12 mol%, fracture propagation only occurs through the glassy film/alumina grain interface plane. However, when the calcia content is greater than 12 mol%, the larger calcia content in the glassy phase disrupts the siloxane bonding and causes fracturing to occur through the interior of the glassy film. This causes the overall fracture strength to decrease. When the intergranular film contains ≥30 mol% of calcia and is at least 1.5nm thick, the interfacial energy becomes negative, causing abnormal grain growth to occur during sintering [9].

2. Experimental

The alumina powder used had a mean particle size of 400nm (Martoxid CS400M). 5 wt% of calcium silicate additive of molar ratio $CaO:SiO_2$ of 10:1 (C10S) and of 2:1 (C2S) was added. The powder was calcined at 750°C, passed through a 100μm sieve and hot-pressed for 10 minutes at 1450°C at a pressure of 20MPa. SEM (CamScan IV, Cambridge instruments) and an image analysis software package (Kontron, KS 400) were used in order to determine the median grain sizes. STEM, EELS and EDX facilities on a FEI CM200 FEG TEM were used for detailed grain boundary and triple pocket analysis.

A wet erosion wear test using alumina slurry was carried out using a modified high torque attritor mill. The wear rate (ms^{-1}) was calculated from the observed weight loss as function of time. Hardness (HV10) and fracture toughness (K_{IC}) were measured using a Vickers indenter at 1 and 10 kg respectively. A Raman optical microprobe was used for the measurement of internal stress.

3. Results and Discussion

The experimental results are summarised in Table 1. SEM analysis (Fig.1) revealed the

Table 1 Properties of the examined samples.

	C10S specimen	C2S specimen
Additive composition	$10CaO.SiO_2$	$2CaO.SiO_2$
Medium grain size [μm]	0.96 ±0.05	1.8 ±0.05
Wear rate [nm/s]	11 ±1 with evidence of grain pull-out wear mechanism.	17 ±2 with evidence of grain pull-out wear mechanism.
Hardness HV10 [GPa]	14.9 ±0.6	14.2 ±0.5
K_{IC} [MPa $m^{1/2}$]	4.4 ±0.1	4.2 ±0.1
Grain boundary (STEM/EDX)	Amorphous film (~ 4nm) of atomic ratio $CaO:SiO_2$=1:3.1	Amorphous film (~ 2nm) of atomic ratio $CaO:SiO_2$=1:2.6
Triple pocket glassy phase (STEM/EDX)	Composition in corundum phase field ($1CaO: 2.6SiO_2: 3.5Al_2O_3$)	Composition in anorthite/mullite phase field ($CaO: 25SiO_2: 3Al_2O_3$)
Crystal structure of crystallized phase	Grossite with composition ($6CaO: SiO_2: 15.4Al_2O_3$)	Gehlenite with composition ($1.1CaO: 1.1SiO_2: 1.2Al_2O_3$)
Residual fluctuating local stress [MPa]	439 ±110	422 ±120
Mean residual compressive stress [MPa]	494 ±100	233 ±100

C10S sample consisted of mainly equiaxed alumina grains with a median grain size of 0.96μm, whereas the C2S sample exhibited a higher median grain with elongated alumina grains, caused by a reduction in interfacial energy due to the presence of calcia. It is interesting to note that during thermal etching the C2S sample experienced grain and glassy phase pull out whereas the C10S specimen was less affected. In both samples, TEM revealed areas under heavy stress, due to the differences in thermal expansion coefficients (K^{-1}): alumina = 11.05 x 10^{-6}, gehlenite (C2S) = 8.3 x 10^{-6}, grossite (C10S) = 4.1 x 10^{-6}. The higher the calcia content in the amorphous film the higher the thermal expansion coefficient of the glass will be. This causes microcracks and dislocations mainly in the coarser-grained C2S. STEM/EDX elemental mapping of both samples revealed that the glassy phase in the both triple pockets and grain boundaries had a higher

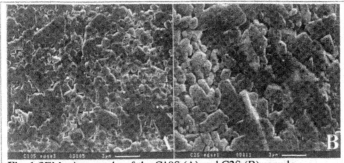

Fig. 1 SEM micrographs of the C10S (A) and C2S (B) sample.

C10S: C2S:

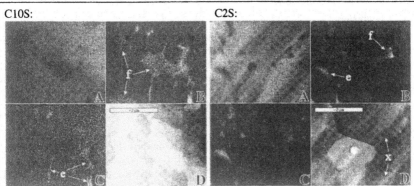

Fig. 2. EDX elemental analysis: (A) Al Kα map, (B) Ca Kα map, (C) Si Kα map, (D) STEM dark field image. Note the glassy phase (e) in C10S contains a larger amount of silica than crystalline (grossite) phase (f). C2S showed no visible differences between the composition of glassy phase (e) and crystalline (gehlenite) phase (f), but structural damage was observed (x).

concentration of silica than calcia (Fig. 2). The triple pockets exhibit inhomogeneity in composition and the glass in the C2S sample (anorthite/mullite phase field) contained considerably less calcia than the C10S glass (corundum phase field) (Table 1). However, the grain boundary films in both samples contained a significant amount of calcia (Table 1). Surprisingly, a larger concentration of calcia in the grain boundary was observed in the C2S sample, which could be explained by crystallisation of calcia rich grossite phase in triple pockets of the C10S specimen. Abnormal grain growth was observed in C2S probably because of a larger concentration of calcia in the grain boundary. The high aluminium content, shown in the amorphous phase by EDX mapping, is from the neighbouring alumina grains, and is included because of beam broadening effects. HRTEM revealed thicker grain boundary films in the C10S sample than in the C2S sample (Figure3a, 3b).

The differences seen between the two samples in terms of the glass composition and the crystalline secondary phases, resulting in abnormal grain growth and residual stresses, can explain the difference in the wear rates. Both samples experienced grain pull out during the wear process. However the C10S sample, because of its smaller grain size, experienced less pull out than the C2S sample.

276

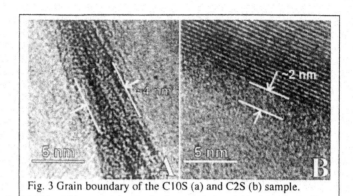

Fig. 3 Grain boundary of the C10S (a) and C2S (b) sample.

4. Conclusion

This work investigates why there are differences between the wear rates of C10S and C2S specimens, where C10S is finer-grained. It is unlikely that grain size alone results in such a dramatic difference in wear rates. There are clear differences in the composition of the secondary phases present in the materials. The interactions of these phases with the alumina on wear rates have been discussed. Clearly, sintering additive chemistry plays a significant role in alumina grain growth and wear rate.

Acknowledgements

The support of this work by the Marie Curie Fellowship (for PS) under the contract number HPMF-CT-2002-01878, National Slovak Grant Agency under the contract N° 2/3101/23, the NATO Science for Peace Programme, EPSRC (for a studentship to FL) and the project N° 97 41 22, from the Alexander von Humboldt Foundation (for DG) is gratefully acknowledged. The authors wish to thank Imperial College, UK for the residual stress measurements.

References

[1]   R. Brydson, P. C. Twigg F. Loughran and F. L. Riley, J. Mater. Res., 16, No.3, 652 (2001).
[2]   D. Galusek, R. Brydson, P. C. Twigg and F. L. Riley, A. Atkinson and Y. Zhang, J. Am. Ceram. Soc., 84 [8], 1767 (2001).
[3]   D. Galusek, P. C. Twigg and F. L. Riley, Wear 233, 588 (1999).
[4]   R. Brydson, S. Chen, X. Pan, F. L. Riley, S. J. Milne, X. Pan and M. Ruhle, J. Am. Ceram. Soc., 81, [2] 369 (1998).
[5]   M. Miranda-Martinez, R. W. Davidge and F. L. Riley, Wear 172, 41 (1994).
[6]   P. Svancarek, D. Galusek, C. Calvert, F. Loughran, Andy Brown, R. Brydson and F. Riley J. Eur. Ceram. Soc., Submited
[7]   C. A. Powell-Dogan, A. H. Heuer, M. J. Ready and K. Merriam, J. Am. Ceram. Soc., 74, [3] 646 (1991).
[8]   S. Blonski and S. H. Garofalini, J. Phys. Chem. 100, 2201 (1996).
[9]   S. Blonski and S. H. Garofalini, J. Am. Ceram. Soc., 80 [8], 1997 (1997).

Inst. Phys. Conf. Ser. No 179: Section 6
Paper presented at Electron Microscopy and Analysis Group Conf. EMAG2003, Oxford, 2003
©2003 IOP Publishing Ltd

A proposed new structure for GPB2/ S" in Al-Cu-Mg alloys

S C Wang and M J Starink

Materials Research Group, School of Engineering Sciences, The University of Southampton, Southampton SO17 1BJ

Abstract: Based on previously published HREM work, we propose a new structure for the ordered GPB2/ S" phase with a composition of $Al_{10}Cu_3Mg_3$, which is aluminium-rich compared to S phase (Al_2CuMg). The proposed structure is coherent with the f.c.c. Al matrix and is formed by the replacement of some Al atoms with Cu/Mg, and has a orthorhombic structure (space group Imm2) and the lattice parameters of $a = 0.405$ nm, $b = 1.62$ nm and $c = 0.405$ nm. The HREM simulation and reflection intensities based on this structure match the experimental image and diffraction patterns.

1. Introduction

In the early 50s, Bagaryatsky [1] first proposed a 4 stage precipitation sequence for the ageing of Al-Cu-Mg alloys:

SSS → GPB zone → S" (GPB2) → S' → S (CuMgAl₂).

where SSS stands for supersaturated solid solution and GPB was termed as Guinier-Preston-Bagaryatsky by Silcock [2]. Fifty years later, there are still disputes on the structure of all these phases, and especially on whether S" (GPB2) truly exists as a phase separate and distinct from the equilibrium S phase. According to Shchegoleva and Buinov [3], the S" is monoclinic crystal with $\alpha=88.6°$ which slightly distorted S structures. Various other claims for the presence and the structure of a distinct S" have been made. For example, in Silcock's X-ray work [2], she did not observe a phase resembling the S" phase reported by Bagaryatsky [1], rather she suggested the existence of a structure rich in copper, more likely to be related to the compound $Al_5Cu_5Mg_2$ with cubic structure and $a = 0.827$ nm. Based on TEM diffraction, Cuisiat et al. [4] suggested S" as a orthorhombic structure $a = 0.405$ nm, $b = 0.405$nm and $c = 0.81$nm. Shih et al. [5] proposed a partially ordered so-called GPB2 zone (S") which has a tetragonal structure and lattice parameters of $a = 0.58$nm, $c = 0.81$nm. Recently, by calculations of formation enthalpies for GPB / complex precipitates in Al alloys using first-principles, Wolverton [6] predicted a new structure of GPB2 zone as a tetragonal structure with $a = 0.401$nm and $c = 0.81$nm. However, none of the above structures have been independently confirmed. Other researchers (e.g., Wilson and Partridge [7] and Ringer et al. [8]) were unable to confirm the presence of the S" phase.

In recent HREM work on an Al-2.03wt%Cu-1.28wt%Mg alloy, Charai et al [9] claimed to have obtained new evidence for an S" phase, which was determined as a primitive monoclinic structure. The purpose of this paper is to re-examine the data in Charai et al's work and reconsider the evidence for existence of S", in that and other papers.

2. Experimental HREM data

In HREM work on an Al-2.03wt%Cu-1.28wt%Mg alloy (Fig.1a) after ageing 200°C for 4 hours, Charai et al [9] found the reflections A-D (Fig.1b) by Fourier transformations from the area indicated in Fig. 1a. These authors measured the spacings of OA/OD and OB/OC to be 0.25 and 0.32nm, which are similar to $\{112\}_S$ and $\{111\}_S$ respectively. As $\{111\}_S$ should be not observed in $[001]_{Al}$ according to the orientation relationship between S phase Al matrix, and none of the reported S" structures in [1,2,4,5,6] could explain these reflections, Charai et al [9] claimed a new S" phase which has a primitive monoclinic structure with $a = 0.32$ nm, $b = 0.405$ nm, $c = 0.254$ nm, $\beta=91.7°$. However, their suggested structure could not explain why one S" variant in Fig. 1a produced two sets of diffraction patterns in $[001]_{Al}$ (Fig. 1b), and no HREM simulation supported such a structure. We reanalysed their results and propose a new structure which fits well with the experimental results.

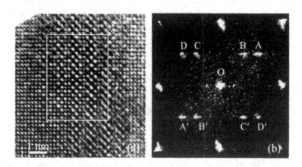

Fig. 1a and b are HREM images and diffraction pattern by Fourier transformation in $[100]_{Al}$ after 200°C for 4h in an Al-2.03wt%Cu-1.28wt%Mg (by courtesy of Prof. A. Charai [9])

3. Reinterpretation of HREM data

As the diffraction patterns are very similar to those of S'/S phase, one possibility is that the reflection A-D are from a variant of S'/S phase. However, no match can be found between the experimental HREM images and images of S phase (for details see [10]). According to Bagaryatsky [1], the S" phase is a slightly distorted S phase, and therefore his model cannot elucidate the images and diffractions of Fig.1.

To elucidate the HREM and diffraction in Fig.1, we performed simulations of these images for reported types of GPB2/S" [4,6] using the EMS on-line software, but again none of them could be found to match Fig.1 [10]. Considering the coherent between the precipitate and Al matrix, we propose a new structure in which the positions of Cu and Mg atoms are mixed 50%Cu/50%Mg as shown in Fig. 2a. The simulated HREM along $[001]_{S"}$

(Fig. 2b) matches the experimental image in Fig. 1a for 4nm of thickness and defocus at 68 nm (200 kV, Cs=0.5mm). Fig. 2c shows the corresponding diffraction patterns in [001]$_{S''}$ based on the model of Fig. 2a, which again satisfies the reflections of Fig. 1b. The orientation relationship between S'' and Al matrix satisfies: $<100>_{S''}$ // $<100>_{Al}$, $<010>_{S''}$ // $<010>_{Al}$. On calculation of its structural factors, we can predict the diffraction patterns for all 6 independent variants of S'' precipitates in [001]$_{Al}$ [10]. Accordingly, some previously observed diffraction patterns can be elucidated. Simulations of diffraction patterns from our proposed structure shows that the reflections around {110}$_{Al}$ could be elucidated well combination of two [100]$_{S''}$ variants of the proposed GPB2/S structure {variants 1 and 2 in [10]) (see Fig. 3a & 3b). The extra diffractions observed for Al-Cu-Mg alloys [11] and [12] are consistent with two variants of the structure we propose (variants 3 and 4 in [10]), see Fig. 3c & 3d. (Note that in Ref 12 the pattern in Fig 3c was ascribed to an oxide layer with α– or γ-Al$_2$O$_3$ structure.)

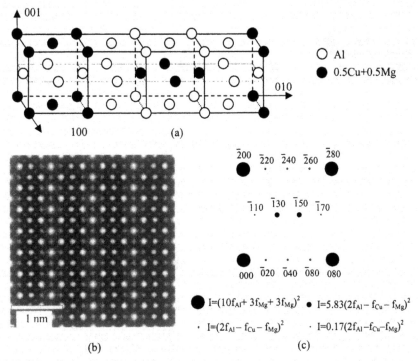

Fig.2
a) The new proposed structure
b) Simulated HREM along [001]$_{S''}$ based on the above structure, with defocus of 68 nm and thickness of 4 nm, matching the pattern in Fig.1a
c) Corresponding diffraction pattern of [001]$_{S''}$. The spots with size are proportional to the diffraction intensities (I) in which f$_{Al}$, f$_{Cu}$ and f$_{Mg}$ are the atomic scattering amplitudes.

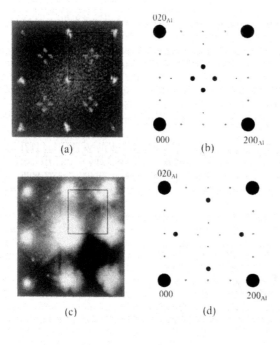

(a)

(b)

(c)

(d)

020_{Al}

000

200_{Al}

020_{Al}

000

200_{Al}

Fig. 3
a) The $[001]_{Al}$ diffraction pattern from the clusters by Fourier transformation after 200°C for 4h in an Al-2.03wt%Cu-1.28wt%Mg adopted from [9]
b) Simulated patterns using variants of the proposed GPB2/S", showing a good correspondence with a) (see text).
c) The observed $[001]_{Al}$ diffraction pattern from the disordered zones aging at 180°C for 34 h in an Al-0.6wt%Cu-4.2wt%Mg alloy (by courtesy of Dr P. Ratchev [11])
d) Simulated patterns using variants of the proposed GPB2/S", showing a good correspondence with c) (see text).

4. Conclusions

We propose a new structure for GPB2 which is coherent with the matrix and matches the experimental results well. Its composition is $Al_{10}Cu_3Mg_3$ which is between that of Cu-Mg clusters which have about 90% Al and S phase (Al_2CuMg).

References

[1] Bagaryatshy Y A, 1952 Dokl. Akad. S.S.S.R. 87 397 & 559
[2] Silcock J M 1960-61 Journal of the Institute of Metals 89 203
[3] Shchegoleva T V and Buinov N N 1967 Sov. Phy. Cryst. 12 552
[4] Cuisiat F, Duval P and Graf R 1984 Scripta Met. 18 1051
[5] Shih H, Ho N and Huang J C 1996 Metall. & Materials Trans. A 27 2479
[6] Wolverton C 2001 Acta Mater. 49 3129-3142
[7] Wilson R N and Partridge P G 1965 Acta Met. 13 1321
[8] Ringer S P, Hono K, Polmear I J and Sakurai T 1996 Acta Mater. 44 1883-98
[9] Charai A, Walther T, Alfonso C, Zahra A M and Zahra C Y 2000 Acta Mater. 48 2751
[10] Wang S C and Starink M J, to be published.
[11] Ratchev P, Verlinden B, de Smet P and van Houtte P 1998 Acta Mater. 46 3523
[12] Ringer S P 2000 Materials Forum 24 59-94

Inst. Phys. Conf. Ser. No 179: Section 6
Paper presented at Electron Microscopy and Analysis Group Conf. EMAG2003, Oxford, 2003
©2003 IOP Publishing Ltd

Microstructural Observation of Overload Effect on the Fatigue Crack Growth of 2024-T3 Al-Alloy

D Turan[1], A Karcı[1] and S Turan[2]

[1] Anadolu University, Civil Aviation School, Iki Eylul Campus, 26555, Eskisehir, Turkey
[2] Anadolu University, Department of Materials Science and Engineering, Iki Eylul Campus, 26555, Eskisehir, Turkey

ABSTRACT: Microstructural investigation of fracture surfaces after fatigue crack growth experiments on 2024-T3 centrally cracked specimens was carried out. Variable amplitude loadings were used with periodic overload cycles added to constant amplitude cycles. During fatigue tests when R=0.1, for every 1000 cycles 1, 2, 3, 4 and 5 overload was applied. The fatigue fracture surfaces were examined in the scanning electron microscopy (SEM) to obtain more detailed information on crack growth contributions of different overload cycles and it was shown that fatigue crack growth retardation occurred when periodic overload cycles were applied after every 1000 cycles.

1. Introduction

Engineering components and structures often operate under variable amplitude loading rather than constant amplitude loading conditions. Continued safe operation of structures subjected to variable amplitude loading may depend on understanding the behaviour of fatigue crack under such conditions. To understand crack growth behaviour under such conditions, several studies have been carried out (e.g., Skorupa 1999) and it was found that significant accelerations and/or retardations in crack growth rate can occur as a result of these load variations under either single load (Vardar 1988) or multiple overloads (Lang and Marci 1999). The retardation is affected by spacing between overload (Pommier and De Freitas 2002, Turan et al 2001), peak of the overload (Skorupa 1999) and by number of overloads (Pommier and De Freitas, 2002). However, prediction of fatigue crack growth is a very difficult task to fulfil due to the complexity of the fatigue damage processes involved. Such verifications are made by comparing predicted and experimental results (Schijve 1999). If the predicted relation and the experimental are similar, the prediction model could be supposed to be accurate. Thus, an accurate prediction of fatigue life requires an adequate evaluation and detection of these load interaction effects using microscopy techniques (Schijve 1999, Lang and Marci 1999, Turan et al 2001). Schijve (1999) investigated the fatigue fracture surfaces in the SEM and found that striation patterns of the fractographic pictures could easily be related to the load history, whereas Turan et al (2001) observed the retardation effect of block overloading by using SEM images. Therefore, in this study, it was aimed to study the effects of single and periodic overloads by using SEM micrographs to explain the experimental results in detail.

2. Experimental procedure

Fatigue crack growth tests were carried out on 2024-T3 aluminium alloy centre crack tension (CCT) specimens. The width, length and thickness of the specimens were 90, 300, 6 mm, respectively. These specimens were machined according to ASTM E 647 recommendation (Roberta 1985). The fatigue tests were conducted with computer controlled electro hydraulic machine (Instron 8501) under load control with the program MAX using a sinusoidal waveform with a frequency of 5 Hz. The crack length was monitored with a travelling microscope (Fatigue Technology Inc.). Periodic and single tensile overload fatigue tests were performed by superimposing 1, 2, 3, 4 and 5 overloads on constant amplitude cycles (after every 1000) with R=0.1 at the same overload ratio of 1.7. For observing overload effect on the fatigue crack growth, fatigue fracture surfaces were examined visually and in a SEM (CamScan S4).

3. Results and discussion

The fatigue crack growth curves of the constant amplitude loading tests with single and periodic overloads are given Fig.1. for 2024-T3 aluminium alloy. Constant amplitude (CA) tests were carried out as a reference to see whether crack growth retardation have occurred.

As shown by the results in Fig. 1, the highest crack growth rates were obtained with the CA loading. This implies that in all other tests crack growth retardation have occurred. The retarding effect of overloads appears to be more significant for periodic tensile overloads in comparison to single overload. The retardation is also higher for 1 periodic overload compared to other periodic block overloads.

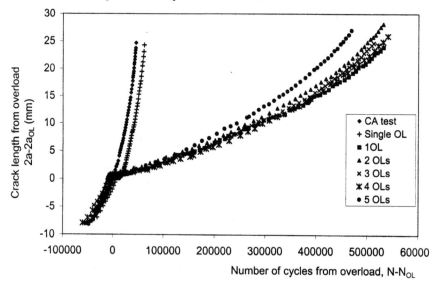

Figure 1. Effect of the single overload and different number of periodic overloads on the fatigue crack growth.

283

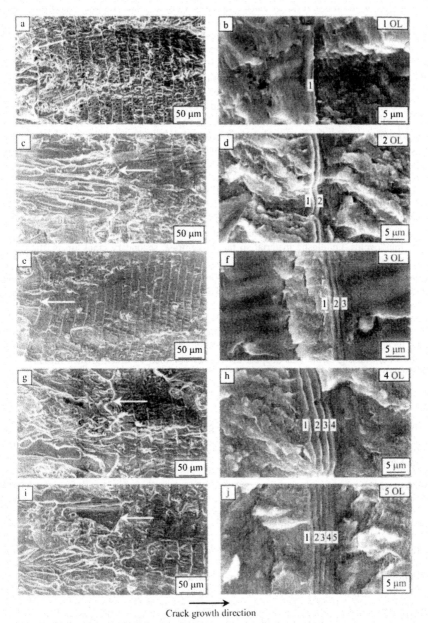

Crack growth direction

Figure 2. Fracture surface micrographs clearly show the bands corresponding to 1, 2, 3, 4 or 5 periodic overload cycles applied after every 1000 cycles. Magnified images shown in b, d, f, h and j also show how many overloads are applied. Starting point of the overloads is indicated with an arrow in a, c, e, g and i.

284

The fatigue crack surfaces macroscopically showed a flat fracture surface, with the crack growth direction perpendicular to the loading direction. Starting points of overloads on the specimen were visible whereas the bands between overloads were not visible. Therefore, microstructure of fractured samples was examined in the SEM (Fig. 2). The bands observed in the SEM micrographs correspond to either 1, 2, 3, 4 or 5 overload cycles. The band spacing was calculated by measuring the distance between a number of bands of periodic block overloads as shown in Fig.2 and these results were used to calculate the local crack growth rate, da/dN.

Calculations from SEM micrographs in Fig 2a showed that da/dN decreased from 0.029 μm/cycle after the first overload to 0.016 μm/cycle in the following overloads for the single periodic overloads. This observation is important for the evaluation of crack growth prediction models. Older crack closure models predict an immediate maximum retardation, because it is assumed that the crack closure level goes down to its maximum immediately after the first overload. This is in conflict with the SEM observations. This conflict can be explained with the concept of primary and secondary plasticity at the crack tip (Koning and Doughorty 1989). Primary crack tip plasticity is plastic deformation occurring in material that has not yet been plastically deformed by previous cycles, i.e. elastic material so far. In contrast, secondary crack tip plasticity occurs in material that had already plastically deformed in advance. Therefore, during primary crack tip plasticity the incremental crack extension for the same K-increment is expected to be larger than during secondary crack tip plasticity as calculated from SEM images.

For the 5 periodic overloads, the da/dN value was 0.034 μm/cycle after the first overload (Fig. 2i). When the da/dN values for the 1 and 5 overloads are compared, it can be concluded that 1 periodic overload is more effective than 5 periodic overload to retard the crack growth. This support the results obtained by crack growth measured using a travelling microscope (Fig 1).

4. Conclusions

The studies on the effect of single and multiple overloads on fatigue crack growth by measuring the bandwidth in the SEM micrographs showed that delayed retardation have occurred after overloads and as the periodic overload number increases fatigue crack growth rate increases.

Acknowledgements

We would like to acknowledge the support by Anadolu University Research Foundation under a contract no of 011516.

References

Koning AU, Dougherty DJ 1989 In Fatigue Crack Growth Under Variable Amplitude Loading, Elsevier Science, Amsterdam, p208.
Lang M, Marci G 1999 Fatigue & Fract. of Engng. Mater. & Struct. 22, 257.
Pommier S, De Freitas M 2002 Fatigue & Fract. of Engng. Mater. & Struct. 25, 709.
Schijve J 1999 Fatigue & Fract. of Engng. Mater. & Struct. 22, 87.
Skorupa M 1998 Fatigue & Fract. of Engng. Mater. & Struct. 21, 987.
Turan D, Turan, S, Ay N 2001 Inst. Phys. Conf. Ser. No 168, 227.
Vardar O 1988 Engineering Fracture Mechanics 36, 71.
Roberta A 1985 Annual Book of ASTM, p739.

Inst. Phys. Conf. Ser. No 179: Section 7
Paper presented at Electron Microscopy and Analysis Group Conf. EMAG2003, Oxford, 2003
©*2003 IOP Publishing Ltd*

Perspectives in Nanoanalysis

A J Craven

Department of Physics and Astronomy, University of Glasgow, Glasgow, G12 8QQ, Scotland UK

Abstract. Advances in aberration correction, parallel detection, spectrum imaging and modelling have revolutionised nanoanalysis to the point where real systems containing both light and heavy elements can be analysed at true atomic resolution and approaching single atom sensitivity.

1. Introduction

The SiO_2 gate dielectric in state-of-the-art CMOS devices is only 1.2nm thick. Layers in magnetic sensors have similar thickness. Even in something as traditional as steel, the early stage of precipitate formation is controlled by atomic scale effects. Thus it is crucial that techniques to characterise such systems on the atomic scale are available.

Rapid progress in instrumentation and software, combined with improved understanding of the beam-specimen interaction, has led to very significant progress in such nanoanalysis. Conventional TEM/STEM columns with Schottky sources can give a probe which has an energy spread of ~0.7eV and a resolution better than 0.14nm using high angle annular dark field (HAADF) imaging. Using aberration correction and cold field emission, instruments such as SuperSTEM will soon achieve sub 0.1nm HAADF resolution with a higher probe current and an energy spread of ~0.3eV. Improved detectors on spectrometers and improved stability provide higher quality data and allow significantly longer acquisition times. With monochromation, sub 0.1eV energy resolution can be obtained, albeit with some loss of spatial resolution due to reduced probe current. Electron energy loss near edge structure (ELNES) in electron energy loss spectroscopy (EELS) allows not only local composition but also local chemistry to be studied. The ability to record nanoanalytical data under computer control as a spectrum image together with processing software has further transformed nanoanalysis.

2. The probe

Two key parameters determining the spatial resolution in nanoanalysis are the size of the incident probe and the current it contains. In a system based on round lenses, the ultimate limit is set by the balance between the effects of diffraction and spherical aberration. The probe current is then determined by the brightness of the source and the contribution that the geometric image of the finite source makes to the overall probe size. Thus a lens with low spherical aberration is required to give the smallest probe size but such a lens has little space in the pole piece, restricting the facilities available in the stage. Correction of three-fold astigmatism is essential to get the smallest probes. However, effects such as mechanical vibration, acoustic noise, earth loops, external AC magnetic fields and instability in the supplies to the alignment coils often limit the probe size in the short term with drift dominating in the long term. Drift is sensitive to air and water temperature, airflow patterns and even atmospheric pressure. Muller gives an excellent discussion of how to minimise such effects [1] and has demonstrated the resolution of single Sb atoms in a silicon lattice [2]. It is clear that modern TEM/STEM columns would benefit greatly from reduced sensitivity to such effects.

Instruments with correction of third order aberrations are now available (e.g. SuperSTEM1 [3]) and ones with correction up to fifth order will soon be available (e.g. SuperSTEM2 [4]). As shown schematically in Fig. 1, chromatic aberration assumes increasing significance as the geometric aberrations are corrected. However, to achieve a sub-0.1nm diameter probe and to carry out nanoanalysis at this spatial resolution requires either the elimination of the sources of non-fundamental limits to the probe size noted above or the de-sensitisation of the column to their presence.

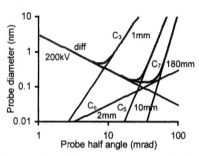

Figure 1. Probe size vs angle for limits set by 3rd, 5th and 7th order aberrations.

One way of using aberration correction is to achieve the highest spatial resolution possible. For example, using HAADF imaging, SuperSTEM 1 has already demonstrated the existence of a predicted 2×1 reconstruction at the interface between Si(001) and NiSi$_2$ [5]. However, another approach is to obtain the smallest probe size currently available in a non-aberration corrected instrument but with more probe current and/or greater flexibility in the specimen region by the use of a lens with a bigger gap.

Even in a relatively thick sample, the HAADF image can show atomic resolution information from the region near the entrance surface. However, spectroscopic techniques such as EELS give information averaged over the volume excited by the probe. The zeroth approximation to the maximum specimen thickness is the geometric depth of field. An Airy disc with a diameter of 0.2 nm at 100 keV has a probe half angle of 22 mrad. To keep the maximum geometric diameter of the extremal rays below 0.2 nm with the probe focussed in the centre requires the specimen to be less than 9 nm thick. An Airy disc of 0.1nm diameter requires the extremal rays to have half the diameter with twice the angle so that the specimen thickness has to drop by 4×. Detailed calculations of the probe-specimen interaction are required for a full understanding of this aspect of the process e.g. [6].

3. Energy dispersive x-ray spectroscopy

Energy dispersive x-ray (EDX) spectroscopy, based on Si(Li) detectors, is still a key nanoanalytical technique but the energy resolution, while sufficient to separate most characteristic peaks, is inadequate to give information on chemical bonding. Since the characteristic x-rays excited in a specimen are emitted isotropically, the detection efficiency is also low with a typical 30mm^2 detector only subtending ~0.13mrad at the specimen so that only ~1% of the x-rays are detected. Care is also required to ensure that spurious contributions to both characteristic peaks and the background are minimised by careful design [7]. Systems with a total solid angle of 0.3 mrad have been built and give excellent results e.g. [8] but the absolute collection efficiency remains low.

New energy dispersive detectors, based on both microcalorimeters and superconducting tunnel junctions have shown energy resolutions below 10 eV.

However, the area of an individual detector is only ~0.04 mm^2. Nonetheless, arrays of such detectors offer great potential in the future e.g. [9].

4. Electron energy loss spectroscopy

4.1 Spectrometers and detectors

EELS has made major advances because of the commercial availability of parallel detection [10] allowing much greater detection efficiency, shorter acquisition times and improved energy resolution. This development continues with charge coupled devices (CCDs) optimised for spectroscopy now available in spectrometers. These have faster read out speeds, flexible binning modes for the cells, low channel to channel gain variations and on-chip amplification to match the analogue signal of the CCD to the input of a 16-bit analogue to digital converter giving better dynamic range and signal to noise ratio. An improved method of coupling the scintillator to the CCD has also been implemented so that the point spread function is also significantly reduced [11].

4.2 Coupling the spectrometer to the optical column

One of the key factors determining the sensitivity of EELS is the fraction of the available signal entering the spectrometer. Inelastic scattering has a Lorentzian distribution around the incident direction with a characteristic angle, ϑ_E. In the classical limit, ϑ_E is given by $E/2E_o$, where E is the energy loss and E_o is the incident energy. More than half of the inelastic scattering lies outside $10\vartheta_E$ [12]. Thus, at a minimum, the spectrometer should accept an angle equal to the probe angle to ensure that the forward scattering is detected and the acceptance angle should be increased with increasing energy loss to take account of the width of the inelastic scattering distribution. Since $10\vartheta_E$ for a 2 keV loss at a 100 keV is 100 mrad, large acceptance angles have always been required for high energy losses. With aberration correction, a large angle of collection will be required whatever the energy loss since a 0.1nm diameter probe at 100keV requires a probe half angle of 44 mrad It is not a trivial matter to couple such large angles of scattering into the spectrometer without introducing loss of energy resolution, defocus of the spectrum as a function of energy loss and change of collection efficiency with energy loss and this will be a challenge for the designers of such instruments.

Elastic scattering can also prevent electrons from entering the spectrometer. Thus strong diffraction conditions should be avoided, a situation that also simplifies quantitative interpretation. However, when performing simultaneous spectroscopy and atomic resolution HAADF imaging, the probe must travel down a high symmetry direction so that strong dynamical effects will occur. This requires a more detailed understanding of the beam specimen interaction to give quantitative results e.g. [13].

4.3 Energy resolution

Energy resolution is fundamentally limited by the energy spread of the source with stability playing a major role in practice. Without monochromation, energy spreads of sources used at their highest brightness are typically $1.0 - 1.2$ eV (thermionic W), $0.7 - 0.9$eV (Schottky zirconiated 100 W) and 0.3eV (cold field emission from 310 W). Even in systems without special stabilisation, a resolution of better than 0.5eV can be obtained e.g. resolution of the spin-orbit splitting of the Si $L_{2,3}$-edges in α-quartz using an HB5 equipped with a Gatan 666 spectrometer, as shown in Fig. 2 [14]. With specialised retarding spectrometers like the Wien filter of Batson, the energy resolution is set by the source itself and an absolute calibration can be maintained over a long period [15]. The energy spread can be reduced by introducing a monochromator into the electron gun, albeit with some loss of probe current, which normally translates into some loss of spatial resolution. With good design of both the high tension system and the spectrometer, an energy resolution of ~0.1eV is possible over 1 sec and ~0.15eV over minutes even with a sector magnet based system [16].

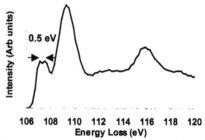

4.4   Energy loss near edge structure

In the dipole approximation, ELNES is essentially controlled by the site and symmetry projected unoccupied density of states on the excited atom. As such ELNES is sensitive to the local chemical

Figure 2. Si $L_{2,3}$ edge of α-quartz recorded using an HB5 and Gatan 666 spectrometer.

environment. However, the level of detail that can be observed is limited by the broadening resulting from the lifetimes of the initial and final states. The lifetime of the initial state depends on the atomic number of the species and the particular shell excited, in general increasing with binding energy. Values are tabulated and the broadening is less than 0.3 eV for the K shells of elements up to Al and the L_3 shells of elements up to Ti [17]. The effect of final state lifetime is to introduce a broadening dependent on the square of the energy above the edge threshold [18] so that fine detail is only present close to the threshold. However, this is a region which is sensitive to local environment. Thus with sub-eV instrumental energy resolution, ELNES provides key information.

Figure 3 shows the O K-edges from three samples of yttria stabilised zirconia, each with a different crystal structure. The detailed shape for each phase is sufficiently characteristic that the phase can be identified from the shape. Moreover, the energy separation of the first two peaks is linearly dependent on the yttria fraction so that a single O K-edge allows both phase and composition to be determined [19].

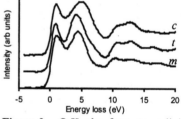

Figure 3. O K-edge from monoclinic (*m*), tetragonal (*t*) and cubic (*c*) Y_2O_3 stabilised ZrO_2 obtained with an HB5.

When analysing overlying phases, ELNES can be used to determine how much of the signal from a particular atomic species comes from each phase. This approach proved invaluable when analysing sub 10nm precipitates extracted from steel using carbon replicas. After the majority of the amorphous carbon had been removed using a plasma cleaner, it was possible to determine the carbon content of the precipitate by comparing the C K-edge shapes obtained from the precipitate to linear combinations of the shapes from amorphous carbon and a carbonitride standard [20].

Methods of modelling ELNES have advanced rapidly in recent years and good agreement with experiment can be obtained from software packages that can be used by experimentalists. For calculations based on density functional theory, WIEN2K and CASTEP are popular while FEFF8 gives good results from a real space multiple scattering approach. Care is needed when there is a high density of states at the Fermi level so that transition metal $L_{2,3}$ edges and $M_{2,3}$ edges of transition metals and $N_{4,5}$-edges of rare earths are best treated by an atomic multiplet approach e.g. [21].

A high density of states at the Fermi level also offers the possibility of lowering the total energy if the system spin polarises. Fig. 4 shows the experimental N K-edge from CrN and the calculated N p-DOS with and without spin polarisation. The agreement is significantly better with the spin polarised calculation [22]. The energy reduction per formula unit from spin polarisation is sufficiently large that spin polarisation effects persist well above the Néel or Curie temperature, where the long range order disappears e.g. in $MgCr_2O_4$, spin polarisation effects are present at 300K even though the Néel temperature is 16K [23].

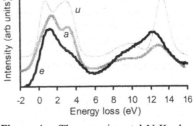

Figure 4. The experimental N K-edge of CrN (e) compared to the N p-DOS for no spin polarisation (u) and for anti-ferromagnetic spin polarisation (a).

5 Spectrum Imaging

Another major advance for nanoanalysis has been the commercial availability of systems for spectrum imaging, first proposed in 1988 [24]. Here, computer control is used to position the probe and orchestrate the signal acquisition at a set of pre-determined points. In its simplest form, a single spectrum is acquired at each point resulting in a 3-D data cube. The key to success is software which both acquires and processes the data efficiently and effectively. This is now available and continues to develop. In principle, simultaneously available signals can all be included in the data set e.g. an EELS spectrum, an x-ray spectrum and the HAADF detector signal. However, acquiring spectrum images take a significant time. The computer can use reference images and spectra to correct for drifts in the system. However, such corrections introduce significant overheads and the better the system stability the less correction is necessary and the more rapidly the data sets are acquired. Reducing hardware and software overheads is an area where progress is required.

Fig. 5 shows a line spectrum image across a high-k dielectric gate stack on a Si substrate recorded on a Tecnai F20 with a Gatan ENFINA spectrometer. The stack consists of 3.5nm of HfO_2, 1nm of HfSiO (an HfO_2 rich co-deposition of SiO_2 and HfO_2) and 250nm of poly-Si. In Fig. 5, the energy loss is displayed horizontally and the position vertically. The energy loss range includes the Hf $M_{4,5}$-edges and the Si K-edge.

Figure 5. Line spectrum image of a high-k gate stack with background subtracted prior to the Hf $M_{4,5}$-edges. Energy loss is horizontal and position vertical.

Plotting the intensity along the energy axis at constant position gives the spectrum from that point on the specimen. It is clear that the Si K-edge has two forms. The shape of the one with the lower threshold energy matches that from crystalline Si while the one with the higher threshold matches the shape from SiO_2. Using these two shapes and that for the Hf $M_{4,5}$-edges, a multiple linear least squares fit can be made to the spectrum at each point on the specimen and hence the distribution of the three phases across the gate stack found [25]. Similar analysis can be performed on the O K-edge and the Si $L_{2,3}$-edges. Thus not only composition but also chemistry can be mapped in complex systems containing a mixture of light and heavy elements. Experimental investigations of the fundamental limits to the spatial resolution that can be obtained in EELS are currently under way e.g. [26].

290

6. Conclusions

With advances in parallel detection in EELS, spectrum imaging software and aberration correction, nanoanalysis of real systems at true atomic resolution and approaching single atom sensitivity is becoming a reality. However, limitations are not only determined by the instrumentation but also by the specimen itself, both in regard to the need for artifact free preparation of very thin specimens and to their sensitivity to the electron beam.

References

[1]   Muller D A and Grazul J 2001 J. Electron. Microsc. 50 219-226
[2]   Voyles P M, Muller D A, Grazul J L, Citrin P H, Gossmann H J L 2002 Nature 416 826-829
[3]   Bleloch A, Brown L M, Brydson R, Craven A J, Goodhew P J and Kiely C 2002 Microscopy and Microanalysis Proceedings 8 (suppl. 2) 470-471
[4]   Krivanek O L, Nellist P D, Dellby N, Murfitt M F and Szilagyi Z 2003 Ultramicrosc. 96 229-237
[5]   Falke U, Bleloch A and Falke M 2003 *submitted* Phys. Rev. Lett.
[6]   Möbus G and Nufer S 2003 Ultramicrosc. 96 285-298
[7]   Nicholson W A P, Gray C C, Chapman J N and, Robertson B W 1982 J. Microsc. 125 25-40
[8]   Papworth A J, Knorr D B and Williams D B 2003 Scripta Mat. 48 1301-1305
[9]   Friedrich S, Mears C A, Nideröst B, Hiller L J, Frank M, Labov S E, Barfknecht A T and Cramer S P 1999 Microscopy and Analysis 4 616-621
[10]   Krivanek O L, Ahn C C and Keeney R B 1987 Ultramicrosc. 22 103-116
[11]   Hunt J A, Dickerson F E, Abbott A A, Szantai G and Mooney P E 2001 Microscopy and Microanalysis 2001 Proceedings 7 (suppl. 3) 1132-1133
[12]   Egerton R F 1986 Electron Energy Loss Spectroscopy in the Transmission Electron Microscope 1st edn. (New York: Plenum) 158
[13]   Dwyer C and Etheridge J 2003 Ultramicrosc. 96 343-360
[14]   Garvie L A J, Rez P, Alvarez J R, Buseck P R, Craven A J and Brydson R 2000 American Mineralogist 85 732-738
[15]   Batson P E 1986 Rev. Sci. Inst. 57 43-48
[16]   Mitterbauer C, Kothleitner G, Grogger W, Zanbergen H, Freitag B, Tiemeijer P, Hofer F 2003 Ultramicrosc. 96 469-480
[17]   Stöhr J 1992 NEXAFS Spectroscopy Springer Series in Surface Sciences 25 (Berlin: Springer) 14
[18]   Muller D A, Singh J and Silcox J 1998 Phys. Rev. B 57 8181-8202
[19]   Vlachos D, Craven A J and McComb D W 2001 J. Phys.:Condens. Matt. 13 10799-10809
[20]   Wilson J A and Craven A J 2003 Ultramicrosc. 94 197-207
[21]   de Groot F M F 1994 J. Electron. Spect. and Related Phenom. 67 529-622
[22]   Paxton A T, van Schilfgaarde M, MacKenzie M and Craven A J 2000 J. Phys.: Condens. Matt. 12 729-750
[23]   McComb D W, Craven A J, Chioncel L, Lichtenstein A J, and Docherty F T 2003 *submitted* Phys. Rev. B
[24]   Jeanguillaume C and Colliex C 1989 Ultramicrosc. 28 252-257
[25]   Craven A J, MacKenzie M, McComb D W and Hamilton D A 2003 IOP Conf. Ser. 180 *in press*
[26]   Lupini A R and Pennycook S J 2003 Ultramicrosc. 96 313-322

Inst. Phys. Conf. Ser. No 179: Section 7
Paper presented at Electron Microscopy and Analysis Group Conf. EMAG2003, Oxford, 2003
©2003 IOP Publishing Ltd

Composition profiles from TiN/NbN multilayers: elastic and inelastic electron scattering techniques compared

S J Lloyd and J M Molina-Aldareguia*

Department of Materials Science and Metallurgy, University of Cambridge, Pembroke Street, Cambridge, CB2 3QZ, UK.

*CEIT, P. Manuel Lardizabal 15, 20018 San Sebastian, Spain.

Abstract. A TiN/NbN multilayer has been examined by Fresnel contrast analysis and energy filtered image series in the TEM for a range of foil thickness. Good agreement between the composition profile determined by the two techniques was obtained if the lower resolution of the EFTEM images was taken into account.

1. Introduction

Inelastically scattered electrons in the transmission electron microscope (TEM) are most commonly used to characterise sample composition, e.g. electron energy loss spectroscopy (EELS). However elastically scattered electrons also contain compositional information that in some cases will offer a more accurate analysis, particularly for features less than a few nanometers in size [1]. For example, Fresnel contrast analysis measures the local change in mean inner potential (ΔV_0) via the acquisition of a through focal series of images and comparison with simulations. Recently this technique has become much easier to implement through the use of energy filtering (removing the majority of the inelastic scattering whose effects on the image are difficult to model) and programs that automatically fit potential profiles to the experimental contrast [2].

In this paper we compare the composition profiles obtained from a TiN/NbN multilayer using Fresnel analysis and an extended energy-filtered (EFTEM) image series [3]. This multilayer system has aroused much interest for its potential 'ultrahard' properties. Models of the deformation in this system suggest that the amplitude and interface width of the composition profile strongly affect the hardness [4]. The structure of this multilayer was also characterised by X-ray diffraction [5].

2. Experimental

The TiN/NbN multilayer (period 13.6nm) was grown on a MgO (001) substrate by reactive magnetron sputter deposition as described in more detail elsewhere [6]. TEM specimens were prepared using a FEI200 focused ion beam workstation to make a series of electron transparent windows of differing thickness. A wedge thickness profile (3.5° angle) was also machined. This allowed the inelastic mean free path of the multilayer to be calibrated so that the foil thickness could be determined accurately from the ratio of elastic and inelastic scattering in any region [5]. Fresnel and EFTEM series from identical areas of specimen were acquired at 297keV in a Philips CM300 FEG TEM.

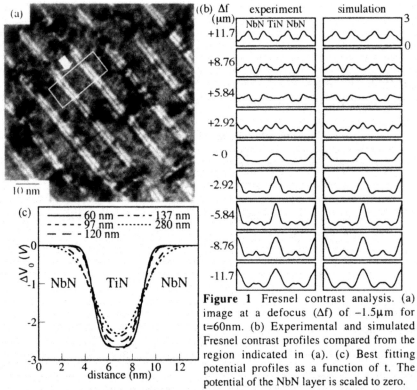

Figure 1 Fresnel contrast analysis. (a) image at a defocus (Δf) of $-1.5\mu m$ for t=60nm. (b) Experimental and simulated Fresnel contrast profiles compared from the region indicated in (a). (c) Best fitting potential profiles as a function of t. The potential of the NbN layer is scaled to zero.

The sample was tilted so that the layers were parallel to the beam and weakly diffracting. Fresnel images were energy filtered and an automated fitting program was used to match the experimental contrast to a potential profile [2]. The magnification of the images was calibrated using the multilayer period determined by X-ray diffraction. EFTEM series were acquired over an electron energy loss range of 0-590eV in 10eV steps. Fourier-log deconvolution was used to remove the plural scattering from these series [3].

3. Results and discussion

Fig.1a shows a filtered image, at a small underfocus for a foil thickness, t, of 60nm. The area analysed is marked with a box and the intensity was projected in the direction indicated to give the profiles of layer contrast as a function of defocus shown in fig.1b, after averaging and symmetrising, in comparison with the profiles from the best fitting simulation. Fitted potential profiles for a range of foil thickness are shown in fig.1c.

The projected potential profile is flat in the center of the NbN and TiN layers for t=60nm (fig.1c) indicating a uniform composition in the center of the layers. The FWHM of the TiN layer is 4.2nm, identical to the value from X-ray diffraction, but more precisely determined [5]. However, the interface width (defined by 0.1 and 0.9 of the profile height) is about 1.5nm, greater than the width of 0.8nm from X-ray diffraction probably due to layer waviness through the thickness of the foil or because the layers are imaged at a slight tilt with respect to the beam. For greater t the potential profile is more rounded and broadened with the area under the profile remaining approximately constant. This is indicative of layer roughness causing some compositional averaging through

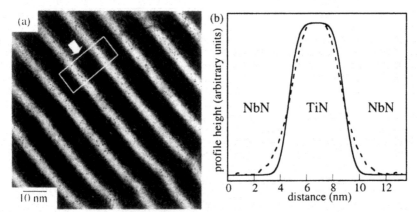

Figure 2 (a) Ti map from the EFTEM series acquired from the same region of multilayer as shown in fig.1a. (b) Composition profiles from the Fresnel analysis (full line) and EFTEM (dotted line) compared for t=60nm. The Fresnel potential profile has been inverted and scaled to the same height as the EFTEM profile.

the thickness of the foil. In this sample flat regions of multilayer are separated by column boundaries which displace the layer interfaces slightly (e.g. see the bottom right hand corner of fig.1a.) The lateral spacing of the columns is around 50-100nm, consistent with the range of foil thickness over which the roughness is observed.

Assuming neutral atom scattering factors [7] and lattice parameters of 0.424 and 0.439nm for TiN and NbN respectively the predicted ΔV_0 is 1.74V. However for this multilayer period the layers are likely to be coherently strained. For a fully coherent structure ΔV_0=3.17V. Experimentally for t=60 nm, ΔV_0=2.66V. This lower experimental value could be accounted for by a combination of bonding effects modifying the scattering factors (increased ionicity would tend to reduce ΔV_0), some incoherent regions being sampled through the foil thickness and/or by some intermixing changing the composition of the layers from pure TiN and NbN.

EFTEM series were acquired from the same regions as the Fresnel profiles for the same range of t, although the series for t=280nm was too thick to be deconvoluted. The Ti map for t=60nm is shown in fig.2a and the projected intensity (after averaging and symmetrising) is shown in fig.2b in comparison with the potential profile determined from the Fresnel analysis. The profiles are similar, but the EFTEM profile is more rounded. Convolving the Fresnel profile with Gaussian profiles of radii between 0.5 and 1nm provided a close match to the EFTEM profile. This is consistent with the approximate resolution expected from the EFTEM under the conditions used due to delocalisation effects, chromatic aberration and specimen drift [5]. Fresnel simulations incorporating a potential with the same shape as that determined from EFTEM did not match the experimental Fresnel contrast, providing further confirmation that the Fresnel analysis gives a more accurate determination of the composition profile.

Fig.3 shows spectra from the EFTEM data set. The area under the background subtracted Nb and Ti edges was measured from the spectra and relative scattering cross-sections were determined by assuming that the Nb/Ti ratio was 2.5 (given by the nominal deposition thickness of the layers) for the total spectrum (a). Masks were used to extract spectra from regions with the maximum Nb signal in the NbN layers (b) and the maximum Ti signal in the TiN layers (c). While this confirmed that the center of the TiN

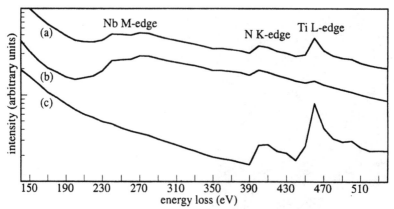

Figure 3 Deconvoluted EEL spectra from multilayer for t=60nm. Spectrum from (a) the whole area shown in fig.2a, (b) the centre of the NbN layers, (c) the centre of the TiN layers. Spectra are plotted on a log scale and offset for clarity.

layers was pure TiN, interestingly the NbN layers contain a uniform composition of about 6% Ti. Given this level of mixing the expected ΔV_0 for a fully coherent structure is reduced to 3.0V, still greater than the experimental value of 2.66V. Assuming neutral atom scattering factors this suggests that the layers are not fully coherent.

4. Conclusion

Fresnel contrast analysis and EFTEM have been used to characterise the composition profile in a TiN/NbN multilayer. Fresnel imaging determines the shape of the composition profile with greater accuracy. Further advantages are that it can be applied reliably to relatively thick foils and it damages the specimen far less. EFTEM identifies the particular elements present and is helpful in reducing ambiguities in the interpretation of the Fresnel potential profile which is sensitive to atomic volume and bonding effects as well as composition. Used together they are a powerful combination that can provide a local quantitative characterisation that complements, and for some parameters improves upon, the 'average' structure determined by X-ray diffraction.

Acknowledgements

We are grateful to the Royal Society, the Basque government and the DTI for financial support. We thank Dr R.E. Dunin-Borkowski Dr M. Weyland and Dr P.J. Thomas for use of their analysis programs and for helpful discussions.

References

[1]   Stobbs W M 1992 in Future developments of metals and ceramics Eds. Charles J A, Greenwood G W and Smith G C (London: Institute of Materials) p 279
[2]   Dunin-Borkowski R E 2000 Ultramicroscopy **83**, 193
[3]   Thomas P J and Midgley P A 2001 Ultramicroscopy **88**, 179 & 187
[4]   Chu X and Barnett S A 1995 J. Appl. Phys. **77**, 4403
[5]   Lloyd S J and Molina-Aldareguia J M 2003, in preparation
[6]   Molina-Aldareguia J M, Lloyd S J, Odén M, Joelsson T, Hultman L and Clegg W J 2002 Philos. Mag. **82**, 1983
[7]   Rez D, Rez P and Grant I 1994 Acta Cryst. A**50**, 481

Inst. Phys. Conf. Ser. No 179: Section 7
Paper presented at Electron Microscopy and Analysis Group Conf. EMAG2003, Oxford, 2003
©2003 IOP Publishing Ltd

EELS ALCHEMI Revisited

R Brydson, A P Brown, S McBride, C Calvert and A J Bell
Institute for Materials Research, University of Leeds, Leeds, LS2 9JT, UK.

Abstract. We have revisited the technique of ALCHEMI, in particular using EELS, for the extraction of site occupancies and local site chemistries in a crystalline material. We have attempted to repeat previous studies on a chromite spinel and also discuss the application of such techniques for the study of complex perovskite materials for ferroelectric applications.

1. Introduction

Atom Location by Channelling Enhanced MIcroanalysis (ALCHEMI) is a technique employed to investigate site occupancies and site symmetries within a particular crystalline unit cell. It relies on the existence of Bloch waves that result from the solution of the Schrodinger equation for an electron in a periodic lattice potential. For the simple case of a crystal at an exact Bragg orientation, corresponding to diffraction from a particular set of (hkl) atomic planes, two standing waves are set up. One standing wave has its maximum amplitude peaked directly on the atomic planes, while the other has its maximum amplitude peaked mid-way between the atomic planes. As the specimen is tilted through the Bragg condition, the relative intensities of these Bloch waves will vary. This variation in intensity will influence the probability of inner shell ionization, and hence the relative intensity of an ionization edge in an EELS spectrum. The exact nature of this variation will also depend on the location within the unit cell of the atoms being ionized, relative to those atoms which compose the atomic planes undergoing Bragg diffraction. As X-ray emission follows inner shell ionization, these channelling effects will also affect the intensities in the X-ray emission (EDX) spectrum.

Although ALCHEMI experiments are relatively common in EDX microanalysis, particularly for the investigation of site occupancies and hence ordering in alloys [1], only limited EELS and ELNES studies have been demonstrated. These EELS studies have employed large scattering angles, designed to maximize the site selectivity of the excitation process [2].

We have attempted to revisit the technique of EELS ALCHEMI for the determination of local valencies and site symmetries from ELNES measurements, and have attempted to reproduce the data of Tafto and Krivanek on a chromite spinel, now some twenty years old [2]. This seminal work is highly important and, if this technique could be routinely applied, it would constitute a powerful structural probe of site occupancy and chemistry. We have also begun to investigate the applicability of this technique to study A and B site occupancies (and in principle valence states of dopant elements) in complex perovskite-based materials (ABO_3) for ferroelectric applications.

2. Method

Powder samples of a normal chromite spinel (nominally $[Fe,Mg][Fe,Cr,Al]_2O_4$) from the South African ore fields and also a lead zirconium titanate ($Pb(Zr_{0.52}Ti_{0.48})O_3$) were dispersed onto holey carbon support films. Thin areas were examined in a FEI CM200 FEGTEM operated at 200 kV and fitted with an Oxford Instruments UTW energy dispersive X-ray (EDX) spectrometer and Gatan Imaging Filter GIF200 using HRTEM, selected area electron diffraction (SAED), EDX and EELS. Further TEM measurements were also performed on a 0.65PMN-0.35PT (PMNT) single crystal sample (PMN = $Pb(Mg_{0.33}Nb_{0.67})O_3$ and PT = $PbTiO_3$), grown by a flux method, which was core drilled, dimpled and ion beam thinned to electron transparency in a Gatan PIPS system.

3. Results and Discussion

The general formula of a so-called 2-3 spinel is XY_2O_4 where X and Y are divalent and trivalent cations respectively. The spinel unit cell is cubic and contains eight formula units, the structure is formed from a cubic close packed array of oxygens which results in two different types of interstitial sites: 32 octahedral and 64 tetrahedral sites. A normal spinel has solely A^{2+} cations on 1/8 of the tetrahedral sites and solely B^{3+} cations on 1/2 of the octahedral sites, whereas in an inverse spinel all the A^{2+} ions and half of the B^{3+} ions are on the octahedral sites with the remaining half of the B^{3+} cations on the tetrahedral sites. Whether a spinel is normal or inverse depends on the octahedral site preference energies (OPSE) of the various divalent and trivalent cations present. For chromite, the octahedral site preference energy of Cr^{3+} is so large that it only occupies octahedral sites. Considering the cubic spinel unit cell, it is clear that the (004) plane contains the oxygen anions plus 2/3 of the cations (those in octahedral interstices), while the (008) plane contains the remaining third of the cations (those in tetrahedral interstices). Thus if we can employ an ALCHEMI technique to localize the ionization process firstly on the (004) planes and secondly, midway between the (004) planes (i.e. on one set of alternate (008) planes), we can then probe the occupancy of the octahedral and tetrahedral sites in the spinel unit cell.

A sample of the chromite spinel was determined by EPMA and XRF to have the composition $Cr_{1.217}Fe_{0.733}Mg_{0.458}Al_{0.577}Ti_{0.016}O_4$. The cubic lattice parameter was determined by X ray diffraction to be 0.8269 nm. From thermodynamic calculations using the Sack and Ghiorso model, it is expected that at temperatures below 800K, the spinel will be normal and the Fe^{2+}/Fe^{3+} ratio is predicted to be 0.553/0.180 = 3.07 [3]. General EELS spectra were recorded in diffraction mode with a collection angle of 10 mrads from a number of chromite grains and quantified using Hydrogenic cross-sections plus a white line correction and a 40 eV integration window. This gave an average Cr/Fe atomic ratio of approximately 1.9, which is slightly higher than that suggested by EPMA perhaps indicating that Mg and Al are somewhat heterogenously distributed. Using an empirical peak fitting procedure specifically developed for the Fe L_3-edge [4], the Fe^{2+}/Fe^{3+} ratio was determined to be 3.0 ± 0.1 in good agreement with the thermodynamic predictions.

ALCHEMI measurements were undertaken by locating a crystallite in the [100] orientation. A schematic electron diffraction pattern is shown in figure 1. Here the crystallite has been tilted away from the zone axis, so it lies close to a systematic row orientation ((horizontal in the figure), in that we are only exciting a row of diffraction spots (in this case the (004), (008) etc. reflections) – this is known as planar ALCHEMI.

If we move from a thin region to a slightly thicker region it is possible to observe Kikuchi lines (KL), particularly when the beam is focused onto the sample area. Observation of the exact position of these KLs as a function of tilt can provide information on the deviation parameter (s) from the exact (004) Bragg orientation. At the exact Bragg orientation (s=0), the dark (deficit) KL passes through the (000) spot, while the bright (excess) KL passes through the (004) spot. When s is positive (KLs moved to the right of the (004) spot in figure 1a), we are localizing the ionization process inbetween the (004) planes and are thus probing the tetrahedral sites. Conversely, when s is negative (KLs moved to the left in figure 1b) we are localizing the ionization process on the (004) planes and are therefore probing the octahedral sites. The position of the spectrometer entrance (collection) aperture is also shown in figure 1, and in practice this can be viewed directly on the GIF TV rate camera. Here we have used a 10 mrad aperture displaced (using the diffraction shift coils) by approximately 20 mrads parallel to the (004) Kikuchi band. The reason for this is that a large scattering angle localizes the excitation onto the atomic planes, here to within about 0.1 nm of the atomic nuclei. However, this constraint does result in very low EELS signals, necessitating long integration times and intense, focused illumination.

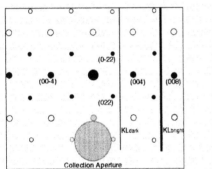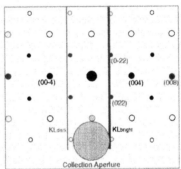

Figure 1. Schematic diagram of diffraction patterns and dark and bright (004) Kikuchi lines (KL) for (a) s > 0 and (b) s < 0. The position of the EELS collection aperture is also marked.

The EELS spectra are shown in Figures 2 and 3. In figure 2, a background has been subtracted before the O K-edge. The Cr $L_{2,3}$ edge at 578 eV is clearly enhanced relative to to Fe $L_{2,3}$-edge at 707 eV for the case of the octahedral site selected spectrum (figure 2b) (Cr/Fe = 3.75 as compared to a vale of 1.9 for a general orientation – figure 2a), additionally the Cr/Fe ratio is decreased (to 1.37) for the tetrahedral site selected spectrum (figure 2c). We estimate the quantification error to be at most 15%.

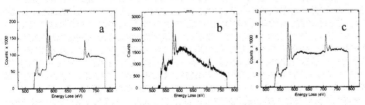

Figure 2. Background subtracted EELS spectra for: (a) a general orientation, (b) octahedral sites selected and (c) tetrahedral sites selected.

298

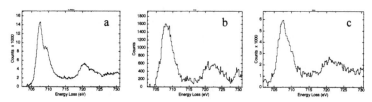

Figure 3. Fe $L_{2,3}$-spectra for: (a) a general orientation, (b) octahedral sites selected and (c) tetrahedral sites selected.

Fe $L_{2,3}$-spectra are shown in figure 3. The general orientation shows two distinct L_3 features at 707.5 and 709.5 eV, known to be associated with Fe^{2+} and Fe^{3+} respectively [4]. Preliminary Fe $L_{2,3}$-spectra for the two ALCHEMI selected sites, tentatively suggest a localization of Fe^{2+} in the tetrahedral sites (as expected in a normal spinel), evidenced by the decreased intensity of the high energy shoulder in the Fe L_3-edge in figure 3(c) relative to the octahedral site-selected signal in figure 3(b). However the data is noisy and the energy resolution and dispersion are somewhat degraded compared to the high quality data of Tafto and Krivanek [2]. Importantly, we did observe, that after prolonged irradiation with a focussed probe, the spinel was sensitive to radiation damage which led to an increase in the Fe^{3+} content, perhaps as a result of exsolution of γ– or even α- Fe_2O_3 by electron beam heating. We believe the spectra in figure 3 to be from an undamaged specimen area. This was not discussed by the previous study. Finally, we also note that the site-selected signal was highly sensitive to the magnitude of the scattering angle used to enhance the signal localization, which we intend to investigate in more detail.

We have also begun to perform similar experiments on PZT and PMNT pervoskites. EDX experiments to extract compositional changes at different sites are relatively straightforward, however EELS suffers from the limited edges which are accessible in the sub-2keV energy range. Furthermore, these materials are highly susceptible to damage, via volatilisation and phase separation of lead oxide and a cooling holder was required which currently restricts our sample tilting capabilities. Single crystal specimens were found to be extremely prone to cracking, perhaps as a result of piezoelectric strain induced by the electron beam charging.

4. Conclusion

Initial EELS ALCHEMI experiments have tentatively reproduced the findings of Tafto and Krivanek on a normal chromite spinel. This study has highlighted the requirement for a cooled, double tilt holder for such investigations, as localized, site selected EELS signals are extremely weak and materials are susceptible to irradiation damage. The use of oriented, single crystal specimens would also greatly aid such measurements.

5. References

[1]   Jones I P, Inst. Phys. Conf. Ser. 168, 231-234, 2001.
[2]   Tafto J and Krivanek OL, Phys. Rev. Lett. 48, 560, 1983.
[3]   Tathavadkar V D, PhD Thesis, IMR, University of Leeds, 2001.
[4]   Calvert C C et al., Inst. Phys. Conf. Ser. 168, 251-254, 2001.

Inst. Phys. Conf. Ser. No 179: Section 7
Paper presented at Electron Microscopy and Analysis Group Conf. EMAG2003, Oxford, 2003
©2003 IOP Publishing Ltd

Spectrum Imaging of High-*k* Dielectric Stacks

M MacKenzie[1], A J Craven[1], D W McComb[2], D A Hamilton[1] and S McFadzean[1]

[1]Department of Physics & Astronomy, University of Glasgow, Glasgow G12 8QQ
[2]Department of Materials, Imperial College London, London SW7 2AZ

Abstract: Electron energy loss spectroscopy combined with spectrum imaging gives a powerful technique that can be used to examine the structural, chemical and electronic properties of materials on a sub-nanometre scale. We have used this technique to examine high-*k* dielectric devices. ELNES from experimental standards have been utilised in the interpretation of the data.

1. Introduction

Spectrum imaging (Jeanguillaume and Colliex, 1989) is a powerful technique with which to investigate the structural, chemical and electronic properties of materials on a sub-nanometre scale. It uses computer control to position an electron beam and to record one or more spectra at each point.

We have used this technique to investigate high-*k* dielectric devices prepared by International SEMATECH. The devices contain a high-*k* dielectric stack sandwiched between a Si substrate and 200nm of Ti-silicided poly-Si. The stack itself consists of 3.5nm of HfO_2 and 1nm of HfSiO (i.e. co-deposited HfO_2 and SiO_2). An isolating oxide film separates the devices.

Here, electron energy loss spectra are recorded and the electron energy loss near edge structure (ELNES) is used to provide information on the chemical and structural environment of the elements at each point. Since the local bonding and coordination of the atom determines the ELNES, it is possible to separate out the contributions to an edge from atoms in different phases. This is achieved by modelling the edge shape as a linear combination of the ELNES from appropriate standards.

2 Method

In order to reduce the likelihood of specimen preparation artefacts, transmission electron microscopy (TEM) specimens were prepared by a number of different techniques. Cross-sectional specimens were prepared from the high-*k* dielectric stacks and isolating oxide using 3 different techniques: (i) FIB lift-out onto holey C film using an FEI Strata FIB 200TEM; (ii) cross-sectional tripod polishing (Muller, 2003); (iii) standard grinding, polishing, dimpling and ion milling. International SEMATECH also deposited 20nm reference layers of HfO_2 and HfSiO on Si wafers and TEM specimens were prepared from these by low-angle cleaving. In order to aid interpretation of the data from the SEMATECH material, EELS analyses were also performed on commercial HfO_2 powder and crystalline hafnon ($HfSiO_4$). TEM specimens were prepared by crushing these materials in propanol using a pestle and mortar and then dispersing the suspension onto a holey C film.

The specimens were examined in an FEI Tecnai F20 equipped with a Gatan ENFINA electron spectrometer. Spectrum imaging was performed using Gatan Digitial Micrograph and DigiScan software. The microscope was operated at 200keV with a probe half-angle of 8.8mrad. The recording conditions depended on the region of the energy loss spectrum being recorded. For the O K-edge (~532eV), typical values were: spot size ~0.5nm; dispersion 0.2eV/ch; collection semi-angle 10 mrad; integration time 10 sec.

Line spectrum images were recorded across the dielectric stack and also across the interface between the Si wafer and the isolating oxide layer i.e. in a region of the wafer where there is no high-k device present. The background under an edge or group of overlapping edges was removed by fitting Ae^{-r} to the smooth background preceding the first edge, extrapolating it under the edges and then subtracting.

3   Results and discussion

Fig. 1 is a conventional bright field image of the high-k dielectric stack. Two layers can be seen at the interface: a light amorphous layer and a dark partially crystallised layer that contains Hf. It is not possible to separately identify the HfO_2 and HfSiO layers expected to be present. The width of the dark layer, ~3nm, is less than the 4.5nm expected for the combined HfO_2 and HfSiO layers.

Fig. 2 is a high angle annular dark field image of the dielectric stack with the contrast expanded. The bright band indicates a high Hf concentration and there is a less intense band on the poly-Si side to the left of it. A fainter, narrower band can also be seen on the Si substrate side. This specimen was prepared using the FIB but similar effects were observed in tripod polished and ion milled specimens.

Fig. 3 is a line spectrum image taken

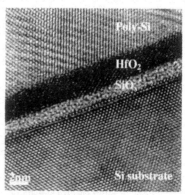

Figure 1. TEM bright field image of high-k dielectric stack.

Figure 2. High angle annular dark field image of high-k dielectric stack

along the line shown in Fig. 2. The horizontal axis corresponds to energy loss while the vertical axis is the position across the oxide layer. The poly-Si is at the top and the Si substrate is at the bottom. Each spectrum has had the background removed prior to the O K-edge so that the intensity is directly related to that of the O K-edge at a particular position and energy loss. The bright area in Fig. 3 corresponds to the O K-edge. Fig. 4 shows two O K-edges extracted by summing the spectra in regions A and B, respectively. It is clear that O is present in at least two chemical environments.

The interface between the Si substrate and the isolating silicon oxide layer was then examined. Fig. 5 is the resulting background subtracted O K-edge line spectrum image. It can be seen that the intensity distribution in the edge from the interface (C) differs from that in the bulk of the isolating oxide material (D). Fig. 6 shows the summed edges from these 2 regions. The O K-edge from the interface region has more intensity close to the threshold. This is characteristic of oxygen deficient SiO_2 (Ramanathan et al, 2002). The edge from the oxygen deficient SiO_2 provides a good match to the edge from region B of the gate stack (also shown in Fig. 6).

Comparison of the edges from the SEMATECH HfO_2 layer and the commercial HfO_2 powder with that from region A of the gate stack (Fig. 7) showed that the edge from the stack approximates well to that from monoclinic crystalline HfO_2. The disagreement above 545eV is a consequence of plural scattering that has not been corrected for.

The O K-edge shape from crystalline $HfSiO_4$ was observed in neither the gate stack nor in the HfSiO reference layer. In fact the O K-edge from the reference HfSiO layer can be modelled as a linear combination of the O K-edges from HfO_2 and SiO_2. (McComb et al, 2003).

Multiple linear least squares (MLLS) fitting was used to model the O K-edge shape at each point in the spectrum image of Fig. 3 using the edge shapes from crystalline HfO_2 and oxygen deficient amorphous SiO_2. The data was fitted in the range 530-545eV in order to minimise the effects of plural scattering. The resulting weights from the fits are shown in Fig. 8. The specimen region primarily associated with the oxygen deficient SiO_2 shape is next to the substrate

301

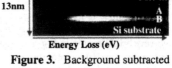

Figure 3. Background subtracted O K-edge spectrum image taken along line indicated in Fig.2.

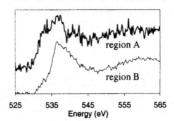

Figure 4. O K-edges taken from regions A and B in the spectrum image in Fig.3.

Figure 5. Background subtracted O K-edge spectrum image taken across interface between Si substrate and isolating oxide layer.

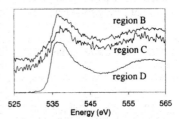

Figure 6. O K-edges from 2 different regions of spectrum image in Fig. 5. Also shown is the O K-edge from region B in the gate stack spectrum image.

302

and corresponds to the light amorphous region in Fig. 1. The main region corresponding to the HfO₂ shape is next to the poly-Si and has the same width as the dark region in Fig. 1 and the very bright region in Fig. 2. There is some tailing of the HfO₂ into the poly-Si region corresponding to the fainter bright band noted above in Fig. 2. Thus, the 1nm HfSiO layer deposited originally appears to have phase separated with the HfO₂ interpenetrating the poly-Si gate electrode material during processing. However, in this particular spectrum image we see no evidence of the SiO₂ in the poly-Si which was observed in spectrum images from other specimens (Craven et al, 2003). Investigation of this is continuing. The apparent HfO₂ signal on the Si substrate side of the stack is probably an artefact. It is likely to be highly O deficient SiO₂ which the MMLS routine fits as HfO₂.

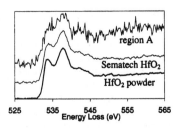

Figure 7. O K-edge from region B in gate stack compared to those from HfO₂ powder and the reference HfO₂ layer.

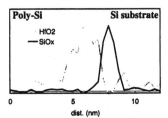

Figure 8. Weights in MMLS fit of O K-edge shapes of SiO_x and HfO_2 to the spectrum image in Fig. 3.

4   Conclusions

Spectrum imaging across the high-k dielectric stack showed the presence of two distinct layers. The O K-edge shapes correspond best to oxygen deficient amorphous SiO₂ and monoclinic HfO₂ but no evidence of crystalline hafnon. Modelling the O K-edge as a linear combination of those from SiO_x and HfO₂ demonstrates the overlap of the phases and suggests that the 1nm HfSiO layer originally deposited has phase separated and interpenetrated the poly-Si gate electrode material during processing. This is currently being investigated further.

5   Acknowledgements

Dr R Murto and his colleagues at International SEMATECH for supplying the SEMATECH materials; Prof J Hanchar, The George Washington University for supplying the crystalline hafnon; Dr P Thomas of Gatan for use of routines for the MLLS analysis; Mr W Smith and Mr B Miller for preparing the TEM specimens.

References

Craven A J, MacKenzie M, McComb D W and Hamilton DA 2003 IOP Conf. Series **180** (in press)
Jeanguillaume C and Colliex C 1989 Ultramicroscopy **28** 252.
McComb D W, Craven A J, Hamilton DA and MacKenzie M 2003 APL (submitted for publication)
Ramanathan S, McIntyre P C, Luning J, Pianetta P and Muller D A 2002 Phil. Mag. Letts. **82** 519.
Muller D A 2003 private communication

Inst. Phys. Conf. Ser. No 179: Section 7
Paper presented at Electron Microscopy and Analysis Group Conf. EMAG2003, Oxford, 2003
©2003 IOP Publishing Ltd

EELS study of near edge fine structure in Al$_x$Ga$_{1-x}$N alloys

G Radtke, P Bayle-Guillemaud and J Thibault

CEA-Grenoble/Département de Recherche Fondamentale sur la Matière Condensée/SP2M
17 rue des Martyrs - 38054 Grenoble Cedex 9, France

E-mail: pbayle@cea.fr

Abstract. N and Al K electron-energy-loss spectra have been recorded in thick layers of Al$_x$Ga$_{1-x}$N alloys (x = 0, 0.15, 0.46, 0.81, 1) in wurtzite structure. The evolution of near-edge-fine-structure (ELNES) with the composition is discussed in terms of modifications of the chemical and geometrical environment around the excited atoms. Some of the experimental results have been confirmed by *ab initio* calculations.

1. Introduction

Group III nitrides (AlN,GaN, InN) are wide band gap semiconductors widely used in opto- and micro-electronics and, in most of these devices, AlGaN alloys play an important role. The ionization edges observed in electron-energy- loss-spectroscopy (EELS) are the result of transitions of core electrons into unoccupied states. In first approximation, the energy-loss-near-edge-fine-structure (ELNES) can be interpreted as an image of the site- and symmetry-projected density of states (PDOS) which depends strongly on the arrangement and the type of atoms located around the excited atom, i.e. its local environment. Through a systematic study of the changes observed near N and Al K edges in Al$_x$Ga$_{1-x}$N alloys (x = 0, 0.15, 0.46, 0.81, 1), the aim of this work is to check how this local environment can influence the near edge fine structure and to determine the relevant parameters which govern these changes.

2. Experimental details

The Al$_x$Ga$_{1-x}$N thick layers ($\approx 1 \mu m$) were grown by molecular beam epitaxy on AlN or GaN single crystal buffers grown on sapphire $\langle 0001 \rangle$ substrates. Cross-sectional TEM specimens were prepared using the standard techniques of mechanical polishing and Ar^+ ion milling. EELS spectra were recorded on a JEOL 3010 LaB_6 equipped with a Gatan Imaging Filter (GIF). They were collected in diffraction mode, the spectrometer aperture centered on the transmitted beam and the crystal oriented with the c axis perpendicular to the optical axis. The collection and convergence semi-angles were respectively 3,66 *mrad* and ≈ 3.5 *mrad*. All the spectra displayed hereafter have been recorded under the same experimental conditions. As a consequence, the variations observed in ELNES between alloys of different compositions cannot be attributed to anisotropy effects. The strong anisotropy which dominates the ELNES in hexagonal (wurtzite) nitrides has already been reported in the literature both in EELS (Radtke *et al* 2003 or Keast *et al* 2002) and X-Ray Absorption Spectroscopy (Lawniczak-Jablonska *et al* 2000). The acquisition time was 40 *s* and the spectra were corrected for dark current and gain variation. The background has been subtracted using the standard power-law procedure and the deconvolution of multiple inelastic scattering has been performed using the

Fourier ratio method. *Ab initio* band structure calculations were performed using the density functional theory in the local density approximation. We used the Wien97 implementation of full potential linearized augmented plane waves (FLAPW) method (Blaha *et al* 1999) and the generalized gradient approximation (Perdew *et al* 1996) for the exchange correlation potential. The maximum momentum for the radial wave functions l_{max} was chosen as 14 and the plane wave cutoff RK_{max} was set to 8. Muffin-tin radii were 1.85 a.u. for Al spheres and 1.65 a.u. for N spheres. ELLS spectra near the Al K edge were calculated according to the formalism described by Hébert-Souche *et al* (2000) accounting for the experimental conditions, i.e. the orientation of the crystal, the collection and the convergence angles. These theoretical spectra were convolved by a 1.5 eV FWHM Gaussian to account for the incident beam energy dispersion. Ground state calculations have been performed in a single AlN cell using 1000 k-points in the first Brillouin zone (1BZ) to get a good convergence of the total density of states. The core hole effects have been included by removing an Al 1s core electron and modeling it as a uniform background negative charge. The spatial separation between excited centers required the use of a 2x2x1 supercell. In this case, 250 k-points in the 1BZ have been used.

3. Results and discussion

Figures 1 a) and b) show respectively the N and Al K edges obtained from the AlGaN thick layers. The reference spectra of AlN and GaN binary compounds were recorded on the buffers of the same samples. The variation of the chemical composition leads to important changes in

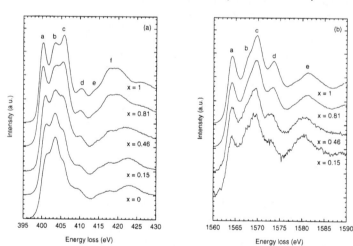

Figure 1. Experimental spectra recorded in $Al_xGa_{1-x}N$ ternary alloys : (a) N K edge with x = 0, 0.15, 0.46, 0.81, 1 and (b) Al K edge with x = 0.15, 0.46, 0.81, 1.

the fine structure of the N K edge. If the general shape of this edge, constituted by a first triplet of preeminent peaks (labeled (a), (b) and (c) in the figure) followed by smoother features ((d), (e) and (f)), remains the same, the relative intensity and energy position of the different peaks show a strong variation with the Ga concentration. The energy distance between peaks (a) and (b) progressively decreases from AlN to GaN whereas the relative intensity of peaks (a) and (c) with respect to the central (b) peak varies. Comparatively, only small changes occur in the Al K ELNES when the Ga concentration increases : figure 1 b) shows that the energy position and the relative intensity of the peaks are almost independent of the composition of the alloy. These behaviors can be easily understood. In hexagonal AlN, each atom has a tetrahedral coordination : N atoms are located in a tetrahedron of Al and Al atoms are located in a tetrahedron of N. In AlGaN ternary alloys, Al and Ga atoms share the same type of sites in the crystal. It means that Ga atoms appear as first nearest neighbors for N atoms but only as second nearest neighbors for Al atoms. It can be concluded that in these alloys the ELNES of N and Al K edges are dominated by the chemical bonding of the excited atom with its nearest neighbors. However, a weak effect can be observed in the Al K edge which consists in a small

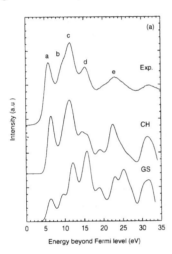
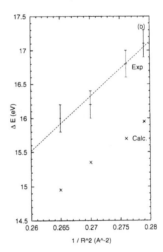

Figure 2. (a) Comparison between experimental (Exp.) and theoretical Al K edge in AlN with (CH) and without (GS) a core hole. (b) Evolution of the energy of the resonance peak (e) of Al K edge vs $1/R^2$ both experimental in Al, $Ga_{1-x}N$ alloys (Exp.) and in AlN calculated with Wien97 with different lattice parameters (Calc.).

shift toward lower energy losses of the structures with respect to the edge onset. This effect is particularly visible for the peak labeled (e) in figure 1 b). A possible interpretation of this experimental result can be found in the increase of bond lengths with the Ga concentration : the peaks recorded in this energy range can be attributed, in first approximation, to interference effects resulting from multiple scattering of the excited electron by neighboring atoms. In this framework, different authors (Lytle et al 1988 or Kurata et al 1993) used the following relation between the energy position of the resonance peak above the threshold (ΔE) and the distance (R) between the excited and the backscattering atoms (Bianconi et al 1983) :

$$\Delta E \, R^2 = const. \tag{1}$$

Figure 2 b) shows a plot of ΔE versus $1/R^2$. The position ΔE of the peak (e) has been measured on the experimental spectra with respect to the first peak labeled (a) in figure 2 a). R is the Al-N interatomic distance calculated in AlGaN alloys by assuming the validity of the Vegard law (from 1.89 Å in AlN to 1.94 Å in $Al_{0.15}Ga_{0.85}N$). The dotted line corresponds to the linear interpolation of the experimental data. The shift is more than 1 eV from an Al composition of 100 to 15 %. From figure 2 b), the correlation between the energy position of peak (e) and the variation of Al-N bond length is clearly established even if the strict *linear* dependence of ΔE with $1/R^2$ is difficult to verify.

In order to confirm the *geometrical* origin of this effect, a series Al K edges has been calculated in pure AlN with different lattice parameters, corresponding to those of the AlGaN alloys. As it can be seen on figure 2 a) in the case of AlN, the agreement between experimental and theoretical spectra is clearly improved when the core hole effects are taken into account in the calculations. For each theoretical spectrum, the energy position of peak (e) with respect to peak (a) has been measured and plotted versus $1/R^2$ in figure 2 b) together with the experimental results. If the absolute value of ΔE is underestimated in the theoretical spectra (of $\approx 1eV$), this discrepancy with the experimental data is almost independent of the composition of the alloy. The same trend is observed and then, confirms that the small changes in Al K fine structure can be partially attributed to an increase of bond length induced by the presence of Ga atoms.

4. Conclusion

This experimental study of N and Al K edges in $Al_xGa_{1-x}N$ alloys allowed us to determine the dominant parameters which guide the evolution of the fine structure with the composition :

i) The remarkable changes near N K edge with the Ga composition, as compared to the weak effect near Al K, is linked to the distance of Ga atoms with respect to the excited atom.

ii) The small changes in the Al K fine structure, characterized by an energy shift of the structures lying at around 15 to 20 eV beyond the edge onset, have been partially attributed to the increase of Al-N bond length due to the introduction of Ga. This effect has been confirmed by *ab initio* calculations.

Acknowledgments

We would like to thank to B. Daudin (CEA-Grenoble) for providing the $Al_xGa_{1-x}N$ layers.

References

Bianconi A, Dell'Ariccia M, Gargano A and Natoli C R 1983 *EXAFS and Near-Edge Structure* ed by Bianconi A *et al* (Springer-Verlag, Berlin) pp 57-61

Blaha P, Schwarz K and Luitz J 1999 *Wien97, a Full Potential Linearized Augmented Plane Wave Package for Calculating Crystal Properties (Computer Program)* (Techn. Universität , Wien, Austria)

Hébert-Souche C, Louf P H, Blaha P, Nelhiebel M, Luitz J, Schattschneider P, Schwarz K and Jouffrey B 2000 *Ultramicroscopy* **83** 9-16

Keast V J, Scott A J, Kappers M J, Foxon C T and Humphreys C J 2002 *Phys. Rev.* B **66** 125319-1-7

Kurata H, Lefèvre E, Colliex C and Brydson R 1993 *Phys. Rev.* B **47** 13763-13768

Lawniczak-Jablonska K, Suski T, Gorczyca I, Christensen N E, Attenkofer K E, Perera R C C, Gullikson E M, Underwood J H, Ederer D L and Liliental Weber Z 2000 *Phys. Rev.* B **61** 16623-16632

Lytle F W, Greegor R B and Panson A J 1988 *Phys. Rev.* B **37** 1550-1562

Perdew J P, Burke S and Ernzerhof M 1996 *Phys. Rev. Lett.* **77** 3865-3868

Radtke G, Epicier T, Bayle-Guillemaud P and Le Bossé J C 2003 *J. Microsc.* **210** 60-65

Inst. Phys. Conf. Ser. No 179: Section 7
Paper presented at Electron Microscopy and Analysis Group Conf. EMAG2003, Oxford, 2003
©2003 IOP Publishing Ltd

Retrieval of anisotropic displacement parameters in Mg from convergent beam electron diffraction experiments

J Friis[†], K Marthinsen[‡] and R Holmestad[†]
[†]Department of Physics, Norwegian University of Science and Technology (NTNU), 7491 Trondheim, Norway
[‡]Department of Materials Technology, Norwegian University of Science and Technology (NTNU), 7491 Trondheim, Norway

Abstract. We present an accurate Wilson plot-like method, based on convergent beam electron diffraction, for measuring the anisotropic displacement parameters in magnesium, and compare it with the method proposed by Saunders *et al.* A generalization of this method to isotropic crystals with more than one type of atoms is also discussed.

1. Introduction

With quantitative convergent beam electron diffraction (QCBED) [1, 2] it is possible to very accurately measure the low order structure factors in small-unit cell crystals. This technique is based on a pixel to pixel comparison between an experimental CBED pattern and a Bloch-wave simulation. The input parameters (such as structure factors, beam direction, etc.) in the Bloch-wave simulation are refined until the best fit is obtained.

In structure factors measured by QCBED, the largest source of errors comes from uncertainty in the thermal displacement parameters (DPs). Even though the DPs might be known as a function of temperature, either from phonon calculations, or from X-ray or neutron measurements, the exact sample temperature is unknown. Several methods have therefore been proposed for measuring the DPs directly by electron diffraction [3-7]. Most of these methods, except [7], require separate experiments for the determination of DPs and structure factors. The same experimental conditions, e.g. sample temperature, can therefore not be guaranteed. We will here investigate a Wilson plot-like method based on the values of the refined electron structure factors, and compare it with the method proposed by [7].

2. CBED experiment and refinement

The experiments were performed using a 120 kV LEO 912B TEM with an in-column Ω-filter and a Gatan CCD camera. The systematic row orientation was used and the sample was cooled to liquid nitrogen temperature. Eleven low order structure factors, listed in Table 1, were measured and refined with the EXTAL program [2]. For Mg the DPs have been measured at different temperatures by neutron diffraction [8]. Hence, good initial estimates of the anisotropic DPs of $\langle u_1^2 \rangle = 0.0076$ Å^2 (**a**-direction, parallel to 2-fold axis) and $\langle u_3^2 \rangle = 0.0085$ Å^2 (**c**-direction, parallel to $\bar{6}$-fold axis) at the experimental temperature (around 110 K) have been used in the refinements.

Table 1. Refinement results. The scattering angles s are in units of Å^{-1} and the electron structure factors U_g are in units of $10^{-4}\ \text{Å}^{-2}$.

hkl	1 0 0	0 0 2	1 0 1	1 0 2	1 1 0	1 0 3	2 0 0	2 1 0	0 0 4	2 0 4	2 2 0
s	0.181	0.193	0.205	0.264	0.313	0.341	0.361	0.385	0.478	0.528	0.626
U_g	184.5	346.9	279.6	115.3	185.8	143.8	77.21	141.0	52.42	44.71	67.0
	±0.7	±0.4	±0.9	±0.5	±0.2	±0.3	±0.15	±0.3	±0.15	±0.16	±0.5

The measured electron structure factors U_g (Fourier coefficients of the crystal potential) are converted to X-ray structure factors F_g (Fourier coefficients of the electron density) with the Mott-Bethe formula [1]

$$F_g = \sum_i Z_i T_{i,g} e^{-2\pi i g \cdot r_i} - \frac{8\pi^2 \varepsilon_0 h^2 \Omega s^2}{\gamma m_e e^2} U_g. \tag{1}$$

The sum goes over all atoms, where Z_i and $T_{i,g}$ are the atomic number and temperature factor of atom i, respectively. Ω is the unit cell volume, $s = \sin\theta/\lambda$ the scattering angle and $\gamma = 1 + E_0/(m_e c^2)$ a relativistic correction, with E_0 being the acceleration voltage of the microscope.

3. Determination of DPs using the method proposed by Saunders *et al.*

A strategy for determination of DPs by comparing the refined structure factors with calculations was proposed by [7]. In short, each refinement is performed for a range of fixed DP values, where the same DPs are used for conversion to X-ray structure factors. If the order of the structure factor is sufficiently high, so that bonding effects can be neglected, the X-ray structure factor obtained with the correct DPs should equal the calculated value from an independent atom model (IAM) [9].

In Fig. 1a this method is applied to the (204) structure factor ($\sin\theta/\lambda = 0.528\ \text{Å}^{-1}$) of Mg. The neutron diffraction measurement [8] is used to relate the two anisotropic DPs for Mg to each other and to assign them to a temperature. The refined structure factors, converted to X-ray values intersect the IAM values at 102 K, corresponding to $\langle u_1^2 \rangle = 0.0074\ \text{Å}^2$ and $\langle u_3^2 \rangle = 0.0082\ \text{Å}^2$.

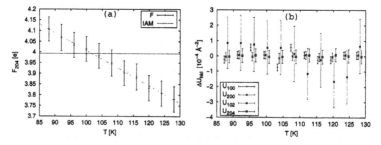

Figure 1. (a) The refined and converted (204) structure factor as a function of the temperature used in the refinement. The dashed line represent the IAM value. (b) The difference between electron structure factors and their mean value as a function of the temperature used in the refinement.

4. Wilson-like method for determination of DPs

The refinement procedure for low order structure factors itself is not very sensitive to the DPs, since they mainly affect the non-refined high order structure factors responsible for HOLZ-lines. This is demonstrated in Fig. 1b, where four structure factors are refined for a range of (fixed) temperatures. No dependence between $U_\mathbf{g}$ and the temperature is observed. However, one should keep in mind that this result is obtained in a systematic row orientation when strong HOLZ-lines are avoided. The case might be different in the zone-axis orientation used in [7]. The conversion to X-ray structure factors is, on the other hand, sensitive to the DPs because of the atomic temperature factors $T_{i,g}$ in Eq. (1). The method for determination of DPs presented here is purely based on this fact.

In the harmonic approximation the atomic temperature factor for Mg is

$$T_\mathbf{g} = \exp[-(h^2 + hk + k^2)P - l^2 Q] \tag{2}$$

where

$$P = 2\pi^2 \frac{4}{3a^2} \langle u_1^2 \rangle \quad \text{and} \quad Q = 2\pi^2 \frac{1}{c^2} \langle u_3^2 \rangle. \tag{3}$$

Since Mg only has one atom species, the X-ray structure factors can be written as $F_\mathbf{g} = f_\mathbf{g} T_\mathbf{g} C_\mathbf{g}$, where $f_\mathbf{g}$ are static lattice scattering factors and $C_\mathbf{g} = \sum_i \exp(2\pi i \mathbf{g} \cdot \mathbf{r}_i)$. This, together with Eq. (1) and (2), gives

$$-\ln T_\mathbf{g} = -\ln \left[\frac{8\pi^2 \varepsilon_0 h^2 \Omega s^2 U_\mathbf{g}}{\gamma m_e e^2 C_\mathbf{g}(Z - f_\mathbf{g})} \right] = (h^2 + hk + k^2)P + l^2 Q. \tag{4}$$

Given $f_\mathbf{g}$, the left hand side of Eq. (4) can be calculated. Hence P and Q can easily be obtained by least square fitting. In Fig. 2 we have plotted the left hand side of Eq. (4) against the right hand side for the fitted values of P and Q. We do therefore expect the points to follow a straight line with a slope of 1. We have used $f_\mathbf{g}$ from both Dirac-Fock [10] and density functional theory (DFT) [11] calculations. The latter includes bonding effects and brings even the low order points nicely onto the line (Fig. 2b). Using the DFT values this gives $\langle u_1^2 \rangle = 0.00777(4)$ Å^2 and $\langle u_3^2 \rangle = 0.0082(1)$ Å^2. These values are so close to the initial values used that new refinements were not performed. However, if the initial DPs were less accurate, one would have to redo the structure factor refinement with the new DPs until convergence is achieved.

This method benefits from the large number (eleven) of measured structure factors. Usually one does not measure that many structure factors with the CBED technique. In

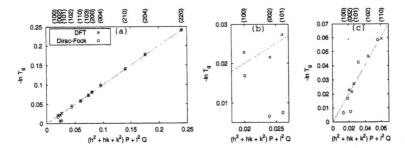

Figure 2. Wilson-like plot for low order data. The static lattice reference scattering factors are taken from both DFT (×) and Dirac-Fock (o) calculations. The slope of the line is 1. (b) shows an enlargement around the three first structure factors in (a), where the disagreement between the two models is pronounced. In (c) the same fit is performed using only the five lowest order structure factors.

310

Fig. 2c we have therefore redone the fit by only considering the five lowest order structure factors. The fitted DPs now become $\langle u_1^2 \rangle = 0.0079(3)$ Å^2 and $\langle u_3^2 \rangle = 0.0082(7)$ Å^2. In this case it is really essential to consider bonding effects in the static lattice reference scattering factors $f_\mathbf{g}$. If even fewer structure factors are measured, it is probably sensible to constrain $\langle u_3^2 \rangle$ to $\langle u_1^2 \rangle$ or use an isotropic model.

This method can be generalized to crystals with more than one type of atoms, in the case where only one independent DP needs to be determined per atom type. This is the case for isotropic DPs, which we will consider here. We can then write $T_i = \exp(-B_i s^2)$, where the B_i are known as Debye-Waller factors. Eq. (1) can now be rewritten as

$$U_\mathbf{g} = \frac{\gamma m_e e^2}{8\pi^2 \varepsilon_0 h^2 \Omega s^2} \sum_i \left(Z_i - f_{\mathbf{g},i} \right) e^{-2\pi i \mathbf{g} \cdot \mathbf{r}_i} e^{-B_i s^2} \tag{5}$$

resulting in a set of equations (one for each $\mathbf{g}$) each containing a sum of exponentials in $-B_i s^2$. These exponents can be determined with non-linear fitting, giving the Debye-Waller factors.

5. Discussion and conclusion

The very good fit in Fig. 2 shows that there is enough information in the CBED data to accurately determine DPs.

We have here compared two different methods for determining the DPs in Mg. The method proposed here differs from the one in [7] in that all structure factor refinements are initially carried out only once with assumed DPs. As seen from Fig. 1b, errors in the DPs will not affect the refined structure factors very much when a HOLZ-line free region is used in the refinement. The determination of DPs in the method proposed here is purely based on the Mott-Bethe formula, and involves all measured structure factors via least square fitting. Another improvement, is that DFT, instead an IAM, is used in the calculations of static lattice reference scattering factors. A minor drawback of this is that a systematic error in the DFT model will bias the experimental structure factors towards the DFT-values, making them less appropriate for testing the DFT model [12]. Nevertheless, we believe that the DPs obtained by this method are the most accurate as long as the number of refined structure factors exceeds the number of fitted DPs with a factor of at least two or three.

Acknowledgments

Helpful discussions with B Jiang and J C H Spence and fundings from the Research Council of Norway (project 135270/410) are gratefully acknowledged.

References

[1] Spence J C H and Zuo J M 1992 *Electron Microdiffraction* (Plenum Press, New York)
[2] Zuo J M 1999 *Microscopy Research and Technique* 46 220–233
[3] Holmestad R, Weickenmeier A L, Zuo J M, Spence J C H, and Horita Z 1993 *Inst. Phys. Conf. Ser.* 138 141–144
[4] Menon E S K and Fox A G 1998 *Philos. Mag.* A77 577–592
[5] Midgley P A, Sleight M E, Saunders M and Vincent R 1998 *Ultramicroscopy* 75 61–67
[6] Nüchter W, Weickenmeier A L and Mayer J 1998 *Acta Cryst.* A54 147–157
[7] Saunders M, Fox A G and Midgley P A 1999 *Acta Cryst.* A55 480–488
[8] Iversen B B, Nielsen S K and Larsen F K 1995 *Philos. Mag.* A72 1357–1380
[9] Doyle P A and Turner P S 1968 *Acta Cryst.* A24 390–397
[10] Su Z and Coppens P 1997 *Acta Cryst.* A53 749–762
[11] Blaha P, Schwarz K, Madsen G K H, Kvasnicka D, and Luitz J 2001 *WIEN2k*
[12] Friis J, Madsen G K H, Larsen F K, Jiang B, Marthinsen K, and Holmestad R 2003 Submitted

Inst. Phys. Conf. Ser. No 179: Section 7
Paper presented at Electron Microscopy and Analysis Group Conf. EMAG2003, Oxford, 2003
©2003 IOP Publishing Ltd

The Effect of Radiation Damage on Anomalous Absorption in the TEM

L A Scruby, J M Titchmarsh and M L Jenkins

Department of Materials, Oxford University, Oxford OX1 3PH, UK

Abstract: *In-situ* electron irradiation of Si at 400keV was performed to determine the variation of two-beam absorption parameters by the method of Hashimoto. Corresponding theoretical sensitivity studies were performed to assess limitations of experimental errors. No significant variations in parameters were revealed. Possible reasons for this are discussed.

1. Introduction

The embrittlement of nuclear reactor pressure vessels as a consequence of neutron irradiation has at least three components. One of these is known simply as 'matrix damage' and is believed to be due to point defect clusters, probably associated with minor alloying or impurity atoms [1]. Such features are currently beyond the direct imaging capability of TEM but they are amenable to investigation by indirect methods in bulk samples. However, if a high density of such features is present then it is feasible that they could affect electron absorption in the TEM and, consequently, influence the contrast of features such as thickness fringes and bend contours. TEM could then be used to quantify variations in damage on a local level. The aim of this experiment was to explore the possibility of detecting and quantifying electron irradiation damage, induced directly in the TEM, by measuring simple 2-beam absorption parameters.

2. Procedures

2.1 Theoretical background

Under 2-beam dynamical diffraction conditions, the absorption of the Bloch waves can be described by the mean and anomalous absorption parameters, ξ_g/ξ_0' and ξ_g/ξ_g', respectively, where ξ_g is the extinction distance for the excited reflection, **g**. These parameters can be derived from values of A_n and B_n (n is the order of the fringe) measured from thickness fringe intensity profiles, as illustrated schematically in Fig.1, and using equations derived by Hashimoto [2]:

$$\xi_g/\xi_0' = (-1/\pi) \ln (A_{n+1} / A_n) \quad \text{and} \quad \xi_g/\xi_g' = (1/\pi) \{\cosh^{-1} (B_{n+1} / A_{n+1}) - \cosh^{-1} (B_n / A_n)\}$$

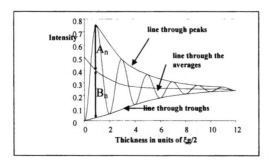

Figure 1. Schematic fringe profile (dark field) illustrating the definition of the parameters A_n and B_n.

The method can be applied to either bright- or dark-field profiles. Surprisingly, there are few reported measurements of absorption parameters but those which have been made suggest that the 'typical' values of ~0.1, often used for intensity calculations, could be significantly in error. In the present study, two-beam profiles were generated assuming various combinations of absorption parameters and then subjected to manual analysis to determine how accurately the parameters could be recovered from the profiles. Sensitivity studies were also performed in this manner on theoretical profiles for a range of tilt deviations from the exact Bragg condition but for fixed values of absorption parameters. Many-beam profiles were also calculated and analysed as for the 2-beam profiles to assess the systematic errors induced in idealising the experimental profiles.

2.2 Experimental method

Silicon was chosen as the material for the experiment because it has a threshold for displacement damage by high-energy electron irradiation of <200keV, is available in large single crystal form, and is rigid. The resistivity of the Si was 4.47Ωcm. Specimens were prepared by the cleavage method to provide clean wedges of angle 55°. The irradiation and examination were performed in a JEOL 4000 TEM and images recorded digitally, allowing rapid quantification of intensity profiles. The specimen was first tilted to excite the required reflection (both 220 and 400 were used), as close as possible to the precise Bragg condition, and BF and DF intensity profiles recorded with a widely spread beam at selected, identifiable locations. The specimen was then tilted to a kinematic orientation, to minimise channelling, before removing the condenser aperture and focusing the beam to a diameter ~ 1μm and irradiating at different locations for varying times up to a maximum of 60 minutes. The beam was again defocused and the specimen tilted back to the Bragg orientation to record more profiles from the newly damaged areas. Further experiments were performed by systematically tilting the specimen through the Bragg position to explore the sensitivity of the measurements to tilting errors. An objective aperture of 2.1mrad diameter was used and great care was taken to position this centrally around the imaging beam. Fringes of order 1 and 2 were discarded because of a potential geometric error [3] and fringes of order >5 were ignored because the fringe contrast variations became low as the thickness increased. Average values of the parameters were then derived from BF and DF images as a function of irradiation time.

The irradiated areas were imaged using weak-beam conditions to explore the nature of any damage but no defects were observed.

3. Results

The analysis of simulated profiles showed that the method was robust for both BF and DF. Mean recovered values of ξ_g/ξ_0' and ξ_g/ξ_g' were within 1% of the values input to generate theoretical profiles, with an error of 5-10%. Both values fell as the deviation parameter was increased. However, mean values remained very close to those for the exact Bragg condition until the deviation parameter $w > 0.4$, which is much greater than any experimental uncertainty in orienting the crystal at the Bragg condition. The effect of increasing the number of beams in the theoretical profiles from 2 to 8 on ξ_g/ξ_0' was less than or similar to the processing errors but a small, significant increase in ξ_g/ξ_g' occurred.

Initial experiments seemed to indicate a significant increase in ξ_g/ξ_g' with irradiation time, relative to the experimental uncertainty, but this was not observed when the experiment was repeated (Fig.2). Average values are presented in Table 1.

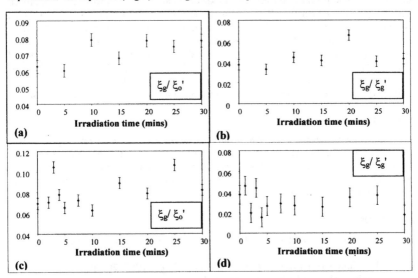

Figure 2. ξ_g/ξ_0' (a and c) and ξ_g/ξ_g' (b and d). (b) and (d) are from a repeated equivalent experiment to (a) and (c). (All BF {220} at 400keV.)

Table 1. Average values of ξ_g/ξ_0' and ξ_g/ξ_g' at 400keV for the {220} reflection.

	ξ_g/ξ_0'	ξ_g/ξ_g'
Present study: 1st experiment	0.062	0.037
Present study: 2nd experiment	0.070	0.032
Reference [3]	0.047	0.002

314

4. Discussion

It is clear from the results in Fig.2 that no significant changes were observed as a function of increasing irradiation time. The sensitivity studies suggest that any errors induced by inaccurate orientation and limitations in assuming two beams rather than many beams are unlikely to have concealed systematic damage increase. The random experimental errors were greater than any trends within the data. It is concluded, therefore, that either insufficient displacement damage occurred for the selected exposures or that any additional contribution from displacement damage to the absorption parameters is small relative to the main contributions from thermal diffuse scattering and electronic excitations. That no defects are observed by weak beam imaging simply confirms that any damage is below the resolution limit, as assumed initially.

The result is somewhat surprising because; (a) the 400keV irradiation exceeds the reported threshold for displacement damage in Si and (b) unintentional irradiation damage was reported in an earlier investigation of absorption parameters [4]. An alternative analysis procedure was adopted in that study whereby several parameters, including absorption parameters, were systematically varied and used to generate theoretical fringe profiles that best fitted experimental profiles. However, that approach sometimes produced very small or even negative values for absorption parameters which were clearly unrealistic. Table 1 compares the results from the present study with values reported in Ref. [4] for the {220} reflection at 400keV. A more feasible explanation is that damage is more likely to occur in heavily doped Si than in the low-doped material used in the present study. While the present results suggest that the method will not be suitable for assessing damage in ferritic steels, it should still be worthwhile to perform additional experiments either on heavily-doped Si, or a steel, to confirm this conclusion.

5. Conclusions

No significant variation in the two-beam absorption parameters in low-doped Si was observed following *in situ* irradiation by 400keV electrons, although such damage has been reported by others. Therefore, measurement of such parameters as a means of assessing the matrix damage component in neutron-irradiated ferritic steel pressure vessels would not appear to be practical.

References

[1]   Akamatsu M, Van Duysen J C, Pareige P, Auger P, *Journal of Nuclear Materials* (1995) **225** 192
[2]   Hashimoto M, Howie A, Whelan M J, *Proc R Soc* (1962) **269** 80
[3]   Goringe M J, *Phil Mag* (1967) **16** 1111
[4]   Walther T, Schaublin R E , Dunin-Borkowski R E, Boothroyd C B, Humphreys C J and Stobbs W M, *Electron Microscopy and Analysis* (1995) **147** 195

Acknowledgements

JMT is grateful to the Royal Academy of Engineering, AEA Technology and the Institute of Nuclear Safety System, Inc. (Japan) for support.

Inst. Phys. Conf. Ser. No 179: Section 7
Paper presented at Electron Microscopy and Analysis Group Conf. EMAG2003, Oxford, 2003
©2003 IOP Publishing Ltd

Sputtering and the formation of nanometre holes in Ni₃Al under intense electron beam irradiation

B B Tang, I P Jones

Department of Metallurgy and Materials, University of Birmingham, Edgbaston, Birmingham B15 2TT, UK

Abstract. The radiation damage of polycrystalline Ni_3Al thin foils of stoichiometric composition by a nanoscale stationary 200 keV FEG electron probe in an FEI Tecnai F20 (S)TEM has been investigated. Nanometre holes are produced in these TEM thin films and EDX spectra from the irradiated volume have been collected with the incident electron beam at both [001] and [110]. From the EDX results, preferential surface sputtering of aluminium from Ni_3Al has been demonstrated. The sputtering rate at the electron-exit surface of a Ni_3Al thin foil has been measured. The film sputtering and hole formation processes have been simulated successfully. All the experimental results agree well with theoretical results from Molecular Dynamics (MD) and Monte Carlo (MC) computer simulations (W. S. Lai & D. J. Bacon, University of Liverpool).

1. Introduction

In recent years, scanning transmission electron microscopes fitted with field emission guns (FEG STEM) have been used to perform high spatial resolution chemical microanalysis. In a FEG STEM, the electron beam can be focused down to ~0.5 nm while still retaining sufficient current that x-ray analysis or electron energy loss spectroscopy can be used to assess the composition of the irradiated volume. Unfortunately, it is becoming clear that with such enormous beam current densities at electron energies between 100 and 300 kV the specimen is damaged and altered via radiation damage. For example, Shang et al. [1] recently suggested that preferential surface sputtering of aluminium from grain boundaries in Ni_3Al leads to spurious segregation profiles. Experimental observations by Bullough [2] using a stationary focused 100 keV high-current-density electron probe in a STEM clearly showed that electron-induced surface sputtering can result in the formation of nanometre voids and holes in an Al foil. Under the same experimental conditions, Muller et al. [3] demonstrated by EDAX the preferential surface sputtering of Al from Ni_3Al.

2. Experimental Procedure

An exact in stoichiometric Ni_3Al alloy was prepared using a radio-frequency induction generator. The alloy was then capsulated in a silica tube and homogenized at 1200°C in a furnace for 30 hours. The polycrystalline Ni_3Al alloy rod was cut, sectioned and mechanically thinned to thin discs of about 0.8 mm in thickness and 3 mm in diameter. Using a Tenupol twin jet polisher, these thin discs were finally electropolished into transmission electron microscope samples.

An FEI Tecnai F20 FEG (S)TEM operating at 200 keV was used in this work, with the vacuum in the microscope column better than 10^{-5} Pa. The current intensity of the incident electron beam was measured from the main fluorescent viewing screen. All energy-dispersive X-ray (EDX) spectra were acquired by an Oxford Link Ultra-thin Be-window Si-Li X-ray detector and were processed and calculated using the Oxford Instruments commercial software package Link ISIS 300.

3. Results and Discussion

With the incident electron beam at [001] direction of Ni_3Al crystal, after 10 minutes' FEG electron probe continuous irradiation, a hole formed as shown in Figure 1 at atomic resolution. Also shown in the figure is its tilted image. At the same time, the composition and thickness changes with time of irradiation are shown in Figure 2.

8 nm

10nm

Figure 1. A typical [001] irradiation hole at atomic resolution and its tilted (30°) projection

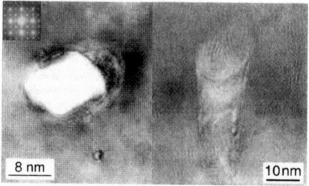

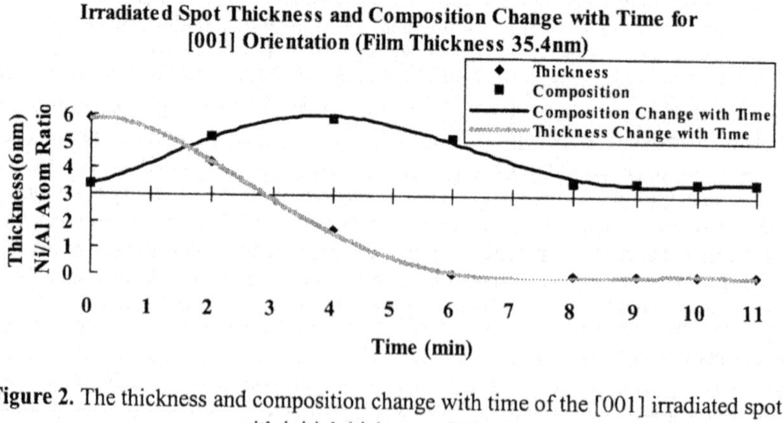

Irradiated Spot Thickness and Composition Change with Time for [001] Orientation (Film Thickness 35.4nm)

Legend:
- ◆ Thickness
- ■ Composition
- —— Composition Change with Time
- ······ Thickness Change with Time

Y-axis: Thickness(6nm) / Ni/Al Atom Ratio (0–6)

X-axis: Time (min) (0–11)

Figure 2. The thickness and composition change with time of the [001] irradiated spot with initial thickness of 35.4 nm

The composition curve shows clearly that Al is preferentially sputtered at least in the early stages of the sputtering process and the thickness curve implies strongly that the hole is through the film after about 6 minutes' irradiation. So in 6 minutes, a volume of Ni_3Al of the hole was sputtered out. Using the method of measuring the intensity of the electron probe, the sputtering yield for [001] irradiation is calculated to be $7.74*10^{-8}$ atoms/electron.

Finer scale experiments of interrupted irradiation of as short time as 10 seconds were conducted for the [001] orientation. Typical results are shown in Figure 3. Notice that the hole size is only 1 or 2 nm in diameter and the hole of 84 nm in length is through within 1 minute. Actually these micrographs reflect the early stage of the hole-forming process in the 10-minute continuous irradiation experiment whose result is actually a long time average one. So the sputtering yield, which is calculated to be $30.9*10^{-8}$ atoms/electron, should be more accurate than that from long time continuous irradiation experiments.

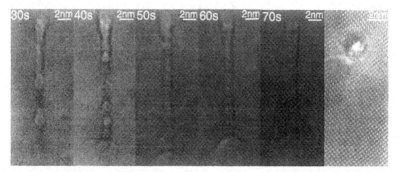

Figure 3. Sequential details (14.4° tilted projection) of the [001] irradiation hole-forming process and the final hole at atomic resolution

Another significance of these finer scale micrographs is that they confirm the observations by Bullough [2] which showed that electron-induced surface sputtering can result in the formation of nanometre voids and holes in an Al foil. The process of hole formation starts as a surface pit and grows by sputtering of the atoms from the electron-exit surface. As the depth of the pit increases, because most of the sputtered atoms from the pit base are recoiled at an angle, they are increasingly attached onto the pit side walls and the pit eventually seals at a point near the pit opening. This leaves a subsurface void which moves under the influence of the electron irradiation along the irradiated volume towards the electron-entrance surface, allowing subsequent pit growth and void formation to repeat at the electron-exit surface. As the process is repeated many times, the arrival of the voids at the electron-entrance surface results in the growth of a pit at that surface, which eventually extends into a hole through the entire sample thickness.

Figure 4 is the theoretical simulation of the [001] finer scale hole forming process, using the surface sputtering energy data from Lai et al. [4], which agrees with our experimental micrographs quite well.

Same experiments as above were also carried out for [110] orientation. Their results confirmed the results of the [001] irradiation experiments. The preferential sputtering of Al and the nanometre voids formation were also found in the early stage of the [110] irradiation hole-forming process. But the sputtering yield for the [110] was calculated to be $4.2*10^{-8}$ atoms/electron, much smaller than the [001] result and shows that the sputtering yield of Ni_3Al crystal is orientation dependent.

[001] Sputtering Simulation

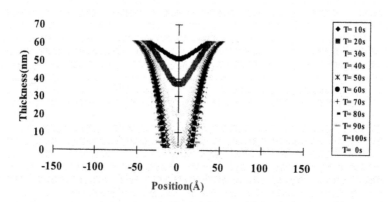

Figure 4. Theoretical simulation of the [001] irradiation hole forming process

4. Conclusions

Electron beam radiation damage in stoichiometric Ni_3Al films was found to cause hole drilling and preferential surface sputtering of Al under conditions typical for microanalysis in an ultra-high-vacuum FEG (S)TEM operating at 200 kV for irradiation directions at low index orientations such as [001] and [110]. The sputtering rate/yield and the changing composition of the irradiated area during the hole-forming process were measured accurately by applying some new (S)TEM methods and utilities. The void mechanism of the hole drilling process, first observed in pure Al, was confirmed in our Ni_3Al experiments at higher resolution. Sputtering yield was found to be orientation dependent: $30.9*10^{-8}$ atoms/electron for [001] irradiation and $4.2*10^{-8}$ atoms/electron for [110] irradiation. Using some known data of Ni/Al sputtering energy from Ni_3Al, the dynamic process of hole drilling was successfully simulated, which coincides quite well with our experimental results.

Acknowledgments

The authors would like to thank Prof. D. J. Bacon and Dr. W. S. Lai of the University of Liverpool for valuable discussions concerning their theoretical simulation results during the course of collaborative work, and the Engineering and Physical Science Research Council for provision of financial support.

References

[1]   Shang P, Keyse R, Jones I P and Smallman R E 1999 Phil. Mag. A 79, 2539
[2]   Bullough T J 1997 Phil. Mag. A 75, 69
[3]   Muller D A and Silcox J 1995 Phil. Mag. A 71, 1375
[4]   Lai W S and Bacon D J private communication

Inst. Phys. Conf. Ser. No 179: Section 7
Paper presented at Electron Microscopy and Analysis Group Conf. EMAG2003, Oxford, 2003
©2003 IOP Publishing Ltd

High Loss EELS at Dislocations in Diamond

A T Kolodzie*
A L Bleloch*[+]

*Cavendish Laboratory, University of Cambridge
[+]SuperSTEM, Daresbury Laboratory, Warrington

Abstract. High Loss EELS acquired at dislocations in natural type IIa brown, colourless, and high pressure, high temperature treated brown diamonds reveal additional electronic energy states associated with the dislocations in the brown diamond. These results are consistent with the outcome of low loss EELS experiments. Comparison of the spectra acquired in the experiments with recent theoretical models of EELS from the various types of dislocations in diamond suggest that the brown diamond may contain shuffle type dislocations.

1. Introduction and Low Loss EELS Results

This work is a continuation of electron energy loss spectroscopy (EELS) research that has been performed on natural, type IIa diamonds in an effort to understand if dislocations are responsible for the colour of brown diamond. Previously, using a scanning transmission electron microscope (STEM), low loss EELS spectra were taken on an untreated colourless diamond, a brown diamond, and a brown diamond that had become colourless after high pressure, high temperature annealing. The spectra were acquired on a dislocation, the electron probe was then moved to a position off the dislocation and spectra were again recorded. The 'off' spectra were subtracted from the 'on' spectra, resulting in 'difference' spectra, which should indicate energy states in the band gap that may be caused by the dislocations.

In a prior low loss experiment [1] the difference spectra for the brown diamond exhibited a broad peak in the band gap which the colourless and annealed diamond did not have. A subsequent improved low loss experiment confirmed the existence of these states associated with dislocations in brown diamond at approximately 5 to 6 eV. Some of the EELS spectra in this latter experiment were acquired with the electron probe oriented along the dislocation core or at very small angles to it. The peak in the difference spectra for the brown diamond, which also appeared in some of the annealed diamond spectra, provides evidence of π-bonded states associated with the dislocations in these samples. These π states are characteristic of the presence of sp^2 bonded material. [2] The spectra from the colourless diamond did not show such states. Significantly, the spectra acquired with the EELS probe travelling close to and parallel to the dislocation core reveal more π states than those taken when the beam was oblique to the dislocation.

2. High Loss Experiment

The most recent experiment examined high loss EELS at the carbon K-edge. EELS spectra were again taken on and off of dislocations and all spectra were acquired with the electron probe oblique to the dislocation lines. The longer acquisition times required in obtaining high loss spectra (~ 1 to 2 seconds) necessitate correction of electronic instability or drift in the electron microscope. In response, an electronic stabiliser, which keeps the zero loss EELS peak in the same position by means of a feedback loop, was developed by Dennis McMullan. [3] Modifications were made to the stabiliser for this experiment.

The stabiliser (Figure 1) applies a triangular wave form to the drift tube voltage so that the EELS zero loss peak (ZLP) is scanned back and forth across a tungsten wire. Electrons backscattered off the wire are detected by a scintillator and their signal is processed to determine if the ZLP has moved. If it has, a correction signal is applied to the drift tube to return the ZLP to its original location. The previous design of the stabiliser placed the scintillator for the stabiliser after the wire and detected the shadow of the wire in the ZLP signal. In the new scintillator configuration, the backscattered signal of the ZLP is many times stronger than the signal from the shadow of the wire and improves the performance of the stabiliser dramatically. Because of the success of the stabiliser, 50 high loss spectra at a time could be acquired on and off dislocations and added together, without any alignment of the spectra needed.

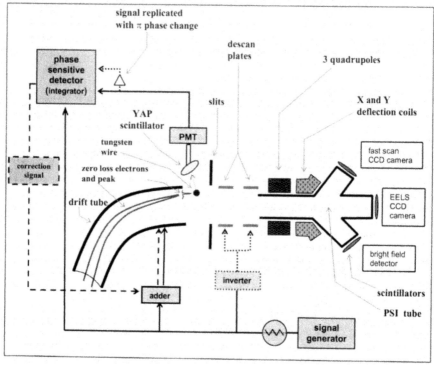

Fig. 1 Schematic overview of McMullan Stabiliser.

3. Results

High loss difference spectra were acquired from 20 or more dislocations in each of the brown, colourless, and annealed diamonds. The difference spectra were averaged and scaled to produce the final results shown in Figure 2.

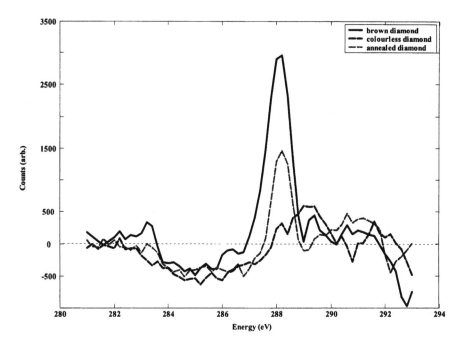

Fig. 2 High loss EELS difference spectra for brown, colourless, and annealed diamond.

The most general observation to be made is that the electronic states associated with dislocations in the three diamonds are different. The deficit in the spectra of all three diamonds from approximately 283.5 to 286.5 eV is attributable to an amorphous carbon surface contribution. [4] The deficits in the graph mean that there was an excess of surface contribution in the spectra off of the dislocations as compared to the 'on' spectra. The physical explanation for this difference in the 'on' and 'off' dislocation surface contribution is likely related to the fact that the majority of the 'on' spectra were recorded before the probe was moved and the 'off' spectra were taken. (This procedure was followed because of considerations arising from use of the stabiliser.) Therefore, the total time that the electron beam was in the vicinity of the 'off' probe position was greater than for the 'on' position, giving surface contamination more time to migrate and accumulate for the 'off' spectra.

Rearrangement of electronic states at the dislocation cores is responsible for the prominent peak at 288 eV. The presence of this excess of states in high loss EELS dislocation 'difference' spectra in the vicinity of 288 eV has been previously reported in type IIb diamond. [4] The small feature in the brown diamond spectrum near 286 eV appears at approximately the energy at which sp^2 bonded carbon produces its pre-carbon K-edge peak. [5] Since the 'off' spectra show an excess of surface contribution over the 'on' spectra, as was just described, this feature at 286 eV is almost certainly not a surface effect, but is a characteristic of the dislocations in the brown diamond. The existence of sp^2 bonded material associated with dislocations in brown diamond is consistent with the low loss EELS experimental results. [1]

4. Modelling of EELS at Dislocations in Diamond

There are two types of dislocations that can exist in diamond, depending on whether the extra plane of atoms of the dislocation terminates at the narrowly spaced ((a/12)<111>) glide plane (called a glide dislocation) or the more widely spaced ((a/4)<111>) shuffle plane (called a shuffle dislocation). [6]   Calculations have shown that dissociation of diamond edge and screw perfect dislocations into two partial dislocations separated by an intrinsic stacking fault (ISF) is energetically favored.   The diamond perfect screw dislocation is predicted to dissociate into two 30° (the angle between the line of the dislocation and its Burgers vector) glide partials separated by an ISF; the perfect 60° glide dislocation is predicted to dissociate into a 30° glide partial dislocation and a 90° glide partial dislocation separated by an ISF; the perfect 60° shuffle dislocation is predicted to dissociate into a 30° glide partial dislocation and a 90° shuffle partial dislocation separated by an ISF. [7]   Both low and high loss EELS spectra from the cores of each of these dislocation types in diamond have been modelled using a pseudopotential calculation within density functional theory (DFT). [8]

The theoretical spectra were compared to those obtained in the low and high loss EELS experiments. To facilitate the comparison, modelled bulk diamond spectra were subtracted from the calculated EELS high loss spectra to produce theoretical high loss 'difference' spectra for the three energetically favored partial dislocations.   The experimental high loss EELS states for the brown diamond appear marginally more similar to modelling of the 90° shuffle dislocation. Additionally, the band gap states present in the two low loss EELS experiments for the brown diamond appear in the calculated low loss EELS spectra for the shuffle type dislocations. [8]

It should be noted, however, that it is very difficult to correlate directly theoretical and experimental spectra. For instance, this model does not take into account excitonic (core hole) effects, which have been shown to substantially influence theoretical diamond K-edge EELS spectra. [9] Thus, the modelling should be viewed as offering a qualitative picture of EELS spectra from the different types of dislocations in diamond. Nonetheless, the experimental results, in combination with the DFT model, provide evidence with which to begin to address the question of whether the differences in the experimental spectra from the three diamonds could be because the diamonds contain different types of dislocations, which might also play a role in the colour of the diamonds.

5. References

[1]   Kolodzie A T, Murfitt M and Bleloch A L 2001 Inst. Phys. Conf. Ser. No. 168 247
[2]   Yuan J, Saeed A, Brown L M, and Gaskell P H 1992 Phil. Mag. B 66 187
[3]   Bleloch A, Brown L M, Marsh M J, McMullan D, Rickard J J and Stolojan V 1999 Inst. Phys.Conf. Ser. 161 195
[4]   Bruley J and Batson P E 1989 Phys. Rev. B 40 9888
[5]   Yuan J and Brown L M 2000 Micron 31 515
[6]   Hirth J P and Lothe J 1982 Theory of Dislocations (New York: John Wiley & Sons)
[7]   Blumenau A T, Heggie M I, Fall C J, Jones R and Frauenheim T 2002 Phys. Rev. B 65 205205
[8]   Fall C J, Blumenau A T, Jones R, *et al.*, 2002 Phys. Rev. B 65 205206
[9]   Pickard C J 1997 Ph.D. thesis, University of Cambridge

Inst. Phys. Conf. Ser. No 179: Section 7
Paper presented at Electron Microscopy and Analysis Group Conf. EMAG2003, Oxford, 2003
©2003 IOP Publishing Ltd

Characterisation of the <500nm fraction of airborne particulates using TEM, EDX and EELS

C C Calvert [1], A P Brown[1], R Brydson [1], J J N Lingard[2],
A S Tomlin[2] and S Smith[3]

[1] Institute for Materials Research, [2] Energy and Resources Research Institute, University of Leeds, Leeds, LS2 9JT UK. [3] School of Health and Life Sciences, King's College, London, SE1 9NN, UK.

Abstract. Current research suggests that atmospheric particles act as a catalyst for a complex set of reactions that involve different oxidation states of transition metals and the generation of reactive oxygen species. Determination of the composition of individual particles, the distribution of transition metals, their oxidation state, and the binding characteristics of transition metals to particles using advanced AEM techniques, such as EELS and EDX will provide new information on the interaction of particles with pulmonary tissues.

1. Introduction

A three-month preliminary study has been carried out to investigate the collection methodology and TEM analysis techniques for the characterisation of the <500nm fraction of urban airborne particulates.

Field emission gun transmission electron microscopy (FEGTEM) in conjunction with high spatial resolution techniques of electron-energy-loss spectroscopy (EELS), energy filtered TEM (EFTEM) and energy dispersive X-ray spectroscopy (EDX) provides new opportunities for determining and elucidating the elemental and molecular structure of individual nano- and micro-sized particles. Of significance is the ability to distinguish between the valency of certain transition metal ions such as Fe^{3+} and Fe^{2+}.

Recent epidemiological research (Pope *et al*., 1995, Bascom *et al*., 1996 and Pope, 2000) has shown that airborne particulate matter has a damaging effect on respiratory and cardiovascular systems of people with a history of heart and lung diseases leading to health deterioration and even death. However the underlying mechanism for these effects is unknown. Toxicological evidence (Wilson *et al*., 2002, Donaldson *et al*., 2000 and Li *et al*., 1999) points towards the fine (<2.5μm) and even ultrafine (<0.1μm) particulate fraction as the most toxic. However, the components responsible within these particulate fractions are again unknown. There is growing evidence that quinoids perhaps in association with transition metal species are responsible for generating free radicals which are related to the production of reactive oxygen species in the lung, in much the same way that has been shown for tobacco smoke.

In light of this there is an urgent need to isolate the damaging components, identify their origins and demonstrate their specific biological reactivity in pulmonary and cardiovascular systems, such that the toxicology can be directly related to the epidemiology. In this quest, there is a need for the development of state-of-the-art measurement systems and characterisation techniques providing quantitative information on submicron-particular components responsible for these health effects.

2. Experiment

2.1. *Collection Methodology*

For the current data set, airborne particulates were collected in two ways:
1) Direct collection of ultrafine particulates onto TEM grids has been achieved by placing 3mm Cu supported holey carbon film grids onto the 0.03, 0.06, 0.108 and 0.17μm stages (1 to 4 respectively) of an Electrical Low Pressure Impactor (ELPI), located on a main road outside the University of Leeds.
2) Extraction of particles from polymer filters used in high volume air samplers placed in Oxford Street, London, by agitation of a section of the filter in de-ionised water to remove the particulates (<10μm in diameter). A few drops of this solution was then pipetted onto holey carbon film TEM grids.

2.2 *Experimental Conditions*

All particulate samples were analysed in an FEI CM200 FEGTEM fitted with a Gatan Imaging Filter (GIF) and an Oxford Instruments ultra thin window EDX detector. For each sample 15 EDX spectra, images and EELS spectra were collected from representative transition metal (TM) bearing particles. All EELS measurements were made in CTEM diffraction mode (image coupling to spectrometer) on ultra thin (<30nm) crystalline areas of ~0.1μm in diameter. TM $L_{2,3}$-, O K-edges and associated low-loss spectra were collected from particles of interest, with total integration times of 40-50 seconds for TM $L_{2,3}$-edges and 20-25 seconds for O K-edges at an energy dispersion of 0.1eV/pixel and a collection angle β of 4.6mrads. The energy resolution for all measurements was 0.7-0.8eV.

3. Results and Discussion

Table 1 EDX data for Leeds and London samples.

Sample/Size	% Fe	% Mn	% Cr	% Zn	% Ti	% Pb	% Zr	% Sb	% Bi
Leeds Stage 1 (0.03μm)	60	7	7	0	20	33	13	7	7
Leeds Stage 2 (0.06μm)	27	7	0	0	7	7	0	0	0
Leeds Stage 3 (0.108μm)	60	7	7	20	7	13	0	0	0
Leeds Stage 4 (0.17μm)	47	20	0	0	7	7	0	0	0
London Oxford St. (<10μm)	53	0	0	7	7	20	0	0	0

Table 1 shows the proportion (in atom %) of metal species within each collection fraction (from EDX analyses of 15 TM-bearing particles per size fraction). The data excludes the dominant C-bearing soot phase, which forms upto 90% of the total particulates. The average particle diameters of the Leeds collection stages 1-4 are 0.03, 0.06, 0.108 and 0.17μm respectively. The London data have a diameter of <10μm.

Soot particles (20-30nm) are the dominant phase (comprising 80-90 % in number of all particles collected) in all size fractions of the airborne particulate samples. These form large aggregations, that can contain TM species, and chains of upto 500nm in diameter/length (figure 1 (A) shows part of a short chain and 1 (B) high resolution detail of the turbostratic or onion-like graphitic layers of soot).

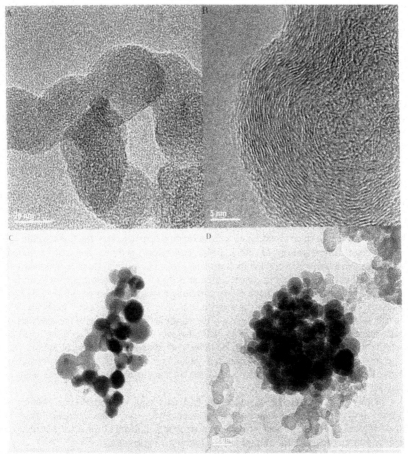

Figure 1 (A) Bright field (BF) TEM Image showing part of a short chain of soot particles; (B) High resolution detail of the soot structure, showing the turbostratic or onion-like graphitic layers; (C) BF TEM image showing an aggregation of Fe-Mn oxide particles and (D) BF TEM showing an aggregation of Fe-oxide particles together with soot.

A range of transition metals (TMs) and other trace metal species has been identified in the current samples (Table 1). Fe was the most commonly occurring TM, forming both discrete oxide particles (figure 1 (C)), often together with Mn (figure 1 (D)), and associated with Mg, Al and Si within large crustal derived mineral fragments. Fe also occurred within large aggregations (>200nm in diameter) of soot. Pb was a common component in both the Leeds and London samples, occurring in small amounts in all sample fractions analysed. From the current EELS data the valence of the Fe present in the particles appears to be mainly Fe^{3+} and the Mn appears to be Mn^{2+}. There was no obvious transition metal segregation between the different size fractions of the Leeds samples, although as the current data set is small (15 analyses) the data is unlikely to be truly representative. However, there are some differences between the Leeds and London data, where the London data showed a more restricted range of TMs present.

4. Preliminary Conclusions

Soot particles (20-30nm) are the dominant phase (comprising 80-90 % of all particles collected) in all size fractions of the airborne particulate samples. TMs from anthropogenic and crustal sources (including Ti, Mn, Cr, Fe and Zn) occur both within agglomerations of soot particles and as discrete particles. Most of the TM-bearing particles are associated with S and Si. Fe (as Fe^{3+}) and Mn (as Mn^{2+}) are the main TMs present are associated with S, O and Si. Pb is also a common component in both the Leeds and London samples.

5. Further Work

It is proposed that further worked be carried out in the following areas to provide comprehensive data on airborne particulates.

FEGSEM and SEM EDX analyses should be carried out to give detailed particle size and size range data and elemental distributions as a function of particle size (micron scale analyses). ICP-MS analyses of bulk particulates will also be carried out to give the average trace metal content for a range of particle size fractions within the PM_{10} range. Investigation of other PM_{10} collection methods are also required, including thermophoretic precipitation directly onto TEM grids, these techniques will run in parallel in order to compare the different methods and so determine the optimum collection method for TEM analysis of particulates. Energy filtered TEM (EFTEM) analyses of particles will also be carried out to provide statistical data for TM distributions within soot aggregations on the nanometre scale. Nanoprobe analyses of individual particles in the FEGTEM will also be carried. These will be used to determine the structure of individual particles and identify any chemical zoning (i.e. core rim effects), utilising a liquid nitrogen cooled holder to minimise beam damage of the particles during EDX and EELS analyses and avoid changes in valence state of TM-bearing phases.

Finally it is intended to correlate the EM data with epidemiological studies at Leeds and KCL. The comparison of ambient with surrogate particles used in toxicological experiments will provide key structural information on the distribution and binding of metals to the surfaces of the two sets of particles.

References

Bascom, R., Bromberg, P. A., Hill, C., Costa, D. L., Devlin, R., Dockery, D. W., Frampton, M. W., Lambert, W, Samet, J. M., Speizer, F. E. and Utell, M. (1996) *American Journal of Respiratory and Critical Care Medicine*, **153 (2)**, 477-498.

Donaldson, K., Stone, V., Gilmour, P. S. Brown, D. M. and MacNee, W., (2000) *Transactions of the Royal Society of London A*, **358**, 2741-2749.

Li, X. Y., Brown, D., Smith, S., MacNee, W. and Donaldson, K. (1999) *Inhalation Toxicology*, **11**, 709-731.

Pope, C. A. (2000) *Journal of Aerosol Medicine-Deposition Clearance and Effects in the Lung*, **13 (4)**, 335-354.

Pope, C. A., Dockery, D. W. and Schwartz, J. (1995) *Inhalation Toxicology*, **7 (1)**, 1-18.

Wilson, M. R., Lightbody, J. H., Donaldson, K., Sales, J. and Stone, V. (2002) *Toxicology and Applied Pharmacology*, **184**, 172-179.

Inst. Phys. Conf. Ser. No 179: Section 7
Paper presented at Electron Microscopy and Analysis Group Conf. EMAG2003, Oxford, 2003
©*2003 IOP Publishing Ltd*

Electron Energy-Loss (EEL) Spectroscopy Observations of Filled Carbon Nanotubes

A Seepujak, A Gutiérrez-Sosa, A J Harvey, U Bangert
Department of Physics, UMIST, PO Box 88, Manchester M60 1QD, UK

V D Blank, B A Kulnitskiy, D V Batov
Technological Institute for Superhard and Novel Carbon Materials, 7a Centralnaya Street, 142092 Troitsk, Moskow, Russia

Abstract. EEL intensity maps have allowed the distribution of $CaCO_3$ and SiO_2 in a filled MWCNT (multi-wall carbon nanotube) to be determined. The spectral weight of nitrogen in the $CaCO_3$ and SiO_2 fillings provides evidence of the inhibition of nitrogen incorporation into the MWCNT in the vicinity of the $CaCO_3$. Fitting a sum-of-Gaussian function to low loss EEL spectra, required fitting of a discrete resonance of energy ~4 eV. This resonance is attributed to varying levels of nitrogen incorporation and/or differing interfacial structures in the filling.

1. Introduction

The solid state properties of semiconductor devices rely on electronic transitions between interband states of metal-oxide-semiconductor media. With semiconducting applications based on MWCNT systems predicted [1], spatially-resolved EEL spectroscopy represents a powerful method for probing electronic transitions of metal filled MWCNTs.

Intraband transitions from 1s core levels to unoccupied π^* and σ^* orbitals are typically probed in core loss EEL spectroscopy of carbon nanotubes, allowing local composition changes to be scrutinised. In contrast, low loss EEL spectroscopy (0-10 eV) probes the joint density of states, with momenta transfers $q \rightarrow 0$ exciting optically allowed transitions. Excitations explicitly identified in the low loss EEL regime include collective volume and surface modes [2] attributed to the electric dipole $\pi \rightarrow \pi^*$ transition. Additionally, boundary conditions [3] predict an additional collective mode in the low loss EEL regime, excited by an electron beam passing close to an interface between two media. The present contribution utilises a Gatan Enfina EEL system to correlate collective interface resonances in the low loss EEL regime, to local compositional changes of a MWCNT filling, as determined by core EEL losses.

2. Experimental

The MWCNTs were formed in a water-cooled HIP apparatus with a carrier gas containing pure nitrogen at a pressure of 70 MPa. Details of the MWCNT production are presented elsewhere [4]. EEL spectra were acquired in a VG-HB601 STEM (scanning transmission

328

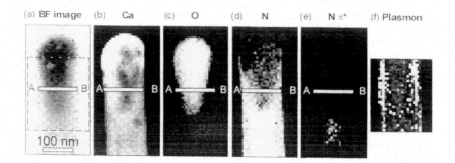

(a) BF image   (b) Ca   (c) O   (d) N   (e) N π^* (f) Plasmon

Figure 1. (a) STEM BF image of the MWCNT studied. The horizontal line AB shows the location of extracted EEL spectra. (b)-(e) EEL intensity maps showing core-losses in the entire area of Figure 1(a). White/black represent the greatest/least intensity. (b) Map of the calcium $L2$-edge. (c) Map of the oxygen K-edge. (d) Map of the nitrogen K-edge. (e) Map of the nitrogen π^* resonance. (f) EEL intensity map, showing low losses in the dashed square defined in Figure 1(a), in the energy range 3.4-3.9 eV

electron microscope) fitted with a Gatan Enfina spectrometer. The Gatan Enfina system enables a two-dimensional (x,y) EEL spectral-array to be acquired from a specimen in a single reading. Providing the array size is small, the reduced acquisition time, compared to the conventional method of acquiring individual EEL spectra, significantly reduces the likelihood of radiation damage and specimen drift.

Interface modes have a very low intensity, compared to the intensity of volume and surface modes, and are are typically obscured by the ZLP (zero-loss peak). The STEM possesses an energy resolution of down to 0.30 eV, given by the FWHM of the ZLP, thus ensuring a sharp cutoff of the ZLP tail. Objective / collector aperture semi-angle of 5.9 / 3.4 mrad were used, providing a compromise between obtaining the best conditions and best statistics.

3. Results and Discussion

One low loss EEL spectral-array, and one core loss EEL spectral-array were acquired. The low loss EEL spectral-array was acquired from the area defined by the dashed square in Figure 1(a), which was then divided into a specified number of pixels. At the centre of each pixel, 10 EEL spectra were acquired and recorded, thus maximising count statistics. Each spectrum was acquired with the maximum dwell-time, without saturating the detector array. A dispersion of 0.01 eV over the detector array composed of 1024 CCDs, provided a energy range of ~10 eV. The core loss EEL spectral-array was acquired by an identical method, apart from spectra being acquired from the entire area of Figure 1(a). The dispersion of 0.3 eV allowed scrutiny of an energy range from before the onset of the carbon K-edge to beyond the O K-edge.

In order to generate the EEL intensity map of a given element, the background before the onset of the core loss edge of the element, was extracted from each pixel of the EEL spectral-array using a power law function. An energy window was then specified.

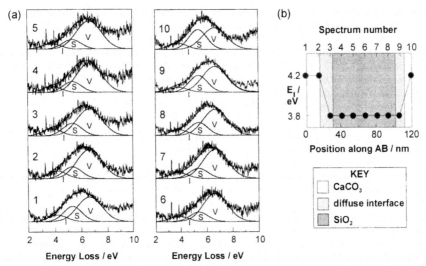

Figure 2. (a) Low loss EEL spectra. Spectra 1 to 10 represent consecutive and equidistant measurements along the line AB, identified in Figure 1(a). Spectra have been smoothed using a function with w=0.05 eV. I, S and V are assigned to the interface, surface and volume modes, respectively. (b) Variation of the interface mode central resonance energy E_I, with position along AB.

from the onset of the core-loss edge to around the end of the near-edge structure. Utilising a greyscale for the number of counts over this energy window, for each pixel of the EEL spectral-array, formed the element-specific EEL intensity map showing the spectral weight of the given element. EDXS (energy dispersive X-ray spectrometry) data (not shown) and the EEL intensity maps illustrated in Figures 1(b) and 1(c), show the MWCNT filling is comprised of SiO_2 diffusely encapsulated by a $CaCO_3$ coating. Core loss spectra of the carbon edge show that the $CaCO_3$ is itself surrounded by a graphitic coating. Comparing the difference in the spectral weight of nitrogen in various regions of the filling (Figure 1(d)), suggests $CaCO_3$ and SiO_2 have played significantly different roles in the incorporation of nitrogen into the present MWCNT. The role of deposits at MWCNTs tips, in the catalytic growth mechanism, has been described in literature [5]. The complete absence of nitrogen in the vicinity of the $CaCO_3$ deposit, suggests $CaCO_3$ has inhibited the incorporation of nitrogen into the graphene layers. The EEL intensity map calculated from assigning an energy widow to the central energy of the nitrogen $1s \rightarrow \pi^*$ transition (Figure 1(e)) confirms the nitrogen incorporation as substitutional. This map demonstrates the π^* cross section is greatest when the direction of the primary beam is oriented perpendicular to the graphene planes. Such an orientation dependent cross section is consistent with pyridinic substitution of nitrogen into the graphitic hexagonal rings [4].

Low loss EEL spectral arrays can similarly be used to show the spatial distribution of interface modes. Interface modes generally appear in the 2-4 eV energy region conventionally utilised to extract the ZLP from experimental spectra. It was therefore critical to extract the ZLP without introducing artefacts, or inadvertently removing structure in this energy region, both of which would produce meaningless trends. A power law function generally providing an adequate signal-to-noise ratio, and artefact free spectra, at energies above ~2 eV. Spectra were deconvolved with a sum-of-Gaussian function, which minimised the normalised residual χ^2. The procedure for deducing the energy of surface and volume modes is described in ref. [2].

Figure 1(f) shows the distribution of a feature in the energy range 3.4-3.9 eV. The intensity of the feature, greatest around the edge of the filling and in the diffuse $CaCO_3$ / SiO_2 region, demonstrates the feature as originating from interfacial effects. The origin is confirmed by Figures 2(a) and 2(b), illustrating a change in energy of the interface feature occurs ~30 nm and ~90 nm along AB, corresponding to $CaCO_3$ / SiO_2 interfaces. The interface feature is most intense, can be distinguished most clearly, at position 3, showing however intense the surface mode is, the fitting function will always require a peak at ~ 4eV.

Literature provides various studies of the interface feature (ref. [5] and refs. therein), which is not observed in pristine MWCNTs. The nature of the feature as being a $\omega_{\pm}$ split surface resonance can be explicitly precluded, due to the mutual coupling suppression of surface modes propagating on a graphite sample [6]. Moreau et al. [8] describe the modelling of interface modes excited at an abrupt planar Si / SiO_2 interface. Utilising a relativistic correction to the model, the interface mode was seen to shift in energy in the vicinity of the interface. No shifting was observed using a non-relativistic approach. The present deconvolved spectra clearly confirm the relativistic prediction of energy downshifting, in the vicinity of an interface. Measurements at a sharp and planer $CaCO_3$ / SiO_2 interface may provide a more gradual change in energy that shown in Figure 2(b).

In summary, modifying the electronic properties of MWCNT can be achieved by the substitution of electron donors or acceptors. EEL intensity maps suggest $CaCO_3$ has inhibited the incorporation of nitrogen into the graphene layers. The excitation of an energy-shifting interface mode in a filled MWCNT, consistent with relativistic predictions, has been observed for the first time.

References

[1]   Dresselhaus M S and Dresselhaus G 1996 Science of Fullerenes and Carbon Nanotubes (London : Academic Press)
[2]   Seepujak A et al. Interpretation of Electron Energy Loss (EEL) Spectra of Multi-Wall Carbon Nanotubes (MWCNTs) In The 2-10 eV Regime. Accepted for publication in Carbon
[3]   Ritchie R H and Howie A 1988 Philos. Mag. A 58(5) 753-767
[4]   Blank V D et al. 2003 Diam. Relat. Mater. 12 864-869
[5]   Blank V D et al. 2002 J Appl Phys 91(3) 1657-1660
[6]   Seepujak A et al. Mutual Decoupling and Dispersion of π - Surface Modes in Electron Energy-Loss (EEL) Spectra of Multi-Wall Carbon Nanotubes. In press
[7]   Raether H 1965 Spring. Tract. Mod. Phys. 38 84-157
[8]   Moreau P et al. 1997 Phys. Rev. B 56 (11) 6774-6780

Inst. Phys. Conf. Ser. No 179: Section 8
Paper presented at Electron Microscopy and Analysis Group Conf. EMAG2003, Oxford, 2003
©*2003 IOP Publishing Ltd*

Indirect Transmission Electron Microscopy; Aberration Measurement and Compensation and Exit Wave Reconstruction

A I Kirkland and R R Meyer
Department of Materials, Parks Road, Oxford, OX1 3PH

Abstract. Improvements in instrumentation and image handling techniques mean the indirect reconstruction is now realising the promise it has long offered. This approach recovers the phase and modulus of the specimen exit plane wavefunction using datasets comprising either through focal or tilt-azimuth series of images. In order to achieve this it is necessary to measure the objective lens aberrations to high accuracy. This paper will review progress in implementing this approach and will present recent reconstructions from a range of materials.

1. Introduction

High Resolution Transmission Electron Microscopy (HRTEM) is now firmly established as one of the most important tools available for studies of the local microstructure and chemistry of many materials at close to 0.1nm resolution. However, conventional HRTEM records only the image intensity, confused by the aberrations present in the objective lens and resolution limited by the imperfect spatial and temporal coherence of the illumination making it difficult to obtain quantitative structural data. Indirect reconstruction offers one solution to this by enabling the recovery of the specimen exit plane wavefunction providing an attractive (but necessarily indirect) route to quantitative structural information at interpretable resolutions beyond those that can be routinely achieved by conventional imaging [1,2].The datasets required for this approach comprise a set of conventional HRTEM images recorded at either varying defocus levels or with different illumination tilt directions. These provide differently aberrated images containing independent information about the exit plane wavefunction which can subsequently be computationally recovered from the overdetermined data.

The experimental realisation of this approach had been limited, until recently by several factors, but a number of technical developments have changed this situation, making off-line reconstruction viable on a routine basis. The development of field-emission electron sources energy-filtered imaging systems and characterized CCD detectors have enabled the acquisition of high quality experimental data. More recently there has also been considerable success in sophisticated direct correction of the objective lens aberrations [3,4] which also brings benefits to indirect reconstruction. Independently of these instrumental developments, substantial progress has been made on the theoretical and computational problems of eliminating the non-linear image intensity components [5] and in the fully automated measurement of the aberration coefficients [6].

There are two acquisition geometries that have been used for indirect reconstruction. The first involves the acquisition of a suitable focal series of images enabling recovery of the exit plane wavefunction to a resolution determined by the effects of limited temporal coherence. Alternatively tilted illumination can be exploited wherein if the resolution limit for axial illumination is limited to a scattering angle α, then tilting the illumination by the same angle allows beams scattered by up to 2α in the opposite direction to pass through the aperture. However, beams at different azimuths are eliminated and hence it is necessary to combine information from several images recorded at different tilt azimuths. This approach has, in practice achieved significant resolution enhancement for a range of materials and in theory using larger illumination tilts the resolution can even be extended to beyond twice the conventional information limit. Crucially, both of these approaches, together with others such as holography and ptychography, have in common the particular advantage of recovering both the phase and modulus of the specimen exit plane wave function, free from artifacts due to the objective lens aberrations. This additional data allows structural inferences about the specimen to be made by comparing experimental and simulated wave functions for trial structures with only one real unknown experimental parameter (the specimen thickness), in contrast to the multitude of unknown parameters involved in an equivalent comparison using conventional (aberrated) images.

2. Exit wave reconstruction

In the linear imaging approximation the Fourier transform of the recorded image contrast is given by

$$c(k) = \psi_i(k) + \psi_i^*(-k) \tag{1}$$

The Fourier transform of the image wave, ψ_i and the object wave, ψ_o leaving the specimen are related by the wave aberration function, γ (see later) and hence

$$\psi_i(k) = \psi_o(k)w(k) \tag{2}$$

where $w(k) = \exp\{-i\gamma(k)\}$. Thus the relationship between the object wave, the wave aberration function and the Fourier transform of the recorded image contrast is given (in the presence of observational noise n) is given as

$$c(k) = \psi_o(k)w(k) + \psi_o^*(-k)w(-k) + n(k) \tag{3}$$

The essence of all reconstructions is therefore to find an estimate, ψ_o' of the object wave given an experimentally recorded set of image contrast transforms and a measurement of their transfer functions, $w(k)$. With data available from several differently aberrated images the Wiener least squares estimate can be used to define an *optimum* solution for ψ_o'. This provides an estimate in the form of a weighted superposition of the image transforms,

$$\psi_o'(k) = \Sigma_i r_i(k)c_i(k) \tag{4}$$

in which the multipliers r_i, called *restoring filters*, depend on the transfer functions for the set of images recorded according to

$$r_i = \frac{(W(-k)w_i^*(k) - C^*(k)w_i(-k)}{W(-k)W(k) - |C(k)|^2 + \upsilon(k)} \tag{5}$$

in which $W(k) = \Sigma_i |w_i(k)|^2$, $C(k) = \Sigma_i w_i(k)w_i(-k)$, and $\upsilon(k)$ is the noise to object power ratio. In the final step of the reconstruction process the object exit plane wavefunction itself is obtained simply by inverse transformation of this estimate. It is clear from equation (5) that this approach is only viable if the imaging parameters for each image and hence the wave aberration functions are known.

3. Aberration measurement

The wave aberration function describes the distance, $W(u, v)$ (where u, v is position in the diffraction plane) between an ideal spherically diffracted wavefront and the real diffracted wavefront in the presence of lens aberrations. These aberrations can be conveniently parameterised in the wave aberration function written in polar form with $w = u + iv = \kappa e^{i\phi}$, $A_n = |A_n| e^{i\alpha_n}$ and $B_n = |B_n| e^{i\beta_n}$ as

$$
\begin{aligned}
W(k, \phi) = & |A_0| \theta \cos(\phi - \alpha_0) \\
& + \tfrac{1}{2} |A_1| \theta^2 \cos 2(\phi - \alpha_1) + \tfrac{1}{2} C_1 \theta^2 \\
& + \tfrac{1}{3} |A_2| \theta^3 \cos 3(\phi - \alpha_2) + \tfrac{1}{3} |B_2| \theta^3 \cos(\phi - \beta_2) \\
& + \tfrac{1}{4} |A_3| \theta^4 \cos 4(\phi - \alpha_3) + \tfrac{1}{4} |B_3| \theta^4 \cos 2(\phi - \beta_3) + \tfrac{1}{4} C_3 \theta^4
\end{aligned}
\tag{6}
$$

in which the azimuthal and radial dependence of the coefficients is clear with terms containing A_1, A_2 and A_3 representing two, three and four fold astigmatism, those containing B_2 and B_3 being the axial coma and axial star and the circularly symmetric terms, C_1 and C_3 being the defocus and spherical aberration. Few of the coefficients in equation (6) are observable under axial illumination and their determination therefore requires measurements taken as a function of known injected beam tilts. Observations that have been made include tilt induced image shifts [7] or changes in C_1 and A_1 (measured from diffractograms) [8] or the orientation of diffractograms [9]. However all of these suffer from practical limitations; shift measurements are inaccurate due to specimen drift and distorted correlation peaks between tilted and untilted images whereas diffractogram measurements require thin amorphous sample areas. For this reason a more general approach has been developed using a Phase Correlation Function (PCF) to determine relative focus levels and image registrations and a Phase Contrast Index (PCI) to calculate the absolute values of A_1 and C_1 (Figure 1) [6]. This approach is optimally implemented using a combined tilt / defocus dataset comprising short focal series recorded at several tilt azimuths. From any of the above sets of measurements the aberration coefficients can be reliably determined by simple least squares fitting of the parameters to be determined to the observations available.

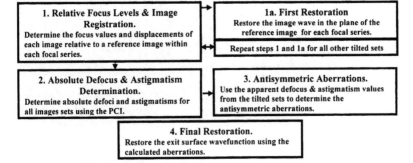

Figure 1. Outline flowchart of the steps involved in aberration determination using a general PCF / PCI approach.

4. Experimental applications

Our first example describes an experimental tilt azimuth series reconstruction which has been used to refine an inorganic perovskite structure. Perovskites of general formula ABO_3 are known for a large variety of metallic cations, A and B and often form complicated layered structures when the oxygen content is increased. This composition dependence allows the engineering of materials with tailored properties and therefore an understanding of their structure/composition relationship is important. In this regard perovskites find applications as dielectric materials, colossal magneto-resistive materials and the perovskite La,Ba_2CuO_4 was the first high temperature oxide superconductor to be discovered. The compound, $Nd_4SrTi_5O_{17}$ investigated here is composed of slabs of perovskite sliced along the [110] plane in which the two excess oxygen anions are accommodated at the interface region between two slabs. This material shows the dramatic effect of the partial substitution of Nd by Sr in the [010] diffraction pattern where the systematic (h0l) rows with h=2n+1 are replaced by continuous oscillating streaks, an effect not present in the parent ternary structure, $Nd_5Ti_5O_{17}$.

The exit plane wavefunction for this material reconstructed from a tilt azimuth series is shown in Figure 2. In this example the aberration coefficients were determined using the PCF/ PCI approach due to a lack of any amorphous material from a dataset recorded with the geometry described earlier. The reconstruction shows improved resolution in both phase and modulus and an enhanced sensitivity to the weakly scattering oxygen sublattice in the reconstructed phase. This allows weak contrast from the O anions to be detected in the reconstructed phase between the Ti sites which are bridged by O. The reconstruction also reveals local distortions in $Nd_4SrTi_5O_{17}$ in which the Nd cations at the outside of the perovskite slabs are displaced by alternate amounts. This improved data has led to the development of a refined model in which alternate displacements of the outer Nd atoms create two possible ways of stacking the slabs. In $Nd_5Ti_5O_{17}$ this stacking always occurs in the same direction but in the $Nd_4SrTi_5O_{17}$ the stacking changes direction (Figure 3).

Figure 2. (a) Modulus and (b) Phase of a tilt azimuth reconstruction of $Nd_4SrTi_5O_{17}$ projected along the [010] direction. showing clear resolution of the cation lattice in both phase and modulus with Nd and Ti sites distinguished. The reconstructed phase also shows the anion sublattice as weak contrast between the Ti cations

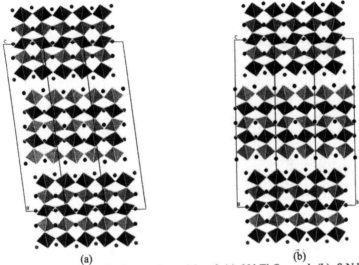

Figure 3. Structural models of (a) $Nd_5Ti_5O_{17}$ and (b) $SrNd_4Ti_5O_{17}$ projected along the [010] direction showing the variation in direction of the stacking sequence between the perovskite slabs.

Figure 4 shows the reconstructed modulus and phase of the complex oxide $K_{0.1}WO_3$ projected along the [001] direction, recovered to a resolution better than 0.11nm. The reconstructed modulus shows the positions of the cation columns in projection with negative contrast at high resolution and importantly remains directly interpretable to a greater specimen thickness than the axial image. We also note that in general the reconstructed modulus is less sensitive to errors in the determination of individual image aberrations than the phase to the extent that even where these are not known to high accuracy it is still possible to identify the basic cation lattice sites in the modulus.

Compared to the modulus, the reconstructed phase shows a more complex contrast. In addition to the strong positive contrast (white, corresponding to a phase advance) located at the cation sites and corresponding directly to the strong negative (black) contrast in the modulus there is additional weak contrast at positions between the cations corresponding to a direct imaging of the anion sublattice (inset Figure 4).

The modulus and phase show also almost constant contrast over the entire crystal suggesting that there is no rapid thickness variation. and the tunnel sites containing the K cations appear brighter in the reconstructed phase than the W sites except in a very thin region at the crystal edge where this is reversed. Multislice simulations of the reconstructed exit wave suggest that this behaviour is consistent with a predominantly flat crystal with a thickness greater than 4 unit cells. Interestingly however, the modulus shows local contrast variations (of *ca.* 10%) at selected "tunnel" sites in which the K are located (Figure 4) which cannot, given the above, be attributed to thickness variations These local variations thus provide evidence for variable tunnel occupancy and / or partial substitution of K by W.

336

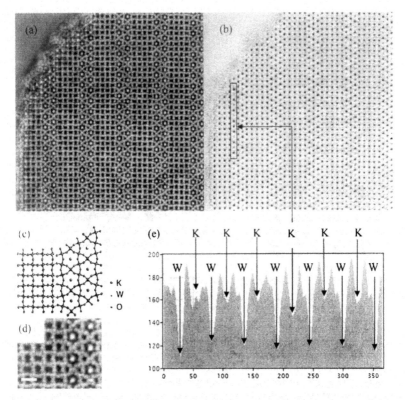

Figure 4. (a) Phase and (b) Modulus reconstructed from a 20 member focal series of the complex non stoichiometric oxide $K_{0.1}WO_3$ together with (c) structural model and (d) an enlargement of the reconstructed phase (d). (e) Line profile obtained from in (b) showing the variation in tunnel site occupancy with a local high occupancy site indicated.

References

[1] Coene W, Janssen G, Op de Beeck M and van Dyck D 1992 Phys. Rev. Lett. **69**, 3743

[2] Kirkland A I, Saxton W O and Chand G 1997 J. Electron Microscopy **1**, 11

[3] Haider M, Rose H, Uhlemann S, Schwan, E Kabius B and Urban K 1998 Ultramicroscopy **75**, 53

[4] Lentzen M, Jahnen B, Jia C L, Thust A, Tillmann K and Urban K 2002 Ultramicroscopy **92**, 233

[5] Coene W, Thust A, Op de Beeck M and Van Dyck D 1996 Ultramicroscopy **64**, 109

[6] Meyer R R, Kirkland A I and Saxton W O 2002 Ultramicroscopy **92**, 89

[7] Koster A A J, van den Bos A and van der Mast K D 1987 Ultramicroscopy **21**, 209

[8] Zemlin F, Weiss K, Schiske P, Kunath K and Herrmann K-H 1978 Ultramicroscopy **3**, 49

[9] Saxton W O 2000 Ultramicroscopy **81**, 41

Inst. Phys. Conf. Ser. No 179: Section 8
Paper presented at Electron Microscopy and Analysis Group Conf. EMAG2003, Oxford, 2003
©2003 IOP Publishing Ltd

Moveable aperture lensless microscopy: a novel phase retrieval algorithm

H M L Faulkner and J M Rodenburg

Department of Electronic and Electrical Engineering, University of Sheffield, UK

Abstract. We discuss a new method of phase retrieval which uses measured diffraction patterns created with a movable aperture assumed to interact multiplicatively with the post-specimen wavefunction. The method combines several useful properties of iterative phase retrieval with the ease of measuring a small number of diffraction patterns. It allows the possibility of lensless microscopy, since a focussed image and beam are not required for the retrieval. The method is demonstrated in simulation and compared with other iterative methods.

1. Introduction

Interest has recently been renewed in the use of the Gerchberg-Saxton [1] and Fienup [2] iterative algorithms to solve non-periodic phase problems in microscopy. The advantage of this is that the diffraction pattern, used as input to these algorithms, can be measured to very large angles, limited only by the size of the detector. For some important types of short wavelength radiation, such as electrons, protons, X-rays, very good quality lenses are not available. Even for good quality lenses, the object may not scatter strongly, or it may only alter the phase of the transmitted wave so the that interpretation of the conventional image is difficult. However if we can record the diffracted intensity to a large angle, and then employ a suitable algorithm to solve for the phase of the diffraction pattern, there is no requirement for a lens at all. This is because the back Fourier transform of the complex-valued diffraction pattern is a perfect representation of the exit wave-field at a resolution determined only by the size (and pixel resolution) of our diffraction plane detector.

In addition, iterative methods are numerically stable and robust given noisy data. They do not suffer from the numerical problems and inaccuracies found in some more direct algebraic methods, such as the need to accurately estimate the derivative of the wavefunction intensity.

However, there are limitations to the success of current iterative methods. The major problem is the common requirement for the specimen to be defined by a distinct support. Experimentally, it is often difficult to align a support-defining aperture accurately over the particular feature of interest, especially at very high magnifications. If an algorithm that does not require a known support is used, it will usually require accurate measurement of one or more images in real space instead. This poses additional experimental difficulties.

We show here that these problems can be solved by allowing several measurements performed with the probe shifted laterally by a known amount to a number of different positions. The method can be effectively used with a moveable sharp aperture. We make one assumption: that the aperture interacts multiplicatively with the object transmission function. For strong, dynamical scattering in thick objects, particularly crystals in electron microscopy, this is not a good approximation. However, our method may nevertheless be useful for a wide range of thin, though possibly strongly scattering, samples.

Probe movement, or the act of shifting a sharp aperture in the specimen plane, gives much more information than the classic Fourier domain phase problem. However as far as iterative methods are concerned we must tackle certain complications: as we move the probe, we illuminate a different region of specimen, and so the variables we are solving for progressively change. The technique we develop applies to all forms of scanning transmission microscopy, although our principal interest is to apply the method experimentally to electron microscopy, where very substantial gains in resolution should be possible.

2. Conventional iterative algorithms

Most iterative phase retrieval algorithms have similar structure. They require images in two or more different planes (for simplicity we restrict ourselves to considering two planes), related by some transformation. At each plane they impose a constraint on the wavefunction, before transforming to the next plane. Example types of information and constraints used are summarised in Table 2.

In the Gerchberg Saxton method, intensity in both the real and diffraction plane is used as input information for the algorithm, and the transform relating these is the Fourier transform. The Fienup method requires only the support shape in the real plane rather than the entire intensity. However to satisfy the Nyquist criterion the support must cover at least half of the image area and must often be much larger to allow practical recovery of the wavefunction. Thus the amount of the object that may be examined using this method is constrained. The Fienup method is also prone to symmetry problems in the phase retrieval and it is necessary to use a known support of a shape that will remove these problems.

Note that it is not essential to use the Fourier transform to relate the data from the two planes in an iterative algorithm. One approach that uses a different transform is the through focal series algorithm of Allen et al. [4], which uses the free space propagator to relate data at different defocii. Our proposed new algorithm uses yet another transform.

Algorithm	Plane 1	Plane 2	Constraint 1	Constraint 2	Transform
Gerchberg-Saxton	Image	Diffraction Patt.	Intensity	Intensity	FFT
Fienup	Image with support	Diffraction Patt.	Support	Intensity	FFT
Through Focal Series	Image	Image	Intensity	Intensity	Free space propagator
Moving Aperture	Diffraction Pattern	Diffraction Pattern	Intensity	Intensity	Aperture shift transform

Table 1. Comparison of iterative algorithms.

3. A new algorithm

A new algorithm is proposed which avoids the problems faced by the previously discussed iterative methods, by using diffraction data. We assume that the intensity of the diffraction pattern is available to us at one or more known probe positions, and that the probe intensity is also known. The algorithm, illustrated in Figure 1 works as follows:

(i) Start with a guess at the object function
(ii) Multiply the current guess at the object function by the probe at the current position, producing the exit wavefunction for that probe position.
(iii) Fourier transform to obtain the guessed diffraction pattern for that probe position.
(iv) Correct the intensities of the guessed diffraction pattern to the known values.
(v) Inverse Fourier transform back to real space to obtain a new and improved guess at the exit wavefunction.

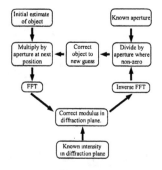

Figure 1. New phase retrieval algorithm.

(vi) Where the probe for that probe position is relatively large, divide the exit wave by the probe, to obtain an improved guess at the object wavefunction for those regions.

(vii) Update the guessed wavefunction in those regions only

(viii) Repeat ii – vii until the sum squared error (SSE) as measured in the diffraction plane is sufficiently small.

When the SSE is small, it means that the object function has been correctly found in the region covered by the different probes. We have no information about the other areas of the object, so do not obtain useful results for those regions.

This algorithm does not require any sort of focussed illumination. Therefore, the probe forming lens may be discarded, as it is not necessary to focus the beam. All that is required is a source of coherent radiation, an aperture, and a method of moving the aperture a known distance across the sample. This last may be achieved with the use of suitable piezo-electric devices. As a result, the equipment requirements of this algorithm are significantly less than those of the conventional iterative algorithms. This method has the potential to revolutionise microscopy and allow truly lensless imaging over a large field of view.

4. Example simulation

Fig. 2(a)-(d) show the test data used for a demonstration of the new algorithm. The test object used is shown in intensity (a) and phase (b). The simple tophat function used to represent the aperture is shown in (c) and (d). These are combined at two different aperture positions to produce the diffraction patterns shown in intensity in Fig 2(e) and (f). These diffraction patterns and the known aperture shape are used as input data for the algorithm.

The retrieval result is shown in Fig 2(g) and (h). The algorithm in this case stagnated after 698 iterations with an SSE in the diffraction plane of $9.44346e - 06$. In the image plane the sum squared error was 0.000401491 in the region of interest, which is that covered by the combined apertures at different positions. As is seen, the object structure is recovered to high accuracy in the area of interest.

The new algorithm was compared with other iterative algorithms using the same simulated data. The Gerchberg Saxton was the most successful in attaining the lowest error $(9.54075e - 11)$ in the fewest iterations (111). However this algorithm cannot be implemented successfully in a lensless configuration because it requires measurement of the focussed image, which is not available if we dispense with the electron lens.

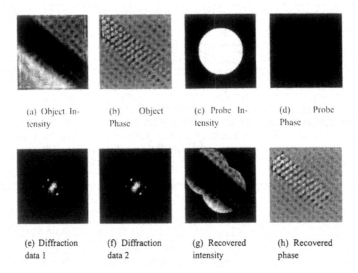

(a) Object Intensity

(b) Object Phase

(c) Probe Intensity

(d) Probe Phase

(e) Diffraction data 1

(f) Diffraction data 2

(g) Recovered intensity

(h) Recovered phase

Figure 2. Simulation of new phase retrieval method.

The Fienup algorithm was the worst performer, stagnating with an SSE of 0.000104994 after 510 iterations. In addition it naturally retrieved a smaller region of interest, whereas the moving aperture algorithm may be used to retrieve part or all of the object, as required, by varying the number of position of aperture positions used.

5. Conclusions

We have demonstrated a new algorithm for retrieval of an object function, based on knowledge of a simple aperture, and diffraction pattern measurements at two or more different aperture positions. The method retrieves the object structure with good levels of accuracy, and is expected to be experimentally simple to apply compared to other iterative techniques.

There are many potential future directions of this research. These include refining the algorithm further by optimising the number and location of the aperture positions used. Other approaches, including using averaging techniques to implement a more parallel algorithm are also under investigation. The relatively simple apparatus required for these techniques has the potential to revolutionise electron microscopy, and permit 'lensless' imaging of three dimensional structure at the atomic level.

[1] R. W. Gerchberg and W. O. Saxton. A practical algorithm for the determination of phase from image and diffraction plane pictures. *Optik*, 35(2):237–246, 1972.
[2] J. R. Fienup. Phase retrieval algorithms, a comparison. *Applied Optics*, 21(15):2758–2769, 1982.
[3] U. Weierstall, Q. Chen, J. C. H. Spence, M. R. Howells, M. Isaacson, and R. R. Panepucci. Image reconstruction from electron and X-ray diffraction patterns using iterative algorithms: experiment and simulation. *Ultramicroscopy*, 90:171–195, 2002.
[4] L. J. Allen, H. M. L. Faulkner, M. P. Oxley, K. A. Nugent, and D. Paganin. Phase retrieval from images in the presence of first-order vortices. *Phys. Rev. E*, 63:037602, 2001.

Inst. Phys. Conf. Ser. No 179: Section 8
Paper presented at Electron Microscopy and Analysis Group Conf. EMAG2003, Oxford, 2003
©2003 IOP Publishing Ltd

Wavelet image analysis in aberration-corrected scanning transmission electron microscopy

QM Ramasse[†] and AL Bleloch[‡]

[†] Cavendish Laboratory, University of Cambridge, Cambridge CB3 0HE
[‡] SuperSTEM, Daresbury Laboratory, Warrington WA4 4AD

Abstract. The correction of aberrations on the VG HB501 instrument of the SuperSTEM facility at Daresbury Laboratory relies on the numerical analysis of a series of electron Ronchigrams of an amorphous specimen. In particular, cross-correlation techniques are used during the auto-tuning procedure to measure the shifts between images recorded at different tilt angles, hence deducing the experimental values of the aberration to be corrected. The use of wavelets and a newly defined wavelet cross-correlation function provide the basis for a powerful new method for image registration that could simplify and systematize the tuning procedure. The method was tested on simulated Ronchigrams of amorphous samples and used in conjunction with edge detection algorithms to measure the movements of fringes observed in Ronchigrams of crystalline samples.

1. Introduction

The advent of aberration correctors has opened up new horizons in electron microscopy. With all optical aberrations of the magnetic lenses used in a STEM corrected up to fifth order, the path to $0.5 \mathring{A}$ resolution seems now clear of any major obstacle [1]. The tremendous progress made in the instrumentation side of aberration correction calls for a parallel effort in the image analysis techniques it uses. It was mentioned in earlier work [2] that wavelets could play an important role there. The localization properties of a wavelet transform as well as the large amount of information about the original image it encompasses in a single mathematical object make it particularly suitable indeed for the analysis of such complex micrographs as electron Ronchigrams. We investigate here the possibility of applying wavelet transforms to improve the image registration and feature tracking algorithms used for tuning the aberration-corrected $100\,keV$ VG HB501 electron microscope of the SuperSTEM facility.

2. From cross-correlation functions to wavelets

The first step of the auto-tuning procedure is an automated diagnosis of all aberrations present up to fifth order [2, 3]. The diagnosis is performed by numerical analysis of several electron Ronchigrams of an amorphous region of the specimen to be studied. An electron Ronchigram, or shadow image, can be thought of as a convergent beam electron diffraction (CBED) pattern recorded in the STEM diffraction plane with a large aperture size. As the aperture increases the diffraction discs observed in the CBED pattern start overlapping and, provided a highly coherent source is being used (as is the case in STEM), they interfere, giving rise to what is customarily called a Ronchigram. This "shadow image" presents rather remarkable features such as circles of infinite magnification and micro-diffraction fringes (in the case of a crystalline sample), described in more details elsewhere [2, 4, 5]. See figure 1a.

342

Shifting the electron probe position across the specimen induces shifts in the image. Cross-correlation techniques are applied to obtain a map of those displacements and the values of the aberration coefficients can subsequently be deduced.

The displacements and features one is looking at in this procedure are intrinsically very localized, observable in specific areas of the image and have particular orientation and scale. Cross-correlation and other Fourier-related numerical techniques are therefore not so well suited for such analysis due to their global nature: Fourier transforming an image is after all equivalent to globally filtering it with a periodic analyzing function, hence giving information averaged over the whole image. Wavelet transforms on the other hand can take advantage of rapidly decaying, hence very localized, analyzing functions. They can be described as "mathematical microscopes" [6] probing the signal at the position and scale the analyzing wavelet itself lives at.

Wavelets seemed a very adequate tool indeed for the analysis of electron micrographs. In order to apply it to our case an analogous procedure to image registration by means of usual cross-correlation functions had to be developed. Structural similarities with Fourier transforms allowed us to do so by defining a "*wavelet cross-correlation function*" whose properties, remarkably, are conserved from those of its non-wavelet counterpart. In other words, the wavelet cross-correlation function of two images shifted with respect to one-another will present a sharp peak at the corresponding displacement. But because wavelet transforms also accept scale and orientation parameters (they are 4-dimensional functions), rotations and dilations between the two analyzed images will also be detected, making the wavelet cross-correlation function a generally useful tool.

3. Fringe tracking algorithm

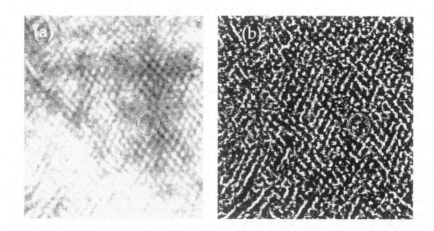

Figure 1. *(a).* Electron Ronchigram of a Silicon (crystalline) sample taken underfocus and showing micro-diffraction fringes. *(b).* The same Ronchigram after edge detection and non-maxima suppression. In both *(a)* and *(b)* the circle merely marks a position on the image.

Successful first tests of the wavelet image registration method on simulated Ronchigrams of an amorphous sample encouraged us to apply it to a more complex problem. The Ronchigrams obtained from a crystalline sample characteristically present many sets of fringes due to the coherent interference between the various Bragg-diffracted discs as they overlap in the diffraction plane. Varying the defocus at which the image is formed introduces a phase change and the observed fringes are shifted. Tracking their displacements with defocus could be a key to the diagnosis of aberrations from crystalline specimens and is an ideal field which to apply wavelet cross-correlations.

As illustrated in figure 1a the observed fringes in experimental images are not well resolved objects as far as computer feature detection is concerned. In order to improve the accuracy of our wavelet image registration scheme it was necessary to make the fringes stand out more noticeably from the rest of the image. An easy way to numerically "detect" fringes in an image is of course using the change of contrast between dark and bright fringes by means of an edge detection algorithm. The algorithm used to produce the results presented in figure 1b is a "simple" Canny edge detector combined with a non-maxima suppression filter. This algorithm, widely used and shown to be optimal for edges corrupted by white noise [7], was a straightforward and easy-to-implement choice to obtain results rapidly. However, wavelet multi-scale edge detectors as defined for example in [8] would probably be more adequate and versatile for the same reasons than exposed above, in particular if different scales of fringes are present in the same image and need to be separated out. We are presently investigating this.

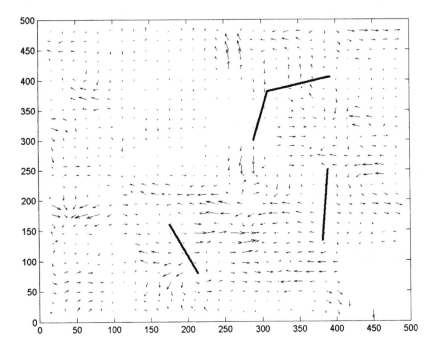

Figure 2. Map of the displacement field output by the algorithm. The vectors indicate displacements in pixels between two consecutive images of 512x512 pixels in a defocus series. The lines mark the locus of defocus-independent points in the image.

The same process was applied to a series of images recorded at increasing values of defocus. The core of an existing particle image velocimetry (PIV) routine [9] was modified to incorporate the wavelet cross-correlation and with this program pairs of consecutive images were then analyzed to measure the fringe displacements. A typical output displacement map thus obtained is presented in figure 2: the arrows represent a displacement measured in pixels between two consecutive images at the position (x, y) in pixels in the original image.

This preliminary result is rather encouraging as it seems to be consistent with the theory on a significant point. It can be shown (cf. [4]) that the intensity recorded in certain regions — actually a set of intersecting lines — of the diffraction plane is not sensitive to the variation of defocus. Or, as observed in figure 2 sets of arrows seem to diverge from / converge towards points where the displacements measured are null and the locus of these points looks quite convincingly like a set of intersecting lines. Some of these lines have been drawn by eye on figure 2.

4. Conclusion and future work

Due to their localization properties and their versatility wavelet transforms are an ideal tool for the digital processing of electron micrographs. We have developed and demonstrated a fringe tracking scheme making extensive use of wavelets. However we have not at present fully exploited the capabilities of wavelet transforms. The algorithms presented in this paper need to be systematized: enabling tracking of rotations or dilations of features, and optimized: better suited sub-algorithms and faster computer routines. It is hoped that this scheme will set the basis of a unifying algorithm for the diagnosis of aberrations in both amorphous and crystalline samples.

Acknowledgments

This work was sponsored by the EPSRC scholarship. Q. Ramasse is also grateful to the Cambridge European Trust and the SuperSTEM project for their support.

References

[1] Krivanek OL, Dellby N, Nellist PD, Murfitt MF and Szilagyi S 2003, *Ultramicroscopy* **96 (3–4)**, 229–227
[2] Lupini AR 2001, Ph.D. thesis, University of Cambridge, UK, p. 180
[3] Krivanek OL, Dellby N and Lupini AR 2000, *Auto-adjusting Charged-particle Probe-forming Apparatus* (U.S. Patent Application)
[4] Spence JCH and Cowley JM 1978, *Optik* **50 (2)**, 120–142
[5] Cowley JM 1986, *J. Electr. Micr. Tech.* **3**, 25–44
[6] Antoine J-P 1999, *Wavelets in Physics* van Den Berg JC ed., (Cambridge: CUP), pp. 9–75
[7] Canny JF 1986, *IEEE Trans. Pattern Anal. Machine Intell.* **8 (6)**, 679–698
[8] Mallat S and Zhong S. 1992, *IEEE Trans. Pattern Anal. Machine Intell.* **14 (7)**, 710–732
[9] mPIV: PIV toolbox in Matlab, $http://homepage2.nifty.com/nobuhito\_mori/mpiv/mpiv\_index.html$

Inst. Phys. Conf. Ser. No 179: Section 8
Paper presented at Electron Microscopy and Analysis Group Conf. EMAG2003, Oxford, 2003
©2003 IOP Publishing Ltd

Calculations of Limited Coherence for High Resolution Electron Microscopy

L Y Chang, R R Meyer and A I Kirkland

Department of Materials, Parks Road, Oxford, OX1 3PH

ABSTRACT: Improvements in instrumentation through the use of monochromators and field emission electron sources suggest that the ways in which the limited coherence effects are calculated must be re-examined. In this paper the effects of these approximations on the accuracy of the resultant image intensity are investigated as a function of both defocus and beam divergence.

1. Introduction

The effects of partial temporal and spatial coherence restrict the information limit of high resolution electron microscopes. In the general case, the resultant image intensity these effects can be calculated as an incoherent summation of the coherent image intensity arising from a suitable distribution of incident beam tilts and energies. In this approach for each coherent intensity, the exit wave has to be re-calculated using a multislice simulation with an appropriate incident beam direction and energy. Several approximations have been reported to make this calculation feasible. Almost universally, the effect of beam tilt and energy variation on the exit wave is neglected. In a further approximation the wave aberration function is locally replaced by its first order Taylor expansion, and a transmission cross coefficient for each pair of beams describes the transfer of their interference term to the image intensity. In order to avoid the computationally demanding summation over all pairs of beams, this approach is further simplified in the weak object approximation using envelope functions.

In this paper the effects of these approximations on the accuracy of the resultant image intensity are investigated as a function of defocus and beam divergence. In order to increase the efficiency of the integration over beam tilts and defoci, a Monte-Carlo integration method is used where the offsets are Gaussian random numbers with standard deviations given by focal spread and beam divergence, respectively. The accuracy of this method is then compared with the above approximations.

2. Theory

The effects of the finite electron source size, and the fluctuations in the electron energy, the objective lens current and the accelerating voltage lead to both the partial spatial and temporal coherence effects in TEM. If the illumination system of the electron microscope is treated as the incoherently filled effective source [1] with an intensity distribution $S(q)$, the partially spatial coherent image intensity can be calculated by incoherent summation of intensities from all the incident angles from the effective source. The energy spread of the electrons leaving the source, the fluctuations of the high voltage, as well as the variance of the objective lens, result in the chromatic aberration, and therefore in a focal spread [2]. This temporal partial coherence effect can also be treated incoherently and is also reasonable to assume that the spatial and temporal distribution of the source have no correlation with each other Under these conditions we can therefore write the image intensity as:

$$I(r) = \sum_q \sum_f S(q)F(f)\left|F^{-1}\left\{\psi(k;q)\exp(-i2\pi\chi(k,D+f,q))\right\}\right|^2 \tag{1}$$

Equation (1) describes the image intensity as the incoherent summation of the coherent image intensities at incident angle q and focus level f weighted by the spatial and temporal intensity distributions $S(q)$, $F(f)$, respectively. The individual coherent image intensities are given as the modulus square of the convolution of the specimen exit wave function $\psi(k,q)$ at an incident angle q and the objective lens aberration function $\chi(k,D+f,q)$ with corresponding beam tilt q and focus $D+f$. Computation of (1) therefore involves a separate dynamical exit wave function calculation for each q.

For a small beam tilt in the incident wave and for reasonable specimen thickness, the specimen exit wave function can be approximated to a single exit wave function at the mean incident angle. A further approximation can also be made by taking the Taylor expansion of the lens aberration to first order in q and f if the beam tilt q and focal spread f are small. Using these approximations the image intensity therefore be written as [3]:

$$I(k) = \sum_k \psi(k+k')\psi(k')T(k+k',k') \tag{2}$$

$$T(k_1,k_2) = \exp(-i2\pi[\chi(k_1) - \chi(k_2)])\exp(-\pi^2 q_0^2[\chi'_{k_1} - \chi'_{k_2}]^2)\exp\left(-\pi^2\Delta^2\left[\left.\frac{\partial\chi_k}{\partial D}\right|_{k=k_1} - \left.\frac{\partial\chi_k}{\partial D}\right|_{k=k_2}\right]^2\right) \tag{3}$$

The Fourier transform of the intensity is given by the integral over the pairs of diffracted beams and T is the transmission cross coefficient which accounts for the lens aberrations and limiting coherence effects. The second exponential term describes the spatial coherence for a Gaussian spatial intensity distribution and assuming an even lens aberration function. This term is a function of the square difference of the gradient of the aberration function at two spatial frequencies. It has the maximum transfer when the slopes of the aberration function are identical at k_1 and k_2. The second exponential term describes the temporal coherence, and is a function of square difference of the derivative of the aberration function with respect to the focus at two spatial frequencies. This function has maximum transfer when $|k_1| = |k_2|$.

If the further assumption of a weak scattering object is made, which only considers the interference between the transmitted beam and diffracted beam, the transmission cross coefficient T can be further simplified as [4]

$$T(k) = \exp(-i2\pi(k))\exp(-\pi^2 q_0^2 \chi'^2_k)\exp(-\pi^2\Delta^2\chi'^2_D) \tag{4}$$

Unlike the complicated behaviour of the two partial coherence functions in equation (3), both spatial and temporal coherence functions for weak objects decrease with increasing spatial frequencies, and therefore are often called the damping envelopes.

3. Discussion

The effect of beam tilt is normally ignored in high resolution image intensity simulations. Although this may be a reasonable approximation for certain materials and thicknesses for small beam tilt it has been shown [5] that images of materials with dynamically forbidden reflections are very sensitive to the beam tilt [6]. Accordingly the dependence of the exit wave functions on beam tilt must be investigated in order to correctly simulate the high resolution image intensities. To investigate this multislice calculations [7] of Si [110] were used to simulate a series of specimen exit wave functions calculated for incident plane waves with beam tilts of 0.4 and 2 mrad through the specimen which were then passed through the objective lens at the Scherzer defocus to obtain the image intensities. To investigate the effect of beam tilt on the exit wave, another set of intensities are simulated under the same condition but using a normal incident wave in the multislice calculation with the beam tilt included only in the wave aberration function.

The results show that for a beam tilt of 0.4mrad, the image intensities show almost no difference for thicknesses up to 543A whereas for a beam tilt of 2mrad, the image intensities show no difference for thicknesses up to 217Å.

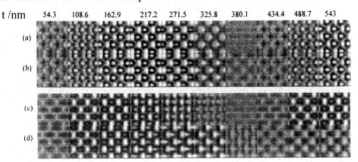

Fig.1, Si[110] image intensities at 300kV accelerating voltage, Cs=0.6 mm defocus =-34 nm. Beam tilt = (a), (b) 0.4mrad (c), (d) 2mrad. (a) and (c) include beam tilt in the multislice exit wave function calculation; (b) and (d) calculate the exit wave function with normal incidence but include beam tilt in the wave aberration function.

A direct comparison of the coherence functions calculated using the three approaches already described is shown in Figure 2.

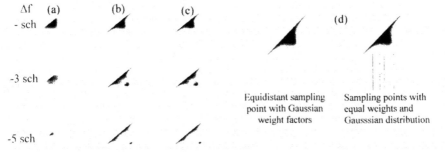

Figure 2. Moduli of coherence functions (300 kV, C_s =0.6 mm, beam divergence angle = 0.2 mrad and focal spread = 3 nm). (a) calculated from equation (4). (b) calculated from equation (3). (c) calculated from equation (1) using Gaussian profiles. (d) Schematic comparison of the Moduli of coherence functions calculated the two sampling schemes described in the text.

The use of the simplified transmission cross coefficient is fast (equation (4)) since the summation over pairs of beams is avoided, but is inaccurate for the interference between scattered beams since $T(k_1)T(k_2)$ is a poor approximation to $T(k_1,k_2)$ (Figure 2(a)). The use of the full transmission cross coefficient is a more accurate approach, but is computationally slow, since summation over all k' is required for all k. Numerical summation (Figure 2(c), equation (1)) gives identical results to those obtained from the full transmission cross coefficient (Figure 2(b)). However when an equal sampling distance is used the total sampling number is necessarily large to avoid artifacts. To overcome this the calculated coherent images can be averaged with equal weights, but using Gaussian random numbers with a standard deviation in the beam divergence angle and focal spread (Figure 2(d)). This allows using fewer sampling points to be used since any artifacts introduced are random rather than systematic. The accuracy of this Monte Carlo Summation (MCS) was investigated by comparison of the rms differences between the image intensities calculated using the full TCC and the MCS which confirms that the rms error is proportional to 1/M where M is the total number of sampling points. This approach can be further refined by using the MCS to calculate the non-linear terms and a standard envelope function to evaluate the linear terms which gives a further 3-4 x improvement in accuracy. In general, we have found that specimens containing heavy atoms have a larger error in the MCS calculation than those with light atoms and that the error shows little focus dependence.

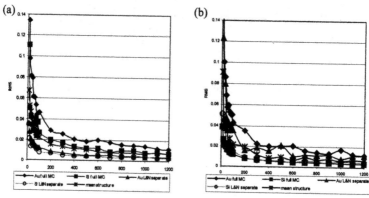

Figure 3. Comparison of rms errors for different calculation and sample types. The curves are calculated for the full MCS for Au[100] and Si[110] structures and for the refined MCS in which the unscattered beams are calculated using an envelope function and the MCS is only used for interference between scattered beams. The curve for the mean structure was calculated for an ideal material giving equal weight to all beams and is the basis for the calculations shown in Figure 2. Simulation conditions were 300kV, C_s =0.6mm, beam divergence =0.2mrad, focal spread =3nm. Defocus (a) –sch and (b) –9sch.

References
[1] Hopkins H H 1953 Proc. Roy. Soc. **A217**, 408
[2] Hanβen K J and Trepte L 1971 Optik **33**, 166
[3] Ishizuka K 1980 Ultramicroscopy **5**, 55
[4] Wade R H and Frank J 1977 Optik **49**, 81
[5] Gjonnes J and Moodie A F 1965 Acta Cryst. **19**, 65
[6] Smith D, Bursill L A and Wood G J 1985 Ultramicroscopy **16**, 19
[7] Cowley J M and Moodie A F 1957 Acta Cryst. **10**, 609

Inst. Phys. Conf. Ser. No 179: Section 8
Paper presented at Electron Microscopy and Analysis Group Conf. EMAG2003, Oxford, 2003
©2003 IOP Publishing Ltd

High tilt, automated, electron tomography of materials by incoherent (HAADF) and inelastic (EFTEM) scattering

M Weyland, L Laffont and P.A. Midgley
[1]Department of Materials Science and Metallurgy, University of Cambridge, Pembroke Street, Cambridge, CB2 3QZ

Abstract. High angle annular dark field (HAADF) scanning transmission electron microscopy (STEM) tomography and energy filtered transmission electron microscopy (EFTEM) tomography are promising techniques for the analysis of nanostructured materials. This paper describes the development of materials tomography methodology using modified holder designs and fully automated acquisition schemes. The details of the improvement achieved are demonstrated, potential pitfalls with EFTEM tomography examined and possible future directions discussed.

1. Introduction

The continuing pace of development in nanostructured materials is presenting serious challenges for transmission electron microscopy (TEM). As the length scales of nanostructured materials approaches, and in many cases goes below, the thickness of an average sample, the projection of objects inherent in TEM, becomes a serious limitation. One approach to tackling this problem is electron tomography in which the three dimensional (3D) structure of objects is reconstructed of from tilt series of 2D projections (images). Conventional electron tomography, based on bright field (BF) imaging, is problematic for materials specimens and recent research has concentrated on proving the validity of 3D reconstruction using alternatives such as HAADF STEM[1] images and elemental maps acquired using EFTEM [2, 3]. The next step is the optimisation of these experimental techniques and methodology, more specifically, in maximising the 3D information available by increasing the maximum tilt range and number of projections acquired.

2. Overcoming limitations to materials tomography

Defining the actual resolution achievable in a tomographic reconstruction is not straightforward and depends on many factors: the degree of structural complexity of the specimen, its orientation relative to the beam direction and the method of the acquisition[4]. However, in general, the resolution is improved by maximising the number of projections and the tilt range over which they are acquired. The former is limited by the degree of automation and the beam sensitivity of the specimen. The latter is limited by the mechanical design of the objective lens pole piece, specimen holder, microscope goniometer and the geometry of the specimen itself.

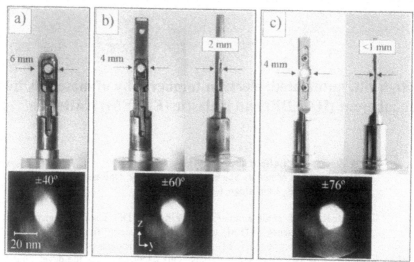

Figure 1. The effect of holder tilt range on the reconstruction of a magnetite crystal from a HAADF STEM tilt series; a single 2D slice (with the tilt range used), from the 3D reconstruction is shown for each holder. a) Philips single tilt holder, tilt limit ±42° due to pole touch. b) Modified EM400 single tilt holder, tilt limit ±60° due to holder shadowing. c) Fischione advanced tomography holder, tilt limit ±80° due goniometer limits.

The tilt range achievable in the Philips CM300 or a Tecnai F20 with a SuperTWIN lens, the microscopes used in this study, is ~±42° using a standard single tilt holder, shown in Fig.1 a). For initial studies into HAADF and EFTEM tomography, a holder from an EM400 was modified to allow a tilt range of ±60°, Fig. 1 b), giving an improved reconstruction. In collaboration with EA Fischione a new specimen holder, specifically for achieving high tilts with minimum shadowing in the SuperTWIN pole piece, was designed allowing tilts up to ±80° limited only by the design of the FEI compustage. The effect of tilt range is shown in Fig. 1, for a reconstruction of a magnetite (Fe_3O_4) crystal. The original STEM tilt series was acquired using the Fischione holder, and the series restricted in tilt range to reconstruct the slices in Fig 1 a) and b). As the tilt range is increased the reconstruction of the crystal becomes markedly less elongated in Z, the hexagonal outline of the crystal becomes increasingly apparent and its intensity more homogeneous.

The successful application of BF tomography in the biological sciences has been enabled by the development of advanced automation procedures[5]. Automation is key to acquiring an increased number of projections as it avoids the fatigue of manual acquisition, speeds up acquisition which in turn minimises beam damage. The adaptation of BF automation techniques for STEM HAADF imaging has been carried out for the Tecnai F20 in collaboration with FEI, as part of the "Xplore3D" software suite. The automation is based on pre-calibration of the specimen stage movements[6], which, although non-linear, are reproducible, and shifting the optic axis of the microscope to the eucentric axis of the stage. This is combined with cross-correlation tracking, for higher magnification acquisitions, and STEM auto-focusing, carried out by analysis of spatial frequencies across a range of defoci. In addition the software allows dynamic focusing, applying a focus ramp as the scanned image is acquired maximising depth of focus.

Tilt series	Number of projections (Range)	Increment	Total time	Number of scanned frames
Manual series from MCM-41 (CM300)	55 (-55° to 60°)	2°	4 hrs	≥3500*
Automatic series from MCM-41 (Tecnai F20)	146 (-71°to 74°)	1°	3hrs 30min	862

Table 1. Comparison of parameters from manual and automatic acquisition schemes, for acquisitions from an MCM-41 based catalyst. * Conservative estimate based on the assumption of 3 minutes for manual titling, repositioning and refocusing of area at each increment with a frame time of 3 sec.

A brief summary of some of the benefits offered by automating STEM tomography is shown in Table 1. A larger number of projections can be acquired in a shorter time and, most importantly, the total number of exposures needed is far lower for the automated series. This is a consequence of the area of interest being imaged constantly throughout manual adjustment of tilt, position and focus. Compared with a small number of acquired frames for the automated tracking and focusing.

3. Limitations of EFTEM tomography

While improvements in tilt range should offer benefits to reconstruction from EFTEM elemental maps they also present additional problems. A rigorous condition for accurate EFTEM mapping is that the thickness of the specimen should be less than 1 mean free path (λ). For a tilt series from a monolithic specimen this presents a clear problem. Areas that are suitable for mapping in 2D quickly become unsuitable due to the increase in projected thickness (2x thicker at 60°, 3x thicker at 70°). The drop in jump ratio as the thickness increases will make any quantitative analysis of the reconstruction difficult as this invalidates the projection requirement for tomographic reconstruction.

An illustration of this problem is shown in Fig 2 for a Cr map tilt series acquired from a grain boundary in a stainless steel. The intensity of two areas of the elemental map series have been examined; area 1 on the vacuum side of the boundary and area 2 on the matrix side. A thickness map has confirmed that area 1 is ~30% thinner than area 2. At low tilts the map intensity in area 1 generally follows the expected relationship with increasing tilt, a higher thickness should give more signal. However there is a marked divergence from this relationship tilts greater than 30° with the intensity in the map decreasing due to the predominantly faster rise in energy loss background compared to the ionisation edge intensity. For the thicker area this divergence from the expected projection relationship is even more rapid with no part of the tilt series conforming to the simple projection relationship. As a consequence the elemental map reconstruction from this series shows significantly lower intensity in the matrix side of the boundary than the vacuum side. This effect makes it is doubtful whether any quantitative EFTEM reconstruction may be possible from monolithic specimens. The case for quantitative results may be more promising for specimens supported on a thin carbon film, where the total thickness of the object is not increasing greatly through tilt.

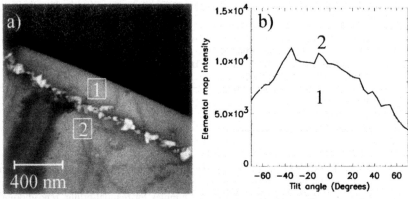

Figure 2. Examination of the effect of increasing specimen thickness on elemental map intensity. a) Zero tilt Cr elemental map showing Cr rich carbides at the grain boundary. b) Line traces of the elemental map intensity from areas 1 and 2 through tilt.

A consequence of increased tilt ranges will be longer acquisition times, assuming a similar angular increment is used. As EFTEM mapping is an inherently high dose technique this will increase the possibility of beam damage, which may change the structure of the object during the series and hence invalidate reconstruction. Automated acquisition of EFTEM tilt series is therefore required not only to increase the number of acquired projections but to minimise the potential for beam damage.

4. Conclusions

The methodology of EFTEM and HAADF STEM tomography has been significantly improved by the use of high tilt holder designs and fully automated acquisition. These allow a significant improvement in both the tilt range and number of projections that may be acquired in a given time, as well as reducing the overall dose on the specimen. However for EFTEM tomography an increased tilt range can causes complications due to the rapid increase in thickness at high tilts, and great care must be used in interpreting results.

It is clear that EFTEM tomography, while a potentially powerful technique, is still problematic and further research into the underlying effects of thickness, diffraction contrast and SNR on reconstruction is required. The technique would also benefit from the development of automated acquisition to maximise resolution and minimise dose.

References

[1]   Midgley P A et al. 2001 Chemical Communications 907-908
[2]   Weyland M and Midgley P 2003 Microscopy & Microanalysis *In press*
[3]   Mobus G and Inkson B J 2001 Applied Physics Letters 79 1369-1371
[4]   Van Aert S et al. 2002 Journal of Structural Biology 138 21-33
[5]   Dierksen K et al. 1992 Ultramicroscopy 40 71-87
[6]   Ziese U et al. 2002 Journal of Microscopy 205 187-200

Inst. Phys. Conf. Ser. No 179: Section 8
Paper presented at Electron Microscopy and Analysis Group Conf. EMAG2003, Oxford, 2003
©2003 IOP Publishing Ltd

Improved visibility of thin amorphous intergranular films using Fourier filtering

Ian MacLaren

Institute for Materials Science, Darmstadt University of Technology, Petersenstr. 23, 64287 Darmstadt, Germany

Abstract. It is shown that the visibility of thin intergranular films or other non-periodic features in HRTEM micrographs can be enhanced by Fourier-filtering out the periodic information in the image. This is particularly useful in the case of the thinnest films where the visibility is very low in the raw HRTEM images. It is shown that the film thickness is typically underestimated in measurements from HRTEM images, whereas it is much more accurately represented in Fourier-filtered images, so long as the boundary tilt is minimised.

1. Introduction

Thin intergranular films are a common feature of ceramic materials, occurring in a wide variety of different compositions and have been intensively studied using transmission electron microscopy (TEM) since Clarke (1979) first published a comparison of different TEM imaging strategies for such films. The three main methods are the use of dark field imaging with a part of the diffuse diffracted ring from the amorphous phase, the use of Fresnel fringes in out-of-focus images of the edge-on boundary, and the direct high-resolution TEM (HRTEM) imaging of the edge-on boundary. For quantitative film thickness measurements, the latter two methods have generally been preferred (Cinibulk *et al.*, 1991; Jin *et al.*, 1998).

Fourier filtering has been extensively used for image processing of TEM images since the advent of powerful desktop computers. It has generally been used to enhance the visibility of features in HRTEM micrographs, for example: small clusters or crystals (Faust *et al.*, 1991; Nihoul, 1991), strain and displacement (Hÿtch, 1997, Snoeck *et al.*, 1998), and dislocations (Snoeck *et al.*, 1998).

These methods use Fourier filtering to enhance periodic information or small deviations from periodicity (e.g. dislocations). In the present work, an alternative approach is proposed whereby Fourier filtering is used to remove the periodic lattice fringes and thereby enhance the non-periodic contrast from the intergranular film. This both enhances visibility and enables more accurate thickness measurement.

2. Experimental Procedure

An alumina ceramic doped with 500 wt. ppm Y_2O_3 was prepared as described previously (Gülgün *et al.*, 2002) and was sintered at 1450 °C for 96 h and then annealed at 1650 °C for 12 h. As a result, abnormal growth of some grains was observed as has been reported in detail elsewhere (MacLaren *et al.*, 2003). TEM specimens were prepared using a standard procedure of slicing, disc cutting, mechanical polishing, dimpling and ion milling, including low energy ion milling (500 eV) to reduce surface damage. The sample was examined using a JEOL3010 UHR transmission electron microscope operated at 297 kV; high resolution images were recorded using the CCD camera of the attached Gatan image filter. Processing of these images was performed using the Digital Micrograph software (Gatan Inc.).

3. Results

The method is in principle very simple. Fourier filtering is used to remove periodic information, not to enhance it. Such an idea was originally proposed by Coene *et al.* (1988) but has apparently received little attention since. Figure 1a shows a HRTEM image of a grain boundary at which there is a very thin amorphous film. This image was fast Fourier transformed to produce the image of Figure 1b. The bright spots corresponding to crystalline periodicities were then blanked out with circular masks to yield Figure 1c. This was then finally reverse transformed to yield Figure 1d, the Fourier-filtered image. In this last image all non-periodic features are strongly enhanced: both the amorphous film as well as surface relief on the specimen. A second example of this procedure is shown in Figure 2, this time with the images cut just to show the grain boundary (square images are better for the FFT). Figure 2a shows the raw HRTEM image of the grain boundary and Figure 2b the Fourier-filtered image.

It may be noted that the apparent film thickness is somewhat larger in the Fourier-filtered HRTEM (FF-HRTEM) images than in the raw HRTEM images. There are a number of effects which can contribute to such an observation. Firstly, it can be difficult to orient a grain boundary exactly in the "edge-on" condition: there are always likely to be small errors, partly from human error, and partly from the non-planarity of the grain boundary. In the HRTEM image, the periodic information will tend to obscure the non-periodic resulting in an apparent width less than the true width. When this periodic contribution is filtered out, the apparent width will then overestimate the true width to a similar degree. It can easily be shown by geometry that the apparent film thicknesses in the raw HRTEM and in the FF-HRTEM images are (in the small angle approximation):

$$w_1 = w - t \tan \theta \qquad \text{and} \qquad w_2 = w + t \tan \theta \qquad (1)$$

where w is the true width, t the specimen thickness, and θ the tilt away from edge-on (in radians). Thus for $\theta = 0.5°$, $t = 20$ nm and $w = 1$ nm, this would give $w_1 = 0.83$ nm and $w_2 = 1.17$ nm. Thus for fairly normal specimens, even a small tilt of less than 0.5 ° may have a significant effect on the apparent film thickness.

A second important effect is that of fringe delocalisation. This may lead to crystalline fringes extending into the amorphous film and thus reducing the apparent width. Such an effect is undoubtedly more serious for highly coherent FEG-TEMs but also has an undue effect in conventional TEMs with thermionic filaments (Krivanek *et al.*, 1979; Clarke, 1979; Cinibulk *et al.*, 1991). This effect is minimised at a certain defocus calculated by Lichte (1991) and Coene and Janssen (1992):

$$\Delta f_{opt} = - M C_s \lambda^2 u_{max}^2 \qquad (2)$$

where M is between 0.75 and 1, C_s is the spherical aberration, λ the electron wavelength and u_{max} is the maximum spatial frequency in the image. This is typically a small negative defocus, but is not simply related to the Scherzer defocus. It can only be truly

355

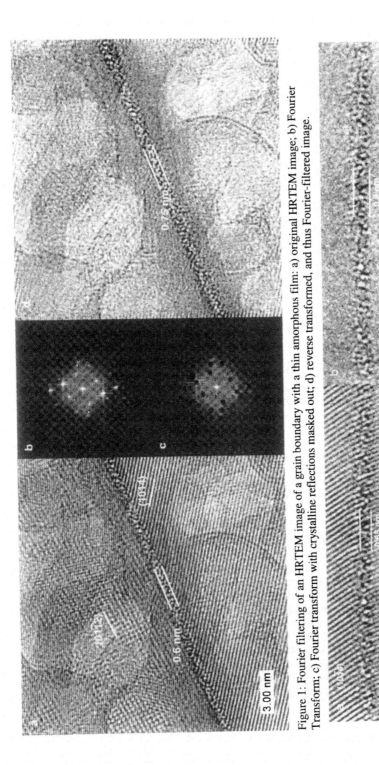

Figure 1: Fourier filtering of an HRTEM image of a grain boundary with a thin amorphous film: a) original HRTEM image; b) Fourier Transform; c) Fourier transform with crystalline reflections masked out; d) reverse transformed, and thus Fourier-filtered image.

Figure 2: Fourier filtering of a HRTEM image of a second grain boundary: a) original HRTEM image; b) Fourier filtered image.

eliminated by correction of C_s. Calculations of the effects of delocalisation in the microscope used in this work are presented elsewhere (MacLaren, 2003), and show that this will in typical cases reduce the apparent film thickness by 1-2 Å. As a general rule, only low values of defocus should be used, this has the additional advantage of minimising the appearance of Fresnel fringes, which could otherwise complicate the thickness measurement.

Finally, film roughness or waviness and the fundamental uncertainty of which atom belongs to the glass and which to the crystal at the amorphous-crystalline interface leaves a certain fundamental uncertainty. Thus, to some degree, it is only possible to define an average film thickness. In the overlap region the crystal fringes will tend to dominate the image, also leading to an underestimate in thickness. On the contrary, the FF-HRTEM image will better represent the true average thickness, and may even show some of the local variations.

Thus to sum up, the film thickness will almost always be underestimated in raw HRTEM images by perhaps 1-3 Å in typical cases. This is a significant discrepancy for the very thinnest films (≤ 1 nm) and may be one reason why there are sometimes discrepancies between thicknesses measured by HRTEM and by Fresnel fringe analysis (Clarke, 1979; Cinibulk et al., 1991). In contrast to this, if the boundary tilt can be minimised, the FF-HRTEM image will represent the film thickness very well.

4. Acknowledgements

The author is thankful for helpful discussions with Profs. M. Rühle, R.M. Cannon and H. Fuess, and Drs. M.A. Gülgün and G. Miehe, as well as useful comments from reviewers of a related journal article. Dr R. Voytovych is gratefully acknowledged for the preparation of the alumina ceramic used in the work, and Mrs M. Sycha for the preparation of the TEM specimen.

References
Cinibulk M K, Kleebe H-J and Rühle M 1991 J Am Ceram Soc 76 426
Clarke D R 1979 Ultramicroscopy 4 33
Coene W, Janssen A 1992 Scanning Microscopy Supplement 6 379
Coene W, de Jong A F, van Dyck D, van Tendeloo G and van Landuyt J 1988 Phys Stat Sol (a) 107 521.
Faust P, Brandstättner M and Ding A 1991 Z Phys D 21 285
Gülgün M A, Voytovych R, MacLaren I, Rühle M and Cannon R M, 2002 Interface Sci. 10 99
Hÿtch M J 1997 Microsc Microanal Microstruct 8 41
Jin Q, Wilkinson D S and Weatherly G C 1998 J Eur Ceram Soc 18 2281
Krivanek O L, Shaw T M and Thomas G 1979 J Appl Phys 50 4223
Lichte H 1991 Ultramicroscopy 38 13
MacLaren I, 2003 Ultramicroscopy under review
MacLaren I, Cannon R M, Voytovych R, Gülgün M A, Popescu-Pogrion N, Scheu C, Täffner U and Rühle M 2002 J Am Ceram Soc 86 650
Nihoul G 1991 Microsc Microanal Microstruct 2 637
Snoeck E, Warot B, Ardhuin H, Rocher A, Casanove M J, Kilaas R and Hÿtch M J, 1998 Thin Solid Films 319 157

Inst. Phys. Conf. Ser. No 179: Section 9
Paper presented at Electron Microscopy and Analysis Group Conf. EMAG2003, Oxford, 2003
©2003 IOP Publishing Ltd

Modern Approaches to the Preparation of TEM Samples

S B Newcomb

Sonsam Ltd, Glebe Laboratories, Newport, Co. Tipperary, Ireland

Abstract. Some of the methods by which focused ion beam thinning can be used to prepare samples for microstructural examination are outlined. Typical applications of the preparative method are described and some of its limitations are discussed in·relation to the development of techniques that can be used to counter them.

1. Introduction

The recent advances that have been made in both instrumentation and the performance of modern transmission electron microscopes (TEM) have been largely matched by improvements to the methods used for the preparation of samples for TEM examination. Much of the need for the continued development of TEM specimen preparation techniques not only stems from the advent of new structures of diminishing dimensions but also from the recognition of the importance of the examination of specific locations that are property determining. For a number of years, microstructural characterisation was limited to the examination of extraction replicas [1], or of samples made using techniques such as jet electropolishing [2] or broad beam milling [3]. Much of the current scenario originates from the use of focused ion beam (FIB) thinning to prepare TEM samples containing highly localised microstructures. Some of the methods used for the FIB preparation of TEM samples are described and examples given of their characterisation that not only demonstrate the adaptability of the technique but also the relative ease with which different types of sections can be made. The need for the further development of preparative methods is discussed particularly in relation to some of the modifications that can take place at the surfaces of FIB prepared thin foil samples.

2. Focused Ion Beam Milling

The key advantages of FIB usage lie in the high spatial accuracy with which samples can be taken as well as the overall speed of the preparative procedure. Samples for TEM examination are made in an ion microscope that generally uses a Ga liquid metal source to generate ions that are accelerated at voltages of between 10 and 50 keV [4]. Focusing is achieved through the use of two electrostatic lenses, and the beam current can be varied using apertures for which the spot size is some 7nm at 1pA and approximately 250nm at 12nA [5]. Gas assisted etching can be achieved by introducing a reactive gas such as iodine whilst metals can be locally deposited in the presence of Pt or W based organo-metallic precursor gases [6].

358

2.1 *Basic Approaches*

There are two general approaches that have been developed for the FIB preparation of TEM samples and that are described in detail elsewhere [e.g. 7, 8]. For the trench technique [7], a small rectangular 'slab' of material that contains the site to be examined and measuring some 3mm in length and 20-50μm in width is prepared using standard metallographic procedures. The slab is fixed to a horseshoe shaped support grid and the site of interest is located in the FIB microscope before its upper surface is protected by the deposition of a high atomic number 'strap'. Trenches are cut to the immediate sides of the target site at a high beam current, and the beam current subsequently decreased as the thickness of the slice is reduced to the point at which it is electron transparent [9]. Final milling is performed using incident ion angles of some 1-2° and electron transparent regions measuring some 25 by 6μm are typically obtained, the total procedure generally taking less than 2 hours. The shape of the trenches can be optimised to enable the sample to be tilted to high angles in the TEM thereby also allowing standard probe methods to be used to determine local compositions. Similar approaches can be used to prepare samples for tomography whilst a wide range of characterisation methods have been used to examine FIB prepared sections including high resolution electron microscopy (HREM) [10], loss imaging [11] and holography [12]. It is critical that surface structures be protected from direct exposure to the ion beam, and this can be achieved through the use of a sacrificial thin film deposit. The second approach uses a lift-out technique to transfer an electron transparent section onto a support grid and does not involve loss of the integrity of the bulk sample through pre-FIB cutting, grinding and polishing [8]. The downside of the technique is the difficulty of the micromanipulation although methods have been developed whereby sections of intermediate thickness can be extracted and subsequently stuck onto 'dummy' substrates [13]. This approach is more attractive than the conventional lift-out method given the typical need to perform repeat milling sequences.

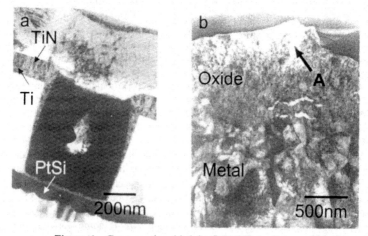

Figure 1. Cross-sectional bright field TEM micrographs of a) a via and b) a metal-oxide interface formed near the surface of a ZrNb alloy.

Two examples of cross-sectional samples that were prepared in an FEI FIB200 Workstation using the trench technique are shown in figure 1. Figure 1a is a bright field image that demonstrates some of the high spatial selectivity with which samples can be made on a routine basis. A number of different layers can be distinguished and it is apparent that only limited differential milling of the different phases has taken place. It is interesting to note that the columnar grained Ti and TiN making up the barrier layer are both fully continuous but that there is a thin band of oxide located at the interface between the lower part of the barrier layer and the underlying PtSi. The presence of this anomalous zone was found to be controlling the poor resistivity of the via. Figure 1b gives a second example of an FIB prepared TEM slice and shows part of a ZrO_2 layer formed at the surface of a ZrNb alloy used in the nuclear power industry. Such interfaces are notoriously difficult to retain in samples prepared using broad beam Ar ion milling procedures partly because of the propensity of the metal to absorb H during the milling [14]. No such difficulties are encountered in the FIB, and figure 1b shows part of a small nodule in the oxide that was found to be associated with a localised increase in the compressive stress of the oxide that has been relieved with the formation of a vertical crack (at A).

Plan-view membranes can be prepared from both site specific and blanket coverage samples and the general approach is based on the trench technique described above. Metallographic methods are used to mechanically thin a slab to within some 2-5μm of the target site, only this time the sample is oriented side-on for FIB milling. From here it is relatively easy to prepare plan-view sections at any given depth within a structure although it is emphasised that protective Pt straps tend not to be deposited for slices taken from upper surface regions. Electron transparent areas measuring some 25 by 20μm can be prepared with ease, and tapered sections and multiple slices taken from single test pieces [15]. Figure 2a shows part of a plan-view section containing an array of magnetic nanodots and it is noted that the section is of generally higher thickness uniformity than an Ar milled sample. Unwanted areas lying outside the TEM section have been milled away and the sample examined using electron holography, which has shown that the magnetic microstructure is dominated by strong interactions between the Co dots [16]. Figure 2b is a low magnification bright field micrograph that was taken from part of a site specific plan-view section and it shows five contacts and a poly-Si/silicide rail lying to their sides. A fine stringer was observed between one of the contacts and the adjacent rail and its presence was found to be related to the poor electrical performance of the device.

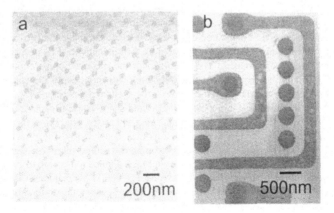

Figure 2. Bright field TEM micrographs of plan-view sections of a) magnetic nanodots and b) part of a device structure.

360

2.2 *Artefacts and the Use of Modified FIB Approaches*

A critical aspect of TEM sample preparation is the need for adaptability. Small samples, for example, can be stuck onto horseshoe support grids for FIB milling and a similar method used to prepare membranes from cleaved materials such as SiC or alumina. Figure 3a shows a secondary electron image of a small triangular shaped piece of alumina attached to a Cu support grid, the shape of the sample allowing short milling times to be attained by targeting one of the 'corner' regions marked. The microstructural changes occurring at the surface of the alumina brought about by Zr implantation are of interest with regard to the potential biomedical application of the material and figure 3b shows a bright field micrograph of a typical surface region of the sample where the peak concentration of Zr was found to lie at approximately the centre of a well developed band of amorphous material. Such observations have been used to rationalise the effects of different implant treatments on the mechanical properties of a range of substrates [17].

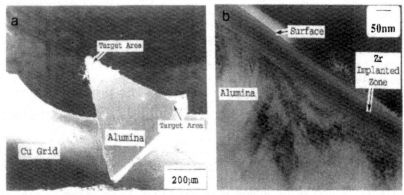

Figure 3. a) Secondary electron micrograph of an alumina sample for FIB milling and b) a cross-section bright field TEM micrograph of the surface amorphous layer formed during Zr ion implantation.

FIB is suited to the preparation of samples for the examination of near surface microstructures but rather less so for structures that are either 'buried' or those with high depth dimensions. Preparation of the latter type of sample is likely to involve long milling times and can be further complicated by the tendency for FIB-prepared samples to be somewhat wedge shaped. The versatility of the specimen preparation procedure, however, is underlined by the fact that structures with high depth dimensions can be prepared with relative ease when the region to be sectioned lies edge-on to the incident ion beam and at one end of a mechanically ground slab. The approach draws parallels with plan-view sections although it should be emphasised that some microstructural modification of the very near surface regions may occur during the prior mechanical stages of the sample preparation. Buried microstructural features such as stress corrosion cracks (SCC) may require the combined application of broad beam Ar and focused ion milling. An approach developed for the TEM characterisation of SCC phenomena in Al based alloys [18] involves the use of conventional broad beam milling to thin a disc containing SCC to the point at which the target crack or crack tip lies approximately 10-20μm from the perforated edge of the sample. FIB milling is subsequently used to 'chop' away material lying between the crack (marked at A in figure 4a) and the edge of the sample, as demonstrated by the secondary electron image shown in figure 4a where the FIB cuts have been marked at C, D and E, so that the target area can then be milled to electron transparency once the sample has been turned through 90°. Figure 4b is a bright field micrograph of a typical SCC seen in a sample prepared using the method described and shows a grain boundary region that has undergone corrosion. The boundary itself was found to be rich in solute elements making it anodic relative to the surrounding precipitate free zones whilst the degradation was notable for being found not to be stress assisted. A further method for the examination of SCC has been described elsewhere [19].

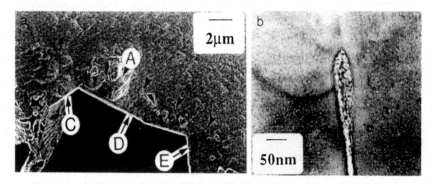

Figure 4. a) Secondary electron micrograph of a part prepared TEM sample containing a SCC and b) a bright field TEM image of a SCC.

Any differential milling that occurs during the FIB preparation of samples is not as significant as that frequently taking place during broad beam Ar milling. Small thickness variations do however develop in the material located beneath and to the sides of structures such as W based contacts, and it is these variations in specimen thickness that can dominate the observed phase contrast in electron holograms. Such curtaining effects can be avoided by performing the critical final stages of the milling from the substrate side of the disc (the sample having been rotated through 180°), the upper parts of the contacts similarly being milled away so that the region of interest lies at an appropriate distance from vacuum and within a carefully optimised thickness of the sample [12].

The upper and lower surfaces of TEM samples made using FIB are amorphised during ion erosion although the thickness of such damage layers is clearly related to the type of material being sectioned. The thicknesses of the damage layers have been measured directly on a range of semiconductors and metals [20] and found to be as high as some 28nm for InP by comparison with a value of approximately 21nm for Si. This is in itself points to one of the inherent limitations of the FIB method for making samples for examination using methods such as HREM and holography. For HREM, there is clearly a finite thickness beyond which the sample will be fully amorphous whilst the application of holography has been found to be impaired by the electrically 'dead' surface damage layers [12]. It has, however, been shown that the damage formation can be reduced by performing the final stages of FIB milling at high angles (4-8°), or the damage can itself largely be removed through the careful use of post FIB broad beam milling [20].

3. Summary

Focused ion beam milling has made a significant impact on the range of structures that can be prepared for TEM examination on a routine basis. Much of the attraction of the method lies in both the speed and spatial selectivity with which samples can be made whilst the favourable control of the thinning characteristics is underlined by the range of TEM techniques that can be used to examine FIB prepared samples. An inherent problem of the method remains the magnitude of the damage zones formed during milling that require the use of other preparative techniques for their part removal. The use of such methods as well as the further understanding of the effects of changes to the gas chemistry during FIB milling are both areas of increasing importance, particularly as the critical dimensions of so many structures of interest continues to fall. To this end, established techniques such as jet polishing and broad beam milling will retain crucial roles in the electron microscopists' repertoire of preparative procedures.

Acknowledgements

It is a pleasure to acknowledge the enthusiastic support and input from Drs David Foord, David Sutton and Rafal Dunin-Borkowski during the period in which some of the methods described were developed and applied.

References

[1] Wilson J H M and Rowe A J 1980 in Practical Methods in Electron Microscopy (Ed. A M Glauert) (Amsterdam: North Holland)
[2] Goodhew P J 1985 Thin Foil Preparation for Electron Microscopy (New York: Elsevier)
[3] Newcomb S B Boothroyd CB and Stobbs W M 1985 J. Microsc. 140 195-207
[4] Young R J 1993 Vacuum 44 353-361
[5] Bischoff L and Teichert J 1997 Materials Science Forum 248 445-454
[6] Prewett P D 1993 Vacuum 44 345-352
[7] Park K H 1990 Mater. Res. Symp. Proc. 199 271-280
[8] Overwijk M H F van den Heuvel F C and Bulle-Lieuwma C W T 1993 J. Vac. Sci. Technol. B11 2021-2024
[9] Szot J Hornsey R Ohnishi T and Minagawa J 1992 J. Vac. Sci. Technol. B10 575-579
[10] Brown P D and Newcomb S B 1998 ICEM 14, Mexico p399
[11] Pirila N Weyland M and Newcomb S B 2000 12th EUREM, Czech Republic IP567
[12] Dunin-Borkowski D B Newcomb S B Doyle D Deignan A and McCartney M R Ibid I64
[13] Kamino T 2001 Workshop on Advances in Focused Ion Beam Microscopy: FIB 2001, Oxford
[14] Newcomb S B Warr B D and Stobbs W M 1994 ICEM 13, France p1103
[15] Newcomb S B and Quinton W A J 2000 12th EUREM, Czech Republic P575
[16] Dunin-Borkowski D B McCartney M R Ross C A Farhoud M and Newcomb S B 2001 Proc. 5th MCEM, Lecce p429
[17] Murphy M Laugier M T Beake B D Sutton D and Newcomb S B 2002 J. Mat. Sci. 37 2053-2062
[18] Deshais G and Newcomb S B 2000 12th EUREM, Czech Republic P573
[19] Lozano-Perez S Huang Y Langford R and Titchmarsh J M 2001 Inst. Phys. Conf. Ser. 168 (Bristol: IOP) p191
[20] Sutton D Parle S M and Newcomb S B 2001 Ibid p 377

Inst. Phys. Conf. Ser. No 179: Section 9
Paper presented at Electron Microscopy and Analysis Group Conf. EMAG2003, Oxford, 2003
©2003 IOP Publishing Ltd

The application of tripod polishing and focused ion beam milling to the TEM specimen preparation of HVOF thermally sprayed coatings

Gaoning Kong*, D Graham McCartney and Paul D Brown

School of Mechanical, Materials, Manufacturing Engineering and Management, The University of Nottingham, University Park, Nottingham, NG7 2RD, UK.

*Now at: School of Metallurgy and Materials, The University of Birmingham, Edgbaston, Birmingham, B15 2TT, UK.

ABSTRACT: The microstructure of high velocity oxy fuel thermally sprayed coatings is highly anisotropic and inhomogeneous. Tripod polishing has enabled the preparation of samples with up to 0.5mm diameter electron transparent areas, where a statistically significant number of features could be examined. Conversely, FIB has been used to prepare TEM samples for site-specific analysis of sub-micron regions of interest, e.g. for the interface characterisation between metallic coatings and the substrate, or the study of secondary precipitation on pre-existing phases in cermet coatings.

1. INTRODUCTION

High velocity oxy-fuel (HVOF) thermally sprayed coatings are produced by the successive deposition of individual splats from melted or partially melted powder feedstock. Because of the high particle temperatures and velocities involved in this process, and the high cooling rate on impact, HVOF coating structures are extremely heterogeneous [Smith, 1995]. TEM characterisation is required to fully understand the microstructure of such thermally sprayed coatings. However, the preparation of large area, electron transparent samples that truly represent the microstructural features of such coatings is quite a challenge for the traditional techniques of electropolishing or dimpling and broad ion beam milling, in addition to the investigation of site-specific features within the coatings. Accordingly, tripod polishing [Benedict, 1992] and focused ion beam (FIB) milling [Ishitani, 1991] have been applied to investigate HVOF thermally sprayed coatings. Representative results are presented enabling an appraisal of the advantages and disadvantages of each approach.

2. EXPERIMENTAL

An HVOF sprayed sample of composition Co-28Cr-4.5W-3.0Fe-3.0Ni-1.2Si-1.1C (wt%) is reported on here. TEM foils were first sequentially polished down to ~ 10μm thickness using a Testbourne Model 590W tripod polisher, then perforated using a Fischione 1010 low angle, argon ion beam thinning machine. An FEI FIB200 workstation was also used in this study for the examination of site-specific features within the HVOF coatings. Both

364

plan-view and cross-sectional TEM samples were prepared using the 'lift-out' technique [Hull, 1997]. An FEI XL30 FEG-SEM operated at 20 kV and a Jeol 2000fx TEM fitted with an energy dispersive X-ray (EDX) detector operated at 200 kV were used to perform the microstructural characterisation.

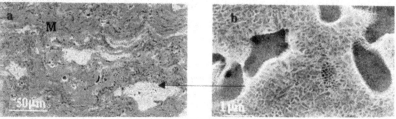

Fig. 1. SEM micrographs of an etched, HVOF sprayed Stellite 6 coating. (a) Cross-sectional image; (b) plan view image at higher magnification.

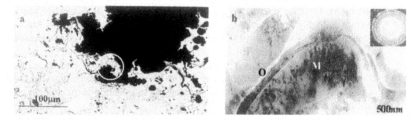

Fig. 2. (a) BSE image of TEM foil; (b) TEM image of circled region in 2a showing the layered structure of Stellite 6 coating with metallic matrix, M, and oxide layers, O. The inset SADP was taken from both the matrix and oxide.

3. RESULTS AND DISCUSSION

3.1 The application of tripod polishing and low angle argon ion milling

A Stellite 6 coating built up layer by layer onto a mild steel substrate with the formation of splats, intersplat oxides and some porosity is shown in the SEM image of Fig. 1a. The regions of lighter contrast that exhibit fine-scale dendritic features are attributed to powder particles that had not completely melted prior to impact on the substrate. Fig. 1b shows a higher magnification plan view micrograph of the area indicated in Fig. 1a. The rounded regions of darker contrast are the unmelted dendritic cores whilst the fine scale cellular structure is considered to have formed by the rapid solidification of the interdendritic liquid. Regions such as M (Fig. 1a) which appear relatively featureless in the SEM are believed to have formed from powder that was fully molten at the time of impact and then rapidly re-solidified at a high cooling rate.

The application of tripod polishing enabled TEM samples with ~ 0.5 mm diameter electron transparent areas to be prepared and directly correlated with the features identified in SEM. Fig. 2a shows a low magnification BSE image of one particular TEM foil. The morphology of the circled region corresponds to the coating region shown in Fig. 2b that had been fully melted, thereby, demonstrating the effectiveness of this sample preparation approach. This TEM image shows several micrometer lengths of string-like oxide, denoted O, which partition the microcrystalline Co-based matrix regions, M. The oxide layers were 100 to 200nm in thickness and electron diffraction demonstrated them to be the spinel oxide $CoCr_2O_4$. The associated selected area diffraction pattern (SADP) corresponding to fcc Co and the spinel oxide is inset.

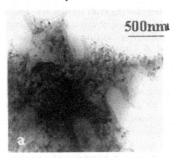 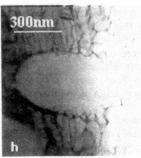

Fig. 3. (a) Bright field TEM image of a partially melted powder particle within the coating; (b) a high magnification image of a similar feature, showing remnant dendritic cores surrounded by a cellurlar structure.

Using the same approach, a partially melted powder particle within a coating was investigated, as illustrated by the low-magnification, plan-view image of Fig. 3a. The ~ 500 nm-sized elliptical regions correspond to unmelted dendrite cores and these are surrounded by material exhibiting a cellular morphology (Fig. 3b). In this instance, the cell size of ~ 100 nm is of a similar size to that revealed by SEM (Fig. 1b). EDX analysis indicated that the elliptical dendrite cores are ~ 15% richer in Co than the surrounding cell structure. It is considered that this compositional difference is the reason why these Co-rich dendrites were locally left unmelted during powder heating in the HVOF spray gun.

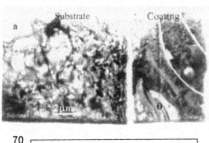 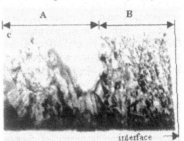

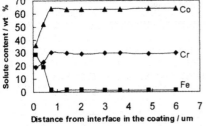

Fig. 4. Cross sectional TEM bright field image of sample prepared by FIB. (a) Sample foil across substrate / coating interface; (b) EDX analysis of elemental distribution in the coating part of the sample (a); (c) sample foil from substrate region next to the interface.

3.2 The application of FIB milling

An example of a cross-sectional Stellite 6 specimen prepared by the FIB "lift-out" technique is shown in the TEM image of Fig. 4a. Thin spinel oxide layers partitioning the fully-melted splats are identified within the coating, e.g. the feature labelled O. EDX analysis across the interface shows no trace of the coating elements within the substrate. However, a ~ 1 μm wide band of Fe was detected in the coating layer next to the interface (Fig. 4b). This suggests that very limited melting has occurred at the coating / substrate interface in this instance. At the moment a molten / semimolten coating powder impacts on the mild steel substrate, a limited amount of Fe will dissolve into the coating material. However, it is considered that the impact momentum and temperature of the coating particles is insufficient to fully melt the substrate which remains solid. Accordingly, no coating species is able to diffuse into the surface of the substrate. It is also interesting to note how the microstructure of this mild steel substrate gradually changes with distance from the interface deeper into the substrate. This phenomenon is more clearly illustrated by Fig. 4c which shows a TEM membrane cut from a region of the mild steel substrate close to the interface. It shows a highly deformed band A and a recrystallisation band B as a consequence of thermal spikes associated with the spray deposition process. Because a grit blasting surface roughening treatment was applied to the substrate before thermal spraying, a certain amount of deformation was built up in the top surface layer. During the subsequent process of thermal spray coating, it is believed that the temperature at the interface is sufficient to induce recrystallisation recovery within this deformed substrate layer, hence the microstructural change observed in the substrate. As limited melting occurred between the substrate and coating, the bonding between them is considered to be dominated by mechanical coupling.

Tripod polishing is undoubtedly the preferred choice for the general microstructural mapping of HVOF coatings because of the large size of electron transparent areas produced. The ability to minimise artefacts and surface amorphisation damage also make this the preferred route for preparing samples requiring chemical microanalysis. However, these advantages are sometimes offset by the problem of preferential argon ion milling of a multi-phase coatings. Conversely, the ability to perform site-specific analysis of a coating / substrate interface, or an interphase reaction within these thermally sprayed coatings, by the FIB approach clearly offers an advantage. However, it is noted that there is always some degree of remnant amorphisation at the two side-walls of these FIB prepared sample foils, the thickness of which is dependent on the material and the milling parameters.

Acknowledgements

This project was funded by the EPSRC (ROPA GR/R40432/01). The authors gratefully acknowledge the provision of focused ion beam facilities by the Dept. of Materials, University of Oxford, and the assistance of Dr. RM Langford.

References

Smith R. W, Knight R, *JOM*, 47 (1995) 32.

Benedict J. P, Anderson R. M, and Klepeis S. J, Mat. Res.Soc. Sypm. Proc. 254 (1992) 121.

Ishitani T, Ohnishi T, Madakoro Y, Kawanami Y, *J. Vacuum. Sci. Technol.* **B9** (1991) 2633.

Hull R, Banck D, Stevie F. A, Koszi L. A, Chu S. N. G, *Appl. Phys. Lett.* **62** (1997) 3408.

Inst. Phys. Conf. Ser. No 179: Section 9
Paper presented at Electron Microscopy and Analysis Group Conf. EMAG2003, Oxford, 2003
©2003 IOP Publishing Ltd

The effects of sample preparation on electron holography of semiconductor devices

A C Twitchett, R E Dunin-Borkowski and P A Midgley

Department of Materials Science and Metallurgy, University of Cambridge, Pembroke Street, Cambridge, CB2 3QZ, U.K.

Abstract. Focused ion beam (FIB) milling and small angle cleaving have been used to prepare semiconductor specimens for examination using electron holography. The experimental results illustrate the effects of sample preparation on the electrostatic potential distribution in a silicon p-n junction sample. The importance of biasing experiments when examining FIB-prepared semiconductor samples is highlighted.

1.   Introduction

Sample preparation is known to have a major effect on phase contrast recorded from a doped semiconductor sample using electron holography [1, 2]. Focused ion beam (FIB) milling is a particularly attractive sample preparation technique as it provides the site-specificity required for characterising future device generations. However, the effect on the electrostatic potential in the sample of the physical damage and Ga implantation associated with this technique [3] must be assessed before holograms obtained from FIB-prepared semiconductors can be interpreted with confidence. In comparison to FIB milling, cleaving provides specimen surfaces undamaged by ion beam milling, which may be crystallographically defined. Furthermore, *small-angle* cleaving allows wedge-shaped samples with shallow thickness profiles to be prepared. Here, we address the effect of sample preparation using FIB milling and small-angle cleaving on the contrast recorded from a silicon p-n junction, which is examined with and without an applied reverse bias.

2.   Sample preparation

The examination of a semiconductor device using off-axis electron holography requires the preparation of a sample that has a region of optimal thickness (~200-500 nm) [4], which is close enough to vacuum to allow overlap of the object and reference waves required for the formation of the hologram. Assuming an overlap width of 1 µm, this optimal thickness range can be maintained across the overlap region for a wedge-shaped specimen that has an angle of between 15 and 30°. FIB milling can be used to provide a sample that has a more flexible geometry, and any desired thickness. A strap of platinum is deposited to protect the area of interest, and further cuts can be made during milling, as shown in Fig. 1, to provide a hole close to the area of interest. Final milling is performed at a low beam current of ~150 pA to minimise Ga implantation.

3.   Off-axis electron holography experiments

The silicon sample examined here contained a p-n junction 2.5 µm below the wafer surface. The nominal dopant concentration was in excess of 10^{18} cm^{-3} on both on both p and n sides, with B and Sb dopants respectively. FIB milling was used to prepare membranes with three different thicknesses in a single window, as shown schematically in Fig. 1. The 'crystalline' thicknesses of the membranes were measured to be 210, 270 and 410 nm using convergent beam electron diffraction (CBED). A small-angle cleaved wedge (SACW) containing the junction was also prepared, with a cleavage angle of $40 \pm 5°$.

3.1.   Unbiased experiments

Off-axis electron holograms were acquired from each sample on a 2048 pixel charge-couple device using a Philips CM300 field emission gun transmission electron microscope operated at 200 kV. The microscope was operated with the objective mini-lens ('Lorentz' lens) used in place of the conventional objective lens. A Möllenstedt-Düker biprism was inserted in the position of the selected area aperture and was used under an applied voltage of 80 V to create interference fringes of spacing 8 nm and an interference width of 0.7 µm. Reference (empty) holograms were acquired immediately after each object hologram to remove artefacts associated with the imaging and recording system.   Phase and amplitude images were reconstructed using the image processing language Semper [5].

3.2.   In-situ electrical biasing experiments

A novel sample geometry based on FIB milling [2] was used in a custom-built biasing holder [6] to examine the p-n junction under applied electrical bias in the electron microscope.   Off-axis electron holograms were acquired with the junction examined under applied reverse bias voltages of 0, 1, 2 and 3 V. The crystalline thickness of the membrane was determined to be 390 nm using CBED.

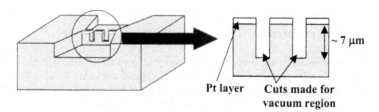

Figure 1. Schematic diagram illustrating the geometry of the FIB-prepared membranes. The three membranes were prepared with different thicknesses.

3.3   Results and analysis

Figs. 2(a) and (b) show phase images obtained from the SACW and FIB prepared samples. In Fig. 2(a), the phase ramp associated with the increase in sample thickness in the SACW sample has been subtracted to show only the phase change arising from the p-n junction. In each image, the p and n sides of the junction appear dark and bright, respectively. In addition, in Fig. 2(b), a grey band, which may correspond to the thickness of a damaged or electrically depleted layer, can be seen at the edge of the phase image. A similar layer was not observed for the cleaved sample.

The step in phase across the p-n junction, $\Delta\varphi$, can be measured from line scans across the junction, as shown in Fig. 2(c). These measurements can be related to the built in voltage across the junction, V_{bi}, projected through the membrane thickness, t, using the relation $\Delta\varphi = C_E V_{bi} t$, where C_E is a microscope-determined constant. This approach assumes that there are no other electric or magnetic fields contributing to the phase change, that there is no diffraction contrast, and that the potential is uniform through the sample thickness. The latter assumption is unlikely to hold as amorphous, electrically 'dead' layers are expected to be present on the membrane surfaces, either due to oxidation or due to FIB damage. Ga implantation may also affect the electrostatic potential in the sample. In order to obtain a better understanding of the effect of sample preparation on the electrostatic potential in the sample, the experimental results were fitted to simulations, allowing parameters such as the electrically active dopant concentration and the junction depletion width W to be measured. Fig. 3(a) shows a schematic diagram of a model used in the simulations, which incorporates the effects of surface passivation by allowing the depletion width to vary through the thickness of the membrane. The increase in depletion width at the sample surfaces is used to approximate the electrical properties of the passivated surface layer.

Measurements obtained from the best-fitting simulations to the experimental phase profiles, which are shown in Figs. 3(b) and (c), provide values for the electrically active dopant concentration on each side of the junction of $8\pm1\times10^{17}$ and $3\pm2\times10^{17}$ cm^{-3} for the SACW and unbiased FIB-prepared samples, respectively. The simulations also provide a measurement for the total electrically 'dead' layer thickness on each surface of the SACW sample of 31 ± 5 nm. For the FIB-prepared samples, in addition to the low electrically active dopant concentration, the holography results and sample thickness measurements revealed the presence of 28 ± 5 nm of crystalline and 25 ± 10 nm of amorphous electrically 'dead' material on each sample surface.

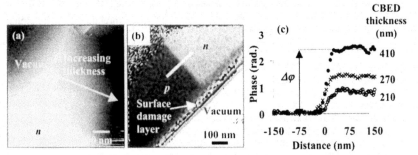

Figure 2. Reconstructed phase images of (a) the SACW sample and (b) the 410 nm thick FIB-prepared sample. The white line in (b) marks the position from which the line traces shown in (c) were extracted.

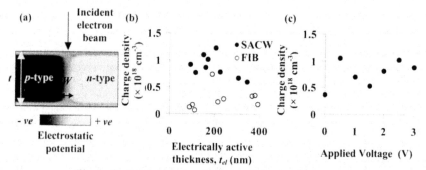

Figure 3. (a) Schematic diagram of the model used to simulate the electrostatic potential in the TEM membranes. Plots of the fitted charge density obtained from (b) the FIB and SACW unbiased experiments and (c) the biased FIB membrane.

EFTEM and CBED thickness measurements reveal the presence of a total of 50 nm of amorphous material in all FIB-prepared membranes. For the SACW samples the best fits to the holographic experimental results revealed a total (crystalline and amorphous) electrically dead thickness of 62 ± 10 nm in the membrane. The fits to the biased results indicate an electrically active thickness of 335 nm, thus suggesting the presence of 28 nm of crystalline, electrically dead material at each membrane surface.

4. Discussion and Conclusions

The off-axis electron holography results illustrate a significant difference between the measured electrical properties of SACW and FIB-prepared samples. The fitted 2-D electrostatic potential distribution is comparable for all samples, but the active dopant concentration is much lower in the FIB-prepared sample. This discrepancy must arise from the sample preparation as the membranes were prepared from the same wafer. The active dopant concentration in the FIB-prepared membranes is seen to increase to a level consistent with the SACW sample when under applied bias, indicating that the application of an electrical bias may 'reactivate' some of the dopants present in the membrane. CBED, EFTEM and holographic measurements of membrane thicknesses have revealed different thickness electrically dead layers for the SACW and FIB-prepared samples. This illustrates the importance of determining the surface properties for each sample preparation method to allow a quantitative interpretation of holographic results.

The authors would like to thank Philips Research Laboratories (Eindhoven) for providing the Si device and the Royal Society, Newnham College, Cambridge and the EPSRC for financial support.

References
[1]   Gribelyuk M A et al. 2002 Phys. Rev. Lett. 89 2 5502-5
[2]   Twitchett A C et al. 2002 Phys. Rev. Lett. 88 23 8302-5
[3]   McCaffrey J P et al. 2001 Ultramicroscopy 87 97-104
[4]   Rau W D et al. 1999 Phys. Rev. Lett. 82 12 2614-2617
[5]   Saxton W O et al. 1979 Ultramicrosocopy 4 343-354
[6]   Twitchett A C et al. 2001 Inst. Phys. Conf. Ser. No. 168 493-496

Inst. Phys. Conf. Ser. No 179: Section 9
Paper presented at Electron Microscopy and Analysis Group Conf. EMAG2003, Oxford, 2003
©2003 IOP Publishing Ltd

TEM analysis of dual column FIB processed Si/SiGe MOSFET device structures

A C K Chang, I M Ross, D J Norris and A G Cullis

University of Sheffield, Department of Electronic and Electrical Engineering, Sheffield, S1 3JD, UK

Abstract. In this report, we present the preparation of specific Si/SiGe MOSFET device structures for TEM analysis, using a new JEOL dual column focussed ion beam (FIB) miller. HREM/STEM imaging has verified the gate oxide thickness and the uniformity of the gate electrode structure. Misfit dislocation bands were observed within the virtual substrate, which found to be associated with steps in the Ge concentration profile.

1. Introduction

In the last decade, the Metal Oxide Semiconductor Field Effect Transistor (MOSFET) industry has been largely dominated by Silicon-based (Si) technologies [1,2]. However, the continuing drive towards the development of smaller and faster conventional Si-based devices is currently, limited by detrimental short-channel effects and subsequent spatial constraints. Consequently, there is an increasing interest in the development of strained Si/SiGe devices, reducing some of these limitations, while economically marrying itself to the current available Si technology [2]. For the purpose of this study, a new generation dual column focused ion beam (FIB) miller has been utilised. This instrument, encompassing a JEOL 6500F field emission gun scanning electron microscope (FEGSEM) equipped with an Orsay Physics ion-beam column and a RAITH Nanolithography System, has been used to prepare cross section TEM specimens from specific areas within a range of technologically important Si/SiGe MOSFET devices. Dual column FIB processing has many advantages over conventional cross-sectional TEM specimen preparation. Not only is this combination able to produce thin (<100 nm) TEM specimens with a uniform thickness from specific regions of the device area, the dual column ability allows the FIB milling process to be continuously monitored at a high resolution via the FEGSEM image [3,4]. Various metal capping layers may also be deposited locally with pinpoint accuracy to protect the milled surface resulting in an improvement in the quality of the subsequent TEM cross-section [5]. In this report we pay particular attention to the configuration of the FIB column and milling parameters to optimise its performance for TEM specimen preparation of the Si/SiGe devices of interest. Three main considerations are addressed in this study. First we consider the deposition of a metallic over-layer to protect the near-surface components of the device during subsequent milling. Secondly, we will describe the parameters used to form an optimum electron-transparent thin section and finally, the HREM/STEM observation and analysis.

2. Experimental

In this study, deposition and milling was achieved with the liquid Ga ion source operating at 30 keV. Precision milling was achieved using a dedicated software package within the RAITH ELPHY *Quantum* Universal SEM/FIB Nanolithography System software. Deposition and milling procedures were monitored in real time via the FEGSEM operating at 15 keV. TEM observation and microanalysis was performed in a JEOL 2010F field emission gun transmission electron microscope (FEGTEM). This instrument is equipped with a Gatan imaging energy filter (GIF), an Oxford LINK/ISIS X-ray energy-dispersive spectrometer (EDS) and a scanning attachment allowing STEM bright and annular dark-field imaging. HREM/STEM analysis was used in this instance to primarily investigate: the dimensional variation and structural uniformity within the device gate oxide layer and, to quantify the Si/Ge concentration profile across the virtual substrate (VS) layers [6]. The devices investigated in this study have a SiGe graded VS (Ge bulk concentrations starting at 5% increasing to 30%) grown on a Si substrate followed by a series of channel layers. The gate oxide sits just under the poly-Si gate electrode, which is then subjected to a poly reoxidation process followed by Au metallization.

3. Results and Discussion

TEM specimen preparation was performed in three stages. Firstly, in order to preserve the near-surface components of the device, a layer of platinum (Pt) was deposited onto the region of interest. Five different materials may be selected (Figure 1a). The appropriate injector, in this case for Pt deposition, was carefully positioned above the area of interest (Figure 1b) and a 1 μm thick layer of Pt deposited using an ion probe current of ~ 200 pA (Figure 2a).

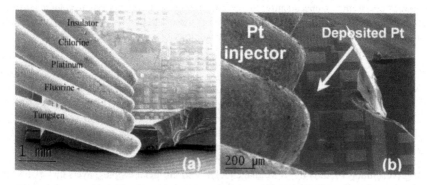

Figure 1. a) Gas injector layout. b) Pt gas injector positioned adjacent to the designated deposition area.

The second stage involves the milling of two 15×7 μm 45° wedges, using a probe current of ~ 190 pA, on either side of the Pt deposited area, leaving a 1-2 μm gap (Figure 2b and c). Finally after lowering the probe current to ~ 40 pA the intended laminar was polished to a thickness <100 nm, by tilting the specimen stage ± 2° to ensure a parallel sided specimen (Figure 2d). Once the final thickness was achieved, the sides and based were milled to sever the remaining laminar (Figure 2e and f). The processed specimen is then lifted out by electro-static charge with a glass needle attached to a micro-manipulator and deposited onto a holy carbon film coated TEM grid ready for analysis.

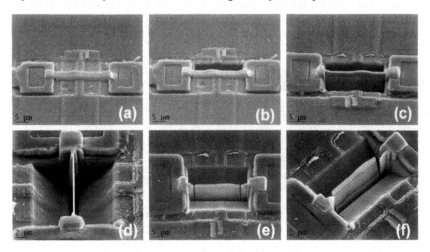

Figure 2. a) Pt deposited on device area. b) 15×7 μm wedge milled on one side. c) 2nd wedge milled on the opposite. d) Final polishing on the intended laminar. e) Sides of laminar milled through. f) Base of laminar milled leaving specimen to rest on side of a wedge.

Figure 3a, shows a [110] bright-field (BF) STEM image confirming the device was grown with the poly-Si gate electrode structure intact and uniform. The gate oxide is also visible and appears planar. No misfit dislocations were observed within the vicinity of the device structure, verifying the intended growth conditions have been met. HREM analysis of the outlined region determined the gate oxide thickness to be 7 nm and confirmed the gate electrode to be poly crystalline (Figure 3b).

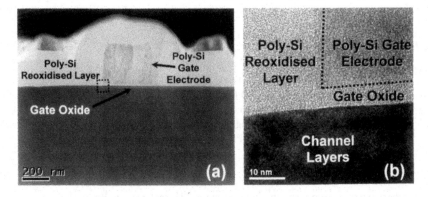

Figure 3. a) HREM [110] image of the gate oxide layer under the poly-Si gate electrode structure. b) High magnification BF STEM [110] image of gate electrode device structure.

A low magnification BF STEM [110] image, Figure 4a, illustrates the graded VS layer showing bands of dislocation density parallel to the growth front. EDS analysis was performed across the VS from the Si substrate to the poly Si reoxidised layer. The Ge concentration profile across the VS is shown in Figure 4b. It is found that the misfit dislocations are observed where major steps in Ge concentration occur, as a result of strain relaxation.

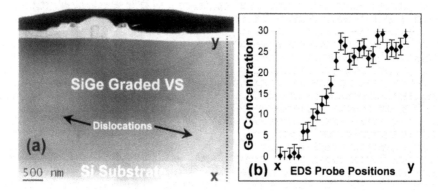

Figure 4. a) Low magnification BF STEM [110] image of device and VS layer. b) Ge concentration profile x to y across the graded VS layer.

4. Conclusions

In conclusion, we have demonstrated that FIB milling enables high precision, site specific TEM specimen preparation. The JEOL 6500F FEGSEM with integrated Orsay Physics ion column allows real time monitoring of the milling process. By subsequent use of HREM and STEM imaging, we are thus able to determine the device gate oxide thickness, and confirm the uniformity of the gate electrode structure with ease. EDS microanalysis verifies, the increase in the Ge concentration within the graded VS layer is as intended. Bands of misfit dislocations are observed within the VS where steps in the Ge concentration occur.

5. References

[1]   Paul D J 1998 Thin Solid Films 321 172-180
[2]   Paul D J 1999 Advanced Materials 11 191pp
[3]   Giannuzzi L A and Stevie F A 1999 Micron 30 197-204
[4]   Krueger R 1999 Micron 30 221-226
[5]   Schraub D M and Rai R S 1998 Prog. Crystal Growth and Charact. 36 99-122
[6]   Ismail K, Meyerson B S and Wang P J 1991 Applied Physics Letters 58 2117

6. Acknowledgements

The authors would like to thank the Engineering & Physical Science Research Council (EPSRC) for their financial support and Willy Scott (JEOL) and Sergio Lozano-Perez (Oxford University) for their experimental assistance.

Inst. Phys. Conf. Ser. No 179: Section 9
Paper presented at Electron Microscopy and Analysis Group Conf. EMAG2003, Oxford, 2003
©2003 IOP Publishing Ltd

Electron Beam Nanofabrication of Amorphous Chalcogenide-Metal Masks

N Nusbar, A G Fitzgerald, R K Debnath and S Persheyev

Carnegie Laboratory of Physics, Electronic Engineering and Physics
Division, University of Dundee, Dundee, DD1 4HN, UK

Abstract. A nanolithography system based on an electron microscope has been employed to fabricate x-ray masks in amorphous chalcogenide-metal bilayers in a single processing stage. As a result of electron beam induced chemical modification (ECM) silver nanometre dimensional patterns can be fabricated in these bilayers. These patterns exhibit a different x-ray absorption behaviour from the regions unexposed to the electron beam. Two types of x-ray mask have been formed under different electron beam accelerating voltage conditions.

1. Introduction

The photodoping of amorphous chalcogenides (e.g. GeSe, As_2Se_3 etc.) with metals and the migration of metals in these materials have potential applications in the formation of high-resolution photoresists [1], optoelectronic devices [2] and just recently as x-ray masks [3].

It has been reported that electron-beam irradiation can induce silver migration in amorphous chalcogenides giving rise to an accumulation of Ag^+ ions in the exposed region. This has been described as electron beam induced chemical modification (ECM) [4,5]. We have studied this effect for a number of years [6-10].

Nanometre dimensional patterns can be produced in amorphous chalcogenide-silver bilayers as a result of ECM. When metal amorphous chalcogenide bilayers are supported on an x-ray transparent membrane, the electron beam patterned regions exhibit a different x-ray absorption behaviour from regions unexposed to the electron beam. Regions where the silver accumulates have higher x-ray absorption. However, under certain conditions the exposed region can be silver deficient to give a higher x-ray transmission.

The present study has focused on developing a novel technique to fabricate x-ray masks in a single processing step using electron beam generated patterns in amorphous chalcogenide/silver bilayers.

2. Experimental

Bilayers of amorphous chalcogenides (a-As_2Se_3 and a-Sb_2S_3) with silver were studied in this experiment. Transparent silicon nitride membranes (Si_3N_4) surrounded by scanning electron microscope (SEM) compatible silicon wafer support frames were used as substrates. The frames consist of a thin film window consisting of the membrane of 100 nm thickness and 1 mm^2 area. The chalcogenide and silver films were evaporated successively onto the membrane to contain silver concentrations between 25 and 50 wt.%. The amorphous chalcogenide films and the silver films ranged from 100 to 530 nm and 30 to 100 nm in thickness respectively. The optimum film thickness for each component of the x-ray mask was calculated using theoretical x-ray transmission data obtained from *URL: http://www-cxro.lbl.gov/*. Good contrast in x-ray transmission curves was observed at 1.2 keV (Cu L line) for a 400 nm arsenic triselenide film and a 100 nm silver film, as shown in figure 1.

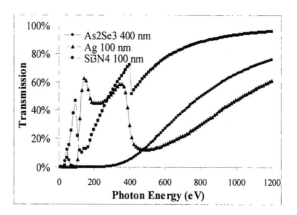

Figure 1. X-ray transmission curves for a 100 nm thick silicon nitride membrane, a 100 nm thick silver film and a 400 nm a-As$_2$Se$_3$ film.

The bilayers were patterned by an electron beam using a modified JEOL JSM-T220 SEM, externally controlled by means of ELPHY Quantum nanolithography software. The effects of exposure parameters such as accelerating voltage, beam current and electron dose were studied to optimize the production of high contrast x-ray masks. Once a suitable beam current and dose were achieved, then the accelerating voltage was varied from 5 to 30 kV. The accelerating voltage was found to determine the type of x-ray mask produced.

The structure and topography of the masks that were fabricated were subsequently viewed and the dimensions measured in an atomic force microscopy (AFM) in which the images were collected in tapping mode with a Nanoscope II multimode scanning apparatus. Energy dispersive x-ray microanalysis (EDX) measurements were carried out to characterise the composition of the mask patterns.

3. Results and Discussion

Two types of X-ray mask were obtained from the a-As$_2$Se$_3$/Ag bilayers (Fig 2). The accelerating voltage was found to be a crucial variable in the formation of these two types of mask. The use of a low accelerating voltage (5 and 10 kV) tended to produce a silver deficient trough-like patterned structure (Fig.2a). These trough-like regions can therefore transmit more x-rays than the surrounding unexposed regions.

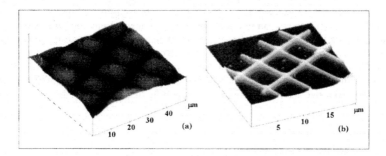

Figure 2. AFM image of a-As$_2$Se$_3$/Ag masks of 300 nm thick that were exposed to the e-beam by using (a) 10 and (b) 30 kV.

However, a protruding patterned structure (Fig.2b) was observed when the electron beam was applied at higher acceleration voltages (e.g. 20, 25 and 30 kV). A small increase in silver accompanied by carbon was detected in these protruding structures when they were analysed by energy dispersive x-ray microanalysis (EDX) in the SEM. It is most likely that the carbon contamination in the protrusions originates from electron beam decomposition of residual hydrocarbon molecules that are present in the high vacuum SEM chamber. Electron beam-induced protruding carbon lines have previously been observed on silicon surfaces [11]. It appears that at 20kV and above when a focused electron beam is scanned across the surface of the amorphous chalcogenide-silver bilayer, an electric field is directed towards the irradiated region that generates the silver ion (Ag$^+$) migration. The electron beam also induces surface deposition of carbon along the path of the electron beam. Our preliminary work has demonstrated that this protruding patterned structure can be used as an x-ray mask [3].

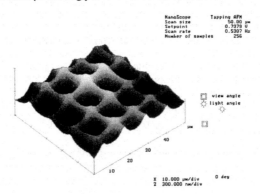

Figure 3. Atomic force micrograph of a-Sb$_2$S$_3$/Ag mask patterns generated with an e-beam using an accelerating voltage of 20 kV and a beam current of 0.4 nA.

X-ray masks formed with a protruding structure, when using accelerating voltages in the 15 to 25 kV range, were also found to form in the a-Sb_2S_3/Ag bilayer system (Fig 3). EDX analysis shows an increase in silver concentration after electron beam irradiation due to electrochemical modification. Unlike the a-As_2S_3/Ag bilayers only a small amount of carbon was detected in these structures. However, it was found that changing the magnitude of the accelerating voltage changed the dimensions of the patterns. Fine patterns could be generated at higher accelerating voltages (30 kV) due to the smaller spot size and the maximum current density that was achieved.

4. Conclusions

It has been demonstrated that accelerating voltage is the main driving factor for the structure and formation of amorphous chalcogenide-silver masks. Under different conditions of the accelerating voltage, two types of x-ray mask can be produced in a-As_2Se_3/Ag bilayers systems. However, only x-ray masks with a protruding structure formed under electron irradiation in a-Sb_2S_3/Ag bilayers. An increase in accelerating voltage was found to improve the contrast of the mask as well as the height and width of the patterns.

References

[1]   Yoshikawa A, Ochi O, Nagai H and Mizushima Y 1997 J. Appl. Phys. Lett. 31 (3) 161-163
[2]   Wagner T, Frumar M, Kasap S O, Vlcek Mir and Vlcek Mil 2001 J. Optoelec. and. Advance. Mat 3 (2) 227-232
[3]   Nusbar N, Fitzgerald A G, Persheyev S, Shaikh W, Hirst G J and Winter D 2003 PREP 2003: Proc. PREP 2003 Conf. 95-96
[4]   Yoshida N, Itoh M and Tanaka K 1996 J. Non-Cryst.Solids. 198–200 749-752
[5]   Tanaka K 1997 J.Appl. Phys. Lett. 70 (2) 261-263
[6]   McHardy C P, Fitzgerald A G, Moir P A and Flynn M 1987 J. Phys. C: Solid State Phys. 20 4055-4075
[7]   Mietzsch K and Fitzgerald A G 2001 Electron Microscopy and Analysis 2001: Inst. Phys. Conf. Ser. 168 477-480
[8]   Nusbar N, Fitzgerald A G and Mietzsch K 2001 Electron Microscopy and Analysis 2001: Inst. Phys. Conf. Ser. 168 510-512
[9]   Nusbar N, Mietzsch K and Fitzgerald A G 2002 ICEM-15: Proc. Int. Conf. Elec. Microsc. 15 967-968
[10]   Romero J S, Fitzgerald A G and Mietzsch K 2002 J. Appl. Phys. 91 (12) 9572-9574
[11]   Djenizian T, Santinacci L and Schmuki P 2001 Appl. Phys. Lett. 78 (19) 2940-2942

Acknowledgements

The authors would like to thank EPSRC for financial support of this research project (Grant No. GR/14682).

Inst. Phys. Conf. Ser. No 179: Section 9
Paper presented at Electron Microscopy and Analysis Group Conf. EMAG2003, Oxford, 2003
©2003 IOP Publishing Ltd

Crystal Silicon Tip Arrays Fabricated by Laser Interference Ablation of Amorphous Silicon Films

Y C Fan, M J Rose, J S Romero, and A G Fitzgerald

Division of Electronic Engineering and Physics, Faculty of Engineering and Physical Sciences, University of Dundee, Dundee DD1 4HN, United Kingdom

ABSTRACT: The crystal silicon tip arrays have been fabricated on the amorphous silicon films by three-laser beam interference ablation process. By atomic force microscopy (AFM) image observation and analysis, the correlations of the tip array morphology and structure with the characteristics of amorphous silicon films, the laser interference configuration and the laser fluence are established. The fabricated silicon tip arrays can be used as cathodes in the future field emission display.

1. INTRODUCTION

Field emission based display, as a promising new technology to fabricate flat panel displays, was hindered in the last few years. One of the main technical difficulties in field emission display is involved in the fabrication of the so-called Spindt type of field emitter arrays (Spindt et al 1968, 1976). Now it is generally accepted that the Spindt type of field emitter array is highly impossible to be used in the future field emission display technology because of its complexity and high cost in the fabrication process. Novel, simple and low cost fabrication processes of field emitter arrays are needed to be developed to revitalise the field emission technology.

The field emission cathode based on amorphous silicon film technology is one of the new progresses in this research direction (Silva et al 1999, Tang et al 2002). In this work, a one-step laser interference ablation process to fabricate crystal silicon tip arrays, which can be used in future field emission display, has been demonstrated. This one-step silicon tip array fabrication process is realised by three-beam interference ablation of amorphous silicon films deposited on metal-coated glass substrates. The novelty of this fabrication process is that the crystal silicon tip array can be formed on amorphous silicon films by using a single laser pulse irradiation without involving any masking or photolithography process.

By atomic force microscopy (AFM) observation and image analysis, the correlation of the characteristics of the fabricated silicon tip arrays with the three-beam interference configuration and the laser ablation fluence are established. Based on the morphology and microstructure analysis results the tip formation mechanism has been discussed.

2.   EXPERIMENTAL

Two kinds of hydrogenated amorphous silicon films (a-Si:H), i.e. low and high concentration hydrogen contained films, were prepared by DC plasma enhanced chemical vapour deposition onto 100 nm Chromium coated 7059 glass substrates. For the deposition of the low concentration hydrogen contained films (designated as L-a-Si:H films), pure SiH_4 gas was used with a gas flow rate of 40 sccm and the gas pressure was kept at 150 mTorr. For the high concentration hydrogen contained films (designated as H-a-Si:H films), 20 sccm SiH_4 and 20 sccm H_2 gas mixture was used and the gas pressure was also kept at 150 mTorr. In both situations, the depositions were performed in a vacuum chamber at a base pressure of typically 10^{-5} Torr, with a DC power of 6.5 W and a substrate temperature at 350 °C. The deposition rate has been calibrated, allowing the prepared films to have a thickness of about 100 nm.

After deposition, the a-Si:H films were structured by two- or three-beam overlapping interference irradiation (Nebel et al 1998, Kelly et al 1998). The laser ablation was performed in air by using a Q-switched and frequency doubled Nd:YAG laser (Surelite, Continuum) at the 532 nm with a seeding injection to obtain sufficiently long coherence lengths ($\cong 3$ m). During the irradiation, the laser machine was operated at 5 Hz and the irradiated sample was scanned in x-direction on a moving stage at a speed of 2 mm per second. The output laser energy was adjusted by changing the Q-switch delay time in respect to the triggering pumping lamp.

AFM morphological observations of the laser irradiated a-Si:H films were carried out by using a Digital Dimension™ 3000 atomic force microscope. The AFM images of the as-deposited and laser irradiated a-Si:H films were captured in tapping mode by using a silicon nitride cantilever tip with a beam length of 125 μm and a resonant frequency of around 300 kHz.

3.   RESULTS AND DISCUSSION

The optimum laser energy density for producing required structure is very much thin film property dependent. Figure.1 shows the representative AFM images taken from two-beam interference irradiated low- and high-concentration hydrogen contained films. A well-defined periodic grating structure can be observed on the laser irradiated L-a-Si:H films as shown in Fig.1(a). This grating structure was formed with a laser beam energy of 120 mJ/pulse before splitting and the energy density on the irradiated area was about 160 mJ/cm². Image section analysis certified that the period p of the grating structure complies with the theoretical formula $p = \lambda/2\sin(\theta/2)$, where λ is the laser wavelength and θ is the incident angle between the two laser beams. For the H-a-Si:H films, the material ablation occurs much easier at a relatively low irradiation energy density. The structure shown in Fig.1(b) was formed with a beam energy

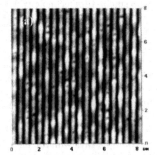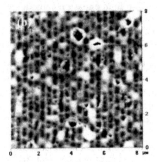

Figure 1 AFM images taken from two-beam interference irradiated (a) L-a-Si:H film (b) H-a-Si:H film. Image scan size: 8 μm

of 80 mJ/pulse and the energy density on the irradiated area was about 100 mJ/cm^2. In most cases, a lot of cavities or pits appear on the irradiated area and the periodic structure can only be dimly recognised.

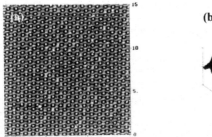 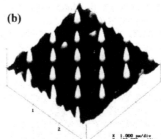

Figure 2 AFM images taken from three-beam interference irradiated L-a-Si:H films showing the crystal silicon tip array structure. (a) Top view AFM image, (b) 3D AFM image.

Figure 2 shows the typical crystal silicon tip arrays which are fabricated on the L-a-Si:H films by the three-beam interference ablation process. The three-beam interference irradiation was accomplished by splitting the main laser beam into A, B and C three beams which are directed towards the sample surface. The overlapped laser beams on the irradiated area form a periodic two dimensional interference pattern. In the optimum conditions the interference maxima can locally raise the temperature above the melting point of the amorphous silicon. The high peak power of the short laser pulse causes a quick melting and bouncing of the melted materials followed by an immediately solidification with a cease of the laser pulse. This quick meting, bouncing and instant solidifying process promotes to form high-aspect ratio crystal silicon tips. The tip array structure illustrated in Fig. 2(a) was fabricated with the intensities of the three split beams A, B and C respectively set at 20%, 30 %, and 50 % of the total beam intensity of 125 mJ/Pulse. In the three-dimensional AFM image Fig. 2(b), it can be seen very clearly that each silicon tip is formed in the centre of a crater. This observation result is consistent with the tip formation mechanism proposed by Evtukh et al (Evtukh et al 2000). The crystalline properties of the silicon tips have been confirmed by the XPS and TEM analysis.

A cross sectional analysis of the tip array is shown in Fig.3. The fabricated silicon tip array has a nearest tip distance of 570 nm, a tip height of about 90 nm and a tip base width of 180 nm. The calculated height to width aspect-ratio of the silicon tips is near to 1:2.

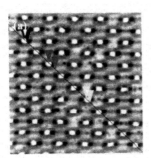 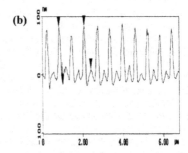

Figure 3 Cross section analysis of a three-beam interference structured L-a-Si:H film, (a) top view AFM image, image scan size 8 μm, (b) section profile along the line indicated in (a).

382

The high concentration hydrogen contained amorphous silicon films were also ablated by the three-beam interference irradiation. The representative AFM images are shown in Fig.4. These structures were fabricated with a laser energy of 70 mJ/pulse before beam splitting. Interestingly, all of the formed silicon cones or tips are cracked on top forming a crater-like structure. It is speculated that these crater-liked structures are probably formed by an explosive hydrogen releasing process.

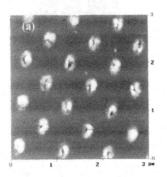 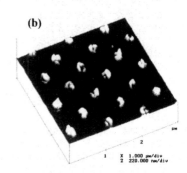

Figure 4 AFM images taken from three-beam interference irradiated H-a-Si:H film. (a) Top view AFM image, (b) 3D AFM image

The field emission characteristics of the fabricated silicon tip arrays have been measured. Preliminary results show that the electron emission threshold value (defined under the conditions that when one nano-ampere current is detected) is in the range of 20 to 50 V/μm. The measured emission threshold value is still too high for the field emission display application. In addition, it is still a difficult task to achieve large area uniformity in using laser interference ablation process to fabricate crystal silicon tip arrays. The non-uniformity of tip arrays is mainly caused by the uneven energy distribution across the laser beam and the interference pattern edging effect. To further improve the surface morphology, structure uniformity and filed emission performance of the silicon tip arrays, more detailed investigations on the thin film deposition and laser irradiation conditions are needed.

Acknowledgements We are indebted to Mr. Stuart Anthony for the preparation of amorphous silicon films.

REFERENCES

Evtukh A A, Kaganovich E B, Litovchenko V G, Litvin Yu M, Fedin D V, Manoilov E G, Svechnikov S V 2000 Semiconductor Physics, Quantum Electronics & Optoelectronics Vol.3 (4) pp 474 -478.

Kelly M K, Rogg J, Nebel C E, Stutzmann M and Katai Sz 1998 Phys. Stat. Sol.(a) **166** pp 651 – 657

Nebel C E, Christiansen S, Strunk H P, Dahlheimer, B, Karrer U and Stutzmann M 1998 Phys. Stat. Sol. (a) **166** pp 667 – 674

Silva S R P, Forrest R D, Shannon J M and Sealy B J 1999 J. Vac Sci. Technol. **B17** (2) pp 596 – 600

Spindt C A 1968 J. Appl. Phys. **39** 3504

Spindt C A, Brodie I, Humphrey L and Westerberg E R 1976 J. Appl. Phys. **47** 5248

Tang Y F, Silva S R P, Boskovic B O, Shannon J M 2002 Appl. Phys. Lett. **80**(22) pp 4154 – 4156

Inst. Phys. Conf. Ser. No 179: Section 10
Paper presented at Electron Microscopy and Analysis Group Conf. EMAG2003, Oxford, 2003
©2003 IOP Publishing Ltd

Microscopy with atomic beams: Contrast in a Scanning Helium Microscope

D A MacLaren and W Allison §

Cavendish Laboratory, University of Cambridge, Madingley Road, Cambridge CB3 0HE, UK

Abstract.
Helium atoms are an established, non-invasive probe of surfaces. The interaction between the atom and the surface, in the thermal energy regime, is predominantly one of elastic scattering that gives surface sensitivity without perturbation to the electronic or physical properties of the surface itself. Recent work has been directed at beam focusing, with the aim of creating a microprobe. All the key atom-optical components necessary for the construction of a scanning helium microscope (SHeM) are now in place and such a microscope is currently under development. We present an overview of the technique, with particular reference to the potential contrast mechanisms.

1. Introduction

Scanning helium microscopy is an emerging imaging technology that will be sensitive to features on the atomic scale, yet will use beam energies of no more than 50 meV. Helium atom scattering (HAS) has proved a uniquely sensitive, non-perturbing probe of surface morphology. The technique is now used routinely in the study of nucleation and growth of thin films, adsorbate diffusion, surface structure, alloying and surface phonons [1, 2, 3]. It is of particular advantage in high-precision structural studies of surfaces and in observation of delicate adsorption systems. Conventional HAS, however, is limited by a lack of spatial resolution, restricting its application to homogeneous and typically single-crystal systems. Recent advances in atom optics allow for the focusing of neutral helium atoms, producing a microprobe that can be scanned across a heterogeneous sample to create an image. Now, a scanning helium microscope (SHeM) is under construction [4]. The technique offers the prospect of a low energy, inert and charge-neutral probe that does not cause sample charging, heating or excitation and exhibits several contrast mechanisms. In this paper, we aim to give an overview of scanned helium microscopy and demonstrate potential contrast mechanisms by reference to existing HAS studies. We then conclude with a brief description of recent techological advances in the field and discuss their incorporation into a prototype SHeM.

2. Contrast and sensitivity

2.1. Helium-sample interactions

Contrast in a SHeM derives ultimately from the nature of the probe-sample interaction, which has been the subject of numerous HAS studies and is well understood [1, 2, 3]. Helium atoms scatter from the surface electronic corrugation or frontier electron orbitals of a sample. The helium-surface interaction is sketched in figure 1a and is well-approximated

§ To whom correspondence should be addressed (wa14@cam.ac.uk)

384

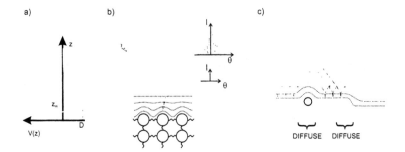

Figure 1. Helium-surface interactions and the origin of SHeM contrast. a) Schematic form of the interaction potential, exhibiting a shallow potential well, then a steep potential wall. b) Scattering mechanisms from a crystalline substrate (see text for details). Dotted lines illustrate the variation in potential corrugation with distance from the surface. Figure after Toennies [1]. c) Coherent scattering from a disordered surface. Aperiodic features such as adatoms and surface steps scatter diffusely and are not observed. Beams scattering from either side of the surface step can interfere to enhance contrast.

by a Lennard-Jones potential [3] acting perpendicular to the surface. At long range, attractive dispersion forces result in a shallow potential well (typically <10 meV) whilst at short range, Pauli exclusion repulsion gives rise to a sharp potential wall. Parallel to the surface, the potential is modulated by a corrugation function, a cross-section of which is suggested by the dotted isopotential lines illustrated in figure 1b. Classically, thermal helium may be considered to scatter from a "hard potential wall" with the locus of turning points - the measured surface corrugation - mapping out a surface of constant electron density, typically 3-4 Å above the surface. As helium approaches a surface it accelerates, "refracting" in the potential (figure 1b) [1]. Subsequent scattering depends upon the nature of the potential and can be diffracted (trajectories 1 and 3) or diffuse (trajectory 2). Elastic, single scattering (trajectory 1) dominates, although time-of-flight techniques are used routinely to resolve inelastic events and study surface phonons [3] or adsorbate dynamics [5]. Elastically-scattered helium diffracts to produce a pattern characteristic of the corrugation and which can be used for structural analysis. For highly corrugated surfaces, transient scattering into resonant states (trajectory 3) may also occur. Such states can give rise to sudden variations in diffraction intensity with beam energy [6], which in themselves could be used generate contrast under favourable conditions. "Diffuse" scattering incorporates both elastic scattering from aperiodic features (which is discussed below) and inelastic events. A major contribution to the diffuse signal arises from Debye-Waller attenuation [3] - inelastic scattering from surface phonons - which is strongly dependent on the sample temperature and composition. Samples of a rigid lattice or high atomic mass will attenuate the diffracted intensity less than amorphous, low-mass specimens [4]. Consequently, heterogenous samples with thermal, structural or compositional variations will give rise to spatial variations in helium reflectivity.

Diffraction and bound-state analysis are most easily observed from crystalline regions of a sample. HAS is also widely used, however, to study aperiodic features on disordered surfaces [2]. Figure 1c illustrates coherent scattering from an isolated adsorbate and surface step on an otherwise flat substrate. Both the adsorbate and step perturb the electron density contours such that impinging helium atoms scatter away from the specular direction (see also trajectory 2, fig 1b). Thus the surface reflectivity is inversely proportional to the surface defect density, or adatom coverage, etc and gives a direct measure of surface disorder. In comparison to other probes, an atom beam is far more sensitive to aperiodic features, as the initial probe-

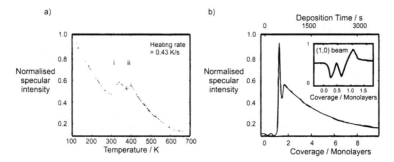

Figure 2. a) Variation of surface reflectivity with temperature for a Ni(100) single crystal with ˜5% surface coverage of molecular CO and H_2. Desorption signatures initate around 290 K and 380 K, superposed on a background thermal attenuation. b) Observation of the metalisation transition for thin-film deposition of K on Si(100), from [9]. The intensity scale for the inset does not relate to that of the main figure.

sample interaction (via the van der Waals force) is relatively long-range. The interaction results in "giant" cross-sections for isolated features, which can exceed the geometric feature size by an order of magnitude. For example, a single surface step appears as a diffuse strip of ˜10 Å width whilst isolated adatoms have typical cross-sections of ˜100 Å ^{2} [2, 3]. Even very small adsorbates, such as atomic hydrogen, which are difficult to see with other techniques, can be detected at very low coverage.

It is the sub-angstrom wavelength of thermal helium that leads to diffraction from the unit cells of a crystalline lattice. That same property also leads to a sensitivity to the vertical morphology of atomically- stepped surfaces, through interference scattering from adjacent terraces. Figure 1c indicates the origin of this sensitivity, which arises from Bragg interference effects between surface regions of different heights. In the figure, the step introduces a relative phase change between beams reflected from upper and lower terraces. Modulation of the perpendicular scattered wavevector - a "specular lattice-rod scan" - allows for determination of step height distributions with extreme accuracy. The technique is used routinely in thin-film growth studies [3, 7], an example of which will be presented in the following section.

2.2. Experimental Contrast

In the context of a SHeM, all of the above mechanisms can give rise to image contrast. Note also that the predominance of elastic single scattering should make image interpretation relatively straight-forward. The discussion of diffuse scattering from disordered surfaces suggests that topological contrast will dominate images of microscopically rough surfaces. In this section, we discuss additional compositional and structural contrast mechanisms by reference to conventional HAS data. All the data discussed derive from single-crystal studies conducted previously in our laboratory. Each study is chosen to demonstrate the signal changes that can arise from small changes in surface structure or imaging conditions and which indicate the likely contrast achievable from inhomogeneous samples.

Close-packed faces of clean transition metals exhibit high helium reflectivity. Subsequent processing of the surface will give rise to a change in reflectivity that, in a heterogeneous system, would produce contrast. Figure 2a demonstrates two such processes: first, annealing and second, desorption of molecular adsorbates. The figure plots the variation of specular reflectivity during annealing of a Ni(100) single crystal that has been dosed with a low

386

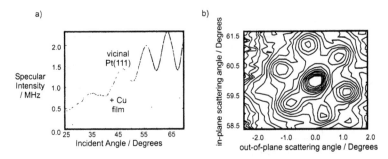

Figure 3. Diffraction contrast from 5ML Cu film grown on vicinal Pt(111) at 300 K. a) Specular lattice-rod scans of the clean and Cu-covered surface. b) Spot-profile of the specular beam at ~60° incidence, indicating a surface dislocation network. Contours are logarithmic.

coverage of coadsorbed CO and H_2. The crystal is annealed from 120 K to 700 K, left to right across the data set [8]. Thermal attenuation reduces the specular reflectivity exponentially with temperature, in accordance with Debye-Waller scattering [3]. Over a temperature range of only 150 K, reflectivity is observed to halve. Samples with inhomogeneous thermal properties may, therefore, give SHeM contrast by altering the sample temperature during scans. The attenuation of figure 2a is then modulated by two separate features, caused by the sequential desorption of molecular hydrogen (i) then carbon monoxide (ii). The total surface coverage is estimated to be of the order of only 5%, yet between ~290 K and ~340 K, surface reflectivity increases by 20%: a strong effect which illustrates the sensitivity to low adsorbate coverages. On an inhomogeneous sample, the effect could also be used to image variations in reactivity - reactive regions could be "titrated" with an adsorbate to cause changes in reflectivity only at selective sites. Surface Debye temperatures can also be fitted to the exponential decays on either side of the desorption peaks, and here are found to differ, in accordance with the change in surface identity.

In contrast to metals, semiconductor surfaces have characteristically low reflectivity. The covalent bonds of semiconductors such as Si cause strong corrugation of the surface potential, causing substantial diffractive scattering. An elegant demonstration of the diffractive contrast resulting from the difference between a semiconducting and a metallic surface is shown by the metalisation transition observed during the deposition of potassium on Si(100) [9]. Figure 2b illustrates the helium reflectivity of the K-Si(100) system during thin film growth. Initially, K adatoms are isolated and retain their atomic character. During submonolayer growth, weak intensity oscillations are observed, in accordance with the variation in surface roughness - a feature used routinely in studies of thin film growth [7]. At a critical coverage just over 1 ML, however, a rapid and dramatic increase in reflectivity is observed. The huge increase in reflectivity is consistent with the onset of metalisation in the K overlayer [9]. The data illustrate that strong contrast can arise from subtle changes in adsorbate density or bonding characteristics. The inset to figure 2b plots the variation in (1,0) diffraction order intensity during the initial phases of growth. Note that the strong reflectivity change in the specular beam is not echoed in the diffraction pattern. Favourable samples, with strong diffraction features, could be monitored in this way to provide "dark-field" SHeM contrast.

Diffraction effects can be used to probe both perpendicular and lateral surface structures and the same mechanisms have the potential to give rise to SHeM contrast. Figure 3 illustrates these two different types of diffraction, both taken from ~5 monolayer copper films grown on vicinal platinum (111) at 300 K. [10]. Figure 3a illustrates a specular lattice rod scan from

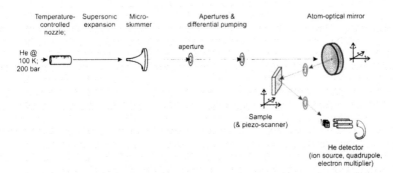

Figure 4. Schematic diagram of a prototype SHeM. A microskimmer collimates a conventional atom beam source. The differentially-pumped beam is then focused by an atom-optical mirror to create a helium microprobe. Micrographs are created by scanning the sample across the microprobe, with the specular reflection detected by a mass-spectrometer.

a vicinal Pt(111) crystal before and after the deposition of copper. Surface steps introduce a relative phase difference between waves scattered from successive terraces, resulting in a fringe pattern as a function of perpendicular wavevector. Copper deposition clearly alters the HAS signature. In both curves, the fringes have a periodicity that is characteristic of the steps on the surface [11]. As copper is deposited, the fringe spacing changes, indicating the presence of Pt-Cu and Cu-Cu steps. In addition to a decrease in net reflectivity (caused by surface roughening in concert with a change in surface Debye temperature) it is clear that under certain kinematic conditions contrast enhancement is possible. For example, at an incident angle of ˜59°, the Bragg minimum for the clean Pt(111) surface coincides with the Bragg maximum for thick Cu films on Pt(111). During Cu growth, therefore, variations of the incident beam wavevector will enhance selectively the reflectivity of different step-structures present in an inhomogeneous system.

Finally, figure 3b illustrates diffraction effects from lateral surface structures that can again impact significantly on the specular reflectivity of a sample. The figure presents a two-dimensional contour plot of the specular beam from the Cu/Pt(111) film. Strong satellite diffraction peaks are evident about the specular beam which are consistent with the formation of a long-range strain-relieving dislocation network in the overlayer [12]. Variations in the spot-profile are observed as a function of deposition rate and temperature as well as during annealing. Each variation causes significant reflectivity variations and is consistent with a structural phase transition, adatom rearrangment or the onset of alloying [10] - all process that will therefore be amenable to a SHeM. The figure illustrates that a SHeM would be sensitive to features on an atomic scale even if the atom probe diameter is not of atomic dimensions.

3. Instrument design

In this section, we will outline briefly the construction of a SHeM. Figure 4 shows the instrument schematically [4]. The microscope operates in a similar manner to other scanning microscopes, where a focused micro-probe is rastered across a sample and the backscattered signal is collected as a function of probe position to generate a micrograph. For the SHeM, the beam is generated by a standard supersonic-expansion atomic beam source [13]. A typical source brightness is 5×10^{22} atoms cm^{-2}s^{-1}sr^{-1} [4]. Alteration of the source temperature provides control over the beam energy, which ranges between 10 meV and 100 meV ($\lambda = 0.45$

- 1.44 Å); $\Delta v/v$ is typically of the order of 1%. A funnel-shaped aperture - the "skimmer" - is then used to select a fine beam from the expansion bubble [13]. It is the skimmer orifice that forms the effective source for all subsequent optics. Conventional skimmers have apertures of ~500 μm, however, apertures as small as 1 μm have been fabricated [14] and will be used in the SHeM. The ultra-fine helium beam is then differentially-pumped and strikes the focusing element: a single-crystal atom-optical mirror. The mirror is fabricated from a hydrogen-passivated silicon single crystal that is deformed under an applied electrostatic pressure - design details are discussed elsewhere [15, 16]. Both the macroscopic mirror structure [15, 16] - which determines the focusing aberrations - and the microscopic mirror structure [17] - which determines the reflectivity - have been optimised for high intensity, low aberration focusing. Crucially, electrostatic control over the pseudo-parabolic mirror profile has been shown to allow for *in-situ* control over mirror aberrations *to fourth-order* [16].

The mirror's specular helium reflection is then focused onto the sample surface. Spatial resolution of the SHeM is therefore limited only by the apparatus geometry and the size of the demagnified source. For example, using a 1 μm-diameter skimmer with source-mirror and mirror-sample distances of 2.0 m and 0.1 m respectively would yield a demagnified probe diameter of 50 nm - several orders of magnitude improvement over conventional HAS, yet with a beam intensity suitable for practicable microscopy. Finally, the sample is rastered across the focused He beam and the back-scattered signal is detected to form a micrograph. It is this final stage that is likely to be the "noise bottleneck" in the system. The ionisation efficiency in a conventional ion source is typically one part per million [18], which will limit image acquisition time. As a consequence, improved detectors, for example using field ionisation from nanotube tips, are currently being designed and tested [19].

In summary, we have outlined recent developments in novel imaging technique: scanning helium microscopy. Several contrast mechanisms have been identified and shown to be strong. A brief description of the microscope's operation has also been presented, from which it is clear that the key technical challenges have been met and that the construction of a SHeM is possible for the first time.

References

[1] Toennies J P 1974 Appl. Phys. 3 91
[2] Poelsema B and Comsa G 1989 Springer Tracts in Modern Physics 115 (Springer-Verlag: Berlin)
[3] Farías D and Rieder K 1998 Rep. Prog. Phys. 61 1575
[4] MacLaren D A, Holst B, Riley D J and Allison W 2003 Surf. Rev. Lett. 10 249
[5] Jardine A P, Ellis J, and Allison W 2002 J. Phys.: Condens. Matter 14, 6173
[6] Buckland J R and Allison W 2000 J. Chem. Phys. 112 970
[7] Dastoor P C and Allison W 2003 Phys. Rev. B. 67 245403
[8] MacLaren D A 2002 PhD Thesis University of Cambridge
[9] Foulias S, Curson N J, Cowen M C and Allison W 1995 Surf. Sci. 331-333 522
[10] MacLaren D A Dastoor P C and Allison W To be published
[11] Green D, MacLaren D A, Allison W and Dastoor P C 2002 J. Phys. D: Appl. Phys. 35 3216
[12] Holst B, Nohlen M, Wandelt K and Allison W 1998 Phys. Rev. B 58 10195
[13] Miller D R 1988 Atomic and Molecular Beam Methods Vol. 2 Ch. 2 (Oxford: Oxford University Press)
[14] Braun J, Day P K, Toennies J P and Witte G 1997 Rev. Sci. Instrum. 68 3001
[15] MacLaren D A, Goldrein H T, Holst B and Allison W 2003 J. Phys. D: Appl. Phys. 36 1842
[16] MacLaren D A, Allison W and Holst B 2000 Rev. Sci. Instrum. 71 2625
[17] MacLaren D A, Curson N J, Atkinson P and Allison W 2001 Surf. Sci. 490 285
[18] Bassi D 1988 Atomic and Molecular Beam Methods Vol. 2, Ch. 6-8 (Oxford: Oxford University Press)
[19] Riley D J , Mann M, MacLaren D A, Dastoor P C, Allison W, Teo K B K, Amaratunga G A J and Milne W 2003 Nano Lett. In press.

Inst. Phys. Conf. Ser. No 179: Section 10
Paper presented at Electron Microscopy and Analysis Group Conf. EMAG2003, Oxford, 2003

SE dopant contrast in LVSEM: The effects of surface and vacuum conditions

G H Jayakody and M M El Gomati

Department of Electronics, University of York, York, YO1O 5DD

ABSTRACT: Secondary electron (SE) imaging in Low Voltage Scanning Electron Microscopy (LVSEM) has been performed on planer p-type doped n-type silicon structures. All imaging has been carried out in a UHV SEM in order to identify the role of the surface contamination on the SE contrast. In-situ thin film metal deposition has been performed on clean silicon surfaces. The experimental results show excellent agreement with the model that explains the SE contrast mechanism to be due to metal to semiconductor structure. Energy band diagrams of the rectifying and ohmic contacts and the energy required for SE emission are also presented.

1. INTRODUCTION

The availability of high resolution FE-SEMs that can be operated at low beam voltages has generated much interest in the imaging of semiconductor devices and structures at low voltages. SE images collected at Low voltages (<5keV) show distinguishable contrasts from doped regions in the semiconductor. Generally, in the SE image, p type doped areas appear brighter and n type doped areas appear darker relative to the undoped areas. Therefore, by combining the high-resolution capability of LVSEM and with the contrast observed from doped regions, it is possible to image doped regions of nanometer scale semiconductor devices.

The majority of reports on LVSEM imaging have been involved in obtaining two dimensional (2D) dopant profiles from imaging cleaved cross section of silicon and superlattice structures (Perovic et al 1995, Elliot et al 2000). However, the use of SE imaging in LVSEM still remains a qualitative technique due to the difficulties in reproducing results. It has also been reported that the image contrast is affected by the carboneous contamination layer present on the sample surface when imaging is carried out in conventional vacuum conditions (El-Gomati and Wells 2001). In this paper, we report on the SE imaging of planer doped Silicon structures carried out in an Ultra High Vacuum (UHV). The use of UHV and in-situ surface cleaning enabled us to clarify the effect of the surface contamination on the SE contrast. In-situ thin metal deposition shows results that supports the metal to semiconductor contact model.

2. EXPERIMENTAL AND RESULTS

Samples used for the experiments were provided by the Institute of Scientific Instruments (ISI, Brno Check Republic) and were patterned with p^+ doped (10^{19} cm^{-3} Boron doped) on the n type (Phosphor doped) Si substrate. Before loading the samples into the UHV microscope they were etched in dilute HF for 3minutes in order to remove the native oxide layer from the surface. The experiments were carried out in a UHV system equipped with analytical instruments such as Cylindrical Mirror Analyser (CMA), Ion gun and a metal evaporator. The SE detector used for imaging is a custom built conventional Everhart-Thornly SE detector. Auger spectra were collected with a 5keV primary electron beam giving 10nA into a spot of ~300nm. For in-situ cleaning, 1 keV Xe$^+$ (90μA) ion beam was used to sputter the surface. Precautionary steps were taken to reduce the contamination sources entering into the UHV system. The Xe gas line was chilled using solid carbon dioxide to reduce contamination from water vapour. The complete filament and source-material unit of the metal evaporator were cooled using a water-cooling jacket in order to reduce the pressure increase during the evaporation. This method thus enables thin films of metals to be deposited at a pressure $< 10^{-8}$ mbar. The pressure in the chamber, during the SE imaging was ~1 x 10^{-10}mbar.

Two types of metals were chosen to be deposited on the ion sputtered Si surface. Chromium (Cr) and Nickel (Ni) are suitable candidates to study the effect of metal to semiconductor contacts with doped Si due to their different work function (ϕ) from that of Si. Cr ($\Phi_{Cr} < \Phi_{Si}$) forms Schottky and Ohmic contacts with p^+ and n type Si respectively. On the other hand Ni ($\Phi_{Ni} > \Phi_{Si}$) forms Ohmic and Schottky contacts with p^+ and n type Si (Sze 1980). Therefore if the SE contrast is affected by the m-s contact formation, a contrast reversal can be observed after Cr deposition. Figure 1 shows SE images collected from a specimen before and after metal deposition. Each sample was bombarded with 1keV Xe$^+$ for 10minutes before the deposition of Cr and Ni. This way the contamination deposited on the surface during the time the samples were handled in air can be removed. All SE images were collected at 1keV electron beam voltage.

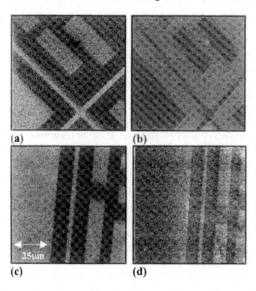

(a)   (b)

(c)   (d)

Fig. 1 a) *SEM image of the sample before Cr deposition b) with ~3nm thick Cr layer (After Cr deposition p-type areas have reversed brightness) c) SEM image of the sample before Ni deposition d) with ~5nm thick Ni layer (Before and after Ni deposition p-type areas appear brighter than the n type substrate. All images were collected with 1keV beam voltage with an E-T SE detector)*

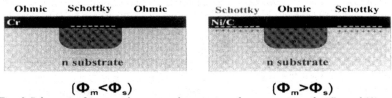

Ohmic Schottky Ohmic   Schottky Ohmic Schottky

Cr   Ni/C

n substrate   n substrate

$$(\Phi_m < \Phi_s)$$ $$(\Phi_m > \Phi_s)$$

Fig. 2 Schematic diagram shows metal to semiconductor contacts for Cr and Ni (or graphite type C) with Si (p^+ and n type). Note that the contacts are reversed for a metal with a work function lower than Si.

As can be seen from fig. 1, the Cr deposition has reversed the contrast from p-type Si areas and after Ni deposition, p-type areas still appear brighter. However, the relative contrast between p^+ and n type areas are reduced after metal deposition of Cr and Ni on each sample, presumably due to the presence of the metal film. After In-situ cleaning both the contrast and brightness were reduced. A schematic representation of the metal to semiconductor contacts of Cr on Si and Ni on Si is shown in fig. 2. Auger Electron Spectra (AES) collected from the as inserted sample, after ion cleaning, Cr deposited and Ni deposited samples are shown in fig. 3. As can be seen the as-inserted sample shows a significant C (KLL) peak. After ion cleaning the Si (LVV) Auger electron peak becomes the dominant peak.

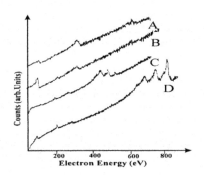

Counts (arb.Units)

200   400   600   800
Electron Energy (eV)

Fig. 3 *AES spectra from A) from as- inserted sample shows dominant C (KLL) peak. B) After 10min, 1keV Xe$^+$ ion bombardment Si (LVV) Auger electron peak at 89eV becomes the dominant peak. C) after 4nm Cr deposition. D) after the 3nm Ni deposition*

3. Discussion of Results

Results from the Cr and Ni deposition experiments show that the SE contrast from the doped Si areas is affected due to metal to semiconductor contact formation. A similar effect can take place when the semiconductor surface is covered with a thin carbon layer. It has been shown that under conventional vacuum conditions the surface of the imaged surface is covered with a carbon layer of about 3-5nm thick and it is of graphite nature (El-Gomati et al 2001). With a graphite type carbon layer ($\phi_C > \phi_{Si}$) metal to semiconductor contacts between p^+ and n type Si would be Ohmic and Schottky respectively. Due to the Schottky barrier that exits between n-type Si and the carbon layer, the energy required for SE emission will be higher than from the p^+ type Si regions. Therefore the p^+ type regions will appear brighter than n-type regions. Similar contacts are formed between Ni and Si; therefore p^+ type regions appear brighter as shown in fig. 1. However, with Cr, metal to semiconductor contacts between p^+ and Cr will be Schottky. Therefore p^+ regions appear less bright than the n-type areas.

Fig. 4 *Metal to Semiconductor contacts formed between p$^+$ type Si and for metal having work functions higher and lower than Si.*

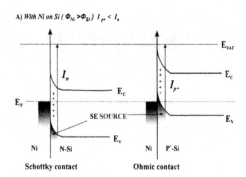

A) With Ni on Si ($\Phi_{Ni} > \Phi_{Si}$) $I_{p^+} < I_n$

Fig.4 depicts the energy required for SE emission (I) for each of these cases. As can be seen the with a thin Ni layer, $I_{p^+} < I_n$ and with Cr I $_{p^+} > I_n$. Therefore, with the Ni layer, the energy required for SE emission is higher for n-type Si compared to p$^+$ type. On the other hand with a Cr layer, the energy required for SE emission is lower for n-type Si compared to the p$^+$ type.

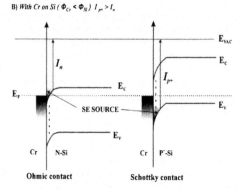

B) With Cr on Si ($\Phi_{Cr} < \Phi_{Si}$) $I_{p^+} > I_n$

4. CONCLUSIONS

SE imaging and thin metal depositions in UHV have been carried out in order to identify the effect of metal to semiconductor contacts on the SE image contrast of doped Si. It has been shown that the SE contrast is affected by Ohmic and Schottky contact formation. The Schottky and Ohmic contacts change the energy required for SE emission and hence change the observed contrast from doped regions. The in-situ ion bombardment caused the contrast and brightness of the SE image to be reduced. This is due to the increased surface states that have been generated. The existing contrast suggests that apart from the metal to semiconductor contacts, the surface potential fields also carry out information regarding the doped areas. In clean Si, this can be explained using the patch field model that has been proposed previously (Sealy et al 2000).

ACKNOWLEDGEMENTS:
The authors would like to thank T. Wells for useful discussions.

REFERENCE
D. D. Perovic, M. R. Castell, A. Howie, C. Lavoie, T. Tiedje and J. S. W. Cole (1995) Ultramicroscopy 58, p104
C. P. Sealy, M. R. Castell and P. R. Wilshaw (2000) J. Electron Microsc. 49, p311
M. M. El-Gomati, T. R. C. Wells and G.H. Jayakody (2001) Microscopy of Semiconducting Materials Conference, Oxford, No169, p435
M. M. El Gomati and T. C. R. Wells (2001) Appl. Phys. Lett. 79, p2931
S. L. Elliot, R. F. Broom and C. J. Humphreys (2002) J. of Appl. Phys. Vol 91, No 11, p9116

Inst. Phys. Conf. Ser. No 179: Section 10
Paper presented at Electron Microscopy and Analysis Group Conf. EMAG2003, Oxford, 2003
©2003 IOP Publishing Ltd

Size dependence of the contact angle of Bi liquid clusters supported on SiO

S Tsukimoto[*][§], S Arai[**] and H Saka[*]

* Department of Quantum Engineering, Nagoya University, Nagoya 464-8603, Japan
§ Now at Department of Materials Science and Engineering,
Kyoto University, Kyoto 606-8501, Japan
** Centre for Integrated Research in Science and Engineering, Nagoya University, Nagoya 464-8603, Japan

Abstract

Crystalline clusters of Bi with the diameter ranging from 70 to 6nm were prepared by in situ evaporation on to an amorphous substrate of SiO in a transmission electron microscope. These Bi clusters were then heated to get molten and the behaviour of the Bi liquid clusters observed in situ in a transmission electron microscope. It was found that the contact angle of a liquid Bi cluster decreases with decreasing the size of the liquid cluster below 7nm

1. Introduction

It is well established that the thermodynamic properties such as melting points depend on the size of fine crystalline clusters, when the size is, say, below 10nm(Takagi 1954, Allen *et al.* 1986, Saka *et al.*1986). By contrast, experimental information about liquid clusters is very limited because of difficulty in preparing isolated liquid clusters on substrates. During the course of in-situ observation of melting processes of fine particles of Bi, it was found that the contact angle of liquid cluster of Bi changes as a function of the size of liquid Bi clusters, which will be described in the present paper.

2. Experimental Procedures

A specimen-heating holder developed by Kamino and Saka (1993) was used, where a fine tungsten filament (25 μ m in diameter), bridged between two electrodes, was used as a heating element, on which a mixture of Si and Bi particles (a few to a few tens μ m in diameter) were mounted directly. The temperature of the heating element is highest at the central part of the span of the heating element. Thus, Bi particles which happened to sit on the central part were molten first during an in-situ heating experiment. From the liquid droplets of Bi, Bi atoms were evaporated and deposited on to the nearby off-central part of the heating element where the temperature was much lower.   Then, the electric current which flew the heating element was increased slightly in such a way the deposited Bi clusters at the off-central part got molten.   Observation was concentrated on Bi clusters deposited on to Si particles. The Si particles used in the present study were covered with a native oxide layer.   Since the native oxide layer on Si consists mainly of SiO (Kim and Carpenter 1990), behaviour of Bi particles supported on SiO substrate was observed.   The specimen chamber was pumped with a 340 L s^{-1} magnetic bearing turbomolecular pump and kept 1 x 10^{-5} Pa during the in situ heating experiments.

3. Results

Figure 1(a) and (b) shows low-magnification micrographs of the Bi particles supported on Si (covered with SiO) before and after melting, respectively.   Before melting (a), the Bi particles were well facetted and some of them showed diffraction contrast.   After melting (b), the particles became perfectly spherical and showed only uniform contrast.   It is noted that the positions of the Bi particles before and after melting remained unchanged and it is very easy to compare the shapes of the individual particles before and after melting.   The contact angles were measured on those particles for which the interfaces between the support and the particles were end-on, an example of which is indicated by arrows.

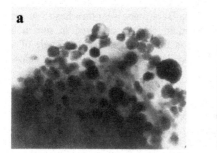

Figure 1. Low-magnification micrographs of Bi particles supported on SiO (a) before and (b) after melting.

To measure the contact angles of smaller particles, or clusters, more accurately, high-resolution electron micrographs (HREM) were taken.   Figure 2(a) and (b) shows an example.   Before melting (a), the cluster shows lattice fringes, the orientations of which are different from one place to another, indicating that this particular cluster was polycrystalline at least when imaged.   After melting (b), the cluster became perfectly spherical and the lattice fringes disappeared.   Furthermore, the projected area of the Bi liquid cluster is smaller than that of solid one.   This is quite reasonable because the volume of Bi shrinks on melting (Faber 1972).   The substrate is amorphous, indicating that the substrate is SiO and not Si.

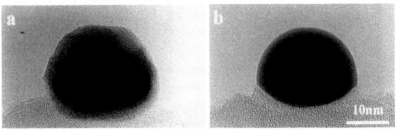

Figure 2. High-resolution electron micrographs (HREM) of a Bi cluster (a) before and (b) after melting.

On further observation, liquid clusters kept shrinking because of continuous evaporation of Bi atoms from the liquid Bi clusters. Figure 3 shows an example. In addition to huge particles denoted by C and D, smaller clusters A and B are sitting on an amorphous SiO substrate. Here, cluster B remained crystalline during observation, although its shape and orientation kept changing. This is what has been observed on crystalline clusters of a variety of metals (Iijma and Ichihashi 1985; 1986). On the other hand, cluster A remained in the liquid state throughout the observation as can be seen from utter lack of the lattice fringes. Furthermore, cluster A kept shrinking from 10nm to 5nm in diameter. The contact angle decreases with decreasing the size of cluster, as shown in fig.4.

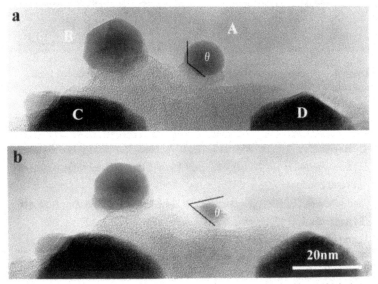

Figure 3. A series of HREM showing dynamical behaviour of Bi clusters with different size. Only cluster A is liquid, while the others B-D remain crystalline.

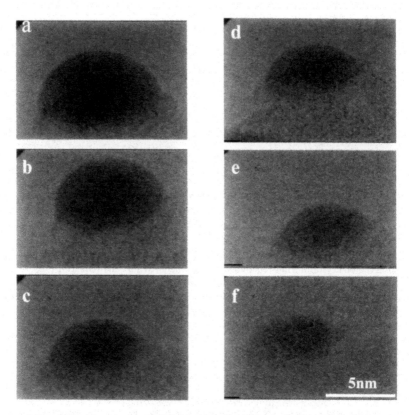

Figure 4. A sequence of shrinking of a liquid Bi cluster reproduced from a video tape. The contact angle decreases with decreasing the size.

References

Allen,G.L., Bayles,R.A., Gile,W.W., and Jesser,W.A.,1986,*Thin Solid Films*,**144**,297.

Fabor,T.E., *Introduction to the Theory of Liquid Metals*, 1972, (Cambridge University Press, London)

Iijma,S., and Ichihashi,T., 1985, *Jpn.J.appl.Phys.*,**24**,L125; 1986,*Phys.Rev.*,B,**56**,616.

Kamino,T., and Saka,H.,1993,*Microsco.Microanal.Microsctruct.*,**4**,127.

Kim,M.J., and Carpenter,P.W., 1990, *J.Mater.Res.*,**5**,347.

Saka.H., Nishikawa,Y., and Imura,T., 1988, *Phil.Mag.*,A,**57**,895.

Takagi, M.,*J.Phys.Soc.Jpn.*,1954,**9**,359.

Inst. Phys. Conf. Ser. No 179: Section 10
Paper presented at Electron Microscopy and Analysis Group Conf. EMAG2003, Oxford, 2003
©2003 IOP Publishing Ltd

ELNES Modelling of Interfaces in Steels

C A Dennis, A J Scott, R Brydson, A P Brown

Institute for Materials Research, University of Leeds, Leeds, LS2 9JT, UK

Abstract. Ab initio calculations, using the plane-wave pseudopotential code CASTEP, have been used to investigate model steel grain boundaries. A $\Sigma=5$ (013), $36.9°$ [100] tilt grain boundary has been constructed and substitutional phosphorus impurity atoms incorporated in order to mimic the segregation of impurity elements at the boundary.

1. Introduction

Impurities in iron that segregate to grain boundaries have a distinct effect on the material's physical and mechanical properties. Electronic structure calculations can be used to study the bonding across interfaces in materials. These calculations allow the mechanism by which segregants change the mechanical properties of materials to be understood. It is possible to use the shape of the d band to study the cohesive nature of the interfaces in iron as well as to determine the number of bonding and anti-bonding electrons present in the d-DOS. Additionally, measures of the total energy of the system can be used to give an understanding of the cohesive energy of the interface.

EELS within the (S)TEM can then be used to detect small amounts of segregants at grain boundaries at high spatial resolution. The Near Edge Fine Structure, ELNES, can be used to measure changes in the unoccupied densities of states (DOS) when a sub-nanometer sized probe is moved across a grain boundary. Band theory can then be used to interpret the energy loss fine structure and relate the observed near edge structure to the local electronic structure in the material. It is then possible to compare these experimental results with theoretical calculations.

This paper presents the results of calculations on the DOS using the plane-wave pseudopotential code CASTEP, which employs density functional theory to simulate the properties of interfaces as well as that of solids and surfaces. Initially, a bcc iron unit cell has been considered and the DOS calculated. A $\Sigma = 5$ (013), $36.9°$ [100] tilt grain boundary has been constructed and the energy of the resulting structure has been minimised. Adhesion at this clean grain boundary has been studied in order to be able to understand the factors affecting adhesion between iron surfaces at the atomic level. The segregation of phosphorus to $\Sigma = 5$ grain boundaries in steel has also been investigated by the addition of phosphorus atoms along the grain boundary plane. The results show that the DOS of ferrite varies considerably with misorientation and the addition of phosphorus clearly causes a decrease in cohesion at the $\Sigma = 5$ Fe/Fe grain boundary.

2. Experimental Procedure

Initial calculations were performed on body centred cubic iron (α-Fe or ferrite) with a lattice parameter of 2.8664 Å. All calculations were carried out using the spin corrected generalised gradient approximation (GGS) so that the magnetic effect of the ferrite is included by explicitly treating the spin of each electron. This is important, as intergranular cohesion along symmetric tilt boundaries in iron has been shown to be strongly dependant on the magnetic structure at the particular interface [1]. A geometry optimisation was carried out to test the value obtained by this approximation. Geometry optimisations using the GGS approximation, ultrasoft pseudopotentials with core correction and an energy cut-off of 300eV gave the lattice parameter within 0.6% of the experimental value for bcc iron.

A model $\Sigma = 5$ (013), [100] tilt grain boundary was constructed by rotating two grains by $36.9°$ about a <100> tilt axis with the boundary plane parallel to {013}.

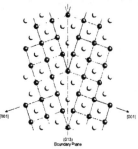

Fig. 1. The structure of the bcc Fe/Fe $\Sigma = 5$ (013), $36.9°$ [100] tilt boundary. The atoms are shown in different greyscales to represent the stacking sequence of the atomic planes.

This particular grain boundary structure was chosen as an example of a real ferrite grain boundary with segregated phosphorus atoms. Experimental work by Lejcek and Hofmann [2, 3] presented experimental values for the segregation of phosphorus at the $\Sigma = 5$ α-Fe grain boundary.

DOS calculations were carried out on the boundary following the complete geometry optimisation of the structure. This geometry optimisation involved firstly relaxing the atomic positions, then the lattice parameters and finally the entire structure.

Phosphorus atoms were then added to the grain boundary and the structure was relaxed again. DOS calculations were carried out in order to determine both the total and partial DOS so that changes in the local electronic structure of the boundary could be determined. The number of substituted atoms was based on the amounts of phosphorus found to be segregated at the $\Sigma = 5$ α-Fe grain boundary by Lejcek.

3. Results and Discussion

3.1 Ferrite unit cell

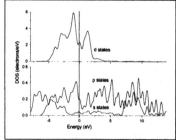 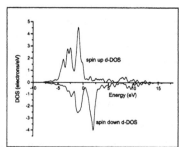

Figs. 2 and 3. The s, p and d-DOS for bulk ferrite and the d-DOS for bulk ferrite split into its spin up and spin down components.

DOS calculations performed on a ferrite unit cell using the spin corrected generalised gradient approximation can be seen in figure 3 above. The structure of the unit cell was relaxed to test the results obtained.

When the cumulative integral of the spin up and spin down curve up to the Fermi energy is determined it is clear that the occupancies of the spin up and spin down components are different. The spin down contribution to the d-DOS has shifted up in energy indicating a net spin and a therefore a spin moment showing that iron is ferromagnetic.

3.2 Σ = 5 ferrite tilt grain boundary

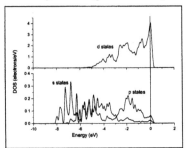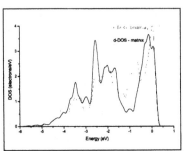

Figs. 4 and 5. The s, p and d-DOS for the atom central to the boundary and the d-DOS for the atom central to the boundary and the atom at the centre of matrix furthest from the boundary.

The relaxation of the model grain boundary caused the two planes of atoms closest to the grain boundary to relax outwards and compress into the surrounding bulk iron.

Integrating under the d-DOS for the Fe atoms on the grain boundary shows that a much higher number of antibonding states are filled than in bulk ferrite. Also the average energy of the d band is higher. Both of these factors indicate an overall decrease in cohesion at the Σ = 5 ferrite tilt grain boundary in comparison to the cohesion in bulk ferrite.

3.3 Σ = 5 ferrite tilt grain boundary with substitutional phosphorus atoms

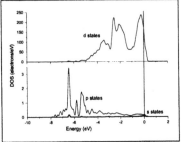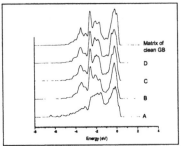

Figs. 6 and 7. The phosphorus s and p-DOS and the iron d-DOS for the atom central to the boundary and the Fe d-DOS for Fe atoms where A is on the plane closest to the phosphorus atom on the boundary and D is the furthest away.

As with the clean Fe/Fe boundary, when the structure was relaxed, the two planes closest to the boundary relaxed outwards. This was more apparent around the substitutional phosphorus atoms. Figure 6 indicates that there is a limited amount of hybridisation between the phosphorus p-states and the iron d-states. The energy of the d-band is higher and there are more antibonding states filled than in the clean $\Sigma = 5$ grain boundary. This suggests a further decrease in the grain boundary cohesion.

Figure 7 shows that the shape of the d-DOS changes rapidly with increasing distance from the grain boundary. It is clear that the effect of the phosphorus on the surrounding iron atoms decreases significantly with increasing distance from the boundary. This is important as it shows that the effect of the phosphorus impurity atoms is localised and the size of the grain boundary supercell is sufficient to prevent interaction between neighbouring grain boundaries.

4. Conclusions

Adhesion between metallic iron interfaces is a very important consideration in many industrial processes. These results clearly show that grain boundary structure and impurity segregation both affect the cohesion at iron grain boundaries. The segregation of phosphorus causes a significant decrease in cohesion at the bcc Fe $\Sigma = 5$ tilt grain boundary.

It is hoped that further work will study the segregation of phosphorus at real grain boundaries in iron. One possible problem however that can occur when studying grain boundaries in the TEM is the difficulty in orientating boundaries so that they are parallel with the electron beam. Most grain boundaries of interest to materials science are general, high angle grain boundaries containing a mixture of tilt and twist which are very difficult to study in the TEM. A possible compromise is to examine grain boundaries that have resulted from standard processing but that also have crystallographic characteristics that allow them to be both imaged and to collect EELS data with greater ease. It is hoped that samples of both iron and iron with phosphorus additions can be studied using EELS within the (S)TEM. This will hopefully allow comparisons to be drawn between experimental data on real grain boundaries and theoretical studies.

Acknowledgements

The authors would like to acknowledge the EPSRC and Corus for funding this research.

References

[1]   Yesilleten D., et al. 1998 Physical Review Letters. 81 (14) 2998-3001.
[2]   Lejcek P. 1994 Analytica Chimica Acta. 297 (1-2) 165-178.
[3]   Lejcek P., Hofmann S. and Krajnikov A. 1997 Materials Science and Engineering a-Structural Materials Properties Microstructure and Processing. 234 283-286.

Inst. Phys. Conf. Ser. No 179: Section 10
Paper presented at Electron Microscopy and Analysis Group Conf. EMAG2003, Oxford, 2003
©2003 IOP Publishing Ltd

Intergranular films in Si_3N_4 studied by TEM

M Döblinger, C D Marsh, D Nguyen-Manh, D Ozkaya and D J H Cockayne

Department of Materials, University of Oxford, Parks Road, Oxford, OX1 3PH

Abstract. Like many ceramic materials, Si_3N_4 contains amorphous intergranular films (IGFs). These films affect or even dominate the properties of the ceramic material. As a result of the sintering process the amorphous IGFs in Si_3N_4 ceramics contain oxides of silicon and other elements. These additional elements are also observed in amorphous pockets where three or more grains meet. To obtain information about the local atomic structure of the IGFs we employ electron diffraction with a convergent probe focussed on the IGF. This method provides the reduced radial distribution function G(r) within the films and thereby a fingerprint of the atomic structure. We have studied Si_3N_4 containing oxygen as an additional element. Our first results show that the order in both pockets and IGFs is closely related to amorphous SiO_2, with little if any nitrogen. This suggests SiO_2 as the main constituent in the pockets as well as in the interfaces. In Y_2O_3 doped material, the G(r) indicates the presence of yttrium in the IGF.

1. Introduction

Si_3N_4 ceramics are an important example [1] of structural ceramics containing intergranular films (IGFs) where the thickness of the IGF resulting from the sintering process is about 1 nm. At high temperature, many physical properties such as strength, creep resistance and oxidation resistance depend on the properties of the IGFs. The almost constant thickness of these films throughout the sample is generally interpreted as meaning that they are thermodynamically stable. Both the pockets at triple junctions and IGFs in Si_3N_4 ceramics contain an amorphous silicate but the IGFs and pockets do not necessarily have the same composition. The solubility of Si_3N_4 in a metastable SiO_2-rich liquid (corresponding to glassy pockets in pure Si_3N_4/SiO_2) has been determined experimentally by EELS to be 1-4 atomic % N/(N+O) [2]. However, a precise determination of the composition of an IGF has not yet been achieved. One of the experimental challenges is to distinguish between the signals from the IGF and the crystalline Si_3N_4 at the grain boundary.

One method to obtain information on both structure and composition of the IGF is to derive the reduced radial distribution function, G(r). G(r) provides information about nearest neighbour distances and thereby also about composition. The derivation of the G(r) for bulk amorphous material by electron diffraction is a well established method [3]. However, to study amorphous IGFs a probe size of about 1 nm is required. Standard G(r) analysis assumes an incoherent beam and the use of parallel illumination. Recently, the implications of a focussed and partially coherent electron beam, as obtained from a field emission gun transmission electron microscope (FEGTEM), have been studied and shown not to be a major obstacle to the derivation of G(r) from nanovolumes [4].

Because reflections from adjacent crystals are excluded from the analysis this method for deriving the G(r) of IGFs inherently considers the amorphous film only. This is a major advantage compared to other techniques where it is always a problem to decide whether the signal being measured comes from the IGF or from the crystal at the boundary.

2. Experimental

Si_3N_4 containing 4 vol % SiO_2 as the only additive and Si_3N_4/SiO_2 also containing 5 vol % Y_2O_3 were produced by hot isostatic pressing [5] [6]. TEM samples were prepared by slicing, grinding, polishing and argon ion milling.

Electron diffraction and HREM were carried out using a JEOL 3000F FEGTEM operating at 297 kV. The diffraction patterns were formed by a focussed beam with FWHM of ~1.5 nm. The diffraction data was energy filtered using a Gatan Imaging Filter (GIF), attached to the base of the FEGTEM column. An energy slit width of 10 eV centred on the zero loss peak was used, with the diffraction patterns recorded on a 2048 x 2048 pixel 7941F/20 MegaScan CCD camera. To avoid saturating the CCD the central spot in the diffraction pattern was placed just off the CCD while the diffraction patterns were recorded. The sampling in reciprocal space was performed out to at least 30 nm^{-1}.

If present, reflections from adjacent crystalline areas in the diffraction patterns from IGFs were used for scaling. These reflections were masked before azimuthal averaging centred on the central spot was performed. Figure 1(a) shows an example of such an experimental diffraction pattern. The one dimensional dataset resulting from azimuthal averaging represents the radial scattering intensity I(s), where s is the distance in reciprocal space from the central diffraction spot. The reduced intensity function $\varphi(s)$ is obtained from I(s) through the formula:

$$\varphi(s) = s\frac{I(s) - Nf(s)^2}{Nf(s)^2}$$

where f(s) is the atomic scattering factor and N is a fitting factor representing the number of atoms. Subsequent Fourier sine transformation results in the reduced radial distribution function G(r). G(r) describes the deviation of the local density $\rho(r)$, at a distance r from an arbitrary atom, from the average density, ρ_0.

$$G(r) = 4\pi r(\rho(r) - \rho_0) = 8\pi \int_0^\infty \varphi(s)\sin(2\pi s r)ds$$

3. Results

Samples of pure Si_3N_4/SiO_2 were found to have IGF thicknesses of approximately 1 nm (figure 1(b)), in agreement with previous reports [7]. On the other hand in the same material, but doped with 5 vol % Y_2O_3, both similarly wide and significantly thicker IGFs were present (figure 1(c)).

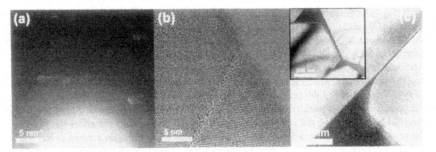

Figure 1. (a) diffraction pattern used for deriving G(r), (b) intergranular film (IGF) in the Si_3N_4/SiO_2 sample and (c) ~1nm thick IGFs and pocket in the Y_2O_3 doped material. Inset – pockets and thick IGF.

The upper two curves in figure 2(a) show the G(r) typical for IGFs and pockets in the pure Si_3N_4/SiO_2 material. These both have a first peak at 0.16 nm and a second peak at 0.27 nm, in agreement with the Si-O and O-O bond lengths respectively in amorphous SiO_2 [8]. A G(r) from a thin IGF in the Y_2O_3 doped material (bottom curve figure 2(a)) also shows a first peak at 0.16 nm, suggesting a large number of Si-O bonds, but the rest of the curve is considerably different from the other two curves. In the sample containing Y_2O_3 the second peak is shifted towards shorter distances and the ratio of the first to second peak heights is much larger, the third peak is also shifted towards smaller distances.

The origin of these differences becomes clear when the theoretical G(r) for SiO_2, Si_2ON_2 and Y_2O_3 are considered (figure 2(b)). The bottom curve in figure 2(b) shows the G(r) of simulated pure amorphous SiO_2 obtained from our own atomistic modelling. In the absence of models for Si_2ON_2 and Y_2O_3 the top two curves in figure 2(b) are derived from published crystal structures of orthorhombic Si_2ON_2 and cubic Y_2O_3. For these crystal structures, the bond lengths and the coordination polyhedra of the cations are expected to be the same as in the corresponding amorphous materials. Hence, the nearest neighbour distances and thereby the first peaks of the G(r) of crystalline and corresponding amorphous material are very similar. It is therefore valid to compare theoretical crystalline curves to experimental amorphous curves as long as only the first peaks are considered. In figure 2 all three curves are convoluted with a Gaussian representing the experimental resolution.

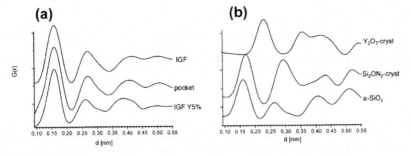

Figure 2. (a) experimentally derived G(r) and (b) simulated G(r).

The positions of the first two peaks in the experimental G(r) of pure Si_3N_4/SiO_2, for both IGFs and pockets (figure 2(a)) are identical to the positions in the theoretical G(r) for amorphous SiO_2 (figure 2(b)), suggesting that the short range order is the same as in amorphous SiO_2 and that SiO_2 is the main constituent in both the pockets and the IGFs. The first peak in a Si_2ON_2 G(r) consists of a superposition of two peaks resulting from the Si-N and the Si-O bonds, which are close in length. This results in an apparent shift of the first peak in the G(r) of Si_2ON_2 (middle curve figure 2(b)) towards larger distances compared to the position of a first peak resulting only from Si-O bonds (bottom curve, figure 2(b)). In the pure Si_3N_4/SiO_2 we did not observe experimentally such a shift for either the IGFs or the pockets (top two curves figure 2(a)). This indicates a low nitrogen content in the IGFs and the pockets. This is in agreement with previous reports that the pockets contain less than 5 atomic % N/(N+O) [2].

In the Y_2O_3 doped material the shift to shorter distances of the second peak in the G(r) of the IGFs (as in figure 2(a)) can be explained by the contribution of Y-O bonds. The first peak in the theoretical G(r) of Y_2O_3 (top curve, figure 2(b)) corresponds to Y-O bond length at 0.23 nm. In figure 2(a), bottom curve, this peak contributes to the second peak at 0.27 nm which is mainly due to the O-O bonds and hence results in an apparent shift of the second peak towards shorter distances. The shift of the third peak towards smaller distances is more complicated to explain because the position of the third peak in the Y_2O_3 doped material is the result of many contributions from yttrium coordination, for example Si-Y.

4. Conclusions

Using a FEGTEM enables the rapid acquisition of electron diffraction data from volumes on the nano-scale. This enables electron diffraction data to be obtained from intergranular films (IGFs) in Si_3N_4 ceramics and hence enables the G(r) spectra from IGFs to be obtained. Our results indicate that in the pure Si_3N_4/SiO_2, the short range order in the IGFs and pockets is very similar to the order in amorphous SiO_2. No evidence for Si-N bond distances was detected in either the IGFs or the pockets, suggesting that nitrogen is a minor constituent in both. The G(r) of IGFs from Y_2O_3 doped material show the presence of yttrium and a major contribution from amorphous SiO_2.

We would like to acknowledge the EU and the NSF for funding under EU contract GRD-CT-2001-00586 and NSF Award DMR-0010062.

References

[1] Petzold G and Herrmann M Silicon Nitride Ceramics, High Performance Non-Oxide Ceramics II, Jansen M (editor), Structure and Bonding, 102, Mingos D M P (series editor), Springer Verlag, 47-167
[2] Gu H, Cannon R M, Seifert H J, Hoffmann M J and Tanaka I, 2002 J. Am. Ceram. Soc. 85 25-32
[3] Cockayne D J H and McKenzie, D R 1998 Acta Crystallographica Section A 44 870
[4] McBride W Cockayne D J H and Nguyen-Manh D, submitted to Ultramicroscopy
[5] Hoffmann M J, IKM, University of Karlsruhe, Germany
[6] Tanaka I, Dept of Materials Science and Engineering, University of Kyoto, Japan
[7] Gu H, Xiaoqing P, Cannon R M and Manfred R, 1998 J. Am. Ceram. Soc. 81 3125-35
[8] Mader W and Heinemann D, 1998 Ultramicroscopy 74 113-122

Inst. Phys. Conf. Ser. No 179: Section 10
Paper presented at Electron Microscopy and Analysis Group Conf. EMAG2003, Oxford, 2003
©2003 IOP Publishing Ltd

The unsuitability of Fresnel fringes for estimating intergranular film thickness in doped-alumina

Ian MacLaren

Institute for Materials Science, Darmstadt University of Technology, Petersenstr. 23, 64287 Darmstadt, Germany

Abstract. Fresnel fringes are found at both film-containing and film-free grain boundaries in Y-doped alumina. For the film-containing boundary the inferred film thickness was significantly smaller than that measured by HRTEM. It is suspected that interface diffuseness or a compositional gradient caused this effect. In the film-free case, the appearance of Fresnel fringes was attributed to a local reduction of the inner potential at the grain boundary due to a dopant-induced grain boundary expansion, in line with a recent theoretical study of the effect of dopants on boundary structure in alumina. It is concluded that the strong dopant segregation to alumina grain boundaries can lead to confusing results in Fresnel fringe analysis.

1. Introduction

Fresnel fringes are shown in out-of-focus images of grain boundaries containing amorphous films as a consequence of the change in internal potential in the material. This has been used for the characterisation of the thickness of intergranular films since the work of Clarke (1979). Assuming that the crystal-film interfaces are sharp and that the potential in the film is lower than that in the crystals, the fringe spacing is given by:

$$w = w_0 + c \, \Delta f^{1/2} \tag{1}$$

where:

$$c = (3\lambda)^{1/2} \tag{2}$$

w is the observed fringe spacing, w_0 is the film thickness, Δf is the defocus, and λ the electron wavelength. A graph is plotted of w versus Δf: w_0 is the intercept at $\Delta f = 0$. This has been widely used for the characterisation of film thicknesses, with perhaps particular success in the analysis of Si_3N_4 (Cinibulk et al., 1991; Jin et al., 1998).

It is however well known that the dependence of fringe spacing on defocus is rarely in agreement with (1) since c is usually $< (3\lambda)^{1/2}$ and there is usually an asymmetry between over- and underfocus (e.g. Cinibulk et al., 1991; Jin et al., 1998). Numerous studies have concentrated on determining the exact form of the potential change between film and crystal, usually finding a diffuse boundary between the two (e.g. Ness et al., 1986, Rasmussen and Carter, 1990, Dunin-Borkowski, 2000). Nevertheless, in some cases, Fresnel fringes are observed from film-free grain boundaries, for example in NiO (Rühle et al., 1984; Merkle et al., 1985) and Al_2O_3 (Simpson et al., 1986). This was believed to occur in all cases because of grain boundary expansion resulting in a lower internal potential. In one case, a dopant which segregated to the boundaries was shown to increase the effect (Rühle et al., 1984).

2. Experimental Procedure

An alumina ceramic doped with 500 wt. ppm Y_2O_3 was prepared as described previously (Gülgün *et al.*, 2002) and was sintered at 1450 °C for 96 h and then annealed at 1650 °C for 12 h. As a result, abnormal growth of some grains was observed as has been reported in detail elsewhere (MacLaren *et al.*, 2003). TEM specimens were prepared using a standard procedure of slicing, disc cutting, mechanical polishing, dimpling and ion milling, including low energy ion milling (500 eV) to reduce surface damage. The sample was examined using a JEOL3010 UHR transmission electron microscope operated at 297 kV; images were recorded onto film or onto the CCD camera of the attached Gatan image filter. Measurement of fringe profiles was performed using the Digital Micrograph software (Gatan Inc.).

3. Results

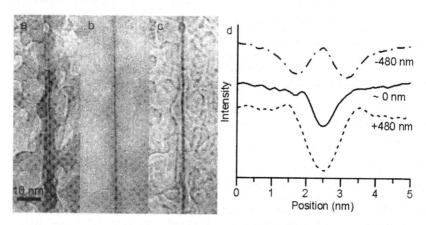

Figure 1: Fresnel fringe images of a film-containing grain boundary in Y-doped Al_2O_3: a) $\Delta f = -480$ nm; b) $\Delta f \approx 0$ nm; c) $\Delta f = 480$ nm; d) profiles across the boundary.

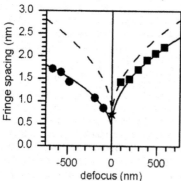

Figure 2: Plot of fringe spacing against defocus for the boundary of Figure 1.

A grain boundary at which HRTEM already showed the presence of an amorphous film was studied by Fresnel fringe imaging. The resulting Fresnel fringes are shown in Figures 1a-c for underfocus, approximate Gaussian focus and overfocus. Fringes are apparent for the boundary and profiles of these fringes averaged for 500 pixels along the boundary are shown in Figure 1d.

The fringe spacings were plotted against defocus in the standard way to yield the plot shown in Figure 2. The measured fringe spacings fall clearly below the theoretical form of (1), as shown in dotted lines. Fitted matches to the points by adjusting c and w_0 are shown in solid lines. Moreover, the commonly seen asymmetry between under- and overfocus is also seen. In this case, this also gave an asymmetry in the measured values of film thickness: 0.35 nm from the underfocus series and 0.5 nm from the overfocus series. Neither of these values agrees well with the measurement from HRTEM images (MacLaren 2003a,b) of 0.75 nm. It is therefore clear that something is going on at the interface not taken into account by the simplistic model of abrupt potential changes at the interface. This may be related to diffuse amorphous-crystalline interfaces or compositional gradients within the amorphous film itself. Whatever the case, it would require much more sophisticated analysis to gain reliable information from this grain boundary by Fresnel fringe analysis, perhaps for example using the method of Dunin-Borkowski (2000).

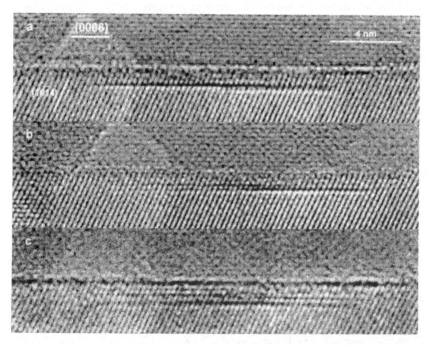

Figure 3: High resolution images of a grain boundary showing Fresnel fringes in under- and over- focus conditions, but containing no detectable amorphous film: a) $\Delta f \approx -48$ nm; b) Approximately Gaussian focus; c) $\Delta f \approx +64$ nm.

Figure 3 shows a series of HRTEM images recorded from another grain boundary at which no amorphous layer could be identified: it is as far as can be seen atomically sharp. Nevertheless, Fresnel fringes can be seen at the boundary in both under- and overfocus conditions. It is therefore clear that something at the boundary is causing a change in internal potential even where no films are present. As noted above, such an effect has previously been seen in alumina (Simpson *et al.* 1986) probably due to reduced density at the grain boundary In the Y-doped alumina of the present work, Y segregates strongly to the grain boundaries (Gülgün *et al.*, 2002). Simulations of the effect of doping on grain boundary structure in alumina show a significant grain boundary expansion of about 6 % perpendicular to the boundary for a Σ13 boundary for similar segregation levels to those seen here (Fabris and Elsässer, 2003). This would as such lead to a significant reduction in density which would probably more than offset the increased potential from the large Z yttrium atoms, thus leading to a reduced inner potential at yttrium doped grain boundaries.

4. Conclusions

Fresnel fringe analysis of a grain boundary containing an amorphous film showed confusing results and apparently smaller film thickness than that measured from HRTEM. It is expected that this could be attributed to diffuse amorphous crystalline interfaces or possibly to compositional gradients across the interface. It is thus clear that a more sophisticated analysis would be required to extract useful information from the Fresnel fringe profiles in this case.

Fresnel fringes were also observed from a film-free grain boundary. It was believed that segregated yttrium atoms had led to a boundary expansion in line with recent theoretical models of segregated grain boundaries in alumina. This would lead to a drop in the inner potential giving rise to the fringes.

It is therefore abundantly clear that the internal chemistry of grain boundaries in doped alumina has a significant influence on the local inner potential, thus significantly complicating any form of Fresnel fringe analysis for this material.

5. Acknowledgements

The author is thankful for helpful discussions with Profs. M. Rühle, R.M. Cannon and H. Fuess, and Drs. M.A. Gülgün and S. Fabris, as well as useful comments from reviewers of a related journal article. Dr R. Voytovych is gratefully acknowledged for the preparation of the alumina ceramic used in the work, and Mrs M. Sycha for the preparation of the TEM specimen.

References
Cinibulk M K, Kleebe H-J and Rühle M 1991 J Am Ceram Soc 76 426
Clarke D R 1979 Ultramicroscopy 4 33
Dunin-Borkowski R E 2000 Ultramicroscopy 83 193
Fabris S and Elsässer C 2003 Acta Materialia 51 71
Gülgün M A, Voytovych R, MacLaren I, Rühle M and Cannon R M, 2002 Interface Sci 10 99
Jin Q, Wilkinson D S and Weatherly G C 1998 J Eur Ceram Soc 18 2281
MacLaren I 2003a Ultramicroscopy under review
MacLaren I 2003b proceedings of EMAG 2003.
MacLaren I, Cannon R M, Voytovych R, Gülgün M A, Popescu-Pogrion N, Scheu C, Täffner U and Rühle M 2002 J Am Ceram Soc 86 650
Merkle K L, Reddy J F and Wiley C L 1985 Ultramicroscopy 18 281
Ness J N, Stobbs W M, and Page T F 1986 Philos Mag A 54 679
Rasmussen D R and Carter C B 1990 Ultramicroscopy 32 337
Rühle M, Bischoff E and David O 1984 Ultramicroscopy 14 37
Simpson Y K, Carter C B, Morrissey K J, Angelini P and Bentley J 1986 J Mater Sci 21 2689

Inst. Phys. Conf. Ser. No 179: Section 10
Paper presented at Electron Microscopy and Analysis Group Conf. EMAG2003, Oxford, 2003
©2003 IOP Publishing Ltd

Compositional characterisation of electrostatic bonds

A T J van Helvoort[1], K M Knowles[2] and R Holmestad[1]

[1]Department of Physics, Norwegian University of Science and Technology (NTNU), Høyskoleringen 5, N-7491, Trondheim, Norway.

[2]Department of Materials Science and Metallurgy, University of Cambridge, Pembroke Street, CB2 3QZ, Cambridge, U.K.

Abstract: Silicon wafers and silicon wafers with a buried oxide have been electrostatically bonded to Pyrex and examined by transmission electron microscopy. In combination with X-ray energy dispersive spectrometry, transmission electron microscopy proves to be a powerful tool for studying the movement of mobile cations during electrostatic bonding. Anodic oxidation of the silicon anode material during bonding can be detected both by using markers in the interfacial region and by detecting boron using high spatial resolution electron energy loss spectrometry.

1. Introduction

Electrostatic bonding is a simple solid state bonding technique used in the production of various silicon-based devices such as pressure sensors, flat screens and fluidic devices. Although it is an established joining technique in industry, the underlying bonding mechanism is surprisingly poorly understood.

Recent bright field transmission electron microscopy (TEM) studies (Xing et al. 2002a, b, Helvoort et al. 2003a, b, c) in combination with X-ray energy dispersive spectrometry (XEDS) have been able to detect a sodium depleted layer in the Pyrex near the bonded interface for both silicon and aluminium as anode materials. Depending on the process conditions and the choice of anode material, the sodium depletion layer varies between 0.5 and 2 μm in width. Less mobile potassium ions pile up close to the bulk Pyrex in this sodium depletion layer.

A permanent bond is formed by anodic oxidation of the anode material during the bonding procedure. This process is less well understood than the cation migration, and to date TEM studies have had limited success in detecting the anodic oxide reaction layer. The anodic oxide reaction layer is only readily detectable using bright field and dark field TEM in special cases where it can be clearly differentiated from the depleted Pyrex. An example of this is found in aluminium-Pyrex electrostatic bonds made at 450°C. In these bonds the reaction product is a 20–25 nm thick layer of crystalline γ-Al_2O_3 (Helvoort et al., 2003a, c). In aluminium-Pyrex bonds made at lower bonding temperatures (200–350°C) the reaction product is amorphous with almost the same average atomic weight as Pyrex and cannot be distinguished from the sodium depletion layer by bright field and dark field imaging techniques (Xing et al. 2002a, Helvoort et al. 2003c).

Here, we have examined cross-sections of electrostatic bonds between Pyrex glass and silicon, and Pyrex glass and silicon with a buried oxide. The bonds have been studied by TEM and scanning transmission electron microscopy (STEM), together with XEDS and electron energy loss spectrometry (EELS). The formation of a sodium depletion layer is discussed for the samples with a buried oxide, but the main focus of this study is on how evidence can be found for the formation of a thin silica amorphous anodic reaction layer adjacent to cation-depleted Pyrex.

2.   Experimental

The bonding procedure, the glass composition and the TEM sample preparation route are described in detail elsewhere (Helvoort, et al., 2003a, b, c). In the work here, n-silicon and silicon with a 400 nm buried oxide layer, starting 40 nm under the silicon surface, were used as anode materials. A 300 kV Philips CM30 (LaB_6) and a 200 kV JEOL 2010F (field emission gun) equipped with XEDS detector, a post-column Gatan Image Filter (GIF) and a STEM unit were used to study the TEM cross-sections.

3.   Results and discussion

A bright field TEM image of an electrostatic bond between the silicon with the buried oxide layer and Pyrex is shown in Fig. 1. XEDS confirms the nature of the various features. The bright layer 1.17 ± 0.03 μm thick visible in the Pyrex adjacent to the silicon-Pyrex interface is the sodium depletion layer.

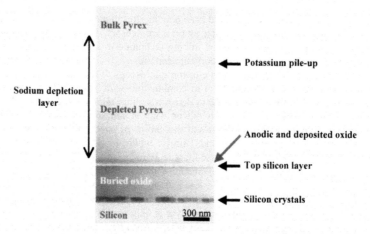

Figure 1. Bright field TEM image of silicon with a buried oxide bonded to Pyrex using electrostatic bonding. The bond was made at 350°C, 1000 V and 60 s.

The black line, 0.26 ± 0.05 μm away from the sodium depletion edge with the bulk Pyrex, is a pile-up region of less mobile potassium ions. This distance away from the bulk Pyrex is slightly higher than in silicon-Pyrex bonds made under similar conditions, but the size of the sodium depletion layer is comparable (Helvoort et al. 2003b).

The silicon with the buried oxide was studied by TEM before and after bonding. The buried oxide layer was made by oxygen ion implantation. The silicon crystals within the buried oxide layer are due to incomplete amorphisation at the lower interface between the oxide and the (bulk) silicon because of the straggle in the ion distribution during ion implantation. A thin layer of silicon 40.4 ± 0.8 nm remained on top of the buried oxide layer after ion implantation. For convenience, we will designate this layer the silicon top layer. On top of this silicon top layer there is a 25.1 ± 1.1 nm protection layer of SiO_2 which has been grown by the device manufacturer.

A comparison of the silicon with a buried oxide examined by TEM unbonded and after bonding shows that the size of the buried oxide is not affected by the electrostatic bonding process. However, the thickness of the top silicon layer is reduced. Here, crystalline silicon is transformed into amorphous silicon oxide. This reduction in thickness of the silicon is interpreted as the thickness of the anodic oxide layer formed during the electrostatic bonding process. For a bond made at 350°C, 1000 V and 60 s, the oxide layer is 4.8 ± 1.2 nm thick. For longer bonding time and a higher bonding temperature (600 s, 450°C) the measured anode oxide is larger, 16 ± 3.8 nm. The oxidation takes place at the interface between the silicon top layer and the anodic oxide layer. The oxygen ions created in the depleted Pyrex move through the already existing oxide layer during the anodic oxidation process, as has previously been suggested (Jorgensen, 1962). However, further bonding conditions need be analysed to characterise fully the effects of the process parameters of time, voltage and temperature on the interfacial reaction process.

The amorphous anodic silica layer cannot be distinguished from the deposited silica layer using bright field TEM, dark field TEM or high resolution TEM and through focus series. Furthermore, in silicon-Pyrex bonds the depleted Pyrex and the anodic silica cannot be distinguished. Both substances are amorphous and the slight compositional difference (7.1 at.% boron and 0.9 at.% aluminium) does not result in a sufficient difference in the average atomic number between the two substances. Hence, the electron absorption behaviour of the two substances is essentially the same. Absorption, and thus contrast, differences are therefore only explicable in terms of density differences between the amorphous substances. Such density differences account for the contrast seen between the depleted Pyrex and the bulk Pyrex in Fig. 1, and also account for the contrast seen in this figure between the depleted Pyrex and the deposited oxide layer on the silicon wafer. The buried oxide in Fig. 1 acts as a marker for the original interface and so the anodic oxide can be indirectly traced because of this marker. Experiments with markers on the interface (deposited inert materials that locally prevent anodic oxidation) or in the surface (channels in the silicon or Pyrex) have been found to be less suitable for detecting the anodic oxide layer.

In silicon-Pyrex electrostatic bonds without interfacial markers the anodic oxidation can only be demonstrated through the detection of small differences in boron content between the amorphous silica reaction layer and the cation-depleted Pyrex. Spot EELS analyses were taken in the interfacial region using a 1 nm diameter STEM probe. Smaller probes led to an unacceptable build-up of carbon on the sample surface and too low a count rate. The results of this study are shown in Fig. 2. The analysis shows that boron could not be detected by EELS in the depleted Pyrex at a

412

distance of 5 nm from the interface. Spot analyses at different distances from the silicon-Pyrex interface show that the thickness of the anodic oxide is 17.5 ± 3.0 nm for a bond made at 450°C, 1000 V and 900 s. Significantly and encouragingly, this is comparable to the oxide thickness determined in the silicon samples with a buried oxide. The spatial resolution is limited due to the probe size and in particular sample drift during the energy loss acquisition. Energy filtered TEM using the two and three window method could not detect the anodic oxide because of the weak boron signal. Mapping the oxide rather than taking spot analyses would require long acquisition times and hence problems with carbon build-up and sample drift.

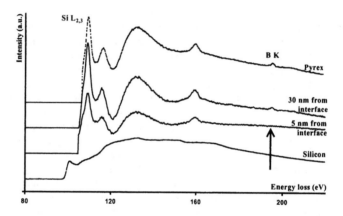

Figure 2. EELS spectra from an interfacial region of a silicon-Pyrex bond made at 450°C, 1000V and 900 s. The probe size was 1 nm, the collection time 8 s and $1 < t/\lambda < 2$. The arrow indicates the position where the boron K edge is absent in the sodium depletion layer close to the silicon-Pyrex interface.

Acknowledgements

We would like to thank the UK Department of Trade & Industry for funding the project through the Postgraduate Training Partnership scheme between the University of Cambridge and TWI, the Norwegian Research Council (NFR) for funding project number 140553/420 and Dr D.G. Hasko for supplying the silicon sample with a buried oxide layer.

References

Helvoort A T J van, Knowles K M and Fernie J A 2003a J. Ceramic Proc. Res. 4 25-29
Helvoort A T J van, Knowles K M and Fernie J A 2003b J. Electrochem. Soc. 150 G624-G629
Helvoort A T J van, Knowles K M and Fernie J A 2003c J. Am. Ceramic Soc. in the press
Jorgensen P J 1962 J. Chem. Phys. 37, 874-877
Xing Q, Sasaki G and Fukunaga H 2002a J. Mat. Sci.: Mat. Electronics 13 83-88
Xing Q F, Yoshida M and Sasaki G 2002b Scripta Materialia 47 577-582

Inst. Phys. Conf. Ser. No 179: Section 10
Paper presented at Electron Microscopy and Analysis Group Conf. EMAG2003, Oxford, 2003
©2003 IOP Publishing Ltd

Oxidation of nanoscale TiAlN/VN multilayer coatings

Z Zhou, W M Rainforth, P Eh Hovsepian[*], W -D Münz[*]

Department of Engineering Materials, University of Sheffield, S1 3JD UK
[*]Materials Research Institute, Sheffield Hallam University, S1 1WB, UK

Abstract. The oxidation behaviour of nanoscale multilayer PVD TiAlN/VN coatings heat-treated in air 600°C for 30minutes were studied using TEM. A duplex surface oxide was found, consisting in a number of phases, with an outer layer rich in V_2O_5, $AlVO_4$ and an inner layer containing V_2O_5 and VO_2 and possibly Al_2TiO_5 with TiO_2 extending across both layers. Oxidation occurred preferentially along columnar grain boundaries, but the oxidation front did not follow the multilayer structure. The oxidation mechanisms are discussed.

1. Introduction

Nanoscale multilayer coatings consisting of alternating layers of two different nitrides (e.g. TiAlN, VN, CrN) exhibit exceptionally high hardness (often referred to as 'superhard'), and are being increasingly used to improve the life of tooling, e.g. milling tools. One example of this family of coatings, TiAlN/VN multilayers, with a layer period of ~3nm, have exhibited excellent sliding wear resistance ($1.26{\times}10^{-17}$ $m^3{\cdot}N^{-1}m^{-1}$), in common with other coatings, but with a lower friction coefficient ($\mu = 0.4$, pin-on-disc test, Al_2O_3 ball counterpart) in comparison to other wear protective coatings [1], e.g. TiAlN/CrN ($\mu = 0.7$-0.9). Laboratory dry sliding wear tests of TiAlN/VN coatings (against an Al_2O_3 ball) yielded wear debris containing V_2O_5 [2], which has inherently low friction, leading to suggestions that formation of this oxide is key to the low frictional behaviour of these coatings.

Thermo-gravimetric analysis (TGA) of TiAlN/VN has shown that the onset of oxidation occurs at ~550°C, while XRD suggested that a significant amount of V_2O_5 was present after 600°C/30minutes, but in conjunction with oxidation products of $AlVO_4$ and TiO_2, the distribution of which could not be evaluated by SEM alone [3]. On the basis of the prior TGA, SEM and XRD investigation, the present work concentrated on a detailed transmission electron microscopy (TEM) analysis of oxidation at 600°C for 30mins.

2. Experimental

TiAlN/VN superlattice coatings were grown on stainless steel substrates in an industrial scale physical vapour deposition coating machine (HTC-1000 ABS, manufactured by Hauzer Techno Coating BV, Venlo, The Netherlands). Details of the deposition process are given elsewhere [1]. Coatings were deposited with a substrate bias voltage of –75V with 3-fold rotation. Since there was no mechanical shielding/shutters during deposition, intermixing between layers occurred, as detailed elsewhere [4].

Cross-sections of the oxidised surfaces were prepared for TEM observation in the conventional manner, as detailed in [5]. A Joel 2010 UHR field emission gun TEM/STEM was used to characterise the coating structure. Scanning transmission electron microscopy (STEM) in conjunction with energy dispersive x-ray spectroscopy (EDS) was employed to provide an overview of the elemental distribution within the coating and oxide [6]. Higher spatial resolution characterisation of the oxide/coating interface was obtained by energy filtered transmission electron microscopy (EFTEM) using a Gatan GIF. Both 3-window elemental maps and jump ratio images were generated using the standard energy slit conditions.

3. Results

An overview of the oxide morphology as a function of temperature and time is given in [3]. Fig. 1a gives a bright field STEM image of the surface, showing the duplex nature of the oxide. The inner layer adjacent to the non-oxidised coating appeared porous with a nano-crystalline structure. The outer layer comprised several different phases with particle and needle morphologies. An overview of the chemical distribution within these layers is given by the STEM/EDS maps in Figs. 1b, c, and d. In the outermost oxide region, angular particles were Al/V rich, identified by electron diffraction as $AlVO_4$. These were surrounded by Ti rich and V rich particles, identified as rutile-TiO_2 and V_2O_5 respectively. The rutile particles were distributed across both inner and outer oxide layers. Similarly, the needle-shaped particles, which were Al and Ti rich, were present across both the inner and outer oxide layers. Unfortunately, this phase could not be identified by electron diffraction. The ring pattern taken from the nano-crystallites in the inner layer was difficult to index, with the best fit being found to a sum of the major diffraction planes of V_2O_5 (200 and 100), VO_2 (100 and 220), and Al_2TiO_5 (230 and 006), but this was not a unique identification. There was no evidence of Al_2O_3.

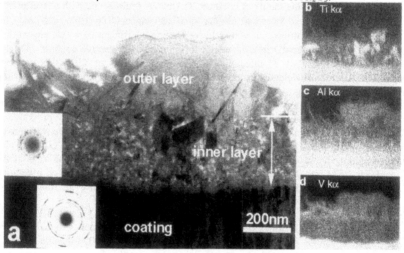

Fig.1 STEM image of the oxidised surface and the associated elemental maps.

Fig. 2 gives a bright field TEM image of oxide/coating interface region. The porous nature of the inner oxide layer is evident. However, careful examination shows a band of fine pores in the oxide immediately adjacent to the interface. The oxidation front clearly did not follow the multilayer structure, but appeared to be strongly determined by the columnar grain boundaries. The Fresnel contrast from the multilayer structure extended right up to the oxide/coating interface, suggesting that inter-diffusion of species between individual layers was not sufficient to remove the multilayer structure prior to oxidation.

Fig. 3 gives a bright field TEM image of the oxide/coating interface and its associated maps from a region indicated by the box in Fig. 2. The thickness map (t/λ typically 0.2-0.3 in the oxide; brighter contrast are thicker regions) confirmed the presence of porosity and also shows that preferential ion beam

Fig. 2. Bright field TEM micrograph of coating/oxide interface.

thinning of the oxide, with respect to the coating, had occurred. Fig. 3 also gives EFTEM maps of the same area using N-K, O-K V-$L_{2,3}$, Ti-$L_{2,3}$, and Al-$L_{2,3}$ edges. N and O maps defined the coating and oxide. The coating front appeared V-rich, followed by a V-depleted band (~40nm) immediately adjacent to the coating/oxide interface. However, the elemental distribution was obviously complex, with V-rich and Ti-rich regions distributed throughout the inner oxide. In contrast, the Al concentration tended to be uniform.

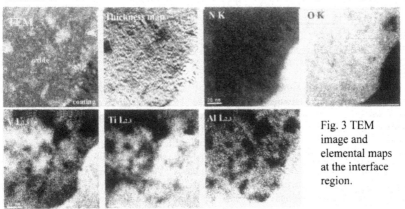

Fig. 3 TEM image and elemental maps at the interface region.

4. Discussion

The observation of a duplex oxide in the present work is consistent with published work on TiAlN, where an outer layer Al-rich oxide and inner layer Ti-rich oxide was observed [7]. In the current work, the distribution of Al and Ti within the oxide layer essentially similar, but important changes resulted from the incorporation of VN.

The inner oxide layer was made up predominantly of TiO_2 and a phase most likely to be Al_2TiO_5, although not uniquely identified. This suggest the reaction: $4TiAlN + 7O_2 \rightarrow 2Al_2TiO_5 + 2TiO_2 + 2N_2$. Although the inner layer did contain some V based oxides, (VO_2, V_2O_5), the majority of the V was found in the outer oxide as V_2O_5 and $AlVO_4$. Several V based oxides can form in the V-O system at 400-600°C, with the phase transformation series from $VO \rightarrow VO_2 \rightarrow V_2O_5$ with major intermediate phases of V_2O_3 and V_4O_9 [8,9]. Only a small proportion of VO_2 was found and no evidence of VO, V_2O_3 or $V_4 O_9$. Thus, much of the V_2O_5 was formed directly, rather than through a sequence of intermediate phases.

Table I gives an indication of the ionic radii, thermal data and molecular volume of Al_2O_3, TiO_2 and V_2O_5. In the current work, no Al_2O_3 was found. However, for the oxidation of TiAlN, the outer layer comprised Al_2O_3 and was believed to arise because of the high mobility of Al^{3+} in rutile TiO_2 (which formed the inner oxide layer) through interstitial sites [7]. Therefore the presence of V prevented the formation of Al_2O_3, with $AlVO_4$ being the major Al bearing phase in the outer layer. Thus, the reaction leading to the formation of $AlVO_4$ appears to have been $Al_2TiO_5 + V_2O_5 \rightarrow 2AlVO_4 + TiO_2$.

The dominance of V_2O_5 in the outer oxide is not expected from the thermodynamic driving force for formation, but is not surprising when all factors are considered. V_2O_5 has a low melting point, and therefore high vapour pressure. Although the diffusion rate of V in TiO_2 or Al_2TiO_5 is not available, V^{5+} has a much smaller ionic radius than Al^{3+} (Table I), and this is believed to provide rapid diffusion of this species to the surface, consistent with the current observations.

Substantial porosity was found in the inner oxide, the size of which depended on distance from the coating/oxide interface. Presumably the porosity arose because of the release of N_2 from oxidation of the coating, and the large change in molar volume on formation of the oxides (Table I).

Table I, Ionic radii, thermal data and molecular volume of Al_2O_3, TiO_2, V_2O_5.[9,10]

	Al_2O_3	TiO_2	V_2O_5
Gold-schmidt Ionic radius	Al^{3+}: 0.57	Ti^{4+}:0.64	V^{5+}:0.4
ΔG_{1000}(kJ), 727°C	-1362.4	-739.5	-1172.8
Melting point: (°C)	2030	1920	674
Molecular volume (cm^3)	25.6(AlN:12.8)	18.8(TiN: 11.8)	54.0(VN:10.7)

5. Summary

Cross-sections of nanoscale multilayer PVD TiAlN/VN coatings heat-treated in air 600°C/30mins were studied using detailed TEM. The surface oxide exhibited a duplex structure, with an outer layer comprised mainly of V_2O_5, rutile-TiO_2, $AlVO_4$ and an Al/Ti rich oxide, while the inner oxide, which had a porous nano-crystalline structure, was believed to consist of TiO_2, Al_2TiO_5 and vanadium sub-oxides. The addition of VN to the coating had prevented the formation of Al_2O_3, which is found for the oxidation for TiAlN, with Al_2TiO_5 the only Al-bearing phases. V was present largely as V_2O_5, primarily on the outer surface, which is ideal for promoting low friction during the operation of these coatings in wear resistant applications. Oxidation occurred preferentially along the columnar grain boundaries, but did not follow the multilayer structure.

Acknowledgement: Financial support from Engineering and Physical Science Research Council (EPSRC), UK, Grant No. GR/N23998/01 is acknowledged.

6. References

[1] Münz WD Lewis DB Hovsepian PE Schönjahn C Ehiasarian A Smith IJ 2001 *Surf. Eng.* 17 15-27.
[2] Constable CP Yarwood J Hovsepian PE Donohue LA Lewis DB Münz WD 2000 *J.Vac.Sci.Tech.*, 18, 1681-1689.
[3] Zhou Z Rainforth WM Lewis DB Creasey SC Hovsepian PE Ehiasarian A Münz WD 2003 *Surf. Coat. Technol.* (in press).
[4] Zhou Z Rainforth WM Rother B, Hovsepian PE Ehiasarian A Münz WD *Surf.Coat.Technol.* (in press).
[5] Zhou Z Reaney IM Hind D Milne SJ Brown AP Brydson R 2002 *J.Mater.Res.* 17 2066-2074.
[6] Lembke MI Lewis DB Münz WD and Titchmarsh JM 2001 *Surf. Eng.* 17, 153-158.
[7] McIntyre D Greene JE Hakansson G Sundgren JE & Münz WD 1990 *J. Appl. Phys.* 67 1542-1553.
[8] Cui J Da D Jiang W 1998 *Appl. Surf. Sci.* 133 225-229.
[9] Kubaschewski O and Hopkins BE 1962 Oxidation of metals and alloys 2nd edition, Butterworths.
[10] Smithells Metal Reference Book 7th edition 1992 Butterworth Heinemann.

Inst. Phys. Conf. Ser. No 179: Section 10
Paper presented at Electron Microscopy and Analysis Group Conf. EMAG2003, Oxford, 2003
©2003 IOP Publishing Ltd

Electrochemical and AFM/SEM Studies of Ni-Cr Electroplated and Sol-Gel Coated Samples

H Xu, R Akid, G Brumpton*, H Wang and J M Rodenburg

Materials Research Institute, Sheffield Hallam University, Sheffield S1 1WB, UK
*HD Sports Ltd., Rutland Road, Sheffield S3 8DG, UK

ABSTRACT: Atomic force microscopy (AFM) and scanning electron microscopy (SEM) have been used to study the surface characteristics and corrosion mechanisms of Ni/Cr electrocoats and Al_2O_3 sol-gel coats before and after accelerated corrosion tests in a 0.5% NaCl solution at 18°C. The results of electrochemical tests show that the corrosion resistance of substrates has been much improved by the application of both coats, and the Al_2O_3 coated sample exhibits the better anti-corrosion performance. Surface morphology observations revealed the presence of highly homogeneous and adherent films formed on the metal substrate during the sol-gel process. These films act as a barrier to prevent the substrate from corrosion in the aggressive environment. This sol-gel coating is therefore seen as a promising replacement for that of Cr^{6+} electroplating.

1. INTRODUCTION

Nickel-Chromium electroplating is a conventional method to apply protective films onto metal substrates which, subsequently, will be used within corrosive environments. Unfortunately, the majority of electroplated chromium is based on the hazardous Cr^{6+} solutions which are toxic, carcinogenic and detrimental to the environment. Exploiting new coating methods to replace hazardous hexavalent chromium electroplating is therefore very important.

It is well known that ceramic alumina coatings can be prepared by various methods but the sol-gel process is an interesting method because this technique is simple, economic, environment friendly and may be used to easily coat any kind of substrate, including sections having complicated shapes. In fact sol-gel processed Al_2O_3 coatings have been used for a broad number of applications [1-3].

In the present work, comparative electrochemical tests have been carried out on the Ni/Cr electroplated and sol-gel (Al_2O_3) coated carbon steel substrates to evaluate their anti-corrosion performance. AFM and SEM have, in addition, been used to observe the surface characteristics of these samples before and after accelerated corrosion testing.

2. EXPERIMENTAL

In this study, Ni and Ni-Cr (VI) coats were electroplated to carbon steel substrates and Al_2O_3 coats were deposited on Ni-plated substrates using a modified sol-gel process in which the substrate was dip-coated four times and subsequently heated at 420°C.

418

Fig. 1 shows a schematic of the 3-electrode cell arrangement used for conducting the electrochemical tests. The tube containing 0.5% NaCl solution was placed on the test sample (working electrode), a saturated calomel reference electrode and a Pt auxiliary electrode were immersed in the electrolyte to complete the 3-cell arrangement. Potential control for the electrochemical tests was monitored by computer controlled (EG&G Instruments) potentiostat.

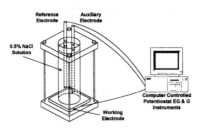

Fig.1. Schematic diagram of electrochemical test equipment

AFM surface characterisation was performed with a Digital Dimension™ 3000 atomic force microscope. The AFM images of the samples were taken in tapping mode with a scan rate of 1 Hz. A Philips XL40 SEM was used for examining the corroded surface of the samples after polarisation tests.

3. RESULTS AND DISCUSSION

3.1 Electrochemical Test Results

Electrochemical tests have been carried out in 0.5% NaCl solution at 18°C to assess the anti-corrosion properties of uncoated and coated samples. Whilst measuring the changes of free corrosion potential E_{corr} with increasing immersion time (up to a maximum of 24 hours), linear polarisation tests were also performed at t = 1 and 30 mins, and 1, 2, 4, 8, 16 and 24 hrs, thereby allowing an assessment of the corrosion rate via the polarisation resistance R_p parameter as a function of immersion time. By convention higher polarisation resistance values correspond to lower corrosion rates. Potentiodynamic polarisation was also used to evaluate the corrosion properties of the samples.

The plot of polarisation resistance of the samples against immersion time in 0.5% NaCl solution at 18°C is exhibited in Fig. 2a. It can be seen that the polarisation resistance of both Ni/Cr plated and sol-gel coated samples is significantly higher than that of the carbon steel substrate. On comparison of the two coated samples, the sol-gel coating possesses greater corrosion resistance. The polarisation curves of carbon steel without and with coatings (Fig. 2b) clearly show that the Al_2O_3 coating has a marked lower current density, implying it has the best anti-corrosion performance amongst all samples. This coincides with the result obtained from linear polarisation tests shown in Fig. 2a. Some researchers [3-5] have also reported that the corrosion resistance of the stainless steel substrates improves upon the application of the Al_2O_3 coating, as deposited by the sol-gel method.

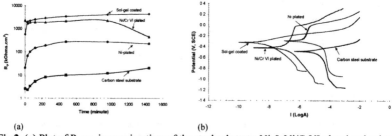

(a)   (b)

Fig.2. (a) Plot of R_p vs. immersion time of the steel substrate, Ni & Ni/CrVI plated and sol-gel coated samples in 0.5% NaCl solution at 18°C; (b) polarisation curves of these samples

3.2 AFM Surface Characterisation

Surface characterisation of the Ni/Cr plated and sol-gel coated samples was carried out using AFM. Both coatings have a mirror-like appearance observed by the naked eye, but observed by AFM as seen in Figs. 3-4 (a) and (b), the CrVI plated sample exhibits a microporous surface and the Al_2O_3 coated surface is more homogeneous than that of the plated one. This is in agreement with Özer et al. [4] who also found the Al_2O_3 sol-gel films possess high homogeneity. Through AFM examination, no visible corrosion damage can be detected (Figs. 3-4 (c)) on the surfaces immersed in 0.5% NaCl solution at open circuit potential (OCP) for 24 hours at 18°C. However, as expected, with exposure to the aggressive environment, the surfaces of samples became increasingly rougher. The Root Mean Square (RMS) roughness data of Ni/Cr plated surface changes from 5.0 nm to 6.4 nm after the sample experienced 1-day immersion in 0.5% NaCl solution and 1.9 nm to 5.8 nm for the sol-gel coating immersed under the same conditions.

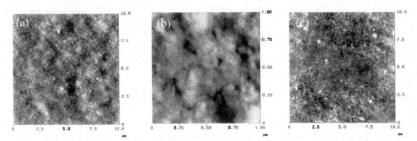

Fig.3. AFM images showing morphologies of (a) & (b) as-received Ni/CrVI plated sample and (c) after 24 hrs immersion in 0.5% NaCl solution at OCP and 18°C

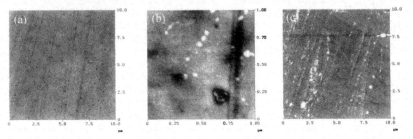

Fig.4. AFM images showing morphologies of (a) & (b) as-received sol-gel coated sample and (c) after 24 hrs immersion in 0.5% NaCl solution at OCP and 18°C

3.3 SEM Surface Characterisation

SEM was used to examine the damaged surface in order to identify the mechanism of corrosion failure. Pitting corrosion was detected on both surfaces although the majority of the surfaces were unaffected, however the extent of corrosion was appreciably different. When comparing Figs. 5(a) and 6(a), an isolated "large" pit having a diameter around 200-300 μm was found on the CrVI plated surface as opposed to a few "small" pits (around 10-50 μm diameter) developed on the Al_2O_3 coated surface. Here localised corrosion has taken place through a micropore or microcrack in the chromium coat where corrosion penetrates to the more noble nickel layer and then spreads laterally. Closer observation of this pit can be seen in Fig. 5(b). Fig. 5(c) shows another example of an individual pit found at another location on the CrVI surface. It is evident from Fig. 5 that following accelerated corrosion small pits containing corrosion products lead to loosening and removal of coatings via delamination and the formation of deep craters.

When focusing on an individual pit on the Al_2O_3 coated surface (Fig. 6c), it is seen that the top layers of coats were peeled off in the accelerated corrosion process, however the coats have not been completely removed except few spots, indicating this sol-gel material has formed highly adherent films on the metal substrate, and the films act as a barrier to diffusion of corrosion initiating species, such as chloride ions and oxygen, thereby lowering the oxidation rate of the metal substrate surface [6]. In the present work the Al_2O_3 coating was heated at 420°C and exists in the form of γ–alumina because of the dried gel transformed from boehmite to γ–alumina between 200-300°C [7] or 350°C [8], this lower processing temperature offers minimal dimensional and weight changes of the protective parts [4] and provides strong adhesion between the sol-gel films and metal substrates.

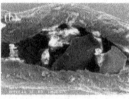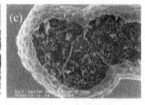

Fig.5. SEM images showing (a) an isolated pit; (b) closer observation at (a) and (c) additional pit on the surface of Ni/CrVI plated sample after accelerated corrosion tests in 0.5% NaCl solution at 18°C

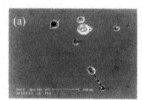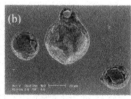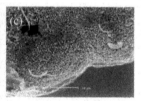

Fig.6. SEM images showing (a) & (b) a few pits on the surface and (c) closer observation at a pit on the surface of sol-gel coated sample after accelerated corrosion tests in 0.5% NaCl solution at 18°C

4. CONCLUSIONS

In conclusions, the improved corrosion resistance of sol-gel coatings is attributed to the formation of a homogeneous and adherent film. This sol-gel coating can be a promising candidate for replacing Cr^{6+} electroplating.

References

[1] Chen Y, Ai X and Huang C 2000 *Mater. Sci. Eng.* **B77** 221-228
[2] Ding D, Rao J, Wang D, Ma Z, Geng L and Yao C 2000 *Mater. Sci. Eng.* **A279** 138-141
[3] Biswas R G, Woodhead J L and Bhattacharaya A K 1997 *J. Mater. Sci. Lett.* **16** 1628-1633
[4] Özer N, Cronin J P, Yao Y and Tomsia A P 1999 *Sol. Energy Mater. & Sol. Cells* **59** 355
[5] Masalski J, Gluszek J, Zabrzeski J, Nitsch K and Gluszek P 1999 *Thin Solid Films* **349** 186
[6] Metroke T L, Parkhill R L and Knobbe E T 2001 *Prog. Org. Coat.* **41** 233-238
[7] Dwivedi R K and Gowda G 1985 *J. Mater. Sci. Lett.* **4** 331-334
[8] Saraswati V, Rao G V N and Rao G V R 1987 *J. Mater. Sci.* **22** 2529-2534

Inst. Phys. Conf. Ser. No 179: Section 10
Paper presented at Electron Microscopy and Analysis Group Conf. EMAG2003, Oxford, 2003
©2003 IOP Publishing Ltd

The Surface Morphology and Nanoscale Electrical Conductivity Studies of the Chemical Etching and Laser Annealing Processed Indium-Tin-Oxide films

Y C Fan[1], W Zheng[2], M J Rose[1], and A G Fitzgerald[1]

[1]Division of Electronic Engineering and Physics, Faculty of Engineering and Physical Sciences, University of Dundee, Dundee DD1 4HN, UK
[2]School of Engineering, Napier University, Edinburgh EH10 5DT, UK

ABSTRACT: The surface morphology and surface conductivity of the chemical etching and excimer laser ablation processed Indium-Tin-Oxide films have been studied by the atomic force microscopy (AFM) and the current sensing atomic force microscopy (CSAFM). AFM and CSAFM image observations show that chemical etching and laser ablation processes can effectively get rid of surface contaminations but these processes cannot reduce the surface roughness.

1. INTRODUCTION

The optical transparent Indium-Tin-Oxide film (ITO) is one of the predominant anode materials used in the light emitting diode (OLED). High performance OLED requires high quality ITO films with nanoscale surface flatness and uniform surface conductivity (Klöppel et al 2000).

In this research work, several types of commercially available ITO substrates have been processed by chemical etching and excimer laser ablation to reduce the surface roughness and improve the uniformity of the thin film surface conductivity. The surface morphology and nanoscale surface conductivity of the as-purchased, chemical etching and laser ablation processed ITO films have been investigated by the atomic force microscopy (AFM) and the current sensing atomic force microscopy (CSAFM). The AFM and CSAFM image observations show that the chemical etching and laser ablation process can get rid of surface contaminations and improve the uniformity of the surface conductivity of the films but these processes cannot reduce the surface roughness.

2. EXPERIMENTAL

The main characteristics of the commercial ITO substrates are summarised in product data sheet. The sheet resistance R_S is measured from 4 to 8 ohms, the nominal transmittance is larger than 76% and the nominal coating thickness is from 150 to 200 nm.

Before surface processing, the as-purchased substrates were first cleaned with acetone and ethanol using an ultrasonic cleaner for 15 minutes in each solvent and then examined by a Dimension™ 3000 atomic force microscope (Digital Instruments) and PicoSPM current sensing atomic force microscope. These substrates were subsequently subjected to chemical etching or laser ablation surface processing.

The chemical etching of the ITO thin films was performed in 18% HCl solution for varying times from 4 to 24 minutes. These reactions were carried out in the clean room at a room temperature of about 20 °C. After a specified etching time, the samples were rinsed with deionised water and blow dried with nitrogen gas.

The laser ablation of the ITO thin films was performed in air by using a 248 nm KrF Excimer laser that was operated at 8 Hz. The laser beam spot on the ablated area has a square shape and the laser energy density can be adjusted by changing the separation distance of the sample from the focus lens. During ablation process, the sample was scanned in x-direction with a scan speed of 2 mm/s.

After chemical etching or laser ablation, the ITO thin films were examined again by AFM and CSAFM to evaluate the thin film surface characteristics.

3.   RESULTS AND DISCUSSION

3.1   Surface Morphologies

Figure 1 shows representative AFM images taken from an as-purchased ITO substrate and the ITO substrates etched in 18% HCl solution. The as-purchased ITO films have a uniform surface morphology and are composed of large patches of flat islands with a size 250 to 600 nm as illustrated in Fig.1a. High magnification AFM images show that these flat islands are consisted of tiny beads with a dimension from 10 to 40 nm. When the substrates are chemically etched, the size of the flat islands was observed to decrease with increasing etching time. For films etched for 4 minutes, the size reduced to a range of 150 to 400 nm (see Fig.1b). Films etched for 8 min (Fig 1c) showed a decrease in island size down to 50 to 250 nm. For the films etched for more than 16 min, the flat island-like features are totally disappeared as shown in Figs. 1e and 1f. These observations suggested that the chemical etching initially occurs at the surrounding edges of the islands. Roughness measurements on the as-purchased and chemical etched films indicate that the adopted chemical etching process cannot reduce surface

Figure 1 AFM images taken from (a) an as-purchased ITO substrate and the chemical etching processed ITO substrates for (b) 4 min, (c) 8 min, (d) 12 min, (e) 16 min and (f) 20 min in 18% HCl solution, Image scan size: 2 μm

roughness. The mean square roughness values measured from images with a scan size of 2 μm from the as-purchased films and the films etched for 4, 8, 12, 16 and 20 minutes are respectively 4.2, 5.5, 6.0, 7.6, 8.5, 9.8 nm.

No obvious surface morphology changes are observed when the etching time is less then 2 minutes. However two minutes etching is good enough to get rid of thin film surface contaminations as will be discussed later.

The surface morphologies of the ITO thin films processed by Excimer laser annealing are shown in Fig.2. When the films are laser annealed, the beads in the flat islands were first observed to merge and then the islands began to coalesce with increasing the laser fluence. These observations suggest that the films have undergone a melting and aggregating process. Once the annealing temperature exceeds the melting temperature of the ITO material, re-crystallization will occur. This melting and re-crystallization process results in forming large flat crystal grains as shown in Fig.1e and 1f. However, some cracks were also observed on the laser annealed films, and particularly on the high laser fluence annealed films. The mean square roughness values measured from the as-purchased film and the films annealed with a laser fluence of 125, 136, 145, 158, and 165 mJ/cm^2 are respectively of 2.8, 4.2, 5.0, 6.6, 7.7, and 8.9 nm.

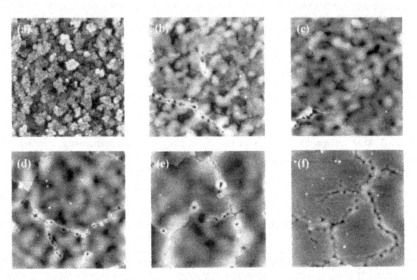

Fig. 2 AFM images taken from (a) an as-purchased ITO substrate and the excimer laser ablation processed ITO substrates at an energy density of (b) 125 mJ/cm^2, (b) 136 mJ/cm^2, (d) 145 mJ/cm^2, (e) 158 mJ/cm^2 and (f) 165 mJ/cm^2, Image scan size: 2 μm.

3.2 Surface Conductivities

The nanoscale surface conductivity of the as-purchased, chemical etching and laser annealing processed ITO substrates were studied by current sensing atomic force microscopy. In CSAFM observation, an electrically conductive probe is scanned over the sample surface in contact mode. While scanning, a dc bias voltage is applied between the tip and sample, and a current flow is generated. This current is detected and used to construct a spatially resolved conductivity image which can be used as a measure of the local conductivity or electrical integrity of the sample (Liau et al 2001, Layson et al 2003).

On the as-purchased ITO thin films, high density of nanosized insulating inclusions have been observed on the CSAFM images as shown in Fig 3d. The CSAFM image illustrated in Fig.3d was captured with a negative bias voltage of -200 mV was applied to the sample surface. In this circumstance, the bright areas in the CSAFM image represent the high resistance regions and the dark areas signify the conductive regions. When a positive bias voltage is applied to the sample surface, the contrast of the CSAFM image will reverse.

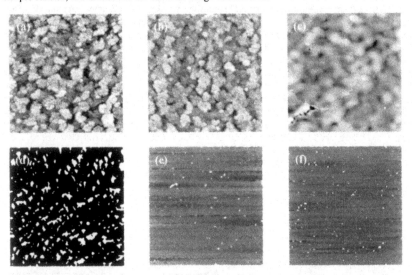

Fig. 3 AFM topographical images of (a), (b) and (c) are taken respectively from an as-purchased, chemical etching and laser ablation processed ITO substrates. (d), (e) and (f) are the corresponding CSAFM conductivity images which were captured with a negative bias voltage of -200 mV was applied to the sample surface. Image scan size 2 μm.

After chemical etching for a short period of time (less than 2 minutes) or laser ablation cleaning at a low laser fluence, the density of the insulating inclusions was substantially reduced (see Fig. 3e and 3f). These observation results indicate that most of these insulating inclusions are surface contaminations. For some ITO thin film samples, even after long time chemical etching and high energy density laser annealing, a few insulating inclusions still exist on the CSAFM images, and in this situation, the insulating inclusions are more possibly formed in the film deposition process and it is buried within the films.

In conclusion we have studied how the surface morphology and conductivity characteristics of the ITO thin films can be modified by chemical etching and laser annealing. These surface processes cannot reduce the film surface roughness. However lightly etching or laser ablation is very effective to remove the surface contamination and improve the uniformity of surface conductivity.

REFERENCES

Klöppel A, Trube J, Hoffmann U, Burroughes J H, Heeks S K, Gunner A, Ramshaw T, Oct 2000 Precision

Liau Y H, Unterreiner A N, Chang Q, Scherer N F, 2001 J. Phys. Chem.B 105(11), pp2135-2142

Layson A, Gadad S, Teeters D 2003 Electrochimica Acta 48 (14-16) pp2207 – 2213.

Inst. Phys. Conf. Ser. No 179: Section 10
Paper presented at Electron Microscopy and Analysis Group Conf. EMAG2003, Oxford, 2003
©2003 IOP Publishing Ltd

Using SEM for Post-Test Analysis of HVOF-Sprayed WC-Cr-Ni and WC/CrC Coatings after Erosion-Corrosion in Concentrated Slurries

H Xu[1] and A Neville[2]

[1]Materials Research Institute, Sheffield Hallam University, Sheffield S1 1WB, UK
[2]School of Engineering and Physical Sciences, Heriot-Watt University, Edinburgh EH14 4AS, UK

ABSTRACT: The scanning electron microscopy (SEM) has been used to study the erosion-corrosion mechanisms of HVOF-sprayed WC-Cr-Ni and WC/CrC coatings after exposure to a concentrated impinging slurry. The surface morphology observations revealed that after the repeated and violent impingement of erodent particles, the coatings were first fractured and then the large and deep craters were formed on the surface, which resulted in the eventual removal of the coatings. The material loss on WC/CrC coating is significantly lower than on WC-Cr-Ni coating, confirming the finding that the erosion-corrosion resistance of coatings is strongly controlled by the composition of the coatings.

1. INTRODUCTION

HVOF sprayed WC-Cr-Ni and WC/CrC coatings were prepared by Greenhey Engineering Services. These coatings are developed to provide high wear resistance and can be applied on pump components. In general, the erosion-corrosion rate of materials in slurry conditions is increased significantly with increasing solid particle concentration [1]. However, it is not fully understood how the slurry impingement process affects deterioration of these coatings.

To evaluate the erosion-corrosion behaviour of these two coatings under conditions relating to the hydro-transport of slurries in the shale oil industry in North America, an experimentally-based programme has been carried out to characterise material loss rates and mechanisms in concentrated 10 wt% slurry erosion. In the study, a weight loss assessment and an optical/microscopical investigation using SEM enables information on degradation of these coatings to be determined.

2. EXPERIMENTAL

The chemical composition of HVOF sprayed WC-Cr-Ni and WC/CrC coatings used in the study is given in Table 1. The surface morphology and microstructural features of the materials were characterised using light optical microscopy and Philips XL40 SEM.

Table 1. Measured composition of two HVOF sprayed coatings

	W	Cr	Ni	Co	C	O
WC-Cr-Ni	60.4	14.6	4.9		13.8	6.3
WC/CrC	74.8	2.6	0.8	8.5	10.6	2.7

426

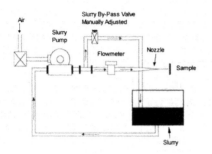

Fig.1. Schematic diagram of slurry jet erosion test rig [2]

This research was carried out using the slurry facility at the Surface Technologies/Tribology Laboratory at the National Research Council, Canada. Jet erosion tests were carried out at room temperature using 10 wt% slurry of silica sand in 3.5% NaCl and tap water. The free slurry jet was directed at the specimens at 15 m/s with a flow rate of 18 l/min, at nominal impingement angle of 90° and at a nozzle-specimen distance of 98 mm. Fig. 1 shows the experimental rig.

3. RESULTS AND DISCUSSION

3.1 Effect of Salinity and Impinging Time

One hour slurry erosion tests in 10% slurry were carried out on WC-Cr-Ni and WC/CrC coatings both in 3.5% NaCl and tap water. It seems that salinity did not affect the material loss rates (Fig. 2a). The weight losses on WC/CrC coating were significantly less than on WC-Cr-Ni coating both in tap water and 3.5% NaCl.

The weight loss on two coatings as a function of impinging time is shown in Fig. 2(b), and a general linear relationship between weight loss and impinging time can be seen. This is in agreement with Karimi et al. who have studied erosion behaviour of some WC coatings in 0.3 wt% sand slurry conditions [3]. The weight loss increased with increasing experimental time as expected, for WC/CrC coating, although the rate of increase was not large. After 15 minutes impingement, the weight loss was 7.0 mg and 14.5 mg after 30 minutes. For the WC-Cr-Ni coating, the increasing rate was much larger due to the coating being partially penetrated after 30 minutes exposure to sand particle impacts. So the weight loss of 33.2 mg after 15 minutes increased to 86.8 mg after 30 minutes. After one hour, the coating was completely removed and stainless steel substrate had been exposed (Fig. 3c) resulting in a higher material loss.

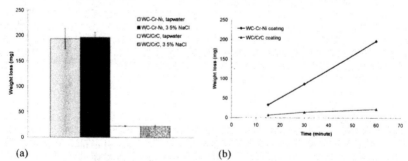

(a)   (b)

Fig.2. (a) Comparison of weight loss on WC-Cr-Ni and WC/CrC coatings in tap water and in 3.5% NaCl solution after one hour slurry erosion at 18°C; (b) weight loss as a function of time on both coatings in slurry erosion at 18°C in 3.5% NaCl solution

3.2 Macrostruture and Microstructure of WC-Cr-Ni and WC/CrC Coatings after One Hour Concentrated Slurry Erosion at 18°C

Macrographs showing the wear scars produced on WC-Cr-Ni and WC/CrC coatings after 15 minutes, 30 minutes and one-hour impingement in 10% slurry erosion at 18°C in 3.5% NaCl are presented in Figs. 3 and 4 (a), (b) and (c) respectively. Obviously, the wear scars for both coatings get deeper as time increases, and material losses on WC-Cr-Ni coating are much higher than that on WC/CrC coating owing to the former being penetrated during one-hour erosion tests.

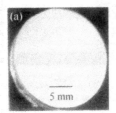 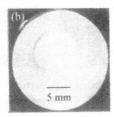 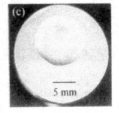

Fig.3. Macrographs showing the wear scars produced on WC-Cr-Ni coating after (a) 15 minutes, (b) 30 minutes and (c) one hour impingement in 10% slurry erosion at 18°C in 3.5% NaCl solution

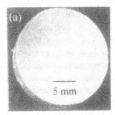 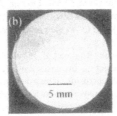 

Fig.4. Macrographs showing the wear scars produced on WC/CrC coating after (a) 15 minutes, (b) 30 minutes and (c) one hour impingement in 10% slurry erosion at 18°C in 3.5% NaCl solution

The microstructure of WC-Cr-Ni coating in the wear scar region after 15 minutes of slurry impingement is presented in Figure 5(a). It shows evidence of some craters formed by the removal of large areas of coating even after short term impingement. More extensive fracture is detected (Figure 5b), even in the region out of wear scar, after one hour impingement. For the WC/CrC coating, the SEM image taken in the wear scar (Figure 5c) shows that less erosion-corrosion damage can be found after one hour in slurry impingement.

Fig.5. SEM images showing the microstructures of (a) WC-Cr-Ni coating in the central area after 15 mins.; (b) WC-Cr-Ni coating in the area out of wear scar after 1-hour and (c) WC/CrC coating in the central area after 1-hour erosion in 3.5% NaCl

428

The examination of cross-sections on these two coatings (Fig. 6) further confirms the erosion-corrosion damage is more severe on WC-Cr-Ni coating than on WC/CrC coating.

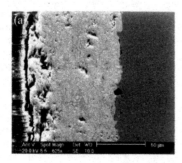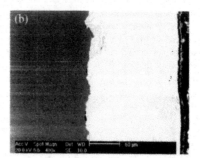

Fig.6. SEM images showing the worn surfaces of (a) WC-Cr-Ni coating after 15 mins. and (b) WC/CrC coating after 1-hour slurry erosion at 18°C in 3.5% NaCl

Rogne et al. [4, 5] found that the erosion-corrosion resistance of coatings is strongly controlled by the composition of the coatings. Generally the Co-based metal phase coatings have better erosion and corrosion properties than the Ni-based metal phase coatings. This is most probably related to the fact that Co has better binding properties to the WC particles than the other alloying elements [4]. However, this will also depend on the erosivity and corrosivity of the medium the coatings are exposed to. In the present work, WC-CrC coating has 8.5% Co and 0.8% Ni, WC-Cr-Ni has 4.9% Ni and no Co (Table 1), so WC-CrC coating should have a better erosion-corrosion resistance. But Rogne et al. [5] also pointed out that increasing the Cr content in the metal phase by replacing Co increases the corrosion resistance of the WC-Co-Cr type coatings. However improved corrosion resistance might result in a reduced erosion resistance depending on the metal composition. At low erosive and high corrosive conditions increased Cr improves the erosion-corrosion resistance, but at high erosive conditions increased Cr has a negative effect. The present experimental results confirmed this conclusion. WC-Cr-Ni coating contains a higher Cr (14.6%) than WC-CrC coating which has only 2.6% Cr content, but in the severe erosive environment, as shown in Fig. 2a, the former possesses very low erosion resistance resulting in the much higher material loss.

ACKNOWLEDGEMENTS

The authors would like to thank Weir Pumps Ltd. for the financial support to H Xu and thanks also go to Dr Hawthorne for provision of test facilities at Tribology Laboratory in National Research Council Canada.

REFERENCES

[1] Postlethwaite J, Tinker E B and Hawrylak M W 1974 *Corrosion* **30**(8) 285-290
[2] Hawthorne H M, Arsenault B, Immarigeon J P, Legoux J G and Parameswaran R 1999 *Wear* **225-229** 825-834
[3] Karimi A, Verdon Ch, Martin J L and Schmid R K 1995 *Wear* **186-187** 480-486
[4] Rogne T, Solem T and Berget J 1997 *Thermal Spray: A United Forum for Scientific and Technological Advances*, Berndt C C (Ed.), published by ASM International, Materials Park, Ohio, USA 113-118
[5] Rogne T, Solem T and Berget J NACE, Houston, Texas, *Corrosion 98* **paper 495,**

Inst. Phys. Conf. Ser. No 179: Section 10
Paper presented at Electron Microscopy and Analysis Group Conf. EMAG2003, Oxford, 2003
©2003 IOP Publishing Ltd

Investigation of the Interfaces between Boride Particles and Aluminium in the As-Received and Re-Melted Ti-B-Al Grain Refiner Master Alloy

P Cizek, B J McKay* and S Lozano-Perez**

IMMPETUS, Department of Engineering Materials, University of Sheffield, Mappin Street, Sheffield S1 3JD, UK
* Manchester Materials Science Centre, UMIST/University of Manchester, Grosvenor Street, Manchester M1 7HS, UK
** Department of Materials, University of Oxford, Oxford OX1 3PH, UK

Abstract. In the Al casting industry, TiB_2 grain refiner particles are added to the melt in order to promote heterogeneous nucleation of α-Al crystallites. It has been suggested that thin Al_3Ti DO_{22} layers, adsorbed on the boride particles, might be necessary for successful α-Al nucleation via the peritectic reaction. In order to elucidate this suggestion, the boride/α-Al interfaces were studied in the as-received Ti-B-Al grain refiner rod, as well as after re-melting and rapid quenching of the same rod. The investigation was performed using the TEM, EDS, STEM and HREM techniques. For the as-received rod, the EDS study failed to show conclusively enrichment of the interfaces in Ti and indicated clearly only a small enrichment in Si and Fe. The HREM investigation found no indication of a possible crystalline layer coating the boride particles. The borides found in the re-melted rod were observed to readily nucleate α-Al on their basal faces with a well-defined low-index orientation relationship but there was again no indication of any layer on the boride particles. Thus, the borides themselves appeared to be potent sites for the heterogeneous nucleation of α-Al crystals during rapid quenching without a necessary presence of the aluminide layers.

1. Introduction

In the Al casting industry, it is common practice to add TiB_2 grain refiner particles to the melt in order to promote heterogeneous nucleation of α-Al crystallites. It has been proposed on the basis of investigations involving Al-based metallic glasses that thin Al_3Ti DO_{22} layers, adsorbed on the surfaces of the boride particles, might be a necessary prerequisite for the successful nucleation of α-Al via the peritectic reaction [1,2].

The aim of the present work was to elucidate whether the aluminide layers suggested in [1,2] might possibly originate directly from grain refiner master alloys and whether they might undergo changes during re-melting of these alloys. For that reason, the local chemical composition, crystallographic characteristics and lattice defects were studied at the interfaces between boride particles and α-Al crystals in the as-received Ti-B-Al 5:1 grain refiner rod, as well as after re-melting and rapid quenching of the same rod using melt spinning.

2. Experimental Methods

A commercial Ti-B-Al 5:1 wt.% grain refiner rod (supplied by London & Scandinavian Metallurgical Co. Ltd.) was used as an experimental material. Rapid solidification of the rod was achieved by melt spinning in an inert helium atmosphere. The charge was heated to 1050°C in a boron nitride crucible and held for 5 minutes before being ejected onto a copper wheel rotating at a speed of 20 ms^{-1}. Both the as-received refiner rod and the melt-spun ribbon were examined using the TEM, EDS, STEM and HREM techniques.

3. Results and Discussion

3.1. As-Received Grain Refiner Rod

TEM analysis revealed that the microstructure of the as-received grain refiner rod contained fine boride and coarse Al$_3$Ti DO$_{22}$ particles randomly distributed within the α-Al matrix. The borides exhibited a facetted, hexagonal platelet morphology with the {001} and {100} planes as their external faces (Figs. 1a-1c). EDS analysis indicated that these particles generally represented a mixed (Ti,Al)B$_2$ phase, with Ti and Al contents changing continuously across the particle dimensions. TiB$_2$ and AlB$_2$ represent isomorphous phases having a hexagonal crystal structure with the 196-Pmmm space group and very similar lattice parameters [2] and thus are difficult to distinguish using TEM diffraction. The diffraction analysis showed that boride particles displayed no apparent low-index orientation relationship (OR) relative to the neighbouring α-Al grains (Figs. 1c, 1d), presumably as a result of recrystallisation of the original as-solidified α-Al matrix occurring during the manufacture of the refiner rod. Diffraction

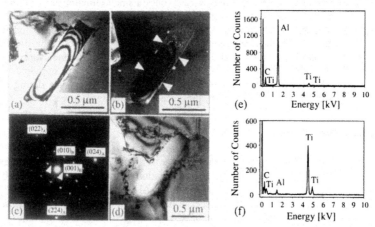

Fig. 1. TEM analysis of a boride particle, composed of an Al-rich core and Ti-rich shell, found within the as-received refiner rod: (a) bright-field image; (b) dark-field image obtained from the streaks located between the centre and {001} spots displaying layer-like contrast effects (arrowed); (c) SAD pattern corresponding to a [100] zone axis showing the streaks (arrowed) (subscripts A and B refer to α-Al and the boride respectively); (d) same as in (a) after re-tilting to obtain the neighbouring α-Al grains in contrast; (e),(f) EDS spectra obtained from the particle core and shell respectively.

patterns obtained from the boride particles showed faint streaks perpendicular to their {001} basal planes (Fig. 1c). Dark-field imaging using the streaks illuminated the basal facets of the borides (Fig. 1b), thus creating layer-like contrast effects with thickness of several nanometres. It has been speculated in [1,2] that such contrast effects might correspond to a physical layer of DO_{22} Al_3Ti phase adsorbed on the surface of boride particles but no conclusive evidence was provided to support this claim. In order to assess the above suggestion, a detailed analysis of the chemistry and atomic structure of the boride/α-Al interfaces was undertaken using the STEM EDS and HRTEM techniques. STEM EDS maps and line scans indicated that the boride particle analysed corresponded to a mixed $(Ti,Al)B_2$ phase (Fig. 2a). The seemingly diffusion-like profile of Al/Ti distribution across the interface in Fig. 2a seemed to be largely a consequence of the interface not being tilted precisely edge-on and several fluctuations in Al and Ti content appeared to result from a small drift of the probe. Thus, the STEM EDS analysis did not provide conclusive information on the possible enrichment of the α-Al/boride interface in Ti suggested in [1,2] and found only evidence of a narrow region at the interface enriched in Si and Fe (Fig. 2b). Moreover, a HREM investigation found no indication of a possible crystalline layer coating the boride particles.

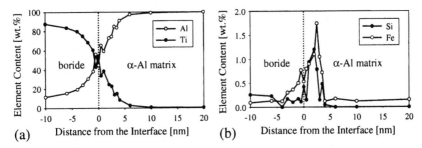

Fig. 2. STEM EDS analysis of the elemental distribution at the boride/α-Al interface within the as-received refiner rod: (a) linescan for Al and Ti; (b) linescan for Si and Fe.

3.2. Melt-Spun Grain Refiner Rod

The melt-spun ribbon microstructure was composed of fine equiaxed α-Al grains. No equilibrium Al_3Ti DO_{22} particles were found, indicating that they had all dissolved prior to melt spinning. A majority of α-Al grains contained small, petal-shaped particles corresponding to a metastable Al_3Ti $L1_2$ phase [3] that systematically displayed a near cube-to-cube OR with α-Al. This suggests that these particles, after their formation from the melt, served as very potent substrates for subsequent heterogeneous nucleation of α-Al, presumably due to an extremely small mismatch between the $L1_2$ Al_3Ti and Al crystal lattices [3]. The boride particles were systematically observed to nucleate α-Al, as well as metastable $L1_2$ Al_3Ti particles (Fig. 3a), each with the same low-index OR found previously [1,2]. In this OR, the close-packed planes and directions for the above phases were approximately parallel to each other (Figs. 3b, 3c). This indicates that α-Al (or $L1_2$ Al_3Ti phase) nucleated heterogeneously on the boride {001} basal faces, which was also directly observed in [1]. A detailed STEM EDS investigation of the boride/α-Al interfaces (Fig. 3d) did not reveal any noticeable enrichment in Ti that would suggest the presence of an aluminide layer on the boride basal faces. A weak-beam dark-field TEM study indicated that these interfaces might have a semi-coherent character, being

432

composed of hexagonal arrays of edge dislocations. Thus, the boride particles without being coated by DO_{22} Al_3Ti layers proved to be very potent sites for heterogeneous nucleation of α-Al during rapid quenching, competing successfully with the $L1_2$ Al_3Ti petal-shaped particles having a negligible lattice mismatch with α-Al.

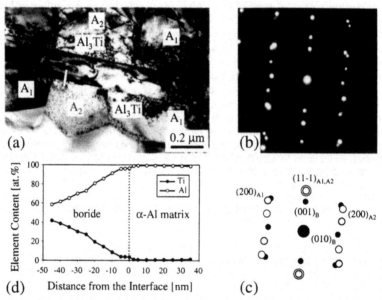

(a)

(b)

(d)   Distance from the Interface [nm]

(c)

Fig. 3. TEM and STEM EDS analysis of a boride particle, found in the melt-spun refiner rod, nucleating several α-Al grains and $L1_2$ Al_3Ti particles with a low-index OR: (a) bright-field micrograph (A_1 and A_2 indicate α-Al grains corresponding to the two possible variants of the observed OR); (b) large-area SAD pattern; (c) indexing of the pattern (zone axes of the boride and α-Al (or $L1_2$ Al_3Ti) are [100] and [011] respectively, subscripts A_1 and A_2 are the same as in (a)); (d) linescan for Ti and Al performed across the boride/α-Al interface as indicated in (a).

4. Conclusions

A detailed study of the interfaces between boride particles and α-Al was performed for the as-received as well as for the re-melted and rapidly quenched Ti-B-Al 5:1 grain refiner master alloy and no conclusive evidence of the adsorbed DO_{22} Al_3Ti layers, suggested in [1,2], was found on the borides. The boride particles themselves appeared to be potent sites for the heterogeneous nucleation of α-Al crystals during rapid quenching without a necessary presence of the aluminide layers.

References

[1]   Schumacher P and Greer A L 1994 Mater. Sci. Eng. A181 1335-1339
[2]   Schumacher P, Greer A L, Worth J, Evans P V, Kearns M A, Fisher P and Green A H 1998 Mater. Sci. Technol. 14 394-404
[3]   Kim W T, Cantor B, Griffith W D and Jolly M R 1992 Int. J. Rapid Solidif. 7 245-254

Inst. Phys. Conf. Ser. No 179: Section 10
Paper presented at Electron Microscopy and Analysis Group Conf. EMAG2003, Oxford, 2003
©*2003 IOP Publishing Ltd*

Understanding interfacial bonding in alumina doped yttria-stabilised tetragonal zirconia using electron energy loss spectroscopy

A J Scott[1], R Brydson[1], C A Dennis[1], I M Ross[2], W M Rainforth[2], C Scheu[3] and M Rühle[3]

[1]Institute for Materials Research, University of Leeds, Leeds LS2 9JT UK
[2]Department of Engineering Materials, University of Sheffield, Sheffield, S1 3JD UK
[3]Max-Planck-Institute for Metallforschung, Stuttgart, D-70174, Germany

Abstract. The addition of trace quantities of alumina to yttria stabilised tetragonal zirconia polycrystal (Y-TZP) leads to enhanced degradation resistance. High spatial resolution EELS measurements were performed on a polycrystalline sample of 3wt% Y_2O_3 - stabilised ZrO_2 + 0.15wt% Al_2O_3; EEL spectra were recorded both on and off grain boundaries. Distinct EELS evidence for Al was obtained at some, but not all boundaries. Al $L_{2,3}$- and Al K-edges were recorded and compared to the corresponding spectra obtained from a polycrystalline α-alumina. The Al K and $L_{2,3}$-ELNES obtained from the modified Y-TZP grain boundaries show a significant decrease in the relative intensity of the two initial peaks lying within about 10 eV of the edge on-set, as compared to the alumina spectra suggesting that this may be interpreted as a lower coordination of oxygen to aluminium compared with the slightly distorted 6-fold coordination found in α-alumina.

1. Introduction

Yttria stabilised tetragonal zirconia polycrystal (Y-TZP) ceramics offer high toughness as a result of the martensitic phase transformation of the metastable tetragonal phase to the monoclinic structure (transformation toughening). However, TZP ceramics undergo hydrothermal degradation, or aging, severely limiting their potential applications. The presence of trace quantities of alumina leads to Y-TZPs with enhanced degradation resistance.

Electron energy loss spectroscopy (EELS) performed within a STEM can be used to understand the bonding and electronic structure at interfaces in ceramic materials [Brydson 1995]. The fine structure in the spectrum within approximately 40 eV of the edge onset (ELNES) reflects the unoccupied densities of states. Characterisation of electronic structure on the nanoscale is therefore possible. Extraction of information from the ELNES is by a combination of comparison with spectra from known bulk materials and theoretical modelling using band structure and multiple scattering methods.

434

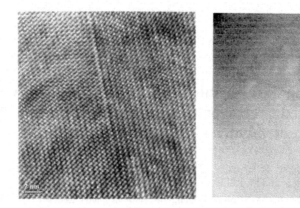

Figure 1. (left) High resolution TEM image of grain boundary
Figure 2. (right) HAADF image of a grain boundary.

2. Experimental

High spatial resolution EELS microanalysis was performed on a UHV VG HB501 STEM, operated at 100 kV and fitted with a Gatan Enfina PEELS. Measurements were made on an ion beam thinned polycrystalline sample of 3wt% Y_2O_3 - stabilised ZrO_2 + 0.15wt% Al_2O_3; EEL spectra were recorded both on and off grain boundaries using a 1 x 10 nm^2 "letter box" scanned area using convergence and collection angles of approximately 12 and 10 mrads respectively.

An FEI CM200 FEG TEM fitted with a Gatan imaging filter (GIF) was used for the high resolution TEM examination of the grain boundaries and for the EELS of the bulk polycrystalline alumina.

3. Results

3.1 Electron energy loss spectroscopy

Previous work by ourselves has revealed the presence of aluminium at the grain boundaries using secondary ion mass spectrometry (SIMS) of a fracture surface and EDX within the TEM [Ross 2001]. HREM of the grain boundaries shows them to be clean - no amorphous or crystalline second phases are observed (fig. 1).

A representative high angle annular dark field image (STEM) from a three grain junction (triple point) is shown in figure 2. The darker contrast apparent at the grain boundaries is consistent with the presence of a lower atomic number element segregated at the boundary, however the possibility of diffraction effects should not be ignored.

Distinct EELS evidence for Al was obtained at some, but not all boundaries. Representative Al $L_{2,3}$- and Al K-edges were recorded. Figures 3 and 4 show a comparison of these edges with the corresponding spectra obtained from a polycrystalline α-alumina.

The Al K and $L_{2,3}$-ELNES obtained from the modified Y-TZP grain boundaries show a significant decrease in the relative intensity of the two initial peaks lying within about 10 eV of the edge on-set, as compared to the alumina spectra. Previous work [Kato 2001; Bouchet 2003] suggests that this may be interpreted as a lower coordination of oxygen to aluminium (possibly fourfold. or perhaps five fold) as compared with the slightly distorted 6-fold coordination found in a-alumina.

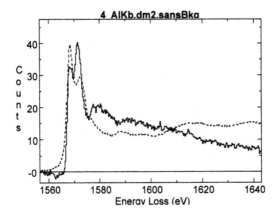

Figure 3. Aluminium K-edge spectrum of reduced area letter box from grain boundary (solid line) compared to the corresponding spectrum from α-alumina (dashed line).

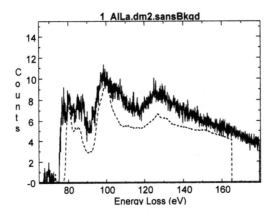

Figure 4. Aluminium $L_{2,3}$-edge spectrum of reduced area letter box from grain boundary (solid line) compared to the corresponding spectrum from α-alumina (dashed line).

3.2 Molecular modelling and electronic structure calculations

In the bulk tetragonal zirconia, there is a distorted 8-fold coordination of the zirconium by oxygen. There are several possible reasons for the lower coordination of aluminium by oxygen at the grain boundaries. Increased yttrium concentration has been noted at the GB. The presence of the trivalent yttrium and aluminium ions, substituting for the tetravalent zirconium ion, leads to increased vacancy concentration at the grain boundary. Study of model GBs reveals the extra space compared with the bulk grain. A model $\Sigma 5$ (310) GB is shown in figure 5. Removal of close O – O contacts together with further

436

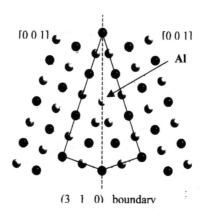

[0 0 1]   [0 0 1]
                         Al

(3 1 0) boundary

Figure 5. Model Σ5 (3 1 0)/ [0 0 1] tetragonal ZrO_2
grain boundary (Zr black, O grey).

anion vacancies will lead to a reduction in the Al coordination.Model GBs are currently being minimised using the density functional theory (DFT) code, CASTEP. Interpretation of the experimental ELNES will then be carried out via comparison with the calculated site and symmetry resolved unoccupied densities of state [Scott 2001].

4. Conclusions

A combination of experimental EELS, molecular modelling and DFT based band structures methods can be used to help us reveal the bonding at grain boundaries in structural ceramic materials. Significant work is required, and is ongoing, to obtain further energy loss spectra from well charaterised grain boundaries and to continue with the GB relaxations and theoretical near edge structure determinations.

References

Bouchet D and Colliex C 2003 *Ultramicroscopy* **96** 139
Brydson R 1995 *J. Microscopy* **180** 238
Kato Y et al 2001 *Phys. Chem. Chem. Phys.* **3** 1925
Ross I M, Rainforth W M, McComb D W, Scott A J and Brydson R 2001 *Scripta Materialia* **45** 653
Scott A J, Brydson R, MacKenzie, M and Craven A J 2001 *Phys. Rev.* B **63** 245105

Acknowledgements

We gratefully acknowledge financial support from the EPSRC (AJS, CD). Thanks are expressed to Alan Real for excellent support with the White Rose Grid computing project.

Inst. Phys. Conf. Ser. No 179: Section 11
Paper presented at Electron Microscopy and Analysis Group Conf. EMAG2003, Oxford, 2003
©2003 IOP Publishing Ltd

Probing electronic states in nanotubes and related-nanoparticles at the nanometer scale

O Stéphan, A Vlandas, R Arenal de la Concha, A Loiseau, S Trasobares, C Colliex

LPS, Bât. 510, Université Paris-Sud, 91405 Orsay, France
LEM, UMR104 Cnrs-Onera, ONERA BP72, 92322 Châtillon, France
Argonne National Laboratory, 9700 S Cass Avenue, Bldg 212-C223, Argonne, IL 60439

Abstract. The combination of the spectrum imaging mode with processing methods such as multiple least square fitting is used for probing electronic states in nanostructures at the nanometer scale. This method is illustrated on two examples: the mapping of boron chemical states for a high accuracy characterization of BN nanotubes samples and the possible detection of five-member rings in highly defective multiwall carbon nanotubes.

1. Introduction

Since their discovery by Iijima [1], the specific morphology of nanotubes with an empty cylindrical core space and the presence of large surface areas on both outer and inner sides of the walls, has captured the interest of many researchers who have demonstrated many types of fillings, coatings or graftings with gases, molecules or solid compounds. In particular, new families of nano-objects have been identified, such as nanowires, nanocontainers, peapods. They offer strong potentialities for exhibiting novel properties and for developing new applications as transport components, storage devices or reaction chambers.

On the other hand, the combination of intense electron probes of near-atomic dimension (in a STEM microscope) with highly efficient 2D detectors for recording electron energy-loss spectra (EELS) within a few milliseconds, has provided quite powerful tools to explore individual nanostructures with unprecedented spatial and energy resolutions. In particular, the identification of single atoms encaged in fullerene molecules along peapods structures was demonstrated [2], and the technique has been extended to the analysis of trapped molecules or single crystals within SWNTs [3]. Here, we demonstrate that beyond chemical mapping, the electronic states in such nanostructures can be locally investigated by mapping the EELS fine structures observed on core edges. Two examples are reported: the mapping of B chemical bond in complex BN nanotubes samples and the detection of pentagonal defects in highly defective multiwall carbon nanotubes.

438

2. Principles of the method

All experiments described here were performed in the context of a STEM-VG HB501 equipped with a field emission gun operated at 100 keV and fitted with a Gatan 666 parallel-EELS spectrometer optically coupled to a CCD camera. Such a STEM instrument delivers a 0.5-nm electron probe of high brightness for local analysis at the nanometer scale and provides 0.5-0.7 eV energy resolution. The high sensitivity home-made detector allows for the acquisition of core-loss spectra within dwell times as short as a few hundreds of milliseconds, making therefore possible the acquisition of collections of several hundreds of spectra within a few minutes, either in a time-resolved mode, either in the spectrum-imaging mode [4]. In the spectrum imaging mode, the probe is scanned over a 2D region on the sample and a whole EELS spectrum is acquired for each position of the probe. A first information to be extracted from a spectrum-image are elemental maps which display the spatial distribution of an integrated intensity over characteristic signals. In this mode, no high energy resolution is required and the acquisition time per spectrum can be further reduced. When elemental mapping is the primary goal, maps as large as 256x256 pixels are now available. This is however, still below the potentiality of filtered imaging which provides images 4 times larger.

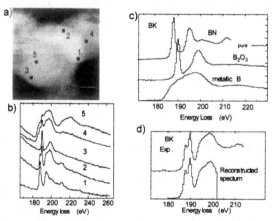

Figure1: Illustration of the NNLS fitting processing applied to a 64x64 spectrum image acquired on a large boron particle. a) HADF image. b) BK edges recorded for the different probe positions marked on the HADF image. c) BK ELNES reference spectra. d) Comparison between one experimental spectrum and a simulated spectrum.

However, the major advantage of the spectrum imaging technique over the filtered imaging approach is associated with the fact that during the data acquisition, the whole energy-loss range of interest is stored on the contrary to the second approach where the spectrum formed after the spectrometer is lost. The mapping of specific spectroscopic features is then possible such as for example the mapping of fine structures associated with a core loss. This requires appropriate methods to process large quantities of data. For example, sophisticated methods involving multivariate analysis have been proposed [5]. Here, we rely on a simpler method, which consists in adjusting (by a mean least square fit) the experimental spectrum with a simulated spectrum from a linear combination of appropriate experimental reference spectra [6]. Such an approach is illustrated in fig. 1 with a 64x64 spectrum image acquired on some large particles in a BN nanotubes sample [7]. Fig.1b shows the BK near-edge fine structures (ELNES) recorded for the 5 different probe positions indicated by some labels on the HADF image – see fig. 1a- (the HADF signal was recorded on line, in parallel to the EELS signal). In fig.1c are shown the 3 different BK ELNES which were chosen as the reference spectra to model the experimental spectrum. They are experimental spectra characteristic of the three following chemical states for boron: boron in hexagonal boron nitride, boron oxide and pure boron. These three materials were expected to be coexisting in the sample. The comparison shown on fig. 1d between an experimental spectrum extracted from the spectrum image and its associated simulated spectrum within a 20 eV fitting window validates such a choice. In general, the relevance in the choice of the references is checked by looking at the difference map, which plots the intensity difference (integrated within the fitting window) between the experimental spectrum and the model spectrum. The output of this fitting procedure is a series of maps plotting the weights associated with the different references in the reconstructed spectrum (hereafter, we will call these maps "NNLS maps" for Non Negative Least Square maps). Provided the intensity of each reference is normalized to one in the fitting window, the weight value then corresponds to an absolute intensity value, and the sum of the different weight values should be equal to the intensity value in the experimental spectrum.

3. Mapping the boron chemical states in a BN nanotube sample

The NNLS fitting method has been applied to complex situations found in some BN nanotubes samples synthesized according to the procedure described in [7]. The unexpected large number of species found in these samples such as B, C, Ca, N, O, made the analysis complicated and the information gained by comparing the different elemental maps appeared to not be sufficient for an accurate characterization of the different boron chemical states.

Fig.2 shows an example of different NNLS maps computed from a 64x64 spectrum image and associated with the different boron states mentioned earlier: B in pure boron (fig.2a), B in hexagonal boron nitride (h-BN) (fig.2b) and B in B_2O_3 boron oxide. The maps provide a clear understanding of the morphology of the investigated area: pure boron particles of two distinct sizes and oxidized at their surface are sitting on a network of BN nanotubes and polyhedral h-BN nanoparticles.

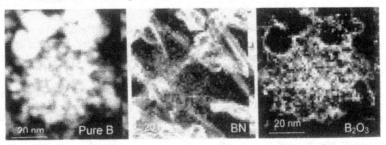

Figure 2: NNLS map deduced from a 64x64 spectrum image acquired on a BN nanotubes sample. a) Elemental boron b) B in hexagonal boron nitride c) Boron in B_2O_3 boron oxide.

Among the advantages of such images, we can mention the high signal to noise ratio despite the short acquisition time, due to the high cross section of boron and in this particular case, the ability for example, to differenciate between boron in the oxide form and elemental boron mixed with CaO impurities also present in high quantities. Furthermore, the quality of the fit accounts for the absence of any boron carbide in the investigated area.

4. Mapping the directionality of the BN unoccupied electronic states

A similar approach used on a single nanoparticle yields interesting insights concerning the growth mechanism of these nanostructures. An example of large particles displaying a concentric morphology is given in fig.3: a hexagonal BN layer encapsulates a pure boron core, widely oxidized at its surface. This type of morphology, often encountered in these samples, seems to suggest a growth mechanism based on a sequence of segregation processes [8].

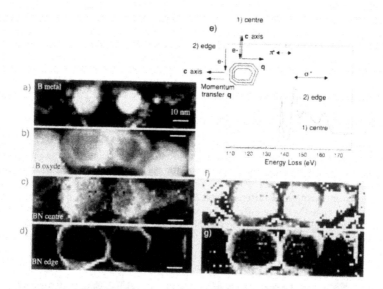

*Figure 3:*NNLS maps from boron particles. a) Pure boron. b) Boron oxide. c) and d) Boron nitride; the maps are calculated using the reference spectra displayed in e). e) anisotropy effects on BK edge in BN. f) and g) Normalised maps as deduced from c) and d).

Contrary to pure boron and boron oxide, hexagonal BN presents the peculiarity to be a highly anisotropic layered material. This results in changes in the BK ELNES as a function of the BN layers orientation with respect to that of the electron beam (or to that of the momentum transfer). Such effects have already been studied in detail for BN and graphite [9, 10]. These anisotropy effects are shown in fig. 3e, where two BK edges are displayed for two orthogonal orientations of the BN layers. Although such effects are reduced in a STEM due to some convergence and collection angles averaging, a difference in the π^*/σ^* ratio is clearly observed between the spectra acquired for a probe position at the edge and at the center of the particle. The NNLS maps displayed in fig 3a to 3d have been calculated using two references for BN, which correspond to the two extreme orientation situations shown in fig.3e. Although a facetting of the BN shell is already clearly seen on fig 3c and 3d, more informative maps are obtained by dividing each BN NNLS map by the sum of both of them. Then, one gets rid of the thickness, and the resulting maps only exhibit the orientation effects. These maps are shown in fig.3f and 3g. The intensity in each pixel ranges between 0 and 1. Each image is complementary to the other, as the sum of both should give a constant map with a pixel value equal to 1. A further step (not shown here) consists in relating the pixels values in each map to that of the inelastic partial cross section value, estimated after summing over the experimental convergence and collection angles, in order get a map in which each pixel is associated to a layer orientation angle. One would then get a very accurate morphology characterization in order to perform a 3D-reconstruction of such particles.

5. Mapping defects in multiwall carbon nanotubes

The last example deals with the detection of structural defects and in particular pentagonal defects in carbon nanotubes. Pentagons induce a positive curvature in the graphitic network and are responsible for example for the tubes end closing. On the opposite, heptagonal defects are responsible for a negative curvature of the network and the two effects cancel out when these defects combine into pairs. Single pentagonal and heptagonal defects have been observed by high -resolution transmission electron microscopy [11]. Some theoretical [12] and experimental [13] scanning tunneling microscopy (STM) and scanning transmission spectroscopy (STS) works have been devoted to identifying the signature of such topological defects in the local density of states. However, so far, experimental studies have shown their limits for a clear assignment of the observed image contrast and spectroscopic signal.

In order to get a spectroscopic signature of defects trapped in the hexagonal network of a nanotube, we have recorded collections of CK ELNES in the spectrum imaging mode by scanning the probe over highly defective multiwall carbon nanotubes. These tubes present a lot of kinks all along their axis (see fig.4b and 4c) which are likely to contain pentagonal defects. In the example shown here, a 50x10 spectrum image was recorded over a 100nm x 20 nm area. Three main component spectra, displayed in fig. 4a, were extracted from the collection of spectra. Two of them were assigned to amorphous and graphitic carbon but the third unknown. In order to realize the mapping of the different ELNES, NNLS processing has been applied. The resulting maps are shown in fig. 4d, 4e and 4f. Fig. 4e and 4f clearly display the graphitic region of the wall of a defective nanotube and its covering by a thin layer of amorphous carbon. The unknown signal is detected for a single position of the electron probe (fig. 4d). Comparison with LDOS calculation (not shown here [14]) pleads for the signature of pentagonal rings while rejecting the possibility of heptagonal rings. We suggest that the scarce detection of such spectroscopic signal in a highly defective nanotube could be explained by a simple geometric argument: several pentagons from adjacent layers need to be aligned vertically along the electron beam to provide a significant spectroscopic contribution to the experimental signal.

Further experiments with a finer sampling should be acquired in order to probe spatially the modification in the LDOS induced by such a defect.

442

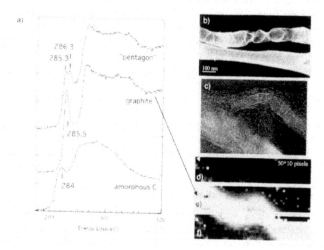

Figure 4: Spectroscopic signature of defects. A) the 3 different identified ELNES components extracted from a 50x10 spectrum image acquired on a multiwall carbon nanotube. b) and c) low and high magnification bright field image from a highly defective nanotube. d) e) and f) NNLS maps associated with the 3 ELNES components displayed in a).

6. Conclusion

The combination of the spectrum imaging mode with processing methods such as multiple least square fitting has been demonstrated to provide a very efficient tool for probing at the very fine scale the electronic states in nanostructures.

7. References

[1] Iijima S 1991 *Nature* **354** 56
[2] Suenaga K *et al.* 2000 *Science* **290** 2280
[3] Stephan O, Doyenette L, Sloan J and Colliex C unpublished
[4] Colliex C, Tencé M, Lefèvre E, Mory C, Gu H, Bouchet D and Jeanguillaume C 1994 *Mikrochimica Acta* **114** 71
[5] Bonnet M, Brun N and Colliex C 1999 *Ultramicroscopy* **77** 97
[6] Tencé M, Quartuccio M and Colliex C 1995 *Ultramicroscopy* **58** 42
[7] Lee R. *et al.* 2001 *Phys. Rev B, Rapid Comm* **64**, 121405.
[8] Arenal de la Concha A *et al.* to be published
[9] Menon HK and Yuan J 1998 *Ultramicroscopy* **74** 83
[10] Gloter A. 2000 PhD Thesis, University of Paris-Sud
[11] Iijima S, Ajayan PM and Ichihashi T 1992 *Phys. Rev. Lett* **69** 3100
[12] Meunier V, Henrard L and Lambin Ph 1998 *Phys. Rev. B* **57** 2596
[13] Ouyang M, Huang JL, Cheung CL and Lieber CM 2001 *Science* **291** 97
[14] El-Barnaby A. *et al.* And Trasobares S. *et al.* to be published

Inst. Phys. Conf. Ser. No 179: Section 11
Paper presented at Electron Microscopy and Analysis Group Conf. EMAG2003, Oxford, 2003
©2003 IOP Publishing Ltd

Energy loss near edge structure study of fullerenes

R J Nicholls[1], D A Pankhurst[1], G A Botton[2,3], S Lazar[3], D J H Cockayne[1]

[1]Department of Materials, University of Oxford, UK

[2]Brockhouse Institute for Materials Research, McMaster University, Ontario, Canada

[3]Laboratory of Materials Science, Delft University of Technology, The Netherlands

Abstract. We have collected carbon K spectra using electron energy loss spectroscopy (EELS) from C_{60} and C_{70} fullerene nanocrystals using TEMs both with and without with a monochromator. Differences in the near edge structure are observed and attributed to the different bonding environments in the two molecules. The importance of high energy resolution for this differentiation is demonstrated.

1. Introduction

Fullerenes have attracted much interest since C_{60} was first discovered by Kroto et al. in 1985 [1]. The number of potential applications of these structures, such as the basis for qubits in a quantum computer [2], is greatly increased if their properties can be controlled. The electronic properties can be altered by doping, using atoms placed inside, outside or within the structure [3].

It is important that these changes in electronic properties are understood if fullerenes are to be used to their full potential. One way of achieving this is through electron energy loss spectroscopy (EELS) as chemical bonding information can be retrieved from samples which are a few nanometres in size and have complex structures. This paper uses the carbon K-edge in EELS to examine nanocrystal samples of pristine fullerenes C_{60} and C_{70}. This edge is sensitive to the unoccupied p-states of carbon. Observed differences between the spectra from the two structures are attributed to the different bonding environments.

2. Theory

EELS measures the energy lost by a beam of electrons after it has passed through a sample. At losses above about 50eV, the energy loss spectrum is a decreasing background with ionisation edges superimposed on it. These edges are caused by excitations of core level electrons into states above the Fermi level. The fine structure which occurs after the onset of an ionisation edge is related to the density of states (DOS) above the Fermi level, whilst the fine changes in the edge energy are related to the valence state of the atom [4]. EELS is, therefore, a suitable tool for monitoring changes in the electronic structure of doped and undoped fullerenes.

444

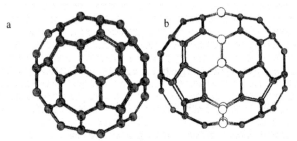

a   b

Figure 1: The structure of (a) C_{60} and (b) C_{70} (pictures supplied by R Scipioni).

The structures of both the C_{60} and C_{70} molecules are shown in Figure 1. The addition of an extra ten atoms to C_{60} to produce C_{70} changes both the shape of the molecule and the bonding environment of the carbon atoms. The atoms are no longer all equivalent and this change in bonding environment has a great effect on the DOS [5]. The relationship between DOS and EELS is not direct as the probability of a transition to a particular state is proportional to the DOS multiplied by the square of a matrix element, as shown in equation 1 (e.g. [6]),

$$P(E) \propto |M|^2 N(E) \qquad (1)$$

where the P(E) is the probability of a transition to a particular state, N(E) is the DOS and M is the matrix element between the initial and final states. For small scattering angles we can consider the matrix element in terms of the dipole approximation. This restricts the change in angular quantum number between the initial and final states to ±1. In the case of transitions from the 1s core level the final states have p-like character.

3. Materials and Methods

The samples used in this study were produced using the arc discharge method (Sigma-Aldrich Ltd). The 99.5% pure powder was dissolved in carbon disulphide and then mixed with an excess of methanol to precipitate fullerene crystals. The precipitate was collected by filtration, re-dispersed in pure methanol and dropped onto a lacey carbon coated copper TEM grid. The methanol was then evaporated. Samples produced in this way have a distribution of crystals of varying shape and thickness as shown in Figure 2.

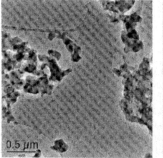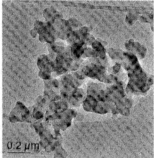

Figure 2: TEM images of C_{70} crystals dispersed on a lacey carbon film obtained in a JEOL 3000F operated at 300kV.

Carbon K-edges and images shown in Figure 2 were obtained on a number of transmission electron microscopes (TEMs) – a JEOL 3000F TEM operating at 300kV, a JEOL 2010F operating at 120kV with a zero loss peak full width half maximum (FWHM) of ~1.4eV and a FEI Tecnai TF200 200kV TEM equipped with a monochromator operating at 120kV with a zero loss peak FWHM of 0.14-0.20eV. Backgrounds were subtracted from all spectra. Spectra were all acquired at 120kV in order to reduce electron beam damage occurring during long the acquisition times necessary to record EELS data.

4. Results and Discussion

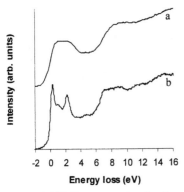

Figure 3: EELS spectrum showing the carbon K-edge of C_{70} obtained on a JEOL 2010F (a) and the Delft Tecnai TF200 (b).

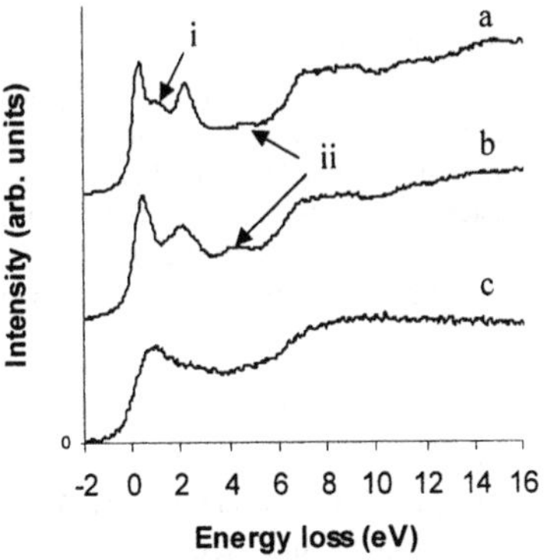

Figure 4: Carbon K-edge of (a) C_{70}, (b) C_{60}, (c) amorphous carbon all taken on the Delft monochromated TEM.

The carbon K-edge of C_{70} obtained on a JEOL 2010F operated at 120kV is shown in figure 3a. Figure 3b shows the carbon K-edge of the same C_{70} sample obtained on the monochromated FEI Tecnai 200kV TEM also operated at 120kV with an energy resolution improvement of a factor of ten. The increase in resolution has a dramatic effect on the detail that can be seen. Figure 4a shows the same C_{70} carbon K-edge along with a C_{60} carbon K-edge (Fig. 4b) and the carbon K-edge from amorphous carbon (Fig. 4c), all obtained on the monochromated Tecnai TEM. Due to the small differences in the fine structure, the high energy resolution is essential to distinguish the different structures and to assess whether beam damage occurs during acquisition of the energy loss spectra. Although our data has been acquired from individual nanocrystals, these spectra agree well with previously published results obtained from thin films and microcrystals [7-9]. Differences between the spectra from the three materials can be seen clearly. The C_{70} spectrum shows a peak [labelled (i)] at about 0.7eV above the edge onset, which is not seen in the C_{60} spectrum. There is also a shift in the position of peak (ii) of around 0.5eV between the two spectra.

The differences between the C_{60} and C_{70} spectra can be explained in terms of the differences between the unoccupied p-DOS for the two molecules. The addition of ten atoms to C_{60} to produce C_{70} changes the shape of the molecule and the bonding environment of the carbon atoms. The symmetry of the molecule is reduced, and additional peaks appear in the spectrum due to additional energy states arising from the different bonding environments. The interpretation of the differences in EELS spectra resulting from a change in the DOS, particularly in the states with p-like character, is confirmed by comparing the spectra to the DOS predicted by density functional theory [10].

5. Summary

The increased resolution of a monochromated TEM makes it possible to distinguish between C_{60} and C_{70} samples using high resolution EELS. The spectra acquired from individual nanocrystals are in agreement with previous results obtained with high energy resolution EELS on large areas of thin films and microcrystals. The spectral differences can be interpreted as differences in the unoccupied DOS of the two molecules due to different bonding environments.

6. Acknowledgements

We would like to thank Andrei Khlobystov for supplying and preparing samples and the Armourers & Brasiers' Company and British Alcan Aluminium for their financial support. The Tecnai TEM used is at Delft University of Technology, the JEOL 2010F at McMaster University, and the 3000F at Oxford University. GAB is grateful to NSERC (Canada), NWO (The Netherlands) and Prof. Zandbergen for financial support, a visiting professorship and for providing access to the Tecnai microscope in Delft.

References

[1]   Kroto H W, Heath J R, O'Brien S C, Curl R F, Smalley R E, Nature 318 162 (1985)

[2]   Briggs G A D, patent number WO03007234 (2003)

[3]   Knupfer M, Surf. Sci. Rep. 42 1 (2001)

[4]   Egerton R F, *Electron Energy-Loss Spectroscopy in the Electron Microscope* (Plenum Press 2nd Ed (1996): New York)

[5]   Pettifor D, *Bonding and Structure of Molecules and Solids* (Oxford University Press (1995): Oxford)

[6]   Gasiorowicz S, *Quantum Physics* (John Wiley & Sons (1994): New York, Chichester)

[7]   Sohmen E, Fink J, Phys. Rev. B 47 (21) 14532 (1993)

[8]   Hansen P L, Fallon P J, Krätschmer W, Chem. Phys. Lett. 181 367 (1991)

[9]   Kuzuo R, Terauchi M, Tanaka M, Saito Y, Shinohara H, Phys. Rev. B 49 (7) 5054 (1994)

[10]   Lee S M, Nguyen-Manh D, private communication.

Inst. Phys. Conf. Ser. No 179: Section 11
Paper presented at Electron Microscopy and Analysis Group Conf. EMAG2003, Oxford, 2003
©2003 IOP Publishing Ltd

447

Structural characterisation of carbon nanotubes in the presence of a Ni catalyst

H K Edwards[1], M W Fay[1], M Bououdina[1], D H Gregory[2], G S Walker[1] and P D Brown[1]

[1]School of Mechanical, Materials, Manufacturing Engineering and Management, and [2]School of Chemistry, University of Nottingham, Nottingham, NG7 2RD, United Kingdom.

Abstract. Multi-wall carbon nanotubes (MWNTs) grown by chemical vapour deposition (CVD) of an ethene / hydrogen gas mixture on alumina supported and unsupported nickel catalysts for 10, 20 and 120 minutes at 600°C and 700°C are compared. Nano-scale nickel particles were identified at the tube ends and also within the main body of the tubes. The particles exhibited comet like structures suggesting the metal to be highly mobile if not molten during processing. HREM observations demonstrate the carbon planes to be parallel to the nickel particle surfaces, while eventually becoming drawn out to define the carbon tube walls. The tubes are generally found to emanate from larger encapsulated nickel particles. A growth mechanism involving the propulsion of mobile nickel through the precipitation of the carbon is discussed.

1. Introduction

The open structures of carbon nanofibres allow for the storage of hydrogen by adsorption and hence there is much interest in developing their application in fuel cells [1]. There is equal interest in developing the properties of multi-wall carbon nanotubes (MWNTs), although hydrogen storage is found to be more difficult using these more closed structures that are presently finding potential applications in nano-scale electrical conduction and nano-composites instead [2]. Here we report on an investigation into the chemical vapour deposition (CVD) synthesis of MWNTs in the presence of a nickel catalyst, assessed using TEM to gain insight into the reaction pathways and growth mechanisms of the nanotubes.

2. Experimental

MWNTs were grown by CVD on alumina supported or unsupported nickel oxide powder in an alumina boat within a tube furnace under flowing argon at temperatures of 600°C and 700°C, established at a ramp rate of 10°C per minute. At the reaction temperature, the gas flow rates were adjusted to achieve a synthesis gas mixture of 80% ethene / 20% hydrogen. Reaction was allowed to proceed in the chamber for 10, 20 or 120 minutes to allow different stages of the growth to be assessed. After growth the furnace temperature was maintained for 15 minutes and then stepped down to room temperature,

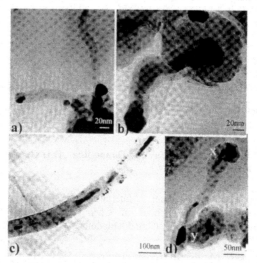

Figure 1 (a,b) Examples of nanotubes grown at 700°C for 10 and 20 minutes, respectively. (c) Nickel catalyst trapped in the hollow core of a nanotube, indicating a highly mobile state has been attained (700°C, 2 hrs). (d) Nanotube with faceted (x) and encapsulated (y) Ni particles at opposite ends (700°C, 20 min).

under flowing argon, to limit further reaction or oxidation of the samples. Powder x-ray diffraction (XRD) confirmed that the initial heating stage was sufficient to fully convert the NiO such that CVD growth proceeded in the presence of a Ni catalyst. Samples of the resultant black powder were removed from the alumina boat and ground or ultrasonically fragmented before being dispersed using acetone onto holey carbon film / copper grids. Specimens were examined using Jeol 2000fx and 4000fx electron microscopes for conventional and high-resolution studies, respectively.

3. Results and Discussion
An initial investigation of tubes grown for a reaction time of 2 hours (using an alumina support) found that those produced at 600°C were more abundant, longer and had larger external diameters than those produced at 700°C for which more carbon encapsulation of the nickel occurred. Tube dimensions were of the order of 1-5 μm in length and 20-100 nm in external diameter at 600°C, and 1-2 μm in length and 15-70nm in external diameter at 700°C, respectively.

Figures 1a,b illustrate MWNTs formed following 10 and 20 minutes of growth at 700°C, respectively. The tubes for these shorter reaction times were typically 100 to 500nm in length, and often found emanating from larger nickel particles encapsulated in carbon. Regardless of reaction time and temperature, highly elongated Ni particles were often situated at both ends of each tube and on occasion distributed along their hollow cores (Figure 1c), the morphology of which might indicate that the catalyst had attained a highly mobile or molten state during synthesis, despite the melting point of Ni being 1453°C.

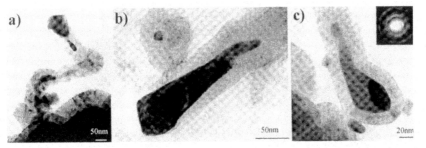

Figures 2a-c. Three varieties of elongated Ni particles characteristic of the nanotube samples (grown at 700°C for 20 min; 600°C for 2hrs; and 700°C for 2 hrs, respectively).

Nanotubes such as those shown in Figure 1d with a faceted particle at one end (x) and a particle encased in carbon at the other end (y) were also frequently observed.

In particular, three distinct varieties of elongated Ni particles were identified, as illustrated by Figures 2a-c. The first type took the form of faceted particles (Fig. 2a, arrowed) with a trailing tail exhibiting clean faces free from carbon at the presumed growth front. It is considered that these clean facets correspond to the catalytically active part of the particle, responsible for adsorbing and absorbing carbon into the metal as part of the mechanism to promote tube growth. The second type of particle observed was 'comet-like' in shape and bounded by carbon walls of graded thickness (Fig. 2b). Thirdly, loosely faceted particles were identified encased in a thick continuous layer of carbon (Fig. 2c). The first variety of particle was more prevalent at shorter reaction times, whereas the latter two types were more abundant in samples grown for 2 hours. Intriguingly, in Fig. 2b the implication is of a graded build up of carbon walls from the trailing edge of the particle, suggesting the release of carbon from the particle is part of the tube growth mechanism.

Observation of the graphitic planes using HREM around the catalysts and along the tube walls confirmed that the nanotubes displayed varying levels of crystallinity. As a general observation the carbon walls followed the shape of the catalyst but then transformed with increasing distance from the catalyst into the parallel-sided walls of the NTs, whilst also become more disrupted and broken up.

Starting from the initial transformation of NiO to Ni, these combined observations implicate the following mechanism in the formation of MWNTs trailing behind clean faceted Ni particles (Figure 3):

- On the surface of a catalyst particle, ethene decomposes to carbon that is adsorbed, whilst hydrogen is released. A resultant carbon monolayer presumably covers the exposed faces of the particle (Figures 3i,ii).
- It is suggested that the Ni faces that have a high carbon diffusion rate preferentially absorb the carbon atoms, creating a low C-content Ni-alloy. (The combined TEM and XRD evidence confirmed that a distinct nickel carbide had not formed.)
- A concentration gradient of carbon will presumably develop within the particle between the catalyst-vapour surface (Fig. 3ii(x)) and the Ni-support interface (Fig. 3ii(y)). The absorbed carbon will thus undergo bulk diffusion through the Ni to the precipitating planes where individual atomic planes of carbon are produced and sequentially build up parallel to the trailing surface of the Ni particle.

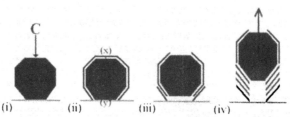

Figure 3. Proposed nanotube growth mechanism including the adsorption, absorption, diffusion and precipitation of carbon.

- This precipitation of carbon causes the propulsion of the Ni particle away from its original location (Fig .3iii).
- As the catalytic reaction proceeds more carbon is adsorbed, absorbed and diffused through the Ni particle to be precipitated and perpetuate the growth of the nanotube (Fig. 3iv). During this process the nickel apparently approaches an almost liquid state.

While it is possible that there is some surface diffusion of carbon around each particle, the evidence suggests that the absorption planes remain clean or only lightly carbon covered throughout the nanotube growth (Figs. 2a,b), whilst thick encapsulated particles are possibly indicative of continued carbon growth at the end of the reaction process (Fig. 2c). The diffusion and precipitation of carbon causes elongation and separation of the particles resulting in Ni being left at the nucleating end of the nanotube or distributed along the core body, before growth eventually terminates and the propagating particle becomes encapsulated. It is suggested that the small particle size with its large surface area to volume ratio, combined with the formation of a Ni-C alloy, promote the molten-like behaviour of the Ni particles observed. To support this suggestion it is noted that a Co-C system theoretically displays molten-like behaviour at elevated temperatures (but well below its melting point) [3]. The details of carbon plane reorientation from the Ni particle to the walls of the tube are not fully understood, however, the formation of tube walls of constant thickness suggests a uniform rate of carbon precipitation.

4. Summary
The initial stages of multi-wall nanotube growth by CVD at 600°C and 700°C have been characterised. Elongated Ni catalyst particles initiating or terminating tube growth, or distributed within the body of the tubes, implicate a highly mobile or liquid state of the Ni that actively promotes the mechanism of tube growth.

5. References
[1]   Cheng H, Yang Q and Liu C 2001 Carbon 39 1447.
[2]   Chen W, Tu J, Wang L, Gan H, Xu Z and Zhang X 2003 Carbon 41 215.
[3]   Gorbunov A, Jost O, Pompe W and Graff A 2002 Carbon 40 113.

Acknowledgements
With thanks to DSTL and EPSRC for funding and to Keith Dinsdale for microscopy support.

Inst. Phys. Conf. Ser. No 179: Section 11
Paper presented at Electron Microscopy and Analysis Group Conf. EMAG2003, Oxford, 2003
©2003 IOP Publishing Ltd

Off-axis electron holography and image spectroscopy of ferromagnetic FeNi nanoparticles

R K K Chong[1], R E Dunin-Borkowski[1], T Kasama[1,2], M J Hÿtch[3] and M R McCartney[4]

[1] Department of Materials Science and Metallurgy, University of Cambridge, Cambridge CB2 3QZ, UK

[2] RIKEN (The Institute of Physical and Chemical Research), 2-1 Hirosawa, Wako, Saitama 351-0198, Japan

[3] Centre d'Etudes de Chimie Métallurgique - CNRS, 15 rue G. Urbain, 94407 Vitry-sur-Seine, France

[4] Center For Solid State Science, Arizona State University, Tempe, AZ 85287-1704, USA

Abstract. The magnetic remanent states and chemical compositions of chains of crystalline Fe_xNi_{1-x} nanoparticles, which have an average size of 50 nm, have been characterised using off-axis electron holography and image spectroscopy. The critical sizes at which the particles can support magnetic vortices are correlated with their compositions. A significant increase in magnetic vortex core diameter is observed when vortices in the nanoparticles are aligned parallel to the chain axes.

1. Introduction

The magnetic properties of closely-spaced nanoparticles are of considerable interest, both fundamentally and for the design of ultra-high-density magnetic devices. However, the factors that influence the critical sizes at which such particles begin to support complicated domain structures, rather than single magnetic domains, are poorly understood. Such effects are difficult to predict theoretically because of the large number of variable parameters that need to be included in simulations of the magnetic microstructure of a distribution of nanoparticles, which is sensitive not only to magnetostatic interactions between each particle and its neighbours and to the magnetic history of the specimen but also to the morphologies and the compositions of the individual particles. Here, we use image spectroscopy[1] and off-axis electron holography[2] to characterise linear chains of crystalline nanoparticles that have an average diameter of 50 nm, nominal compositions of $Fe_{55}Ni_{45}$, $Fe_{30}Ni_{70}$ or $Fe_{10}Ni_{90}$, and 4-nm-thick oxide layers surrounding their metal cores. We begin by assessing the accuracy and the precision to which the nanoparticle compositions can be measured using image spectroscopy. We then correlate these measurements with the magnetic microstructures of individual nanoparticle chains, which are measured using electron holography.

452

2. Image Spectroscopy

Image spectroscopy involves the analysis of an extended series of energy-loss images, and allows the formation of chemical maps of materials with better signal to noise than using three-window elemental mapping. Approaches that are normally applied to electron energy-loss spectra can then be used to improve the fitting parameters for the energy loss background at each image pixel, as well as allowing the deconvolution of plural scattering effects and the use of large energy-selecting windows without introducing additional chromatic blurring. Figure 1a shows part of a series of energy-loss images of two $Fe_{55}Ni_{45}$ nanoparticle chains acquired at 300 kV on a Philips CM300 field emission gun transmission electron microscope (FEGTEM) equipped with a Gatan imaging filter. Energy losses of 0-1200 eV were used, with an energy-selecting slit width of 20 eV and an interval of 20 eV between images. The pixel size is 2.4 nm. Figure 1b shows the corresponding energy-loss spectrum averaged over all image pixels, while Fig. 1c shows O and Fe chemical maps derived from the O K and Fe $L_{2,3}$ edges using optimal pre and post edge windows. The 4 nm oxide shell is visible around each particle in the O map, while diffraction contrast is visible in both signals. Figure 1e shows the Fe/Ni ratios in several nanoparticles in each sample, calculated using a 60 eV pre-edge window, several post-edge window sizes and Hartree-Slater cross-sections[3].

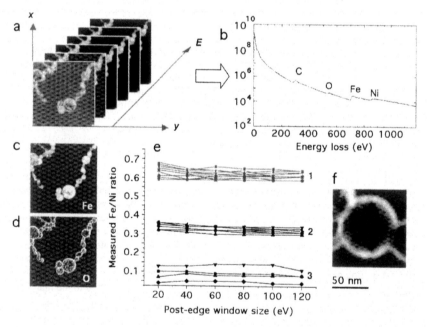

Figure 1 a) Image spectroscopy data from $Fe_{55}Ni_{45}$ nanoparticles. b) Energy-loss spectrum derived from a). c) - d) Elemental maps derived from the Fe $L_{2,3}$ and O K edges. e) Image spectroscopy measurements of Fe/Ni ratios in several nanoparticles in samples of nominal composition: 1. $Fe_{55}Ni_{45}$, 2. $Fe_{30}Ni_{70}$ and 3. $Fe_{10}Ni_{90}$, plotted vs. post-edge window size starting at the Fe and Ni $L_{2,3}$ edge thresholds. f) Oxygen map, showing faceting around a 75 nm nanoparticle, after sub-pixel alignment of the images.

The consistency between the measurements from different nanoparticles in the same sample is better than expected, presumably because some of the possible sources of random and systematic error in the measurements, such as those associated with diffraction contrast and errors in the calculated scattering factors, cancel out when a ratio between the Fe and Ni signals is calculated. The measured Fe/Ni ratios are on average 0.63 ± 0.03, 0.33 ± 0.03 and 0.09 ± 0.03 for the samples that have nominal ratios of 0.55, 0.30 and 0.10, respectively. A slight decrease in these values is sometimes observed with increasing post-edge window size, perhaps because of the inadequate treatment of Fe and Ni white lines in the calculated cross-sections. Although our results confirm the broad trend in the particle compositions, further systematic errors, which are presently being investigated, almost certainly remain. The errors may be associated with the oxide shell around each particle having a different composition from the core, with the presence of the Fe L_1 edge at the position of the Ni $L_{2,3}$ edge in the energy-loss spectrum, and with plural scattering in the largest particles. A final difficulty results from the need to perform sub-pixel alignment of the original images. Only after this procedure is performed is faceting of the particles revealed in the O maps (Fig.1e).

3. Off-Axis Electron Holography

Off-axis electron holograms of the nanoparticle chains were acquired either at 300 kV using a Philips CM300ST FEGTEM or at 200 kV using a Philips CM200ST FEGTEM. Each microscope was equipped with a 'Lorentz' lens, which allows the magnetic microstructure in the samples to be recorded in field-free conditions at high spatial resolution. The technique involves applying a voltage to an electron biprism in order to overlap a coherent electron wave that has passed through the sample with a part of the same electron wave that has passed only through vacuum. Analysis of the resulting interference pattern allows the phase shift of the electron wave to be recovered quantitatively and non-invasively at a spatial resolution close to the nm scale [2].

Figures 2a and b show the magnetic remanent states of two chains of $Fe_{55}Ni_{45}$ particles, which have been measured using electron holography. The density of the contours is proportional to the in-plane induction in the sample integrated in the electron beam direction, and the spatial resolution of the magnetic information is estimated to be approximately 10 nm. The unwanted mean inner potential (MIP) contribution to the phase shift (which provides a value of 21.5 ± 0.3 V for the MIP in each sample) has been subtracted from the images [2]. The remanent state of a 75 nm $Fe_{55}Ni_{45}$ particle sandwiched between two smaller particles is shown in Fig. 2a. Closely-spaced contours run through all three particles in a channel of width 22 ± 4 nm. Comparisons with simulations suggest that the largest particle contains a vortex whose axis lies parallel to the chain axis, as shown schematically in Fig. 2c. The lack of contours on either side of the flux channel that forms the vortex core results from the induction in these regions being substantially out-of-plane. In Fig. 2b, a vortex can be seen end-on in a 71 nm particle at the end of a chain. The positions of the particle's neighbors determine the chirality of the vortex (Fig. 2d), whose core, which is now perpendicular to the chain axis, is only 9 ± 2 nm in diameter. The larger value in Fig. 2a results from dipole-dipole interactions along the chain. These results can be compared with a similar induction map obtained from a chain of $Fe_{10}Ni_{90}$ particles, which is shown in Fig. 2e. The magnetic microstructure in this sample, which has a higher Ni concentration, is closer to that expected for a chain of single-domain spheres. Flux channels, which are 70 nm in diameter, form in this sample only when the particles are above ~100 nm in size.

454

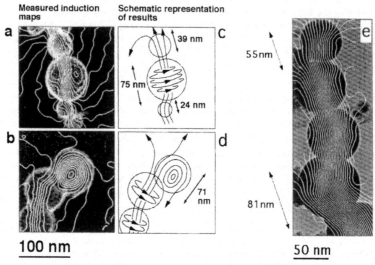

Figure 2 a)-b) Phase contours, which have been overlaid onto O maps, showing the in-plane induction (integrated in the electron beam direction) in chains of $Fe_{55}Ni_{45}$ particles, recorded with the microscope objective lens switched off. The contour spacings are 0.083 and 0.2 radians for a) and b), respectively. The MIP contribution to the phase has been removed from each image. c) and d) show schematic diagrams of the magnetic microstructure in the chains, in which vortices spin around the chain axes. A vortex perpendicular to the chain is also visible in d). e) shows a similar map (spacing 0.125 radians, overlaid onto a bright-field image) from a chain of $Fe_{10}Ni_{90}$ particles.

4. Discussion and Conclusions

The magnetic remanent states shown in Fig. 2 demonstrate the sensitivity of the magnetic microstructure in the FeNi nanoparticle chains to composition, as well as to the sizes and the positions of the particles. Despite the clear trend in the magnetic microstructure with increasing Ni concentration, simple predictions are still complicated by the fact that the critical sizes at which particles cease to be single domain depend not only on their compositions (which affect the exchange length in the alloy) but also on the sizes, compositions and positions of their neighbours. Both high quality magnetic and compositional data and micromagnetic simulations are therefore necessary to understand the magnetic remanent states and reversal mechanisms in such systems fully.

This work was carried out in the framework of the European Research Group "Quantification and measurement in transmission electron microscopy". We thank the Royal Society and Chartered Semiconductor Manufacturing for financial support, and we acknowledge the use of facilities at the Center for High Resolution Electron Microscopy at Arizona State University.

References

[1] P. J. Thomas and P. A. Midgley, Ultramicroscopy **88** (3), 179-186 (2001).
[2] R. E. Dunin-Borkowski and M. R. McCartney, in *Magnetic Nanostructures*, edited by H. S. Nalwa (American Scientific Publishers, 2002), pp. 299-325.
[3] P. Rez, Ultramicroscopy **9** (3), 283-287 (1982).

Inst. Phys. Conf. Ser. No 179: Section 11
Paper presented at Electron Microscopy and Analysis Group Conf. EMAG2003, Oxford, 2003
©2003 IOP Publishing Ltd

LaI$_2$@(18,3)SWNT: the crystallisation behaviour of a LaI$_2$ fragment, confined within a single-walled carbon nanotube

S Friedrichs, E Philp, R R Meyer*, J Sloan, A I Kirkland*, J L Hutchison* and M L H Green

Inorganic Chemistry Laboratory, South Parks Road, Oxford OX1 3QR, UK.
*Department of Materials, Parks Road, Oxford OX1 3PH, UK

Abstract. The encapsulation of the simple binary compound LaI$_3$ within the inner cylindrical bore of a single-walled carbon nanotube (SWNT) yielded an unprecedented structure of LaI$_2$. The filling material was encapsulated within SWNTs by heating bulk LaI$_3$ *in vacuo* in the presence of SWNTs. The obtained encapsulation composite was analysed using energy-dispersive X-ray microanalysis (EDX) and focal series restoration approaches of high-resolution transmission electron microscopy (HRTEM) images, to reveal a one-dimensional LaI$_2$ fragment, 'crystallising' in the structure of LaI$_3$ with 1/3 of the iodine positions unoccupied, whilst simultaneously identifying the chirality of the surrounding SWNT.

1. Introduction

The restoration of a focal series of HRTEM images represents a powerful tool for the complete characterisation of novel nanometre-sized materials, such as SWNT inclusion composites, in which the encapsulated 'crystal' fragments can be as thin as one atomic layer in projection. For the characterisation of SWNTs and their encapsulation composites, however, the conventional image restoration approach was modified, in order to account for rotational movements of the tubular material (Friedrichs 2001). In this modification, the imaged encapsulation composite is restored in several individually subregions. An additional achievement of this modification is the determination of the inclination angle β of the encapsulation composite with respect to the image plane (Meyer 2003). Knowledge of this inclination angle can support the assignment of the precise structure and orientation of the encapsulated lanthanum iodide 'crystal' fragment and furthermore enable the prediction of the electron transfer properties of the host SWNT, which can be either metallic or semiconducting depending on its chiral vector $C_h \equiv (n,m)$.

2. Results

2.1 LaI$_2$@SWNT

Figure 1(a) shows the recombined phase of the encapsulation composite, synthesised by heating SWNTs *in vacuo* in the presence of LaI$_3$, restored from a 20-membered focal series of HRTEM images. Figure 1(b) displays a diagram of the absolute focus values of the individual subregions plotted against their respective centre displacement values along the SWNT axis. A linear fit of the resulting data points (dashed line graph) gives an inclination angle of $\beta = -17° \pm 10°$. This inclination angle is consistent with the observation of different projected C-C spacings in each of the SWNT walls, illustrated in the single-pixel line traces in Figure 1(c). Calculation of the cross correlation coefficients (XCFs) between the power spectrum of the restored SWNT and the power spectra of a large number of simulated SWNTs under different inclination angles β yielded a chiral vector $C_h = (18,3)$, predicting the host SWNT to be a metallic conductor (Meyer 2003).

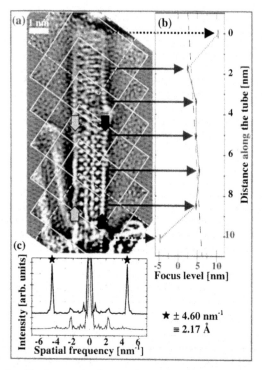

Figure 1. (a) Restored phase of the imaged encapsulation composite 'LaI$_2$@SWNT', obtained by recombination of six individually restored subregions (white frames). (b) Plot of the focus value of each subregion *versus* its centre displacement along the tube axis. (c) Power spectra of line profiles along each of the SWNT walls, indicated by the arrows in Figure 1(a).

The structure and stoichiometry of the encapsulated 'crystal' fragment can be interpreted intuitively. Figure 2(c) displays a selected region of the restored phase, in which the white spots correspond to the projection of one atomic layer of LaI$_x$. The white lines mark the directions of individual single-pixel line profiles, shown in Figures 2(c1) and (c2). The structural motif of the encapsulated LaI$_3$ fragment can be described as a series of three-membered rows, running perpendicular to the tube axis, alternatively aligned with the left and the and right tube wall. Applying a 'weak-phase object'-approximation, the restored phase shift of each projected atom in the imaged object can be regarded proportional to the atomic potential of the inclusion composite. Computational modelling was applied to find a model corresponding to both the structural motif and the contrast intensities of each column. The final structural model was based on the crystallographic data of bulk LaI_3 (Zachariasen 1948), with 1/3 of the iodine atoms omitted to give a 'crystal' fragment with a LaI_2 stoichiometry, while the remaining atoms maintain their positions in the now 'reduced' structure, as indicated in Figures 2(a) and 2(b). Figure 2(d) displays a simulated phase of the proposed $LaI_2@(18,3)SWNT$ encapsulation composite, as modelled in Figure 2(d), illustrating the excellent match of the simulated image regarding both the variation of contrast intensities and projected structural motif for the encapsulated LaI_2 'crystal' fragment and the surrounding (18,3) SWNT.

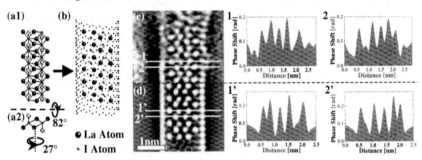

Figure 2. Composite picture displaying the modelling (Figures (a) and (b)), (c) the restoration and (d) the simulation of the observed $LaI_2@(18,3)SWNT$ inclusion composite.

This novel 'reduced' structure of the LaI_2 fragment is closely related to the crystal structure of known bulk LaI_2, which could, until now, only be obtained by careful reaction of LaI_3 with lanthanum metal (Beck 1992). In bulk LaI_2, the iodine atoms form a regular octahedral coordination sphere around the lanthanum, while in the 'reduced' LaI_2 lattice, the coordination sphere is distorted octahedral. In both bulk LaI_3 and bulk LaI_2 the metal adopts a +III oxidation state, since bulk LaI_2, a metallic conductor, is known to have the composition $\{La^{3+}[(I^-)_2(e^-)]\}$.

2.2 $LaI_3@SWNT$

Further investigation of the SWNT encapsulation composite identified a continuously filled SWNT, displaying different arrangements of lanthanum iodide in different parts of the composite. Figure 3(a) shows a composite picture of the restored $LaI_x@SWNT$ inclusion compound and Figures 3(b) and 3(c) display enlarged selected regions of the differently arranged LaI_x fragments, including single-pixel line profiles, obtained from each of the restored regions (Plots 3.1 to 3.4). The structural arrangement seen in Figure 3(b) is similar to the previously observed arrangement, characterised as a LaI_2 fragment in the crystal structure of LaI_3 (Figure2). In contrast to the line profiles shown in Figure 2(c), however, line profiles 3.1 and 3.2 display a very high centre peak. This pronounced phase shift can only correspond to two superimposed iodine atoms, indicating that the encapsulated 'crystal' fragment has the stoichiometry of LaI_3.

Analysis of the second 'crystal' arrangement, displayed in Figure 3(c), supports the above result. The single-pixel line profiles given in plots 3.3 and 3.4 show a high peak, close to the SWNT wall in either one of the two alternating rows. Similar to the

458

interpretation of the arrangement in Figure 3(b), these large phase shifts are assumed to correspond to two superimposed iodine atoms in projection, interpreting the imaged fragment to be LaI₃. Figures 3(d) and 3(e) show end-on views of the proposed models for both arrangement (b) and (c), respectively.

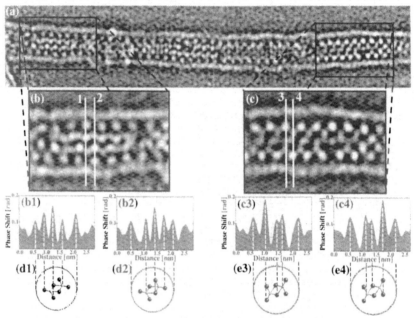

Figure 3. Composite picture displaying (a) the restored phase of a continuously filled SWNT, (a) and (b) selected subregions with different 'crystal' arrangements, incl. single-pixel line traces obtained from each of the subregions and (d) and (e) the end-on views of the proposed LaI₃@SWNT model.

3. Conclusions

We have demonstrated that the focal series restoration of HRTEM images is a powerful technique to completely characterise both the structure and the stoichiometry of the encapsulated 'crystal' fragment and the chirality of the surrounding SWNT. This achievement enabled us to get an insight into the possible physical properties of the encapsulation composite: The host SWNT could be identified as a metallic conductor and the lanthanum iodide 'crystal' fragment was interpreted to adopt LaI₂ stoichiometry, and should therefore also be likely to show metallic conductance.

The author is grateful to Hertford College and the HEFCE fund for financial support.

References
Beck H P 1992 J. Solid State Chem. 100 301.
Friedrichs S 2001 Phys. Rev. B. 64 045406.
Meyer R R 2003 J. Microsc. in print.
Zachariasen W H 1948 Acta Crystallogr. 1 265.

Inst. Phys. Conf. Ser. No 179: Section 11
Paper presented at Electron Microscopy and Analysis Group Conf. EMAG2003, Oxford, 2003
©2003 IOP Publishing Ltd

Electron tomography of heterogeneous catalysts

L Laffont[1], M Weyland[1], R Raja[2], J M Thomas[1,3] and P A Midgley[1]

[1]Department of Materials Science and Metallurgy, University of Cambridge, Pembroke Street, Cambridge, CB2 3QZ
[2]University Chemical Laboratories, University of Cambridge, Lensfield Road, Cambridge, CB2 1EW
[3]The Royal Institution of Great Britain, Davy Faraday Research Laboratory, 21 Albermarle Street, London, WS1 4 BS

Abstract. Heterogeneous catalysts, in which 'heavy' (high Z) nanoparticles are supported on 'light' (low Z) supports, are extremely important for production of bulk chemicals. The efficiency of these catalysts is closely controlled by their three dimensional (3D) structure. This paper will examine the 3D distribution of bimetallic nanoparticles embedded within a mesoporous silica support, MCM-41, using high angle annular dark field (HAADF) scanning transmission electron microscopy (STEM) tomography. This technique has shown itself to be more suitable than conventional TEM electron tomography for these specimens as it offers the advantage of a reduced damage rate and a high contrast incoherent signal. Analysis of 3D structure of these catalysts is demonstrated in both a qualitative and quantitative manner.

1. Introduction

This paper describes the analysis of the heterogeneous catalyst $Ru_{10}Pt_2$ bimetallic nanoparticles, ~1nm in diameter, embedded within MCM-41, a mesoporous silica support with a pore width of ~3nm. This structure has recently been shown to catalyse the conversion of trans, trans-muconic acid, derived from glucose, to adipic acid, widely used in the production of nylon[1]. As this catalyst is based on a renewable feedstock it is an environmentally friendly alternative to traditional, fossil fuel based, production routes.

Although TEM has been used extensively to study heterogeneous catalysts, it is only recently that electron tomography has been applied to determine the 3D structure of such systems[2]. In particular, STEM-HAADF tomography has been shown to be an ideal technique to investigate the 3D distribution of nanoparticles within a mesoporous support[3]. This arises because (i) in STEM mode the silica damages at a rate which is two to three orders of magnitude lower than in conventional TEM mode, (ii) the STEM image shows strong atomic number contrast and (iii) shows little sign of the coherent effects (e.g. Fresnel contrast) that hinder the use of BF TEM images for tomography.

2. Principles of Tomography

Electron tomography consists of three main steps: the acquisition of a tilt series of projections, the alignment of these projections about a common tilt axis and the generation of a reconstruction. A tilt series is acquired by tilting the specimen about a single axis at regular increments, correcting the specimen position and the focus before acquiring the 2D image (projection). The 3D resolution of the reconstruction depends on the number of 2D images acquired in the tilt series and the extent of the tilt range. As such both need to be maximised, yet beam damage, available time and mechanical constraints limit both factors and hence the achievable resolution. Some of these constraints have been overcome by applying automated acquisition (FEI Xplore3D) and by the use of a special high tilt holder (Fischione model 2020). A typical dataset is now composed of 76 images over the tilt range: ± 74°.

The alignment of projections onto a common tilt axis is carried about by a cross-correlation algorithm applied sequentially, starting with the centre projection, to images stretched by an inverse cosθ function. After determination of the exact tilt axis, the reconstruction is carried on consecutive 2D slices using weighted backprojection and/or iterative backprojection using the SIRT algorithm[4].

3. STEM tomography

A STEM HAADF tilt series was acquired from a small region of catalyst, approximately 150nm in diameter. The series consisted of 72 images with a 2° increment from –71° to +71°, a typical image is shown in Figure 1a. After alignment a 3D reconstruction was achieved by weighted backprojection. Contrast selected voxel projections of the reconstruction datasets, Figure 1b and c, show clearly that both mesopores and nanoparticles are resolved in perpendicular directions. In addition, the reconstruction shows that a large proportion of the nanoparticles are sitting within the silica framework.

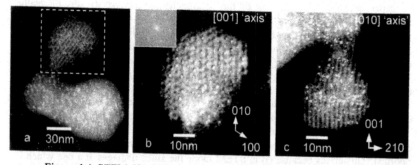

Figure 1a) STEM HAADF image of a typical area of MCM-41 silica support containing $Ru_{10}Pt_2$. b) and c) Two perpendicular contrast selected voxel projections from part of the reconstruction (boxed in Figure 1a) in c) showing clear reconstruction of the hexagonal channels of the MCM-41. The power spectrum shown as an inset in (b) indicates the hexagonal order of the catalyst.

Determining the actual 3D resolution of a reconstruction is far from straightforward [5]but our results seem to indicate that a resolution of ~1nm in all directions is achievable, given a sufficient number of projections within a wide tilt range and a relatively small reconstruction volume (around 150nm^2 for the demonstrated dataset).

Previous results presented on similar specimens[3] have been qualitative in nature. Here we present a more quantitative approach. Two voxel projections in Figure 2 show the hexagonal shape of the silica support, MCM-41 and the hexagonal repartition of the nanoparticles. The ability to separate these components, by the large difference in reconstruction intensity, allows statistical analysis of a small volume of the reconstruction. This is carried out by marking the position and size of each pore in the on-axis voxel projection giving the 2D co-ordinates for the pores. Line traces through the volume aw it is possible to segment the 3D data set into the support in (a) and the nanoparticles in (b). By analysing the sample volume it is possible to investigate the distribution of the particles within the mesopores. The distribution for a sample of 60 pores is shown in Figure 2 (c). In addition by counting the particles as a function of internal pore surface area the catalyst 'loading' was determined as 15 µg/m^2, only 3% of the expected loading, when calculated during synthesis.

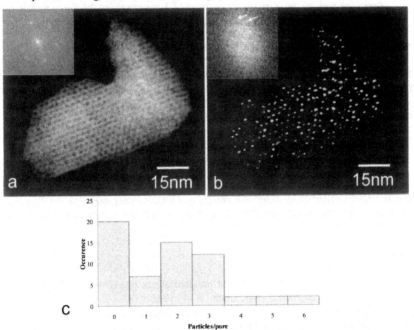

Figure 2 Segmented voxel projections of the reconstruction showing (a) the MCM-41 silica and (b) the nanoparticles, respectively. The 2D power spectra of these projections show the hexagonal shape of the MCM-41 and the ordering of the nanoparticles. (c) Histogram of the nanoparticle population of mesopores.

A complete characterisation of the catalyst structure will include the composition of the nanoparticles, in addition to their positions. An example of a possible approach to such an analysis, using an energy dispersive x-ray (EDX) detector is shown in Figure

4(a) for three nanoparticles in the silica. The Cliff-Lorimer equation [6]applied to the nanoparticles (k factor=1.0 ± 0.2) confirms the composition of each particle to be $Ru_{10}Pt_2$ within experimental error.

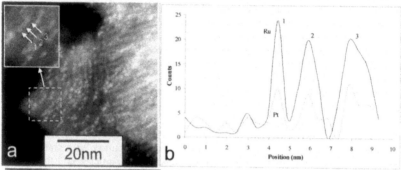

Figure 3(a) STEM HAADF image from the $Ru_{10}Pt_2$ MCM-41 catalyst. A box shows the path of the X-ray line spectrum of 3 nanoparticles. b) X-ray profile of Ru and Pt along the line of particles marked 1,2,3.

4. Conclusions

Previous research has shown that STEM HAADF tomography is an ideal technique for the 3D analysis of heterogeneous catalysts. Results presented here have built upon early work by using automated, high tilt, acquisition to yield a high fidelity, high resolution, reconstruction. Further, it has been demonstrated that reconstructions generated by this technique can be analysed in a quantitative statistical fashion, giving valuable insight into the true nature of these specimens and providing unique feedback into both the synthesis cycle and research into catalytic activity.

The data mined from the 3D reconstructions has the potential to be used in far more complex statistical analyses, and it may well be ideal as initial data for structure and functional calculations. Further combining HAADF tomography and EDX raises the possibility of an almost complete 3D analysis of catalyst structure and chemistry at nanometer resolution.

References

[1]   Thomas J M et al. 2003 Chemical Communications 10 1126-1127
[2]   Koster A J et al. 2000 J. Phys. Chem. B 104 9368-9370
[3]   Midgley P A et al. 2001 Chemical Communications 907-908
[4]   Gilbert P 1972 Journal of Theoretical Biology 36 105-117
[5]   Van Aert S et al. 2002 Journal of Structural Biology 138 21-33
[6]   Williams D, B. and Carter C B 1996 Transmission electron microscopy : a textbook for materials science, (New York ; London: Plenum Press)

Inst. Phys. Conf. Ser. No 179: Section 11
Paper presented at Electron Microscopy and Analysis Group Conf. EMAG2003, Oxford, 2003
©2003 IOP Publishing Ltd

High Resolution TEM investigation of CeO_2/Sm_2O_3 - supported Pd catalysts for methanol reforming in intermediate temperature Solid Oxide Fuel Cells

R T Baker[1], L M Gómez Sainero[2], I S Metcalfe[3], M Sahibzada[4]

[1] Division of Physical and Inorganic Chemistry, University of Dundee, Dundee DD1 4HN, UK. r.t.baker@dundee.ac.uk

[2] Area de Ingeniería Química, Facultad de Ciencias, Universidad Autónoma de Madrid, Cantoblanco, 8049 Madrid, Spain.

[3] Department of Chemical Engineering, UMIST, PO Box 88, Manchester M60 1QD, UK.

[4] Department of Materials, Imperial College, Exhibition Road, London SW7 2AZ, UK.

Abstract. High Resolution TEM (HRTEM) images strongly suggest that Pd catalyst materials containing Sm_2O_3, but not CeO_2, undergo the dissolution and re-precipitation of the oxide support as an semi-amorphous phase during the Pd impregnation step. The Pd appears to be intimately associated with this semi-amorphous phase and is seen to emerge from it following reductive and thermal pretreatments to form identifiable Pd particles. This behaviour can be used to understand the activity of this group of catalysts for methanol reforming.

1. Introduction

Solid Oxide Fuel Cells (SOFCs) are a promising technology for efficient and clean power generation and have significant potential for application in electric motor vehicles. However, SOFCs presently operate at high temperatures (around 1000 °C) leading to materials constraints, high cost of manufacture and problems of long-term stability. Lowering the temperature to about 700 °C would make the fabrication of SOFCs much more cost-effective and allow their integration with conventional balance-of-plant materials [1]. Moreover, the high operating temperature of present designs requires significant energy input at start-up, with an associated, undesirable time delay. By reducing the operating temperature to 500 °C these problems would become manageable for electric motor vehicle applications [2]. Several fuels can be used in SOFCs but, when applied in electric motor vehicles, the use of a liquid fuel such as methanol is of interest. Methanol is easily stored and transported. It can be externally reformed to provide the H_2-rich gas feed to the fuel cell but this elevates the cost of the overall process, so that the internal reforming of methanol, i.e. within the SOFC itself, is a more attractive option. One of the limiting factors in the operation of SOFCs at reduced temperature is the low catalytic activity of the anode for methanol reforming [3]. Transition metals such

464

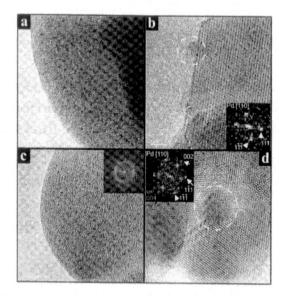

Figure 1. HRTEM images of (a) As-prepared Pd/CS sample; (b) Typical Pd particles in a sample of Pd/CeO$_2$-15; (c) Pd/Sm$_2$O$_3$-15 sample, with DDP inset; (d) Pd particle observed in a sample of Pd/Sm$_2$O$_3$-15, with DDP inset.

as Pd are good catalysts for reforming [3]. CeO$_2$-Sm$_2$O$_3$ mixtures have significant potential as SOFC anodes [4] since they are mixed ionic and electronic conductors at high temperatures. In recent work, a direct methanol high temperature SOFC with SC electrodes performed better than cells with conventional anodes [5]

In this paper, three catalysts with a loading of 2 weight % Pd on CeO$_2$, Sm$_2$O$_3$ and (CeO$_2$)$_{0.8}$(Sm$_2$O$_3$)$_{0.2}$ (CS) supports have been investigated with a view to their application as anodes in intermediate temperature Direct Methanol SOFCs. Here, the relationship between the nanostructure of the catalysts and their activity is considered.

2. Experimental

Pd/CeO$_2$, Pd/Sm$_2$O$_3$ and Pd/(CeO$_2$)$_{0.8}$(Sm$_2$O$_3$)$_{0.2}$ (denoted Pd/CS below) catalysts were prepared with a 2 weight % loading of Pd. Pure CeO$_2$ and Sm$_2$O$_3$ (99.9 %) were used for the preparation of the supports. The (CeO$_2$)$_{0.8}$(Sm$_2$O$_3$)$_{0.2}$ support was prepared by physical mixture of the two components in a mill for 24h. All the supports were heated to 800°C over 5h and calcined at this temperature for 2h. The Pd was incorporated onto the catalyst supports by the incipient wetness impregnation method using PdCl$_2$ in an acid medium (pH<1) as the Pd precursor. The resulting catalyst precursors were dried overnight and calcined at 400°C for 2h. As well as these as-prepared samples, pre-reduced samples were prepared for study in the electron microscope. Samples of all three catalyst compositions were reduced at 150°C in flowing pure, dry H$_2$ for 1h and purged in pure, dry He for 1h at 500°C (indicated with a suffix −15, e.g. CeO$_2$-15). In addition, two further samples of Pd/CS were prepared: (i) by performing the reduction and purging at 500°C (denoted Pd/CS-55) and (ii) by reduction at 500°C and purging at 900°C (denoted Pd/CS-59). Electron microscopy was performed using the 200keV JEOL 2011 TEM instrument equipped with Gatan digital camera and ISIS EDX analysis at St Andrews University. Samples were prepared for the microscope on holey carbon-coated copper grids from the ultrasonicated suspension of the catalyst powders in hexane.

3. Results

Both the CeO_2 and the Sm_2O_3 starting powders employed here consisted of highly crystalline particles. In contrast, the Pd/CS sample is largely made up of particles with little long-range order. A typical example of such a semi-amorphous particle is shown in Figure 1(a). Almost no Pd particles were identified in this sample. The Pd/CeO_2-15 sample contained many well-dispersed Pd particles in the size range 1-6 nm and highly crystalline particles of the support oxide. The Pd particles were aligned with the support oxide as shown in the image and Digital Diffraction Pattern (DDP) presented in Figure 1(b). In complete contrast, the Pd/Sm_2O_3-15 sample (Figure 1(c)) was very largely composed of a semi-amorphous phase similar to that seen in the as-prepared Pd/CS material. There was a minority of crystalline areas, however, and the few Pd particles observed tended to be present in these rather than on the semi-amorphous phase. Such a particle is imaged along the [110] zone axis in Figure 1(d). The particle is orientated such that the Pd (002) planes are aligned with the (00-4) planes of the oxide support (assuming the cubic crystallography for the Sm_2O_3). Figure 2 presents images of Pd/CS samples after the different pretreatments. Images and DDPs, such as those presented in Figure 2(a) show that, even after reduction at 150°C and purging at 500°C, a large quantity of the semi-amorphous phase persisted. In these regions, only very few Pd Particles were observed. However, crystalline areas were also observed in this sample and increasingly as pre-reduction and purge temperatures were increased on going to the Pd/SC-55 and Pd/SC-59 samples. The incidence of Pd particles also followed this trend, these generally being associated with the crystalline regions of the support. These particles were often polycrystalline, misaligned with respect to the support, associated with small support particles or apparently partially encapsulated by the support, as in Figure 2(b). All of these phenomena indicate that the Pd particles had formed during the crystallisation of the semi-amorphous support material. Since this support is associated with the Sm_2O_3 material after impregnation of the Pd precursor, it appears that the Pd is strongly associated with, and finely dispersed within, the Sm-containing semi-amorphous phase. This is in agreement with previous work on Rh/Sm_2O_3 catalysts after Rh impregnation in an acidic medium [6]. On reduction and heat treatment of the samples, the Pd emerges from this semi-amorphous phase to form particles which are still associated with the newly-crystallised support material, as in Figure 2(c). As the severity of the pre-treatment increases, Pd particle size appears to increase gradually until relatively large particles of apparently pure Pd were formed for the most severe pretreatment, as shown in the image and EDX spectrum in Figure 2(d). The difference between the behaviour of CeO_2- and Sm_2O_3-supported catalysts and the appearance of Pd particles in the Sm-containing samples can explain the activity of the Pd/CS catalyst. This was almost inactive after reduction at 300°C but was extremely active once reduced at 400°C [7]. This 'step-change' behaviour can be related to the presence of dispersed Pd particles in the Sm-containing phase which become accessible to the reactants at these reduction temperatures.

466

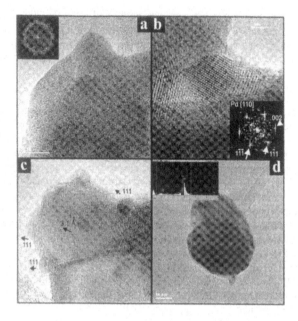

Figure 2. HRTEM images of (a) Pd/CS-15 sample, with inset DDP; (b) A Pd particle in a sample of Pd/CS-55, with DDP inset; (c) several Pd particles in a sample of Pd/CS-55, with alignments of planes indicated; (d) a large particle of Pd observed in a sample of Pd/CS-59 with partial EDX spectrum inset.

4. Summary

Pd catalyst materials containing Sm_2O_3 undergo dissolution and re-precipitation of the oxide as a semi-amorphous phase during the Pd impregnation step. The Pd appears to be intimately associated with this phase and to emerge from it as it recrystallizes after reductive and thermal pretreatments to form identifiable Pd particles. This does not occur with the CeO_2-only support. Appearance of these particles coincides with the onset of catalytic activity for methanol reforming.

References

1   Steele B C H 2000, Solid State Ionics **134**(1-2), 3
2   Yamamoto O 2000, Electrochimica Acta **45** 2423
3   Sahibzada M, Steele B C H, Hellgardt K, Barth D, Effendi A, Mantzavinos D, Metcalfe I S 2000, Chemical Engineering Science **55**(16): 3077
4   Uchida H, Osuga T and Watanabe M 1999, J. Electrochem. Soc. **146**(5) 1677
5   Ohnishi R, Wang W L and Ichikawa M 1994, Appl. Catal. A: General **113** 29
6   Bernal S, Botana F J, Calvino J J, Cifredo G A, García R and Rodriguez-Izquierdo 1990, Ultramicroscopy **34** 60
7   Gómez Sainero L M, Metcalfe I S, Sahibzada M, Baker R T, in preparation

Inst. Phys. Conf. Ser. No 179: Section 11
Paper presented at Electron Microscopy and Analysis Group Conf. EMAG2003, Oxford, 2003
©2003 IOP Publishing Ltd

Self-alignment of CdS nanoclusters observed by HRTEM

D Zhi, M Wei, D W Pashley[1] and T S Jones[2]

Department of Materials Science and Metallurgy, University of
Cambridge, Cambridge, CB2 3QZ, UK

[1]Department of Materials, Imperial College, London, SW7 2BP, UK

[2]Department of Chemistry, Imperial College, London, SW7 2BP, UK

Abstract. II-VI semiconductor nanocrystalline particles, which exhibit
effects of quantum confinement of electrons and phonons in various
optical and electronic properties, have been investigated extensively for
recent years. The interaction between the nanoparticles is certainly of
great importance for their potential application. In this work, uniformly
distributed CdS nanoparticles, with an average size of ~3-5 nm, were
characterised using high-resolution transmission electron microscopy
(HRTEM). The specimens for HRTEM were obtained by varying the
sampling conditions, such as the suspension concentration and
evaporation time. It has been found that the CdS nanoparticles can be
attracted to each other to form either loosely or closely aggregated nanoclusters with an average size of 30-100nm. The closely packed
nanoclusters were self-aligned along 'c' axis, which was determined by
phase-contrast lattice imaging and selected area electron diffraction. Low-
angle grain boundaries (LGBs) were observed between original
nanoparticles. This self-alignment phenomenon is believed to be
significant for both nanoparticle handling and future device fabrication.

1. Introduction

II-VI semiconductors with dimensions in nanometer realm are important because of
their quantum size effect. In these nanocrystalline semiconducting particles, or quantum
dots, the size dependent effective bandgap can give photoluminescence that is tunable
according to the dot size. Due to the small size involved, these systems have a very high
surface to volume ratio and hence it is essential to carefully control the surface states
along with the size. Nanosized II-VI particles, such as PbS, CdS, and CdSe, show lattice
distortions from their bulk counterparts and the presence of residual strain (Murray et al
1993). CdS nanoparticles have also been observed melt at a temperature which is
substantially below that of the bulk material (Goldstein et al 1992).

Besides its fundamental theoretical interest, interaction, e.g. adhesion, between
nanoparticles is a basic and complicated phenomenon, which has rarely been interpreted
on a microscopic scale (Yao and Thölen 1999). This issue is certainly very important
since one is subjected to it in many fabrication processes, applications, and even

handling. In this paper, we report on a high-resolution transmission electron microscopy (HRTEM) study on the interaction between CdS nanoparticles with an average size of 4-6nm and the observation of aligned agglomeration of these CdS particles.

2. Experimental

CdS nanosized particles were synthesised using chemical bath method. After the synthesis, a size selective precipitation enable further narrowing of size distribution. The obtained CdS nanoparticles were suspended in toluene solvent. The nanoparticle suspensions were treated in different condensation and aging conditions before TEM observation. The condensed suspensions were dropped onto a holey carbon-coated copper grid. The distribution and agglomeration of CdS particles were then characterised by a HRTEM (JEOL 2010).

3. Results and discussions

Figure 1 shows TEM images of nanocrystalline CdS particles before and after condensation and aging. For the sample without condensation in Figure 1(a), we can see the monodispersed particles with uniform size of 3-5nm and narrow size distribution. A mixture of the hexagonal close-packed wurtzite showing ABAB stacking and the closely related cubic close-packed showing ABCABC stacking is often seen, although in these very small regions with mixed stacking it is not relevant to distinguish between the two structure. However, for the samples after the condensation of particle suspension solution as shown in Figures 1(b) and 1(c), we can see that different sized nanoclusters were formed. These clusters were formed likely through the coalescence of adjacent nanoparticles. Figure 2 shows one of the nanocluster formed after condensation at 90°C without further aging, in which the cluster was loosely packed by a number of previous independent nanoparticles and the coalescence happened between only some of the particles.

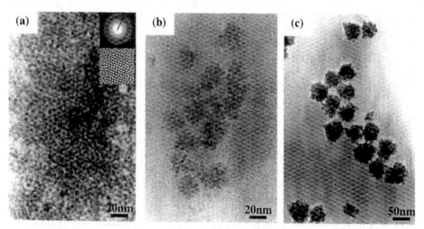

Figure 1. (a) monodispersed CdS nanocrystalline particles with two insets showing selected area diffraction pattern one particle with mixed stacking sequence respectively, (b) 20-30nm nanoclusters formed after 90°C condensation for 60 mintues, and (c) 50-70nm nanoclusters formed after 90°C condensation for 120 mintues and further 48 hours aging at room temperature.

Figure 2. HRTEM image of CdS nanocluster formed after 90°C condensation without further aging.

Figure 3 shows one of the cluster formed after 90°C condensation and 48 hours aging, a unique self-aligned structure was formed along the 'c' axis, as revealed by the inset Fourier transform spectrum taken from the squared area, which shows the hexagonal symmetry. It can also be found that this cluster is formed by joining the original adjacent particle through the low-angle grain boundaries (LGBs) as revealed by further analysis on the squared area in Figure 3. Three of these LGBs were indicated by oval in Figure 4(a) and 4(b) accordingly.

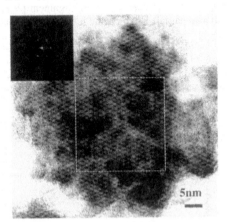

Figure 3. HRTEM image showing one self-aligned nanocluster in the sample after 90°C condensation and further 48 hours aging. The inset FT spectrum taken from the indicated area.

470

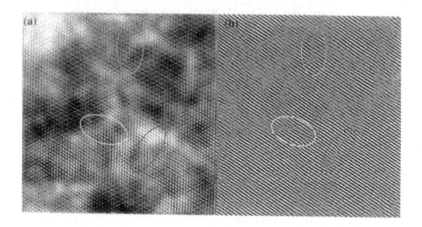

Figure 4. (a) HRTEM image of the squared area in Figure 2. (b) reconstructed image using one side band in the FFT spectrum. The ovals indicate some of the low angle grain boundaries.

4. Conclusions

CdS nanoparticles, which were obtained by varying the sampling conditions such as the suspension concentration and evaporation time, were studied by HRTEM. Either loosely packed or closely packed cluster with different size were observed, which depends on the sampling condition. The closely packed nanoclusters formed after condensation and aging were self-aligned along 'c' axis, which was determined by phase-contrast lattice imaging and selected area electron diffraction. Low-angle grain boundaries (LGBs) were observed between original nanoparticles. The self-alignment phenomenon is believed to be significant for both nanoparticle handling and future device fabrication.

References

Goldstein A N, Echer C M and Alivisatos 1992 Science **256** 1425
Murray C B, Norris D J and Bawendi M G 1993 J. Am. Chem. Soc. **115** 8706
Yao Y and Thölén A R 1999 Nanostructured Materials **12** 661

Inst. Phys. Conf. Ser. No 179: Section 11
Paper presented at Electron Microscopy and Analysis Group Conf. EMAG2003, Oxford, 2003
©*2003 IOP Publishing Ltd*

The effects of catalyst preparation techniques and synthesis temperature on the production of carbon nanotubes by the CVD method

Zabeada Aslam[1], Xuanke Li[2], Rik Brydson[1], Brian Rand[1]

[1] IMR, SPEME, University of Leeds, LS2 9JT, UK
[2] The Hubei Province Key Laboratory of Ceramics and Refractories,
Wuhan University of Science & Technology, Wuhan, Hubei 430081, P. R.
China

Abstract. The effects of synthesis temperature and catalyst preparation conditions for the production of carbon nanotubes were investigated using the CVD method with CH_4 as carbon feedstock. Two Al_2O_3-Fe catalyst aerogels were prepared using the same starting materials (aluminium nitrate and ferric sulphate) and ratios; aerogel 1 was dried normally while aerogel 2 was supercritically dried. CNTs were produced by direct pyrolysis of methane over the catalyst for 30 minutes at a range of different temperatures. Aerogel 1 produced better quality and higher yield of MWCNTS and bundles of SWCNTs at lower temperatures than aerogel 2 catalyst. The yield of NTs for aerogel 2 was observed to increase with temperature.

1. Introduction

Since their discovery by Iijima [1] in 1991, nanotubes have generated a great deal of interest as a result of their unique structural, electronic and mechanical properties. Nanotubes (NTs) can be produced by three different methods, laser ablation, electric arc discharge and chemical vapour deposition. Both laser ablation and electric arc discharge can produce good quality and high yield single-walled carbon nanotubes (SWCNTs); however, these methods are expensive and difficult to scale up to an industrial level. The CVD method provides the most cost effective and efficient route for CNT production. In the CVD method, the size and type of the catalyst used as well as the synthesis temperature, are the most critical variables. The most effective catalysts are transition metals, their oxides and their mixtures. The interactions of the supporting substrate with the catalyst also need to be taken into account as the size of the metal catalyst particles formed on the surface of the substrate is related to the bonding between the substrate and the catalyst. In order to increase the yield of NTs produced, the surface area of the substrate needs to be as high as possible and as a result nanoporous aerogels are increasingly being investigated. Aerogels are known to have high surface areas, high porosity and very low densities [2]. The ability to increase the NT yield using aerogel substrates has been demonstrated by a number of researchers. Piao et al [3] produced MWCNTs 10-20nm in diameter using an alumina supported nickel aerogel catalyst and found that both the reaction and catalyst reduction conditions had strong influences on the morphology of the CNTs produced. Using methane as the carbon feedstock, SWNCTs were produced by Su et al using an aerogel supported Fe/Mo catalyst [4]. The total amount of high quality SWCNTs was greater than 200% the weight of the catalyst with nearly no amorphous carbon being formed. Rul et al [5] have been able to synthesise a majority of SWCNTs from an oxide solid solution foam that increased the production four-fold.

In this paper, work carried out on alumina aerogels with Fe as the metal catalyst is reported.

2. Experimental Method

2.1. Catalyst Preparation and CNT Growth

The catalysts were prepared from the same starting materials and ratios but different preparation conditions. The starting materials for alumina/Fe catalysts were $Al(NO_4)_3 \bullet 9H_2O$ and $Fe_2(SO_3)_3 \bullet xH_2O$. 0.1M aluminium nitrate was dissolved in 70ml of a ferric sulphate solution saturated with ethanol. This solution was then supercritically dried at 7.5MPa and 260°C for 30 minutes to produce aerogel 2. However, the exact ratio of alumina to Fe is not known. For aerogel 1, 0.1M nanosized alumina was impregnated with 70ml of an ethanol solution of ferric sulphate and then dried to obtain the catalyst aerogel.

Aluminium nitrate melts at 73.5°C and decomposes at 150°C, while ferric sulphate does not melt but decomposes at 450°C. On heating, the catalysts lose water and form metallic nm-Fe particles supported in a possible alumina matrix, under a reducing atmosphere.

CNTs were produced using these catalysts via direct pyrolysis of methane for 30 minutes for a range of different temperatures. For aerogel 1, the temperatures ranged from 860°C to 880°C, while for aerogel 2, the temperatures ranged from 880°C to 970°C, (Table 1a).

2.2. Characterisation

The morphology and microstructure of the catalysts and CNTs were characterised using Philips CM200 FEGTEM operated at 200kV and fitted with a Gatan GIF 200 Imaging filter, and also a LEO 1530 FEGSEM. For the FEGTEM, the samples were ultrasonically dispersed in methanol for approximately 5 minutes and a drop placed on a TEM grid. The FEGSEM samples were ultrasonically dispersed for approximately 5 minutes in either acetone or methanol and coated in either gold or chromium. Gas adsorption studies were carried out using a Quantachrome Autosorb 1 with N_2 as the adsorbate.

3. Results and Discussion

3.1. Aerogels

As can be seen from the FEGSEM images, the morphology of the aerogels is quite different. Aerogel 1 particles are smooth and almost spherical, whereas aerogel 2 particles are jagged and plate-like, indicative of a higher surface area. From BET

Fig 1. FEGSEM images of aerogel 1 (left) and aerogel 2 (right).

analysis, the specific surface areas for aerogel 1 and 2 are $81m^2/g$ and $142m^2/g$, respectively, which appears to be consistent with the morphology observed in the FEGSEM images. For aerogel 1, the average pore diameter was calculated to be 4.8nm. The pore size distribution analysis shows the greatest number of pores with pore sizes of 5nm with a large number of 3-4nm pores and a few pores with diameters greater than 10nm. For aerogel 2, the average pore diameter is 22.0nm. The pore size distribution analysis shows the presence of a large number of pores of 8nm diameter, with a large number of pores ranging between 3nm to 18nm. There is also a significant amount of pores with diameters 30nm to 200nm.

3.2. Carbon content

In order to measure the weight of carbon deposited onto the catalyst, the weight loss of the catalyst during heating had to be determined. The weight loss could represent the loss of water as well as the decomposition of the starting materials with the formation of a possible Fe-aluminate spinel. Hence, heat treatment of the aerogel catalysts was carried out at 900°C for 1h, weight losses for which are given in Table 1c. The weight gain of carbon was then given by (weight of product after growing CNTs)-(weight after 900°C heat treatment) divided by the weight of the catalyst after 900°C treatment. The gain in weight for some of the samples is given in Table 1b.

Table 1. Tables showing the temperatures at which pyrolysis was carried out, the weight gained, and the weight loss of aerogel catalysts on heating.

1a	Sample	Temperature (°C)	1b	Sample	Weight gain %
	Aerogel 1			A1.1	21
	A1.1	860		A1.2	24
	A1.2	880		A2.1	22
	Aerogel 2			A2.2	27
	A2.1	880		A2.3	31
	A2.2	900			
	A2.3	920	1c	Catalyst	Weight loss %
	A2.4	940		Aerogel 1	68
	A2.5	960		Aerogel 2	51
	A2.6	970			

3.3. Nanotube Morphology

From FEGTEM observations (Figure 2), good quality MWCNTs and bundles of SWCNTs are produced from the aerogel 1 series at both 860°C and 880°C. However,

474

Fig 2. TEM images of samples A1.1, A1.2, A2.4, A2.5 and A2.6, from left to right.

for aerogel 2 no NTs were observed at 880°C, although graphitic layer formation was observed. In aerogel 2, NTs were first observed at 940°C with the yield increasing with temperature. Since the only difference between samples 1.2 and 2.1 is the catalyst, it is clear that the catalyst preparation conditions affect the Fe-catalyst significantly, delaying the growth of NTs in aerogel and requiring a much higher temperature to activate CNT growth. Comparing A1.2 and A2.6, a 90°C increase for aerogel 2 still does not produce CNTs of the same quality or yield as aerogel 1.

From FEGTEM images (Figure 2), sample A2.4 shows a NT at an early stage of formation. Sample A2.5 shows NTs with a large amount of amorphous material on the surface, while A2.6 shows NTs with less amorphous carbon. This indicates that for aerogel 2, the best conditions for nanotube production in this study are as those in sample A2.6, i.e. at the highest temperature.

4. Conclusions and Future Work

The SEM images show a large difference in the morphology of aerogel catalysts depending on the drying conditions. It is reasonable to assume that the Fe-catalyst will also be different, which in-turn may affect NT production. The different preparation conditions also affect the onset of NT production; supercritical drying results in a delay of at least 90°C. These results indicate that substantial investigation is still required for the completion of this study.

The catalysts will be analysed in order to obtain the size distribution of the nm-sized Fe particles, with a view to identifying the growth mechanism involved for SWCNT production. Studies of morphological changes in the catalyst following reduction will be undertaken using an ex-situ TEM environmental cell. The temperature range of SWCNT production will be investigated more thoroughly for aerogel 1 in order to determine the onset of NT production.

Dynamic in-situ studies of NT growth will be carried out in cooperation with Stig Helveg of Haldor Topsoe company in Lyngby, Denmark.

Acknowledgements

This work has been supported by an EPSRC and Morgan Crucibles studentship.

References
1. Iijima S, Nature, **354**, pp56, 1991
2. Husing N, Schubert U, Angew. Chem. Int. Ed. **37**, pp22, 1998
3. Piao L et al, Catalysis Today, **74**, pp145-55, 2002
4. Su M, Zheng B, Liu J, Chemical Physics Letters, **322**, pp 321-326, 2000
5. Rul S, Laurent Ch, Peigney A, Rousset A, Journal of the European Ceramic Society, **23**, pp1233-41, 2003

Inst. Phys. Conf. Ser. No 179: Section 11
Paper presented at Electron Microscopy and Analysis Group Conf. EMAG2003, Oxford, 2003
©2003 IOP Publishing Ltd

EFTEM investigation of silica nanotubes produced using designed self assembling β-sheet peptide fibrils as templates

**J E Meegan, A Aggeli, N Boden, R Brydson[*], A P Brown[*],
L Carrick, and R J Ansell**

Department of Chemistry, University of Leeds, Leeds, LS2 9JT, UK
[*] Institute for Materials Research, University of Leeds, Leeds, LS2 9JT, UK

ABSTRACT: Sol-gel hydrolysis and polymerisation of tetraethoxy silane (TEOS) in the presence of rationally designed self assembling positively charged peptide fibrils leads to the formation of silica nanotubes. The nanotubes possess a central pore of ~3.5nm diameter determined by the external diameter of the peptide fibril template, an external tube diameter of ~20nm and a length of several hundred nm.

An array of techniques has been used to characterise the nanotubes and their templates. Bright field TEM was used to characterise the template before addition of silane and also the silica nanotubes. Elemental maps acquired using EFTEM were used to demonstrate the efficiency of template removal. EDX was used to characterise the bulk sample.

1. Introduction

Silica nanostructures with various morphologies have been previously produced using supramolecular organic assemblies as templates (Raman et al., 1996). Of particular interest are hollow nano or micro tubules which may have applications in separations, catalysis or nanooptics/electronics; these tubes are mostly non chiral.

Recently Shinkai et al have prepared silica tubes with pore diameters 5-20nm using chiral organogelating templates assembled from low molecular weight gelators (cholesterol and sugar based) in predominantly organic solvents. The tubes have been shown to behave as enantioselective catalysts (Shinkai et al., 2003) giving yields of up to 97% enantioselectivity when employed in the transfer of an alkyl group from diisoproplyzinc to an aromatic aldehyde.

A new class of self-assembling β-sheet peptides which are in some ways analogous to the organogelators of Shinkai et al provide a potentially very versatile range of chiral templates (Aggeli et al., 1997b, Aggeli et al., 1997a) which, by careful manipulation of the amino acid sequence of the monomer can be made to self assemble in a variety of solvents and pH's through an established hierarchy of structures (Aggeli et al., 2001).

2. Experimental

L-DN1Q (*MeCO-QQRFQWQFQQQ-NH₂, 1.25μmol*) was dissolved in aqeous NaF (*6.17mM,324μl*) along with EtOH (*40μl*) in a sealed microtube (final pH 5.2). The peptide solution was allowed to self-assemble and align in a 7.05 T Oxford Instruments widebore NMR magnet with a superconducting solenoid at RT for 72h after which a 20μl sample was withdrawn for characterisation of the template.

To the remaining equilibrated solution TEOS (*161μmol*) was added and the solution left to polymerise for a further 24h inside the magnet and 4h at RT outside the magnet. The sample was then exposed to air to allow any EtOH or MeOH to evaporate, freeze dried and finally calcined at 600°C for 15h in air.

Bright field TEM analysis of the template was carried out on a Philips CM10 equipped with a tungsten filament operating at 80-100kV. The sample (*20μl*) was diluted 15 fold with distilled water and supported on a 400 mesh Cu grid (Agar) before being negatively stained with uranyl acetate solution (*1% w/v*).

HRTEM and EFTEM analysis of the silica nanotubes was carried out on a Philips CM200 FEG TEM operating at 200keV and equipped with an Oxford Instruments ultrathin window energy dispersive X-ray analyzer and a Gatan Imaging Filter GIF 200. For EFTEM analysis of the nanotubes, three window elemental maps were used, the N K (400eV), C K (285eV) and the Si L (100eV) ionisation edges were used for analysis with slit widths of 20, 20 and 10eV respectively. The sample was prepared by suspending 1-2mg of silica in acetone and briefly sonicating. 300μl of the suspension was then dropped onto a holey carbon 400 mesh Cu grid and left to dry.

3. Results and Discussion

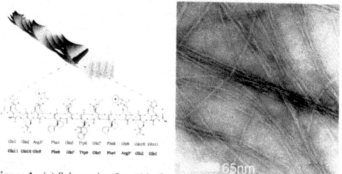

Figure 1. (a) Schematic of peptide fibril and (b) Bright Field TEM image of L-DN1Q Template

The bright field micrograph (Fig. 1b.) of the solution prior to addition of TEOS shows the presence of 3-4nm peptide fibrils (Fig 1a).

HRTEM images (Fig 2.) of the uncalcined sample show the presence of silica nanostructures with clearly defined central 'pores' running the full length of the nanotubes, where the central 'pores' are expected to be filled with template. The EFTEM C K elemental map clearly shows a high intensity region running down the centre of the nanotube corresponding to the central pore (we believe the white edges were caused by uncorrectable differences in the 3 EFTEM images due to the tubes drifting in the electron beam).

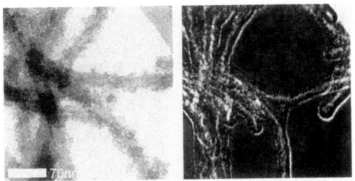

Figure 2. HRTEM (left) EFTEM C *K* Map(right) of uncalcined silica nanotubes

HRTEM and EFTEM (Fig 4.) also provided direct evidence for the alignment of the template by the magnetic field. A raft like area of nanotubes was found in the sample and due to the raft being less susceptible to damage/shrinkage in the electron beam it was possible to obtain C, N and Si maps for the area.

It can be seen that there are high degrees of correlation between the position of the pores in the HRTEM image and the strong C/N signal in the EFTEM elemental map, and the position of the weak Si signals.

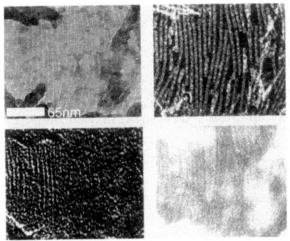

Figure 3. (Clockwise from top left) HRTEM, EFTEM C map, EFTEM N *K* map and EFTEM Si *L* map of the aligned area

An HRTEM investigation of the calcined sample showed tubes with similar dimensions to those observed in the uncalcined sample. No strong C signal could be detected running down the central pore (again we believe the white edges are caused by the tubes damaging/shrinking in the electron beam), therefore we conclude that calcination is an excellent method for removal of the template.

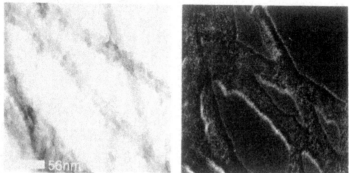

Figure 4. HRTEM (left) and EFTEM C map (right) of calcined silica nanotubes

4.   Conclusion

Hollow silica nanotubes consisting of 89% Q^4 and 10% Q^3 type silica (from ^{29}Si MAS NMR investigation) have been prepared using a sol-gel polymerisation of TEOS and a self-assembling peptide fibril template. HRTEM and EFTEM investigations indicate the existence of central pores in the nanotubes that contain the peptide template prior to calcination and that the template is successfully removed by calcination.

Aggeli, A., Bell, M., Boden, N., Keen, J. N., Knowles, P. F., McLeish, T. C. B., Pitkeathly, M. and Radford, S. E. (1997a) *Nature,* **386,** 259-262.

Aggeli, A., Bell, M., Boden, N., Keen, J. N., McLeish, T. C. B., Nyrkova, I., Radford, S. E. and Semenov, A. (1997b) *Journal of Materials Chemistry,* **7,** 1135-1145.

Aggeli, A., Nyrkova, I. A., Bell, M., Harding, R., Carrick, L., McLeish, T. C. B., Semenov, A. N. and Boden, N. (2001) *Proceedings of the National Academy of Sciences of the United States of America,* **98,** 11857-11862.

Raman, N. K., Anderson, M. T. and Brinker, C. J. (1996) *Chemistry of Materials,* **8,** 1682-1701.

Shinkai, S., Sato, I., Kadowaki, K., Urabe, H., Jung, J. H., Ono, Y. and Soai, K. (2003) *Tetrahedron Letters,* **44,** 721-724.

Inst. Phys. Conf. Ser. No 179: Section 11
Paper presented at Electron Microscopy and Analysis Group Conf. EMAG2003, Oxford, 2003
©2003 IOP Publishing Ltd

HRTEM Characterisation of Surface Effects in Iron Oxide Nanoparticles

G R Lovely[1], A P Brown[1], S D Evans[2], R Brydson[1]

[1]Institute for Materials Research, University of Leeds, Leeds, LS2 9JT, UK
[2]Department of Physics and Astronomy, University of Leeds.

Abstract. Mixed phase iron oxide nanoparticles have been fabricated by colloidal routes. HRTEM images of the nanoparticles show the presence of facets that terminate with a layer of dark contrast, which has been suggested to be caused by a surface cation layer. Work has been undertaken to characterise further this phenomenon.

1. Introduction

Iron oxide nanoparticles are suitable for many applications, including catalysis, high density data storage and magnetic resonance imaging (MRI). As with all nanoparticles, the high surface area to volume ratio can lead to enhanced and often modified surface effects. For example, iron nanoparticles with diameters of ~ 3nm have more than 50% of their atoms at the surface of the particle [1]. Nanoparticles are known to be more reactive than microparticles due to the higher relative concentration of crystallographic areas where atoms exist with low co-ordination numbers. Iron oxide nanoparticles are known to be superparamagnetic, and there is great interest in characterising their properties.

A study of the surface properties and characteristics of iron oxide nanoparticles is therefore both interesting and worthwhile. Over the last few decades, the number of papers published in the field of well characterised metal-oxide surfaces has greatly increased with corresponding advances in characterisation techniques.

This work was undertaken to investigate surface effects observed in HRTEM images of mixed phase iron oxide nanoparticles first reported by D.Jefferson [2].

2. Method

Iron oxide nanoparticles were fabricated using a standard route, by reducing iron chlorides with ammonia under a nitrogen atmosphere [3]. The size of the nanoparticles can be adjusted by altering the rate of addition of ammonia. These nanoparticles were either left bare, or coated with a lauric acid surfactant (to stabilise against nanoparticle aggregation), and then suspended as a colloid in chloroform.

A Philips CM200 FEGTEM operated at 200 kV with a Gatan Imaging Filter (GIF 200) attached, was used for the nanoparticle characterisation. Analytical techniques included selected area electron diffraction (SAED) and electron energy loss spectra (EELS), as well as the acquisition of high resolution images. Zone axis indexing of the individual nanoparticles was completed using the Gatan Digimicrograph software.

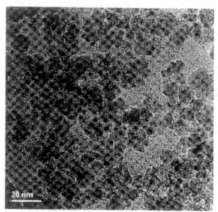

Figure 1. A micrograph of a lauric acid coated sample on a carbon film, showing the degree of agglomeration (despite the coating), and the individual particle size (3-4.5 nm).

3. Results

HRTEM images show that the particles were single crystal and often faceted in nature. They also tend to agglomerate, although the coated particles show this to a slighter extent, as expected due to increased steric repulsion effects of the acid (Figure 1). The SAED and EELS techniques show the samples to be composed of either inter-, or intra-mixed phase nanoparticles of Fe_3O_4 spinel and γ-Fe_2O_3 spinel crystals. The particles had typical diameters of 3-4.5 nm and an extremely narrow size distribution (Figure 2). The micrographs of the samples show that the faceted edges of the nanoparticles frequently terminate with a layer of dark contrast (see Figure 3). Since the images are taken at or near Scherzer defocus, we can assume the darker contrast in the projected image of the crystal structure is caused by iron cations.

In order to identify the Miller indices of the terminating crystal plane, a fast fourier

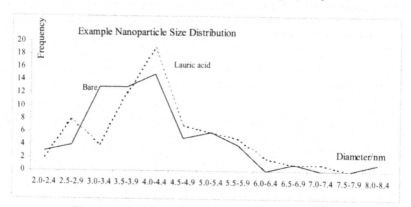

Figure 2. Plot showing typical particle size distributions for surfactant coated and bare nanoparticles.

481

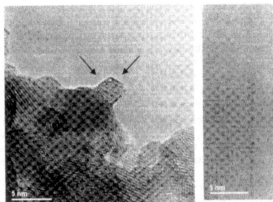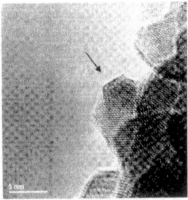

Figure 3. Two examples of micrographs showing nanoparticles exhibiting a dark contrast layer at the terminating edges of particular facets (indicated). The samples shown here are uncoated nanoparticles.

transform (FFT) was taken of the indicated image area. An example is given in Figure 4. It appears that a metal-ion plane has formed on the (1 1̄ 1) facet of the nanoparticle in Figure 4a. It is interesting to note that no such feature is visible on the (311) plane.

4. Discussion

The exposed face at a particle's surface is created by a balance between surface packing, and the need to maintain chemical stoichiometry. Non-cleavage planes have been found at crystal surfaces, but these tend to reform to contain combinations of two more stable cleavage surfaces. It has been theorised that if a low index plane terminates at a nanocrystal surface, then the charge balance of the crystal as a whole is dependent on whether the surface is terminated by anions or cations [2, 4] because of the relative high number of atoms at the surface of a nanoparticle. Fe_3O_4 nanoparticles, for example, tend

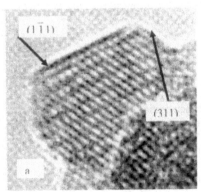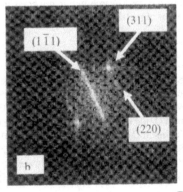

Figure 4a. The dark contrast edge of the particle is found to be the (1 1̄ 1) axis. b. An FFT of the particle shown in figure 4a.

482

to form into octahedral, or cube octahedral morphologies, which are shapes with low surface area to volume ratios. Fe_3O_4 is an inverse spinel and consists of an fcc close packed lattice of oxygen anions. This lattice provides tetragonal (A) and octahedral (B) interstitial sites so the Fe^{2+} cations sit in the B sites, whilst Fe^{3+} cations are equally distributed between the A and B sites, although not all sites are filled. The low index planes of magnetite consist of either anions or cations, but not both. Whereas $\gamma\text{-}Fe_2O_3$ is a defect spinel structure similar to Fe_3O_4 but with more cation vacancies, where the site vacancies depend on the degree of non-stoichiometry (i.e. there is a solid solution between Fe_3O_4 and $\gamma\text{-}Fe_2O_3$. So, if the edges of a pure Fe_3O_4 or a $\gamma\text{-}Fe_2O_3$ nanoparticle were to terminate in a cation layer, cation vacancies might have to be introduced below the terminating layer in order to maintain the overall stoichiometry of the particle, whilst retaining the anion sublattice. This would create a high defect $\gamma\text{-}Fe_2O_3$ structure below the surface.

Multislice imaging techniques [4] require that all the octahedrally coordinated B sites at the surface would need to be filled with cations to create contrast effects comparable to experimental images. However, as oxygen anions do not contribute greatly to the overall particle image contrast, a non-stoichiometric cation lattice would be observed, and should appear as an amorphous layer below the dark contrast surface in the high resolution micrographs. It is also possible that these vacancies may aggregate, enhancing this amorphous contrast. Although this defect structure was observed in micrographs and simulations taken by D.A. Jefferson [4], they were not clearly visible in this work (Figure 3), which may be due to the possible non-stoichiometry of the present samples [6].

5. Conclusion

Observations shown here seem to indicate that iron oxide nanoparticles can form with very different morphologies than those found in bulk specimens. Surface terminating planes exhibiting dark contrast, which are believed to be due to cation surface segregation, were observed at low index surface planes. Further studies will be completed to explore the physical reasons for this phenomena and to investigate whether the effect has a size dependency. Work will also be carried out to determine the precise structure of the terminating layers. The results of this work will have potential implications for research into many areas including the chemical reactivity and magnetic properties of metal oxide nanoparticles.

6. References

[1] Klabunde KJ, editor, Nanoscale Materials in Chemistry, J Wiley and Sons, 2001.
[2] Jefferson DA et al, Micro 90, IOP Publishing, 1990.
[3] Moore RGC, Chemical and Electronic Characterisation of Surfactant Stabilised Iron Oxide nanoparticles, Thesis, University of Leeds, 2001.
[4] Jefferson DA, Phil. Trans. R. Soc. Lond. A, 358 p2683, 2000.
[5] Knauth P and Schoonman J, Nanocrystalline Metals and Oxides: Selected Properties and Applications, Kluwer Academic Publishers, 2002.
[6] A.P.Brown et al, EMAG 2001 IOP Conference Series.

The authors would like to acknowledge the help of Dr .C. Hammond[1], and Dr D.A. Jefferson, Cambridge University, for their help in discussing this work.

Author Index

Subject Index

490